Recipes for Continuation

Computational Science & Engineering

The SIAM series on Computational Science and Engineering publishes research monographs, advanced undergraduate- or graduate-level textbooks, and other volumes of interest to an interdisciplinary CS&E community of computational mathematicians, computer scientists, scientists, and engineers. The series includes both introductory volumes aimed at a broad audience of mathematically motivated readers interested in understanding methods and applications within computational science and engineering and monographs reporting on the most recent developments in the field. The series also includes volumes addressed to specific groups of professionals whose work relies extensively on computational science and engineering.

SIAM created the CS&E series to support access to the rapid and far-ranging advances in computer modeling and simulation of complex problems in science and engineering, to promote the interdisciplinary culture required to meet these large-scale challenges, and to provide the means to the next generation of computational scientists and engineers.

Series Volumes

Dankowicz, Harry and Schilder, Frank, *Recipes for Continuation*

Mueller, Jennifer L. and Siltanen, Samuli, *Linear and Nonlinear Inverse Problems with Practical Applications*

Shapira, Yair, *Solving PDEs in C++: Numerical Methods in a Unified Object-Oriented Approach, Second Edition*

Borzì, Alfio and Schulz, Volker, *Computational Optimization of Systems Governed by Partial Differential Equations*

Ascher, Uri M. and Greif, Chen, *A First Course in Numerical Methods*

Layton, William, *Introduction to the Numerical Analysis of Incompressible Viscous Flows*

Ascher, Uri M., *Numerical Methods for Evolutionary Differential Equations*

Zohdi, T. I., *An Introduction to Modeling and Simulation of Particulate Flows*

Biegler, Lorenz T., Ghattas, Omar, Heinkenschloss, Matthias, Keyes, David, and van Bloemen Waanders, Bart, Editors, *Real-Time PDE-Constrained Optimization*

Chen, Zhangxin, Huan, Guanren, and Ma, Yuanle, *Computational Methods for Multiphase Flows in Porous Media*

Shapira, Yair, *Solving PDEs in C++: Numerical Methods in a Unified Object-Oriented Approach*

HARRY DANKOWICZ

University of Illinois at Urbana-Champaign
Urbana, Illinois

FRANK SCHILDER

Technical University of Denmark
Lyngby, Denmark

Recipes for Continuation

Society for Industrial and Applied Mathematics
Philadelphia

This material is based on work supported by the National Science Foundation under grants 0237370, 0635469, and 1016467 (HD), by the Engineering and Physical Sciences Research Council under grant EP/D063906/1 (FS), and by the Danish Research Council FTP under project 09-065890/FTP (FS).

Library of Congress Cataloging-in-Publication Data
Dankowicz, Harry.
  Recipes for continuation / Harry Dankowicz, University of Illinois at Urbana-Champaign, Urbana, Illinois, Frank Schilder, Technical University of Denmark, Lyngby, Denmark.
    pages cm – (Computational science and engineering series)
  Includes index.
  ISBN 978-1-611972-56-6
  1. Continuation methods. I. Schilder, Frank. II. Title.
  QA377.D364 2013
  515'.35–dc23

                              2012046499

Partial royalties from the sale of this book are placed in a fund to help students attend SIAM meetings and other SIAM-related activities. This fund is administered by SIAM, and qualified individuals are encouraged to write directly to SIAM for guidelines.

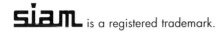 is a registered trademark.

Contents

Preface

Give a man a fish, feed him for a day.
Teach a man to fish, feed him for a lifetime.
                                    – Chinese Proverb

The objective of this text is to present the mathematical methodology known as *parameter continuation* in the context of a treatment that lends equal importance to theoretical rigor, algorithm development, and software engineering. It has been the authors' experience, in the development of the text and the associated software platform, that great mileage comes from accounting for all three of these fundamental pillars of computational science and engineering throughout the design process. Each benefits from careful attention to the other two, collectively ensuring successful and sharable implementations and trustworthy numerical results. In contrast, delayed attention to any one of these aspects likely leads down the road of dead ends, poor documentation, and incompatible computational and theoretical formulations. Although well-trodden, this is hardly a path of progress. If nothing else, we hope that the reader takes this message to heart in all subsequent pursuits in applied mathematics.

The theoretical concept of parameter continuation, as described in the text, is based on the simple observation that solutions to parameterized mathematical equations often belong to solution families, in turn parameterized by the problem parameters. Indeed, the exceptions to this situation continue to form a fertile field of study for generations of applied mathematicians. To a degree, it is possible to suggest that much of applied mathematics emphasizes those exceptional, singular points, where the simple picture breaks down; and that the subject then proceeds to investigate the solution behavior on neighborhoods thereof. Parameter continuation is the complementary tool to such a study, whereby solutions located near exceptional points in parameter space may be traced far outside of their immediate neighborhoods, in many cases until other singular points are reached and the analysis repeats.

To the beginning student of nonlinear equations, we expect that a truly mature appreciation for the subject will come only with the opportunity

- to explore the character and persistence of solutions across parameter domains, wherein no closed-form mathematical analysis is available, and

- to translate novel problem formulations into algorithms suitable for parameter continuation.

A trifecta of accomplishment is the final encoding of such novel problem formulations within a framework that supports their embedding as elements of more complicated continuation problems. An initiate of the paradigm advocated herein is a master of all three levels of understanding and skill: able to intuit the behavior of solutions, to translate statements about their existence to computational algorithms, and to support their modular interface to other larger formulations.

It is vital to recognize that no amount of computation is a substitute for a careful reality check, grounded in the basic and advanced techniques of algebra, calculus, and numerical analysis. The reader will, likely from experience, recognize the difference between being able to compute—so aptly made possible by an ever-growing set of high-level platforms—and being able to argue that what is computed is indeed the desired solution. This is particularly evident when infinite-dimensional problems are reduced, through some method of discretization, to problems of finite dimension, and ultimately approximated by finite-precision arithmetic in a numerical implementation. Examples include boundary-value problems for ordinary differential equations, partial differential equations, integral equations, and combinations thereof. Suffice it to say that confidence in the results of numerical implementations should come only as a result of combining computational tools with theoretical results about existence and convergence, without which a significant degree of skepticism is surely warranted.

Beyond the promulgation of a fundamental paradigm of applied mathematics and computational science and engineering, this text seeks to enable and stimulate its reader, whether novice or adept. We do this with a combination of explicit recipes for implementing classes of continuation problems within the *Computational Continuation Core* (referred to below by the abbreviation COCO) software framework for MATLAB®, developed by the authors, as well as open-ended continuation projects that challenge the reader to take that imaginative leap and believe in the opportunity to be original and creative. To this end, we cite literature only to a minimum, and only in the context of exercises wherein the reader is requested to repeat or reproduce the treatment from an archival source.

It is widely recognized in the applied mathematics community that tools of continuation afford a desirable alternative method of analysis to the more commonly deployed forward-simulation tools that pervade applied design and optimization. Increasingly, this realization is being appreciated also by those in industry whose products need to accommodate a range of design objectives, while being certifiably safe within a challenging set of performance criteria. As a means of reducing the number of simulations necessary in the design and development cycle, continuation methods, when applied to intelligently formulated continuation problems, are likely to become a staple of engineering design. While the recipes and formulations in this text are not production-ready for this context, we hope they afford the reader the necessary foundation for a skillful deployment in practice.

The intended audience for this text includes students and teachers of nonlinear dynamics and engineering, as well as scientists and engineers engaged in modeling and simulation. Of particular interest to us are potential developers of continuation toolboxes. The development framework mentioned above provides a broad suite of basic functionality, common across a range of continuation contexts, leaving the essential development task to those algorithmic decisions that encode a particular problem class. The text assumes some familiarity with MATLAB programming and a theoretical sophistication expected of upper-level undergraduate or first-year graduate students in an applied mathematics and/or computational science and engineering curriculum. Although no particular experience is required in topics traditionally associated with continuation methods, such as bifurcation analysis or nonlinear dynamics, this certainly could be beneficial. To this end, we highly recommend several comprehensive and quite readable textbooks and edited volumes that capture both the theoretical and computational aspects:

1. W.J.F. Govaerts, *Numerical Methods for Bifurcations of Dynamical Equilibria*, SIAM, Philadelphia, 2000.

2. Y.A. Kuznetsov, *Elements of Applied Bifurcation Theory*, Springer-Verlag, New York, 1998.

3. B. Krauskopf, H.M. Osinga, and J. Galán-Vioque (Eds.), *Numerical Continuation Methods for Dynamical Systems*, Canopus Publishing Ltd., Bristol, UK, 2007.

Krauskopf et al. also contains a range of interesting applications of continuation methods as well as state-of-the-art algorithms for a variety of continuation problems, which are not treated here, or mentioned only in passing.

We have sought to organize this text according to the sequence of steps we expect a majority of continuation projects would take. Thus, in Part I, we introduce a fundamental mathematical paradigm within which the subsequent tool development is framed. We provide an illustration of the utility of parameter continuation in the context of an optimization problem from the calculus of variations in Chap. 1. The relevant terminology and notation are then introduced in Chap. 2. Chap. 3 provides an introduction to the syntax particular to command-line interaction with the COCO framework, as well as in the development of special purpose continuation toolboxes for this platform. The framework is expanded in Chaps. 4 and 5 with a discussion of toolbox development and task embedding as a fundamental approach to continuation problems inspired by software engineering. Mastery of the basic object-oriented principles of these chapters is essential for the subsequent treatment.

Part II of this text presents a sequence of *toolbox templates* that build on the task-embedding paradigm, and that also demonstrate basic algorithms that take advantage of a vectorized formulation of the nonlinear equations. A workhorse throughout this part is the collocation toolbox presented in Chaps. 6 and 7 for discretizing first-order systems of ordinary differential equations. This toolbox appears embedded in a sequence of encapsulating toolboxes for the solution of single-segment boundary-value problems of ordinary differential equations in Chap. 8 and for multisegment boundary-value problems in Chap. 9. The treatment includes toolboxes for continuation of single-segment periodic orbits in smooth dynamical systems, multisegment periodic orbits in hybrid dynamical systems, and quasi-periodic invariant tori in smooth dynamical systems. A generalization of the collocation toolbox to solving the linearized, variational problem along solution trajectories for ordinary differential equations is then shown in Chap. 10. The analysis demonstrates the application of this toolbox, and the task-embedding paradigm, to the continuation of connecting orbits between equilibria and periodic orbits, while characterizing local stability properties of the periodic orbit.

In Part III, emphasis is on the development of *atlas algorithms*, implementations of a finite-state machine for generating a collection of charts that covers a portion of the solution manifold of the continuation problem. Although the COCO framework includes default implementations of general-purpose atlas algorithms, certain problem classes naturally benefit from appropriate design not only of the original continuation problem but also of the algorithm used to cover the corresponding solution manifold. A general theory for atlas algorithms and continuation is presented in Chap. 11. An incremental sequence of implementations of 1-dimensional atlas algorithms is then discussed in Chap. 12. Chap. 13 illustrates the generalization of such implementations to the multidimensional context, with several example implementations for covering 2-dimensional manifolds. Further extensions of the 1-dimensional and 2-dimensional atlas algorithms are provided in Chap. 14 in order to support constraints on the computational domain.

An essential ingredient in parameter continuation is the detection of special points associated with critical properties of the solution or the solution manifold. This concept of *event handling* is the topic of Part IV of this text. Chap. 15 defines event handling in the context of detection and location of special points along curve segments on the solution manifold. The basic syntax for command-line handling of events in COCO is introduced there. The integration of event handling in COCO-compatible atlas algorithms and toolboxes is described in Chap. 16. Finally, we conclude the part with an exploration of the use of *event handlers* for associating particular data or actions with candidate special points. Specifically, Chap. 17 provides template event handlers, using a reverse communication protocol, for enabling a selective treatment of bifurcation points of equilibria and periodic orbits, as well as for automated or semiautomated branch switching associated with special points along the solution manifold.

Part V provides an introductory treatment of the problem of adaptive changes to the discretization of a continuation problem during parameter continuation. We discuss several distinct approaches to adaptation that are supported by the COCO framework and explore their associated computational cost. In Chap. 18, we present two paradigms of adaptation that can be directly accommodated within the framework developed in previous chapters. In the first case, we consider a *brute-force mesh-refinement strategy*, in which continuation is terminated when an estimate of the discretization error exceeds a desired tolerance, and an entirely new continuation problem is formulated in order to accommodate a change in discretization. In the second case, we describe a *comoving-mesh strategy*, in which variables defining the discretization are included among the unknowns and solved for as part of the continuation problem without a change in their number. Chap. 19 describes the use of adaptive changes to the order of truncation in a Fourier approximation of a periodic orbit of a smooth dynamical system. Finally, *moving mesh strategies*, in which adaptive changes to the discretization order and discretization parameters are integrated with an atlas algorithm, are considered in Chap. 20. The analysis concludes by a numerical comparison between the different adaptive discretization strategies when applied to the approximation of homoclinic orbits, as well as in the continuation of canard orbits in a slow-fast dynamical system.

At the end of each chapter, we collect exercises for use in self-study or as course assignments. These range from reflections on the theoretical content of a chapter to implementations in the COCO framework of algorithms and toolboxes that generalize the treatment in the text to broader problem classes. As suggested above, some of the latter take the form of open-ended projects that could be used as an alternative source for summative assessment. The epilogue, Part VI, includes a number of proposed development projects for students and junior investigators.

The COCO framework source code is freely available from http://www.siam.org/books/cs11. This site also includes a user manual as well as extensive complete examples. The example code explicitly included in this text can also be downloaded from this site and reused or developed further without explicit permission from the authors. We ask simply that the source be acknowledged in any derivative work.

This text and the COCO framework are a result of a joint effort by the authors since the inception of their collaboration, following the 2007 workshop on *Advanced Algorithms and Numerical Software for the Bifurcation Analysis of Dynamical Systems* organized in honor of Eusebius J. Doedel's 60th birthday at the University of Montréal in Montréal, Canada. We are indebted for the inspiration for this effort to the creativity of our fellow attendees at this and other similar meetings and hope that this text will contribute to the impact that their work has in the scientific community. Among our immediate colleagues, we kindly acknowledge Claudia Wulff and Jens Starke for their patience with this long-term effort, and Michael Henderson for his hospitality and help with recreating MULTIFARIO within COCO. We also gratefully acknowledge David Barton and Mehdi Saghafi, who as developers of COCO-compatible toolboxes have provided invaluable feedback to us. Our sincere thanks also go to Barbara Brickman for her help with editing the text for language and format. As with any work of this magnitude, the countless hours of debugging, editing, and imagining would not be possible without the endless support of family and friends, especially Helena, Melanie, Even, Lia, and Zachary Dankowicz and Bärbel, Anette, and Karsten Schilder.

Champaign, Illinois, and Lyngby, Denmark, October 2012

Notation

We collect below a summary of notational conventions and terminology relied upon in this text, in the event that these are not detailed in the main body. As anyone who has ever attempted authorship of a text of this sort will appreciate, notational consistency is a monumental, but invaluable task. If anything, this summary has been useful in reminding us of the intended meaning of our chosen notation.

Index sets

An index set is a finite sequence of nonrepeating, positive integers. We use the capital letters $\mathbb{I}$, $\mathbb{J}$, $\mathbb{K}$, and $\mathbb{L}$ to represent index sets. The capital letters $\mathbb{R}$ and $\mathbb{Z}$, on the other hand, are used to denote the spaces of all reals and all integers, respectively. The capital letter $\mathbb{S} := \mathbb{R}/2\pi$ denotes the quotient set on $\mathbb{R}$ obtained by defining two numbers as equivalent if they differ by a multiple of 2π.

Denote by $\{k\}_{k=1}^{n} := \{1,\dots,n\}$ the sequence of positive consecutive integers between 1 and n, for some positive integer n, and let $\{1,\dots,0\} = \emptyset$. Then $\mathbb{K} \subseteq \{1,\dots,n\}$ provided that the elements of $\mathbb{K}$ all belong to $\{1,\dots,n\}$. In contrast to $\{1,\dots,n\}$, the elements of $\mathbb{K}$ need not be in increasing order.

Given an index set $\mathbb{K}$, the notation $|\mathbb{K}|$ denotes its *cardinality*, i.e., the number of integers contained in $\mathbb{K}$. It follows that

$$\mathbb{K} := \{k_i\}_{i=1}^{|\mathbb{K}|}, \tag{1}$$

where $k_i \in \mathbb{Z}$ for all $i \in \{1,\dots,|\mathbb{K}|\}$. Given an index set $\mathbb{K} \subseteq \{1,\dots,n\}$, we define the set

$$\{1,\dots,n\} \setminus \mathbb{K} \tag{2}$$

as the index set obtained by omitting the elements of $\mathbb{K}$ from the sequence $\{1,\dots,n\}$.

An index set can be used to extract a subsequence from a given sequence of scalars. For example, if $u \in \mathbb{R}^n$ and $\mathbb{K} = \{k_i\}_{i=1}^{|\mathbb{K}|} \subseteq \{1, \ldots, n\}$, then

$$u_{\mathbb{K}} = \{u_k\}_{k \in \mathbb{K}} := \left\{u_{k_i}\right\}_{i=1}^{|\mathbb{K}|}. \tag{3}$$

Functions

Throughout this text, we identify functions in terms of the nature of their domain and range spaces, as in $f : \mathbb{R}^n \to \mathbb{R}^m$, or in terms of their explicit evaluation, as in

$$f : u \mapsto \left(\begin{array}{c} u_1^2 - u_2 \\ u_1 \end{array} \right) \tag{4}$$

or

$$f(y, p) := \left(\begin{array}{c} p_1 y_1 - p_2 y_2 \\ 1 + y_2^2 \end{array} \right). \tag{5}$$

We often reference functions only by their name, as in f, rather than their value, as in $f(x)$, unless the latter is preferred by the context. The evaluation of a function f at a particular point $x = x^*$ in its domain is also sometimes denoted by $f|_{x=x^*}$. In figures, we use function names to represent coordinates or as a reference to functions, unless the interpretation is ambiguous.

For derivatives of functions of a single variable, we either use the Leibniz operator notation $\frac{dy}{dx}$ or rely on a superscript to denote differentiation with respect to the independent variable, e.g., $y'(x)$ or $h'(t)$. In the case of functions of multiple variables, we try to use the ∂ symbol consistently in order to distinguish between different partial derivatives. In this context, the notation $\partial_{(x,p)}$ denotes a matrix of derivatives with respect to the components of x and p.

Software

The *Computational Continuation Core* (COCO) framework described above is a package of MATLAB routines, libraries, and classes that implement the core functionality described in this text and that support the command-line and toolbox interface illustrated throughout all its parts. One of the key design features of this framework is the possibility to substitute alternate implementations of various numerical algorithms for those that may be assigned by default. To this end, COCO relies on the use of global and local project settings for identifying the source of key algorithms. As an example, consider the encoding below of the `coco_project_opts` function.

```
function prob = coco_project_opts(prob)

prob = coco_set(prob, 'cont', 'linsolve', 'recipes');
prob = coco_set(prob, 'cont', 'corrector', 'recipes');
prob = coco_set(prob, 'cont', 'atlas_classes', ...
  { [] 'atlas_kd' ; 0 'atlas_0d_recipes' ; 1 'atlas_1d_recipes' });

end
```

This uses COCO core utilities to assign references to the source of the linear and nonlinear solvers and default atlas algorithms for use in the application of the coco entry-point function. In each example given in this text, a function file with this content is included in the same directory as the example files. Where appropriate, we make explicit any overriding of its default references.

Throughout this text, the prompt >> is used to indicate the beginning of a command line in MATLAB. When we speak of an *extract*, we mean a verbatim copy of a portion of the MATLAB command window. In the first several parts of the text, we include these extracts in their full glory, but successively begin to omit these in later chapters, as the novelty wears off. In MATLAB, the use of a single variable name to represent both an input and an output argument takes advantage of a call-by-reference mechanism that guarantees that changes to the corresponding variable survive the execution of the function without temporary duplication of its content within the function. Finally, throughout this text, *string identifiers* are used to reference various primitive and composite objects. Such identifiers must adhere to the rules that apply to MATLAB field names.

Part I

Design Fundamentals

Chapter 1

A Continuation Paradigm

Continuation is a numerical technique for computing implicitly defined manifolds. In this book, we restrict our attention to manifolds defined by systems of nonlinear finite-dimensional equations, including those obtained through discretization of infinite-dimensional problems. The objective of this chapter is to illustrate the application of continuation to a classical problem from the *calculus of variations*.

1.1 Problem formulation

Consider the task of finding a smooth curve $y = f(x) > 0$ on the interval $[0, 1]$ that minimizes the *integral functional*

$$J(f) := 2\pi \int_0^1 f(x) \sqrt{1 + (f'(x))^2} \, dx, \tag{1.1}$$

while satisfying the *boundary conditions* $f(0) = 1$ and $f(1) = Y$ for some known, non-negative constant Y. Such a curve, when one exists, depends implicitly on the value of the ordinate Y. By *continuation* we mean a technique intended to compute individual members of a family of such curves under variations in Y.

From Fig. 1.1, it follows that J equals the area of a surface of revolution about the x axis whose intersection with the upper half of the (x, y) plane is given by $y = f(x)$. It is, of course, not necessarily true that a smooth curve, which minimizes the surface area, exists even in a local sense among some family of nearby curves. If such an *extremal curve* does exist, it must correspond to a *stationary value* of the integral J, although the latter would also be true for a local maximum or a saddle-type extremum. It is clear, from the geometric meaning of the functional, that there is no curve that corresponds to a locally maximal surface area, but the possibility of a saddle cannot be excluded a priori.

1.2 An analytical solution

Define the *action integrand*

$$L : (x, f, f') \mapsto 2\pi f \sqrt{1 + f'^2}, \tag{1.2}$$

3

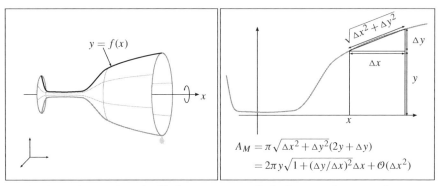

(a) Generating a surface of revolution.   (b) Surface area of a cone section.

Figure 1.1. *A surface of revolution is generated by rotating a planar curve around an axis in the corresponding plane* (a). *If the curve is the graph* $(x, f(x))$ *of a function* f, *its surface area may be approximated by a* Riemann *sum, obtained by constructing a sequence of cone sections and summing their respective areas* (b). *The functional in Eq.* (1.1) *is obtained in the limit as* $\max \Delta x \to 0$, *where the maximum is taken over all cone sections.*

such that

$$J(f) = \int_0^1 L\left(x, f(x), f'(x)\right) dx. \tag{1.3}$$

1.2.1 Extremal curves

From the theory of the calculus of variations, it follows that a function $f : [0,1] \to \mathbb{R}$ corresponds to a stationary value of the integral J if and only if f satisfies the *Euler–Lagrange equation*

$$\left(\frac{\partial L}{\partial f'}\left(x, f(x), f'(x)\right)\right)' - \frac{\partial L}{\partial f}\left(x, f(x), f'(x)\right) = 0 \tag{1.4}$$

for all $x \in (0,1)$. Here, for example,

$$\frac{\partial L}{\partial f} = 2\pi\sqrt{1 + f'^2} \tag{1.5}$$

and

$$\frac{\partial L}{\partial f'} = \frac{2\pi f f'}{\sqrt{1 + f'^2}}. \tag{1.6}$$

Substitution into Eq. (1.4) and further differentiation then yields, after some simplification, the equivalent second-order differential equation

$$1 + f'^2 - f f'' = 0 \tag{1.7}$$

in the unknown function $f(x)$.

More generally, since the function L in Eq. (1.2) is not explicitly dependent on the independent variable x, the vanishing of the left-hand side of the Euler–Lagrange equation in Eq. (1.4) implies the *Beltrami identity*

$$\left(f'(x) \frac{\partial L}{\partial f'} \big(x, f(x), f'(x)\big) - L\big(x, f(x), f'(x)\big) \right)' = 0 \tag{1.8}$$

and thus that the unknown function $f(x)$ satisfies

$$f' \frac{\partial L}{\partial f'} \big(x, f, f'\big) - L\big(x, f, f'\big) = -\frac{2\pi}{a} \tag{1.9}$$

for some constant a. Substitution from Eq. (1.2) then yields, after some simplification, the equivalent first-order differential equation

$$\frac{f}{\sqrt{1 + f'^2}} = \frac{1}{a} \tag{1.10}$$

in the unknown function $f(x)$. Since $1 + \big(f'(x)\big)^2 \geq 1$ and $f(x) > 0$, it follows from Eq. (1.10) that $f(x) \geq \frac{1}{a} > 0$ on $[0, 1]$. In particular, since $f(0) = 1$, it must hold that $a \geq 1$. Separation of variables finally yields the *general solution*

$$f(a, b) : x \mapsto \frac{1}{a} \cosh a (x + b), \tag{1.11}$$

parameterized by the constants a and b.

We reconnect to the original problem formulation by imposing the boundary conditions at $x = 0$ and $x = 1$, respectively, on $f(a, b)$ in Eq. (1.11) to yield

$$\frac{1}{a} \cosh ab = 1 \tag{1.12}$$

and

$$\frac{1}{a} \cosh a (1 + b) = Y. \tag{1.13}$$

These nonlinear equations in the unknown constants a and b must be solved simultaneously for each value of Y. When $Y = \cosh 1$, one such solution is obtained by $a = 1$ and $b = 0$. Collectively, Eqs. (1.12)–(1.13) define the family of viable extremal curves $y = f(x)$ implicitly in terms of the value of Y.

Alternatively, we may characterize the family of viable extremal curves by expressing these in terms of the constant a, in which case we treat b and Y as the unknowns in Eqs. (1.12)–(1.13). In this case, from Eq. (1.12), it follows that

$$b = b_+(a) := \frac{1}{a} \cosh^{-1} a \tag{1.14}$$

and

$$b = b_-(a) := -\frac{1}{a} \cosh^{-1} a \tag{1.15}$$

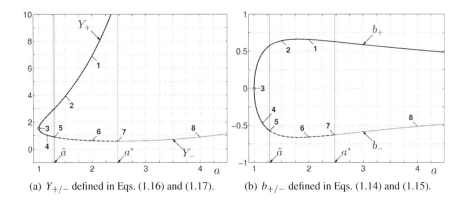

(a) $Y_{+/-}$ defined in Eqs. (1.16) and (1.17).   (b) $b_{+/-}$ defined in Eqs. (1.14) and (1.15).

Figure 1.2. *Values of a, b, and Y corresponding to extremal curves of the integral functional in Eq. (1.3) with action integrand given in Eq. (1.2) and satisfying the given boundary conditions at $x = 0$ and $x = 1$. Points found along the solid black segment correspond to global minimizers, whereas those corresponding only to weak local minimizers are found along the dashed segment $(a, b_-(a), Y_-(a))$, for $a \in (\tilde{a}, a^*)$ (see Sects. 1.2.1 and 1.2.3 for definitions of a^* and $\tilde{a}$, respectively). Selected extremal curves for the points with labels 1–8 are shown in Fig. 1.3.*

both satisfy the boundary condition at $x = 0$ (cf. Fig. 1.2(b)). Substitution from Eq. (1.14) or Eq. (1.15) into Eq. (1.13) then yields the two solutions

$$Y = Y_+(a) := \frac{1}{a} \cosh\left(a + \cosh^{-1} a\right) \tag{1.16}$$

and

$$Y = Y_-(a) := \frac{1}{a} \cosh\left(a - \cosh^{-1} a\right), \tag{1.17}$$

respectively. It is straightforward to show that $Y_+(a)$ is a monotonically increasing, unbounded function of a for $a \geq 1$. Similarly, there exists an $a^* \approx 2.473$, such that $Y_-(a)$ is monotonically decreasing for $1 \leq a < a^*$ and monotonically increasing for $a > a^*$ with a global minimum $Y^* := Y_-(a^*) \approx 0.587$ (cf. Fig. 1.2(a)). From these observations, it follows that the boundary condition at $x = 1$

- cannot be satisfied by any $a \geq 1$ for $Y < Y^*$;

- is satisfied only by $a = a^*$ and $b = b_-(a^*)$ for $Y = Y^*$;

- is satisfied by one value of $a \in (1, a^*)$ with $b = b_-(a)$ and one value of $a > a^*$ with $b = b_-(a)$ for $Y \in (Y^*, \cosh 1)$;

- is satisfied by $a = 1$ and $b = 0$ as well as by one value of $a > a^*$ with $b = b_-(a)$ for $Y = \cosh 1$; and

- is satisfied by one value of a with $b = b_-(a)$ and a different value of a with $b = b_+(a)$ for $Y > \cosh 1$.

These conclusions regarding the existence and number of possible extremal curves of the form in Eq. (1.11) are further illustrated by the number of intersections of the graphs of Y_- and Y_+ with the horizontal grid lines shown in Fig. 1.2(a).

1.2.2 Local minimizers

The functional J is said to have a *weak local minimum* at a function f that satisfies the boundary conditions if there exists a scalar $\varepsilon > 0$ such that

$$J(g) \geq J(f) \tag{1.18}$$

for all smooth functions g that satisfy the given boundary conditions and for which

$$\sup_{x\in(0,1)} |f(x) - g(x)| + \sup_{x\in(0,1)} |f'(x) - g'(x)| < \varepsilon. \tag{1.19}$$

From the theory of the calculus of variations, the extremal function $f(a,b)$ in Eq. (1.11) with $b = b_\pm(a)$ corresponds to a weak local minimum of J if and only if

1. the strict inequality

$$\frac{\partial^2 L}{\partial f'^2}\left(x, [f(a,b)](x), [f(a,b)]'(x)\right) > 0 \tag{1.20}$$

   holds for all $x \in (0,1)$ and, in addition,

2. there exists no $\xi \in (0,1)$ such that the linear combination

$$c_1 \frac{\partial f(a,b)}{\partial a} + c_2 \frac{\partial f(a,b)}{\partial b}\bigg|_{b=b_\pm(a)} \tag{1.21}$$

   vanishes at $x = 0$ and $x = \xi$ for some pair of not simultaneously vanishing quantities c_1 and c_2.

In the case of the function L given in Eq. (1.2) and after substitution of the extremal function in Eq. (1.11), it holds that

$$\frac{\partial^2 L}{\partial f'^2}(x, f, f') = \frac{2\pi f}{\left(1 + f'^2\right)^{3/2}} = \frac{2\pi}{a \cosh^2 a(x + b)} > 0 \tag{1.22}$$

for all x. Moreover,

$$c_1 \frac{\partial f(a,b)}{\partial a} + c_2 \frac{\partial f(a,b)}{\partial b}\bigg|_{b=b_\pm(a)} \tag{1.23}$$

vanishes at $x = 0$ provided that

$$c_1 \left(a - \sqrt{a^2 - 1}\cosh^{-1} a\right) \mp c_2 a^2 \sqrt{a^2 - 1} = 0. \tag{1.24}$$

For example, for $a = 1$, it follows that $c_1 = 0$ and

$$c_1 \frac{\partial f(a,b)}{\partial a} + c_2 \frac{\partial f(a,b)}{\partial b}\bigg|_{b=b_\pm(1)} = c_2 \sinh x, \tag{1.25}$$

which vanishes at an internal point $\xi \in (0,1)$ if and only if $c_2 = 0$. It follows that the extremal curve obtained for $a = 1$ and $b = 0$ is indeed a weak local minimum of J.

For $a > 1$, we assume, without loss of generality, that $c_1 \neq 0$, and consider the linear combinations

$$\varphi_\pm(a) := \sqrt{a^2 - 1}\, \frac{\partial f}{\partial a}(a, b) + \rho \frac{\partial f}{\partial b}(a, b)\bigg|_{b = b_\pm(a),\, \rho = \rho_\pm(a)}, \tag{1.26}$$

where

$$\rho_\pm(a) := \pm \frac{a - \sqrt{a^2 - 1}\, \cosh^{-1} a}{a^2}. \tag{1.27}$$

Here, the slope

$$\frac{d}{dx} \varphi_\pm(a)\bigg|_{x=0} = \pm a. \tag{1.28}$$

Moreover, for $\epsilon \ll 1$,

$$\varphi_\pm\left(1 + \epsilon^2\right) : x \mapsto \pm \sinh x + \mathcal{O}(\epsilon). \tag{1.29}$$

A change in the nature of the corresponding extremal curve then occurs when $\varphi_+(a)|_{x=1}$ changes sign from positive to negative or when $\varphi_-(a)|_{x=1}$ changes sign from negative to positive. Since

$$
\begin{aligned}
\varphi_\pm(a)|_{x=1} &= \frac{\pm a\left(a^2 - 1\right)\cosh a + \left(a^2 \sqrt{a^2 - 1} \pm 1\right)\sinh a}{a^2} \\
&= \sqrt{a^2 - 1}\, \frac{dY_\pm(a)}{da},
\end{aligned}
\tag{1.30}
$$

we conclude that $\varphi_+(a)|_{x=1}$ is positive for all $a > 1$, while $\varphi_-(a)|_{x=1}$ is negative for $1 < a < a^*$ and positive for $a > a^*$. Consequently, the extremal curves for which $b = b_+(a)$ are all weak local minima. In contrast, only those extremal curves for which $b = b_-(a)$ and $1 < a \leq a^*$ are weak local minima of J. It follows from the above derivation that there is a unique weak local minimum for $Y \geq Y^*$ and no smooth extremal curve for $Y < Y^*$. Values of (a, b, Y) corresponding to weak local minima lie along the solid black and dashed curves in Fig. 1.2. Sample local minimizers are shown in Fig. 1.3(b).

1.2.3 Global minimizers

Substitution from Eq. (1.11) into the defining integral yields

$$J\left(f(a, b)|_{b = b_\pm(a)}\right) = J_\pm(a), \tag{1.31}$$

where

$$J_\pm(a) := \frac{\pi}{2a^2}\left(2a \mp \sinh\left(2\cosh^{-1} a\right) + \sinh\left(2a \pm 2\cosh^{-1} a\right)\right). \tag{1.32}$$

The graphs of the functions $J_\pm$ are shown in Fig. 1.4(a).

In order to determine the existence of a *global minimum*, it is necessary to consider the "boundary" curve (see curve 9 in Fig. 1.3(b)) consisting of a line segment from $(0, 1)$ to $(0, 0)$, followed by a line segment from $(0, 0)$ to $(1, 0)$, followed by a line segment from $(1, 0)$ to $(1, Y)$. The area of the corresponding surface of revolution equals

$$\tilde{J}_\pm(a) := \pi\left(1 + Y_\pm^2(a)\right). \tag{1.33}$$

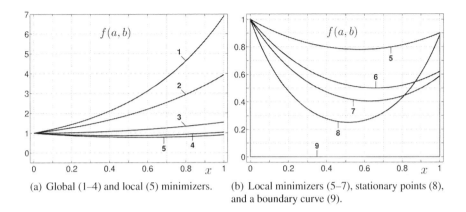

(a) Global (1–4) and local (5) minimizers.   (b) Local minimizers (5–7), stationary points (8), and a boundary curve (9).

Figure 1.3. *Extremal curves of the integral functional in Eq. (1.3) with action integrand given in Eq. (1.2) and satisfying the given boundary conditions at $x = 0$ and $x = 1$. The labels correspond to the points marked in Fig. 1.2. Global minimizers are shown in panel (a). Curves 5, 6, and 7 in panel (b) are weak local minimizers, while curve 8 is a stationary curve, but not a minimizer. The piecewise-straight curve 9 is the boundary curve defined in Sect. 1.2.3. The surfaces of revolution generated by curves 5 and 9 have the same surface area.*

It is straightforward to show that $\tilde{J}_+(a) > J_+(a)$ for all $a \geq 1$. In contrast, $\tilde{J}_-(a) > J_-(a)$ for $1 \leq a < \tilde{a} \approx 1.282$ and $\tilde{J}_-(a) < J_-(a)$ for $a > \tilde{a}$. It follows that the extremal curve corresponding to the unique weak local minimum is a global minimizer only for $Y > Y_-(\tilde{a}) \approx 0.900$. Values of (a, b, Y) corresponding to global minimizers lie along the solid black curves in Fig. 1.2. Sample global minimizers are shown in Fig. 1.3(a). The surfaces of revolution generated by the weak local minimizer for $a = \tilde{a}$ and by the corresponding boundary curve, respectively, are shown in Figs. 1.4(b)–1.4(c).

1.3   A numerical solution

We may approximate the family of extremal functions f by applying techniques of numerical continuation under variations in Y to the algebraic equations in Eqs. (1.12)–(1.13); to a discretization of the Euler–Lagrange equation in Eq. (1.7); or directly to an optimization condition for a discretization of the defining integral in Eq. (1.1). In all cases, we start with the known solution $f : x \mapsto \cosh x$ for $Y = \cosh 1$ and generate iteratively a succession of approximate solutions along a 1-dimensional branch through the starting point.

1.3.1   An algebraic formulation

The result of such an analysis applied to the algebraic equations in Eqs. (1.12)–(1.13) is shown in Fig. 1.5. The initial starting point $(a, b) = (1, 0)$ for the continuation analysis was here obtained from the known solution when $Y = \cosh 1$. The result of the analysis clearly agrees with the analytical prediction, albeit here represented by a finite number of numerical solutions to the coupled equations in Eqs. (1.12)–(1.13).

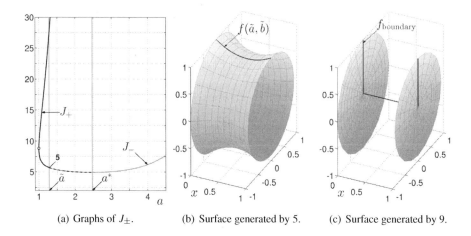

(a) Graphs of $J_{\pm}$.   (b) Surface generated by 5.   (c) Surface generated by 9.

Figure 1.4. *Graphs of the functions* $J_{\pm}$ *defined in Eq.* (1.32) *along the family of stationary solution curves* (a). *Points found along the solid black segment correspond to global minimizers, whereas those corresponding only to weak local minimizers are found along the dashed segment* $(a, J_{-}(a))$, *for* $a \in (\tilde{a}, a^*)$ *(see Sects.* 1.2.1 *and* 1.2.3 *for definitions of* a^* *and* $\tilde{a}$, *respectively). Panels* (b)–(c) *show the surfaces of revolution generated by curves 5 and 9 in Fig.* 1.3, *respectively. The surface generated by the "boundary" curve defined in Sect.* 1.2.3 *consists of two circular disks.*

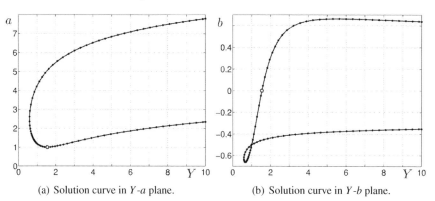

(a) Solution curve in Y-a plane.   (b) Solution curve in Y-b plane.

Figure 1.5. *A branch of solutions of Eqs.* (1.12)–(1.13) *obtained using numerical continuation, as described in Sect.* 1.3.1. *Here, dots indicate computed points, which are connected with straight-line segments. The circle denotes the initial point at* $a = 1$, $b = 0$, *and* $Y = \cosh 1$. *Compare the numerical results with the analysis shown in Fig.* 1.2.

1.3.2   A two-point boundary-value problem

Alternatively, consider the Euler–Lagrange differential equation in Eq. (1.7) written as the following system of two first-order differential equations:

$$\begin{pmatrix} f \\ f' \end{pmatrix}' = \begin{pmatrix} f' \\ (1+f'^2)/f \end{pmatrix}. \tag{1.34}$$

If f is continuously differentiable on the interval $[0,1]$, it follows from Weierstrass's theorem that f and f' may each be approximated arbitrarily closely by polynomials p_1 and p_2, respectively, of sufficiently high degree. Specifically, let

$$p_1(x) := \sum_{k=0}^{m} a_k x^k, \qquad p_2(x) := \sum_{k=0}^{m} b_k x^k. \tag{1.35}$$

We seek to impose conditions on the $2(m+1)$ unknown coefficients $\{a_k\}_{k=0}^{m}$ and $\{b_k\}_{k=0}^{m}$ with the hope that the approximation error may be bounded appropriately and that the sequences of polynomial approximants obtained as $m \to \infty$ converge uniformly on $[0,1]$ to the unknown function f and its derivative f', respectively.

To this end, we translate first the boundary conditions on f into corresponding constraints on p_1, i.e., $p_1(0) = 1$ and $p_1(1) = Y$ or, equivalently,

$$a_0 = 1, \qquad \sum_{k=0}^{m} a_k = Y. \tag{1.36}$$

An additional $2m$ conditions follow by requiring that the polynomial approximants satisfy Eq. (1.34) at m suitably selected *collocation nodes* $\{x_i\}_{i=1}^{m}$ on the interval $[0,1]$, i.e., that

$$\sum_{k=1}^{m} k a_k x_i^{k-1} = \sum_{k=0}^{m} b_k x_i^k \tag{1.37}$$

and

$$\sum_{k=1}^{m} k b_k x_i^{k-1} = \left(1 + \left(\sum_{k=0}^{m} b_k x_i^k \right)^2 \right) \bigg/ \sum_{k=0}^{m} a_k x_i^k \tag{1.38}$$

for $i = 1,\ldots,m$. Provided that these equations may be solved for $\{a_k\}_{k=0}^{m}$ and $\{b_k\}_{k=0}^{m}$, Eqs. (1.36)–(1.38) collectively define a family of polynomial approximants for viable extremal curves $y = f(x)$ implicitly in terms of the value of Y. A comparison between the exact global minimizer given in Eq. (1.11) for $a = 1.5$ and that obtained using a polynomial approximant of degree $m = 2$ is shown in Fig. 1.6. *Spectral convergence* is observed as m grows, as suggested by the analysis in Fig. 1.7.

More generally, we may approximate f and f' separately as continuous functions of x, expressed on each of N equal-sized intervals as polynomials of degree m and parameterized by the unknown values at $m + 1$ equidistant and maximally distributed points. Finally, we impose conditions equivalent to Eq. (1.34) at m suitably chosen collocation nodes in each subinterval for a total of $2Nm$ equations in terms of the $2Nm - 2$ unknown interior values of the piecewise-polynomial approximants for f and f' and the 2 unknown values of the piecewise-polynomial approximant for f' at $x = 0$ and $x = 1$. An example of a family of curves that results from numerical continuation applied to such an approximate formulation is shown in Fig. 1.8(a).

1.3.3 A quadrature approximation

Finally, we consider seeking a locally optimal value for a suitable discretization of the defining integral in Eq. (1.1). Specifically, we substitute for $J(f)$ a *numerical quadrature*

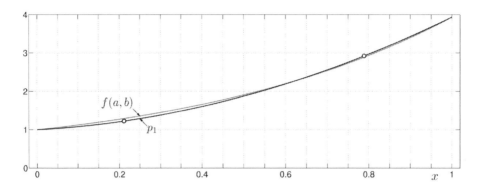

Figure 1.6. *Comparison of the exact solution (gray), obtained from Eq. (1.11), with a numerical solution of the form in Eq. (1.35) of degree $m = 2$ for the case when $b = b_+(a)$, $Y = Y_+(a)$, and $a = 1.5$. The two collocation nodes are marked with circles. For increasing order m, we observe spectral convergence; see Fig. 1.7.*

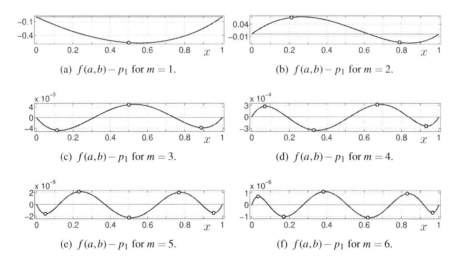

Figure 1.7. *Graphs of the approximation errors $\delta_m := f(a,b) - p_1$ of polynomial approximants p_1 of degree m as in Eq. (1.35), for the case when $b = b_+(a)$, $Y = Y_+(a)$, and $a = 1.5$. In each case, the m collocation nodes are indicated by circles. Panel (b) corresponds to the graphs in Fig. 1.6. We observe spectral convergence, i.e., $\|\delta_m\| < CM^{-m}$ for some $0 < C < \infty$ and $M > 1$.*

of the form

$$I(f) := \sum_{i=1}^{n} w_i \left(2\pi f(x_i) \sqrt{1 + (f'(x_i))^2} \right) \tag{1.39}$$

for some given sequences of *quadrature weights* $\{w_i\}_{i=1}^{n}$ and *quadrature nodes* $\{x_i\}_{i=1}^{n}$ in $[0, 1]$ and some integer n. Suppose again that we approximate the unknown extremal

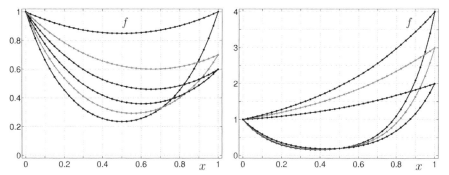

(a) Solutions of the boundary-value problem formulation.

(b) Solutions of the quadrature approximation.

Figure 1.8. *Sample approximate extremal curves of the integral functional in Eq. (1.3) with action integrand given in Eq. (1.2) and satisfying the given boundary conditions at $x = 0$ and $x = 1$. The results shown here were obtained using a collocation method with $N = 10$ and $m = 4$ applied to a two-point boundary-value problem (a) as defined in Sect. 1.3.2, and to a quadrature approximation (b) as defined in Sect. 1.3.3. Each panel shows pairs of stationary curves for each given value of Y.*

function f by a polynomial p of degree m:

$$p(x) := \sum_{k=0}^{m} a_k x^k. \tag{1.40}$$

We seek to impose conditions on the $m+1$ unknown coefficients $\{a_k\}_{k=0}^m$ with the hope that the approximation error may be bounded appropriately and that the sequence of polynomial approximants obtained as $m, n \to \infty$ converges uniformly on $[0,1]$ to the unknown function f.

To this end, we again translate the boundary conditions on f into corresponding constraints on p, i.e., $p(0) = 1$ and $p(1) = Y$ or, equivalently,

$$a_0 = 1, \qquad \sum_{k=0}^{m} a_k = Y. \tag{1.41}$$

We accommodate these constraints on the optimization of the quadrature approximation in Eq. (1.39) by imposing, in addition, the vanishing of the partial derivatives of the function $F(a_0, \ldots, a_m, \lambda_1, \lambda_2) := I(p) - \lambda_1(a_0 - 1) - \lambda_2\left(\sum_{k=0}^m a_k - Y\right)$ with respect to the coefficients $\{a_k\}_{k=0}^m$ or, equivalently, the $m-1$ conditions

$$\left(\frac{\partial}{\partial a_l} - \frac{\partial}{\partial a_m}\right) \sum_{i=1}^{n} w_i \left(2\pi \sum_{k=0}^{m} a_k x_i^k \sqrt{1 + \left(\sum_{j=1}^{m} j a_j x_i^{j-1}\right)^2}\right) = 0 \tag{1.42}$$

for $l = 1, \ldots, m-1$. Provided that these equations may be solved for $\{a_k\}_{k=0}^m$, Eqs. (1.41)–(1.42) collectively define a family of polynomial approximants for viable extremal curves $y = f(x)$ implicitly in terms of the value of Y.

More generally, consider the approximation of f by a continuous function of x, expressed on each of N equal-sized intervals as a polynomial of degree m and parameterized by the unknown values at $m + 1$ equidistant and maximally distributed points, and approximate f' by the (likely discontinuous) piecewise-polynomial derivative of the approximant for f. The functional J in Eq. (1.1) evaluated at the piecewise-polynomial approximant of f may then be approximated by a weighted sum of the values of the integrand at each of n suitably chosen nodes per interval across all N intervals. Finally, we impose the $Nm - 1$ conditions that the derivatives of this quadrature approximation with respect to each of the $Nm - 1$ unknown interior values of the piecewise-polynomial approximant of f should vanish. An example of a family of curves that results from numerical continuation applied to such an approximate formulation is shown in Fig. 1.8(b).

1.4   Conclusions

This chapter has demonstrated the utility of the continuation paradigm in seeking families of approximate solutions to the classical problem of finding a *catenary curve* between two points in the plane. It is not much of a stretch to see the value of this methodology when applied to problems for which an analytical solution is unavailable except, possibly, for isolated values of the problem parameters.

The discussion has also illustrated the possibility of arriving at the sought-after solution through a variety of alternative (and typically approximate) formulations. Each of these belongs to a distinct class of *continuation problems* and requires a separate, and careful, analysis of its numerical accuracy and stability.

In the next chapter, we proceed to introduce the terminology, notation, and fundamental mathematical paradigm associated with a general class of continuation problems that includes all formulations considered in this chapter. The chosen formalism is shown to support natural tasks associated with continuation, including the encapsulation of monitor functions within a continuation problem.

Exercises

1.1. Choose a basic text on the calculus of variations and review the derivation of the Euler–Lagrange equation for an action integrand of the form $L(x, f, f')$. Then derive the equivalent differential equation in Eq. (1.7) for the action integrand given in Eq. (1.2).

1.2. Prove the existence of the *first integral* in Eq. (1.9) for solutions $f(x)$ of the Euler–Lagrange equation in Eq. (1.4), when the action integrand is independent of x.

1.3. Show the explicit derivation of Eq. (1.11).

1.4. Use the implicit function theorem to prove the existence of a local branch of solutions to Eqs. (1.12)–(1.13) through $a = 1$, $b = 0$, and $Y = \cosh 1$.

1.5. Use calculus to draw the conclusions regarding the behavior of $Y_\pm(a)$ in Eqs. (1.16)–(1.17). Hint: show that the derivative of $Y_+(a)$ can never equal zero. Similarly, show that the derivative of $Y_-(a)$ has only one zero (which necessarily must occur for $a > \sqrt{2}$).

1.6. Explain how the observations regarding the behavior of $Y_{\pm}(a)$ are used to draw conclusions as to the number of solutions to Eqs. (1.12)–(1.13).

1.7. The results obtained in Sect. 1.2 show that stationary curves of the integral functional in Eq. (1.1) exist only for $Y \geq Y^*$. Under the transformation $f(x) \mapsto f(1-x)/Y$, however, the boundary conditions

$$f(0) = 1, \, f(1) = Y$$

become

$$f(0) = 1, \, f(1) = \frac{1}{Y}.$$

It follows that for every stationary curve found for $Y > 1$ there exists a stationary curve for $0 < Y^{-1} < 1$, and vice versa. What is wrong with this argument?

1.8. Choose an advanced text on the calculus of variations and review the discussion of the necessary and sufficient conditions for a function f to be a weak local minimum of a functional J. What famous mathematician's name is associated with the condition given in Eq. (1.20)?

1.9. Why is there no loss of generality to assume that $c_1 \neq 0$ in Eq. (1.21) when $a > 1$?

1.10. Given $\varphi_{\pm}(a)$ defined in Eq. (1.26), what is the value of $\varphi_{\pm}(a)|_{x=0}$?

1.11. Show that $[\varphi_{\pm}(a)](x) \approx \pm \sinh x$ for a near 1.

1.12. Use the result of the previous two exercises and the slope at $x = 0$ to explain the conclusion regarding a change in extremum from a local minimum to a saddle accompanying a change in sign of $\varphi_{\pm}(a)|_{x=1}$.

1.13. Derive the expression for $\varphi_{\pm}(a)|_{x=1}$ given in Eq. (1.30).

1.14. Derive the expression for the value of the functional J given in Eq. (1.32).

1.15. Explain the sense in which the curve consisting of a line segment from $(0,1)$ to $(0,0)$, followed by a line segment from $(0,0)$ to $(1,0)$, followed by a line segment from $(1,0)$ to $(1,Y)$ is a "boundary" of the solution space of smooth curves.

1.16. Derive the expression for the limiting value of the functional J given in Eq. (1.33) and use this to verify the conclusions regarding the existence of a global minimizer.

1.17. Choose an introductory text on numerical methods and review the derivation of the formula for the interpolation error

$$f(x) - p_m(x) = \frac{1}{(m+1)!} f^{(m+1)}(\xi) \prod_{k=0}^{m} (x - x_k),$$

where $f \in \mathcal{C}^{m+1}([-1,1])$, p_m is an *interpolating polynomial* of degree m, such that $p_m(x_k) = f(x_k)$ for some sequence $\{x_k\}_{k=0}^{m}$ of nodes in $[-1,1]$, and $\xi \in [-1,1]$. Derive the error estimate

$$\|f - p_m\|_{\infty} \leq \frac{1}{(m+1)!} \left\| f^{(m+1)} \right\|_{\infty} \|w(x_0, \ldots, x_m)\|_{\infty},$$

where

$$w(x_0, \ldots, x_m) := \prod_{k=0}^{m} (x - x_k)$$

and $\| \cdot \|_{\infty}$ is the supremum norm.

1.18. Consider the function $w(x_0, \ldots, x_m)$ introduced in the previous exercise. Find numerical estimates for the supremum norm $\|w(x_0, \ldots, x_m)\|_\infty$ for (i) the uniform sequence $x_k := 2k/m - 1$ and (ii) the Chebyshev sequence $x_k := -\cos(k\pi/m)$ for $k = 0, \ldots, m$ and $m = 1, 2, \ldots, N$ for some $N \geq 10$. In each case, graph the function $w(x_0, \ldots, x_m)$ and compare the graphs. What is the main qualitative difference?

1.19. Let $f \in \mathcal{C}^\infty([-1, 1])$. Formulate a sufficient condition on $f^{(m)}$ to guarantee convergence of the interpolation $p_m \to f$ with rate $\|f - p_m\|_\infty \leq C\|w(x_0, \ldots, x_m)\|_\infty$, where C is a constant independent of m. State examples of functions that satisfy this condition. Compute the actual rate of convergence for these functions.

1.20. Let $f \in \mathcal{C}^\infty([-1, 1])$ and consider the problem of finding an interpolating polynomial of the form $p_m(x) = \sum_{k=0}^m a_k x^k$, given a sequence $\{x_k\}_{k=0}^m$ in $[-1, 1]$. Show that this may be written in the form of a linear equation $C_m \cdot A_m = f_m$ and identify C_m, A_m, and f_m. Investigate numerically the dependence of the condition number of C_m on the order m and the sequence $\{x_k\}_{k=0}^m$. What modification to the form of the interpolating polynomial would improve this condition number?

1.21. Consider polynomial interpolation of the functions $\exp x$, $\sin(\pi x)$, and $\sqrt{1 - x^2}$ on $x \in [-1, 1]$. What is the theoretically predicted rate of convergence? Compute coefficients of interpolation polynomials as the solution of a linear equation of the form $C_m \cdot A_m = f_m$. Use these to compute a numerical estimate of the actual interpolation error and compare with numerical estimates of the theoretical prediction for large $m = 10, 20, 30, 40, \ldots$ for different sequences $\{x_k\}_{k=0}^m$. Explain your observations.

1.22. Consider the general discretization of Eq. (1.34) obtained by

1. approximating f and f' separately as continuous functions of x, expressed on each of N equal-sized intervals as polynomials of degree m, and parameterized by the unknown values at $m + 1$ equidistant and maximally distributed points; and

2. imposing that the piecewise-polynomial approximants satisfy Eq. (1.34) at m suitably chosen collocation nodes in each subinterval.

Show that this formulation results in $2Nm$ equations in terms of the $2Nm - 2$ unknown interior values of the piecewise-polynomial approximants for f and f' and the 2 unknown values of the piecewise-polynomial approximant for f' at $x = 0$ and $x = 1$. Write an explicit set of discrete equations for small integer choices of N and m.

1.23. Consider the general discretization of the problem of finding a local minimum of the functional in Eq. (1.1) obtained by

1. approximating f by a continuous function of x, expressed on each of N equal-sized intervals as a polynomial of degree m, and parameterized by the unknown values at $m + 1$ equidistant and maximally distributed points;

2. approximating f' by the piecewise-polynomial derivative of the approximant for f;

3. approximating the integral functional J in Eq. (1.1) evaluated at the piecewise-polynomial approximant of f by a weighted sum of the values of the integrand at each of n suitably chosen nodes across all N intervals; and

  4. requiring that the derivative of this quadrature approximation, with respect to each of the unknown interior values of the polynomial approximant of f, should vanish.

Show that this formulation results in a total of $Nm - 1$ equations in terms of the $Nm - 1$ interior values of the polynomial approximant of f. Write an explicit set of discrete equations for small integer choices of N and m.

1.24. Let $I(p)$ be obtained from Eq. (1.39) after substitution of Eq. (1.40). Derive Eqs. (1.41)–(1.42) by eliminating the *Lagrange multipliers* λ_1 and λ_2 from the condition that all first-order partial derivatives of

$$F(a_0, \ldots, a_m, \lambda_1, \lambda_2) := I(p) - \lambda_1(a_0 - 1) - \lambda_2 \left(\sum_{k=0}^{m} a_k - Y \right)$$

equal zero.

1.25. Given a polynomial $p := \sum_{k=0}^{m} a_k x^k$, consider the sequence $\{b_k\}_{k=0}^{m}$, where

$$b_0 := a_0,$$

$$b_1 := \sum_{k=0}^{m} a_k,$$

$$b_2 := \left(\frac{\partial}{\partial a_1} - \frac{\partial}{\partial a_m} \right) \sum_{i=1}^{n} w_i \left(2\pi p(x_i) \sqrt{1 + (p'(x_i))^2} \right),$$

$$\vdots$$

$$b_m := \left(\frac{\partial}{\partial a_{m-1}} - \frac{\partial}{\partial a_m} \right) \sum_{i=1}^{n} w_i \left(2\pi p(x_i) \sqrt{1 + (p'(x_i))^2} \right)$$

for some sequence $\{w_i\}_{i=1}^{n}$ of positive scalars and some sequence $\{x_i\}_{i=1}^{n}$ of distinct points on the interval $[0, 1]$. Explore the invertibility of the map $\{a_k\}_{k=0}^{m} \mapsto \{b_k\}_{k=0}^{m}$ for different choices of m and n. Does the result depend on the sequences $\{w_i\}_{i=1}^{n}$ and $\{x_i\}_{i=1}^{n}$? What are the implications of your observations to the discretization method proposed in Sect. 1.3.3?

1.26. Choose a basic text on the calculus of variations and review the treatment of *isoperimetric problems*. Apply your insights to the special case of finding a smooth curve $y = f(x)$ that minimizes the integral

$$J(f) := \rho g \int_0^1 f(x) \sqrt{1 + (f'(x))^2} dx$$

under the integral constraint

$$\int_0^1 \sqrt{1 + (f'(x))^2} dx = L$$

and the boundary conditions $f(0) = 1$ and $f(1) = Y$. Show that a unique solution to this problem when $Y = L = 1$ is given by $f(x) = 1$. Explain how you would apply the numerical techniques discussed in this chapter to the continuation of locally optimal solutions to this problem.

Chapter 2

Encapsulation

Tools that enable the numerical solution of a variety of continuation problems find widespread use in the fundamental and applied analysis of natural or engineered systems. The observations in the previous chapter illustrate an often natural analytical progression. Such an analysis typically proceeds from an initial problem formulation, through a preliminary theoretical study, to a computational implementation. The computational implementation affords the analyst the ability to explore problem solutions across ranges in parameter space, where no closed-form solutions may be available.

At the core of the continuation paradigm is the notion of (local) parameterizations of solutions. In studying the catenary problem in Chap. 1, we recognized the dependence of its solutions on the value of the ordinate Y. We found ranges in Y for which there were no locally optimal curves corresponding to the functional in Eq. (1.1), as well as ranges over which there existed multiple extremal curves (albeit not all corresponding to minima). It was natural to consider Y a priori as a *problem parameter* whose value could be controlled during continuation or extracted from a candidate solution, as in the explicit expressions shown in Eqs. (1.16)–(1.17) in terms of the *problem variable a*. Away from $Y = Y^*$, we found that the solution branches could be locally parameterized by Y, albeit at best implicitly.

As shown in the discussion below, parameters associated with a continuation problem may also arise naturally from an a posteriori characterization of a solution. Access to such parameters can be of great utility in applications. Their proper handling within a computational implementation yields a powerful conceptual approach to studying constrained families of solutions to a continuation problem. We accomplish such a framework in a notion of *encapsulation*, the close association between the nonlinear equations that define problem solutions and the algorithms that encode various computable properties of the problem solutions. As shown in Chap. 4, this form of encapsulation naturally supports the design of toolboxes dedicated to individual classes of continuation problems. With proper attention to the paradigm of encapsulation, especially the assignment of *toolbox data* that parameterizes the contribution of individual toolboxes to a composite continuation problem, the full concept of *task embedding* is then realized in Chap. 5.

2.1   Solution measures and constraints

Consider the task of continuing *periodic solutions* $x : \mathbb{R} \to \mathbb{R}^n$, where $x(t+T) = x(t)$ for some positive scalar T, of an *autonomous dynamical system*

$$\frac{dx}{dt} = f(x, \lambda) \tag{2.1}$$

for some *vector field* $f : \mathbb{R}^n \times \mathbb{R}^s \to \mathbb{R}^n$, parameterized by a vector of *problem parameters* $\lambda \in \mathbb{R}^s$. Periodic solutions may or may not exist for individual values of λ. When such a solution does exist, there is no a priori information in f that determines uniquely the value of the *period* T. Instead, T is a natural parameter characterizing the solution that must be arrived at as part of the continuation analysis.

Following the discretization methodology introduced in the previous chapter, consider approximating the unknown periodic solution $x(t)$ on the yet-to-be-determined interval $[0, T]$ by a polynomial $p : \mathbb{R} \to \mathbb{R}^n$ of degree m:

$$p(t) := \sum_{k=0}^{m} a_k t^k, \, t \in [0, \mathcal{T}]. \tag{2.2}$$

We seek to impose conditions on the $n(m+1)$ unknown coefficients $\{a_k\}_{k=0}^{m}$ and on the unknown period $\mathcal{T}$, with the hope that the approximation error may be bounded appropriately and that the sequences of approximate periods $\mathcal{T}$ and polynomial approximants obtained as $m \to \infty$ converge uniformly to the actual period T and the unknown function x on the interval $[0, T]$, respectively.

To this end, we translate the condition of periodicity of x into a corresponding constraint on p, i.e., $p(0) = p(\mathcal{T})$ or, equivalently,

$$a_0 = \sum_{k=0}^{m} a_k \mathcal{T}^k. \tag{2.3}$$

An additional collection of nm conditions follows by requiring that the polynomial approximant satisfy Eq. (2.1) at m suitably selected collocation nodes $\{\alpha_i \mathcal{T}\}_{i=1}^{m}$ on the interval $[0, \mathcal{T}]$, i.e., that

$$\sum_{k=1}^{m} k a_k (\alpha_i \mathcal{T})^{k-1} = f\left(\sum_{k=0}^{m} a_k (\alpha_i \mathcal{T})^k, \lambda\right) \tag{2.4}$$

for $i = 1, \ldots, m$.

Eqs. (2.3)–(2.4) constitute $n(m+1)$ equations in the $n(m+1)+1$ unknowns, expressed in terms of the vector of problem parameters λ. We may intuit the lack of closure (i.e., the excess of unknowns relative to equations) as associated with a *phase invariance*. Specifically, by the autonomous nature of the vector field, it follows that for every periodic solution $x(t)$ of period T, the function $t \mapsto x(t+\tau)$ is also a periodic solution of period T for arbitrary values of the *phase shift* τ. Each such phase-shifted solution corresponds to a translation of the original solution along the independent variable. The corresponding state-space trajectories, however, are geometrically indistinguishable.

In order to arrive at a closed problem, in which the number of unknowns equals the number of equations, suppose that x^* represents a known approximation to the periodic

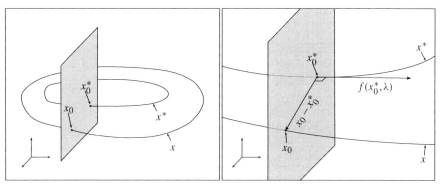

(a) Initial points in hyperplane transversal to x^* and x.

(b) Fixing x_0 in plane through x_0^* and normal to $f(x_0^*, \lambda)$.

Figure 2.1. *Due to translational invariance, a periodic solution of an autonomous ordinary differential equation has a free phase; i.e., the point $x_0 := x(0)$ is not defined uniquely. A typical way to obtain a unique initial point is to require that it lie in a hyperplane* (a), *transversal to the state-space representation of the solution. If the reference function x^* is sufficiently close to the sought-after orbit x, as seen in* (b), *one may use the hyperplane through x_0^* and perpendicular to the tangent vector $f(x_0^*, \lambda)$.*

solution obtained through other means (or, more likely, from a previous step of a continuation algorithm). Then, as suggested in Fig. 2.1, one possible constraint on the polynomial approximant p is the requirement that the deviation of $p(t)$ from $x^*(t)$, for some $t \in [0, \mathcal{T}]$, be orthogonal to the vector field f evaluated at $x^*(t)$, i.e., that

$$\left(x^*(t) - \sum_{k=0}^{m} a_k t^k \right)^T \cdot f\left(x^*(t), \lambda \right) = 0. \tag{2.5}$$

Alternatively, we may consider imposing a vanishing average value of these dot products across the entire interval $[0, \mathcal{T}]$:

$$\int_0^{\mathcal{T}} \left(x^*(t) - \sum_{k=0}^{m} a_k t^k \right)^T \cdot f\left(x^*(t), \lambda \right) dt = 0, \tag{2.6}$$

or, when discretized using numerical quadrature,

$$\sum_{j=1}^{q} w_j \left(x^*(\beta_j \mathcal{T}) - \sum_{k=0}^{m} a_k \left(\beta_j \mathcal{T} \right)^k \right)^T \cdot f\left(x^*(\beta_j \mathcal{T}), \lambda \right) = 0 \tag{2.7}$$

for some sequence of quadrature weights $\{w_j\}_{j=1}^{q}$ and quadrature nodes $\{\beta_j \mathcal{T}\}_{j=1}^{q}$ on the interval $[0, \mathcal{T}]$ for some integer q. Either of these *phase conditions* now provides closure to the continuation problem. Assuming that the closed continuation problem may be solved for the $n(m+1)+1$ unknowns $\{a_k\}_{k=0}^{m}$ and $\mathcal{T}$, its constituent equations collectively define a family of polynomial approximants for viable periodic solutions implicitly in terms of the value of λ.

As alluded to above, it is often advantageous to be able to refer to certain a posteriori numerical characterizations of a given solution. This may be desired for purely descriptive purposes, or because one is interested in constrained families of solutions for which these characterizations take on certain critical values. We have already come across the approximate period $\mathcal{T}$, a solution measure that is used commonly in physical applications. Other relevant measures may involve individual or combinations of Fourier coefficients of the periodic extension of period $\mathcal{T}$ of the approximating polynomial on $[0, \mathcal{T}]$. As an example, consider the mean-squared value of the deviation of the polynomial approximant from its mean on the interval $[0, \mathcal{T}]$:

$$\frac{1}{\mathcal{T}} \int_0^{\mathcal{T}} \left\| p(t) - \frac{1}{\mathcal{T}} \int_0^{\mathcal{T}} p(s)\,ds \right\|^2 dt. \tag{2.8}$$

Now suppose that a family of periodic solutions of the dynamical system in Eq. (2.1) may be locally parameterized by the vector of problem parameters λ. It follows that the approximate period $\mathcal{T}$ and the mean-squared value in Eq. (2.8) are functions of the components of λ. During continuation along this family, their numerical values characterize key properties of the solutions and provide useful measures for quantifying the dependence of the solutions on λ.

Suppose, instead, that it is desirable to continue a constrained family of periodic solutions of Eq. (2.1) for which the approximate period $\mathcal{T}$ or the mean-squared value in Eq. (2.8) is fixed at some preferred value (see, e.g., Fig. 2.2). Having located a candidate solution to either of these problems, we need to append an additional condition to the equations for the polynomial coefficients and the approximate period $\mathcal{T}$ in order to constrain

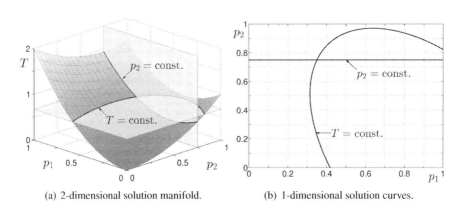

(a) 2-dimensional solution manifold.   (b) 1-dimensional solution curves.

Figure 2.2. *A manifold of periodic orbits of a hypothetical dynamical system depending on two parameters p_1 and p_2, projected onto the hyperplane with coordinates p_1, p_2, and T. Codimension-1 intersections of the manifold with hyperplanes corresponding to constant values of the parameter p_2 or of the period T, respectively, are shown as curves in (a), and projected onto the (p_1, p_2)-parameter plane in (b). In the terminology of Sect. 2.2, each curve is obtained by applying continuation to a suitably restricted continuation problem that is initializable with a 1-dimensional atlas algorithm.*

the solution accordingly. At this stage, however, we find an excess of equations relative to variables. Closure is again obtained only by forgoing the desire to assign values to all the problem parameters λ. Instead, we may choose to fix the values of all but one of these parameters, leaving the remaining parameter as an additional unknown variable to be solved for appropriately. Between the two possible solution constraints and a total of s problem parameters, it is clear that there exist $2s$ possible such continuation problems that could be encountered and for which computational implementations might be sought. For large s, one quickly sees value in not having to encode each of these additional continuation problems separately.

In the remainder of this chapter, we describe the theoretical foundations for encapsulating, in a common formulation, the equations defining problem solutions (e.g., Eqs. (2.3)–(2.5) in the case of periodic solutions) and the algorithms that encode computable properties of these solutions (e.g., Eq. (2.8) for the mean-squared value of the deviations from the mean). A consequence of this framework is the possibility of run-time design of individual continuation problems, as explained in greater detail in Chap. 3.

2.2 Continuation problems

2.2.1 The zero problem

Suppose that the function $\Phi : \mathbb{R}^n \to \mathbb{R}^m$, where $n \geq m \geq 1$, is continuously differentiable. We refer to the equation $\Phi(u) = 0$ as a *(continuation) zero problem* in the vector u of *continuation variables*. In this context, we refer to the components of Φ as *zero functions*.

Let $\partial_u \Phi(u^*) \in \mathbb{R}^{m \times n}$ denote the Jacobian of Φ at some point u^*, for which $\Phi(u^*) = 0$, and suppose that $\partial_u \Phi(u^*)$ has full rank. For $n > m$, the implicit function theorem then implies that there exists a locally unique $(n - m)$-dimensional manifold through u^* of solutions to the zero problem that can be parameterized by some choice of $n - m$ components of u. In this case, the point u^* is a *regular point* of the zero problem and we refer to the manifold dimensionality as the *dimensional deficit* of the zero problem. Solution manifolds of dimension 1 are often called *branches*.

Starting with a single *chart*, i.e., a regular point u^* on the solution manifold together with a local parameterization of the manifold near this point, continuation employs an *atlas algorithm* for computing nearby charts. The process is subsequently repeated for each of the nearby charts. The computed atlas of charts is a *covering* of the manifold.

2.2.2 The extended continuation problem

Denote by $\Psi : \mathbb{R}^n \to \mathbb{R}^r$ a family of continuously differentiable *monitor functions* and let $\mu^* := \Psi(u^*)$. Now consider the *extended continuation problem* $F(u, \mu) = 0$, where

$$F : (u, \mu) \mapsto \begin{pmatrix} \Phi(u) \\ \Psi(u) - \mu \end{pmatrix} \qquad (2.9)$$

and $\mu \in \mathbb{R}^r$ is a vector of *continuation parameters*. By construction, $F(u^*, \mu^*) = 0$. Moreover, F maps $\mathbb{R}^n \times \mathbb{R}^r$ into $\mathbb{R}^m \times \mathbb{R}^r$ and the rank deficit of the Jacobian

$$\partial_{(u,\mu)} F(u^*, \mu^*) = \begin{pmatrix} \partial_u \Phi(u^*) & 0 \\ \partial_u \Psi(u^*) & -I \end{pmatrix} \qquad (2.10)$$

equals that of $\partial_u \Phi(u^*)$. Thus, if u^* is a regular point of the zero problem, it follows that the solution manifold of the extended continuation problem through (u^*, μ^*) is also of dimension $n - m$.

During continuation, each monitor function tracks the value (through the corresponding component of μ) of some nonlinear function of the continuation variables along the solution manifold to the original zero problem. As shown next, the embedding of monitor functions in the extended continuation problem enables the imposition of additional nonlinear constraints on the solution of the corresponding zero problem.

2.2.3   The restricted continuation problem

Let $\mathbb{I} \subseteq \{1, \ldots, r\}$ be an *index set* of cardinality $|\mathbb{I}| \leq n - m$ and denote its complement by $\mathbb{J} := \{1, \ldots, r\} \setminus \mathbb{I}$. Now consider the restriction $G : \mathbb{R}^{n+r-|\mathbb{I}|} \to \mathbb{R}^{m+r}$, where

$$G : (u, \mu_{\mathbb{J}}) \mapsto F(u, \mu)|_{\mu_{\mathbb{I}} = \mu_{\mathbb{I}}^*}. \tag{2.11}$$

Provided that the Jacobian $\partial_{(u, \mu_{\mathbb{J}})} G(u, \mu_{\mathbb{J}})$ has full rank at $(u^*, \mu_{\mathbb{J}}^*)$, the solution manifold of the *restricted continuation problem* $G(u, \mu_{\mathbb{J}}) = 0$ through the point $(u^*, \mu_{\mathbb{J}}^*)$ corresponds to an $(n - m - |\mathbb{I}|)$-dimensional embedded submanifold of the original solution manifold. In this case, u^* is a regular point of the equivalent *reduced continuation problem*

$$\begin{pmatrix} \Phi(u) \\ \Psi_{\mathbb{I}}(u) - \mu_{\mathbb{I}}^* \end{pmatrix} = \begin{pmatrix} 0 \\ 0 \end{pmatrix}. \tag{2.12}$$

2.2.4   Initializability and closure

In practice, we define the restriction in Eq. (2.11) by substituting the ordered subset $\Psi_{\mathbb{I}}(u_0)$, where $u_0 \in \mathbb{R}^n$ is some *initial solution guess*, for $\mu_{\mathbb{I}}^*$. We do this with the expectation that there exists a regular solution $(u^*, \mu_{\mathbb{J}}^*)$ to the corresponding restricted continuation problem and that $u^* \approx u_0$.

The restricted continuation problem corresponding to the index set $\mathbb{I}$ is said to be *initializable with a d-dimensional atlas algorithm* on a neighborhood of a regular point if and only if $d = n - m - |\mathbb{I}|$. In this case, since a d-dimensional atlas algorithm adds d *projection conditions* to the restricted continuation problem, there results a closed set of equations with matching numbers of equations and unknowns. We refer to such a closed set of equations as a *closed continuation problem*.

As a special case, a 0-dimensional atlas algorithm adds no equations to the restricted continuation problem. In the event that $|\mathbb{I}| = n - m$, a regular solution $(u^*, \mu_{\mathbb{J}}^*)$ is then a locally unique solution to the restricted continuation problem. In this case, we often say simply that the restricted continuation problem is *initializable*.

2.2.5   Examples

Example 2.1 To illustrate the formalism introduced above, consider the task of tracing a 1-dimensional manifold passing through the point $u^* = (1, 1)$ and consisting of points $u \in \mathbb{R}^2$ whose Euclidean distances to the point $(0, 1)$ equal 1, i.e., a circle in $\mathbb{R}^2$ of unit radius and centered on $(0, 1)$.

A candidate zero problem whose solutions lie on this manifold is given by the equation $\Phi(u) = 0$, where

$$\Phi : u \mapsto u_1^2 + (u_2 - 1)^2 - 1. \tag{2.13}$$

Here, $m = 1$ and $n = 2$. Moreover, since the Jacobian

$$\partial_u \Phi(u) = \begin{pmatrix} 2u_1 & 2(u_2 - 1) \end{pmatrix} \tag{2.14}$$

has full rank at u^*, it follows that the local solution manifold through u^* is of dimension $n - m = 1$. Indeed, since $\partial_u \Phi(u)$ has full rank everywhere on the solution set, all points on the solution manifold are regular. In the absence of additional monitor functions, the corresponding restricted continuation problem $G(u) = 0$ with

$$G : u \mapsto u_1^2 + (u_2 - 1)^2 - 1 \tag{2.15}$$

is initializable with a 1-dimensional atlas algorithm on a neighborhood of u^*.

Now suppose that $r = 1$ and consider the addition of the single monitor function

$$\Psi : u \mapsto u_1^2 + u_2^2, \tag{2.16}$$

such that $\mathbb{I} = \emptyset$ and $\mathbb{J} = \{1\}$. Clearly, the value of Ψ is constant along each of a family of concentric circles centered at the origin, along each of which it equals the square of the corresponding radius. Let $\mu^* := \Psi(u^*) = 2$ and consider the corresponding restricted continuation problem $G(u, \mu_{\mathbb{J}}) = 0$, where

$$G : (u, \mu_{\mathbb{J}}) \mapsto \begin{pmatrix} u_1^2 + (u_2 - 1)^2 - 1 \\ u_1^2 + u_2^2 - \mu_1 \end{pmatrix}. \tag{2.17}$$

Since the Jacobian

$$\partial_{(u, \mu_{\mathbb{J}})} G(u, \mu_{\mathbb{J}}) = \begin{pmatrix} 2u_1 & 2(u_2 - 1) & 0 \\ 2u_1 & 2u_2 & -1 \end{pmatrix} \tag{2.18}$$

has full rank at the point $(u^*, \mu_{\mathbb{J}}^*)$, it again follows that the solution manifold through $(u^*, \mu_{\mathbb{J}}^*)$ is of dimension $n - m = 1$. The restricted continuation problem is again initializable with a 1-dimensional atlas algorithm on a neighborhood of the point $(u^*, \mu_{\mathbb{J}}^*)$. Solutions to the restricted continuation problem correspond to points along the unit circle centered on $(0, 1)$ together with the square of their distances to the origin.

Suppose, instead, that $\mathbb{I} = \{1\}$ and $\mathbb{J} = \emptyset$, in which case

$$G : u \mapsto \begin{pmatrix} u_1^2 + (u_2 - 1)^2 - 1 \\ u_1^2 + u_2^2 - 2 \end{pmatrix}. \tag{2.19}$$

Since the Jacobian

$$\partial_u G(u) = \begin{pmatrix} 2u_1 & 2(u_2 - 1) \\ 2u_1 & 2u_2 \end{pmatrix} \tag{2.20}$$

has full rank at the point u^*, it follows that u^* is a locally unique solution to the restricted continuation problem $G(u) = 0$. In this case, u^* is the locally unique intersection of the unit circle centered on $(0,1)$ and a circle of radius $\sqrt{2}$ centered on the origin.

Alternatively, let $r = 3$,

$$\Psi : u \mapsto \begin{pmatrix} u_1^2 + u_2^2 \\ u_2 \\ u_1 - u_2 \end{pmatrix}, \tag{2.21}$$

$\mathbb{I} = \{2\}$, and $\mathbb{J} = \{1,3\}$. It follows that $\mu^* := \Psi(u^*) = \begin{pmatrix} 2 & 1 & 0 \end{pmatrix}^T$ and the corresponding restricted continuation problem is given by $G(u, \mu_\mathbb{J}) = 0$, where

$$G : (u, \mu_\mathbb{J}) \mapsto \begin{pmatrix} u_1^2 + (u_2 - 1)^2 - 1 \\ u_1^2 + u_2^2 - \mu_1 \\ u_2 - 1 \\ u_1 - u_2 - \mu_3 \end{pmatrix}. \tag{2.22}$$

Since the Jacobian

$$\partial_{(u,\mu_\mathbb{J})} G(u, \mu_\mathbb{J}) = \begin{pmatrix} 2u_1 & 2(u_2 - 1) & 0 & 0 \\ 2u_1 & 2u_2 & -1 & 0 \\ 0 & 1 & 0 & 0 \\ 1 & -1 & 0 & -1 \end{pmatrix} \tag{2.23}$$

has full rank at the point $(u^*, \mu_\mathbb{J}^*)$, it follows that $(u^*, \mu_\mathbb{J}^*)$ is a locally unique solution of the restricted continuation problem. In this case, u^* is the locally unique intersection of the unit circle centered at $(0,1)$ with the straight line $u_2 = 1$, and μ_1^* and μ_3^* equal the square of the distance to the origin and the difference between u_1^* and u_2^*, respectively. By transferring the index 2 from $\mathbb{I}$ to $\mathbb{J}$, we recover the corresponding extended continuation problem, which is already known to be initializable with a 1-dimensional atlas algorithm on a neighborhood of (u^*, μ^*). ■

Example 2.2 Consider the task of tracing a 2-dimensional manifold passing through the point $u^* = (1, 1, 0)$ and consisting of points $u \in \mathbb{R}^3$ whose Euclidean distances to the axis through the point $(0, 1, 0)$ and parallel to $(0, 0, 1)$ equal 1, i.e., a cylinder of unit radius with symmetry axis parallel to the u_3 axis and running through the point $(0, 1, 0)$.

A candidate zero problem is given by the equation $\Phi(u) = 0$, where

$$\Phi : u \mapsto u_1^2 + (u_2 - 1)^2 - 1. \tag{2.24}$$

Here, $m = 1$ and $n = 3$. Moreover, since the Jacobian

$$\partial_u \Phi(u) = \begin{pmatrix} 2u_1 & 2(u_2 - 1) & 0 \end{pmatrix} \tag{2.25}$$

has full rank at u^*, it follows that the solution manifold through u^* is of dimension $n - m = 2$. In the absence of additional monitor functions, the corresponding restricted continuation problem $G(u) = 0$ with

$$G : u \mapsto u_1^2 + (u_2 - 1)^2 - 1 \tag{2.26}$$

is initializable with a 2-dimensional atlas algorithm on a neighborhood of u^*.

Now let $r = 3$,

$$\Psi : u \mapsto \begin{pmatrix} \sqrt{u_1^2 + u_2^2} \\ u_1 \\ u_3 - u_1 + 0.5 \end{pmatrix}, \tag{2.27}$$

$\mathbb{I} = \{1, 3\}$, and $\mathbb{J} = \{2\}$. It follows that $\mu^* := \Psi(u^*) = (\begin{array}{ccc} \sqrt{2} & 1 & -0.5 \end{array})$ and the corresponding restricted continuation problem is given by $G(u, \mu_{\mathbb{J}}) = 0$, where

$$G : (u, \mu_{\mathbb{J}}) \mapsto \begin{pmatrix} u_1^2 + (u_2 - 1)^2 - 1 \\ \sqrt{u_1^2 + u_2^2} - \sqrt{2} \\ u_1 - \mu_2 \\ u_3 - u_1 + 1 \end{pmatrix}. \tag{2.28}$$

Since the Jacobian

$$\partial_{(u, \mu_{\mathbb{J}})} G(u, \mu_{\mathbb{J}}) = \begin{pmatrix} 2u_1 & 2(u_2 - 1) & 0 & 0 \\ \dfrac{u_1}{\sqrt{u_1^2 + u_2^2}} & \dfrac{u_2}{\sqrt{u_1^2 + u_2^2}} & 0 & 0 \\ 1 & 0 & 0 & -1 \\ -1 & 0 & 1 & 0 \end{pmatrix} \tag{2.29}$$

has full rank at $(u^*, \mu_{\mathbb{J}}^*)$, it follows that $(u^*, \mu_{\mathbb{J}}^*)$ is a locally unique solution of the restricted continuation problem. In this case, u^* is the locally unique intersection between a cylinder of unit radius with symmetry axis parallel to the u_3 axis and running through the point $(0, 1, 0)$; a cylinder of radius $\sqrt{2}$ with symmetry axis parallel to the u_3 axis and running through the origin; and a plane through the point $(1, 0, 0)$ with normal vector $(-1, 0, 1)$.

By transferring the index 1 from $\mathbb{I}$ to $\mathbb{J}$, we obtain the restricted continuation problem $G(u, \mu_{\mathbb{J}}) = 0$ corresponding to the function

$$G : (u, \mu_{\mathbb{J}}) \mapsto \begin{pmatrix} u_1^2 + (u_2 - 1)^2 - 1 \\ \sqrt{u_1^2 + u_2^2} - \mu_1 \\ u_1 - \mu_2 \\ u_3 - u_1 + 1 \end{pmatrix} \tag{2.30}$$

whose Jacobian

$$\partial_{(u, \mu_{\mathbb{J}})} G(u, \mu_{\mathbb{J}}) = \begin{pmatrix} 2u_1 & 2(u_2 - 1) & 0 & 0 & 0 \\ \dfrac{u_1}{\sqrt{u_1^2 + u_2^2}} & \dfrac{u_2}{\sqrt{u_1^2 + u_2^2}} & 0 & -1 & 0 \\ 1 & 0 & 0 & 0 & -1 \\ -1 & 0 & 1 & 0 & 0 \end{pmatrix} \tag{2.31}$$

again has full rank at $(u^*, \mu_{\mathbb{J}}^*)$. It follows that this restricted continuation problem is initializable with a 1-dimensional atlas algorithm on a neighborhood of $(u^*, \mu_{\mathbb{J}}^*)$. The continuation of the corresponding 1-dimensional solution manifold is equivalent to continuation along the embedded submanifold corresponding to the reduced continuation problem

$$\begin{pmatrix} u_1^2 + (u_2 - 1)^2 - 1 \\ u_3 - u_1 + 1 \end{pmatrix} = \begin{pmatrix} 0 \\ 0 \end{pmatrix}. \tag{2.32}$$

Alternatively, if instead $\mathbb{I} = \{1\}$ and $\mathbb{J} = \{2, 3\}$, we obtain the restricted continuation problem $G\left(u, \mu_{\mathbb{J}}\right) = 0$ corresponding to the function

$$G : \left(u, \mu_{\mathbb{J}}\right) \mapsto \begin{pmatrix} u_1^2 + (u_2 - 1)^2 - 1 \\ \sqrt{u_1^2 + u_2^2} - \sqrt{2} \\ u_1 - \mu_2 \\ u_3 - u_1 + 0.5 - \mu_3 \end{pmatrix} \tag{2.33}$$

whose Jacobian

$$\partial_{\left(u, \mu_{\mathbb{J}}\right)} G\left(u, \mu_{\mathbb{J}}\right) = \begin{pmatrix} 2u_1 & 2(u_2 - 1) & 0 & 0 & 0 \\ \dfrac{u_1}{\sqrt{u_1^2 + u_2^2}} & \dfrac{u_2}{\sqrt{u_1^2 + u_2^2}} & 0 & 0 & 0 \\ 1 & 0 & 0 & -1 & 0 \\ -1 & 0 & 1 & 0 & -1 \end{pmatrix} \tag{2.34}$$

again has full rank at $(u^*, \mu_{\mathbb{J}}^*)$. It follows that this restricted continuation problem is initializable with a 1-dimensional atlas algorithm on a neighborhood of $(u^*, \mu_{\mathbb{J}}^*)$. The continuation of the corresponding 1-dimensional solution manifold is equivalent to continuation along the embedded submanifold corresponding to the reduced continuation problem

$$\begin{pmatrix} u_1^2 + (u_2 - 1)^2 - 1 \\ \sqrt{u_1^2 + u_2^2} - \sqrt{2} \end{pmatrix} = \begin{pmatrix} 0 \\ 0 \end{pmatrix}. \quad \blacksquare \tag{2.35}$$

2.3   Algorithm development

As shown above, each restricted continuation problem is equivalent to a reduced continuation problem in the vector of continuation variables corresponding to the imposition of additional constraints on the solutions of the original zero problem. The reduced formulation includes only a minimal set of equations and variables necessary to identify points on the corresponding embedded submanifold. While the restricted continuation problem is given by $m + r$ coupled equations in $n + r - |\mathbb{I}|$ unknowns, the reduced continuation problem corresponds to only $m + |\mathbb{I}|$ coupled equations in n unknowns. When r is large, this appears to be a significant reduction in problem complexity. What then is the advantage of the extended/restricted continuation problem formalism introduced here?

To answer this question, we note that the formulation of an extended continuation problem separates considerations regarding the dimensionality of the solution manifold and the choice of constrained monitor functions from the initial problem construction. Instead, it delegates the imposition or exchange of constraints to a subsequent decision that can be handled in great generality and driven by run-time user preferences. This also allows one to switch between different embedded submanifolds by a simple exchange of elements between the index sets $\mathbb{I}$ and $\mathbb{J}$, thereby making the set-up of involved continuation problems a trivial exercise.

In contrast, a formalism relying solely on the implementation of distinct and predefined reduced continuation problems requires the developer to presuppose the constraints that may be of interest in a subsequent analysis. In particular, one notes that the imposition of k constraints from among a collection of r monitor functions may require the encoding of up to $\binom{r}{k}$ distinct reduced continuation problems. Each such reduced continuation problem presupposes a particular dimensionality of the solution manifold, preventing this choice from being available to a user at run-time.

2.3.1 A return to periodic solutions

The example considered in Sect. 2.1 demonstrates the advantage of the extended continuation formulation. There, we sought a polynomial approximant for a periodic solution to the dynamical system in Eq. (2.1) by requiring that the unknown coefficients $\{a_k\}_{k=0}^m$ and the unknown period $\mathcal{T}$ satisfy the nonlinear equations in Eqs. (2.3)–(2.4) and, for example, the discretized phase condition in Eq. (2.7). We also considered monitoring the values of the period $\mathcal{T}$ and the mean-squared value of the deviation of the polynomial approximant from its mean on the interval $[0,\mathcal{T}]$, as given in Eq. (2.8). Finally, we discussed allowing for the possibility of continuing families of constrained approximate periodic orbits for fixed values of either one of these solution measures.

In the notation of Sect. 2.2, it is now natural to let $u = (a_0,\ldots,a_m,\mathcal{T},\lambda) \in \mathbb{R}^{n(m+1)} \times \mathbb{R} \times \mathbb{R}^s$ constitute the collection of continuation variables, and to define the family of zero functions $\Phi : \mathbb{R}^{n(m+1)+1+s} \to \mathbb{R}^{n(m+1)+1}$, where

$$
\Phi : u \mapsto \begin{pmatrix} a_0 - \sum_{k=0}^m a_k \mathcal{T}^k \\ \sum_{k=1}^m k a_k (\alpha_1 \mathcal{T})^{k-1} - f\left(\sum_{k=0}^m a_k (\alpha_1 \mathcal{T})^k, \lambda\right) \\ \vdots \\ \sum_{k=1}^m k a_k (\alpha_m \mathcal{T})^{k-1} - f\left(\sum_{k=0}^m a_k (\alpha_m \mathcal{T})^k, \lambda\right) \\ \sum_{j=1}^q w_j \left(x^*(\beta_j \mathcal{T}) - \sum_{k=0}^m a_k \left(\beta_j \mathcal{T}\right)^k\right)^T \cdot f\left(x^*(\beta_j \mathcal{T}), \lambda\right) \end{pmatrix}. \tag{2.36}
$$

The dimensional deficit thus equals s, i.e., the dimension of the space of problem parameters. To accommodate the task of monitoring and/or constraining the period and the mean-squared deviation, we may similarly define the family of monitor functions $\Psi : \mathbb{R}^{n(m+1)+1+s} \to \mathbb{R}^{s+2}$, where

$$
\Psi : u \mapsto \begin{pmatrix} \mathcal{T} \\ \frac{1}{\mathcal{T}} \int_0^{\mathcal{T}} \left\| \sum_{k=0}^m a_k t^k - \sum_{k=0}^m \frac{a_k}{k+1} \mathcal{T}^k ds \right\|^2 dt \\ \lambda \end{pmatrix}. \tag{2.37}
$$

With the introduction of a corresponding set of continuation parameters $\mu \in \mathbb{R}^{s+2}$, it follows from the theory of this chapter that we may assign at most s of the integers in $\{1,\ldots,s+2\}$ to $\mathbb{I}$, and thus must retain at least two of these integers in $\mathbb{J}$ in order to obtain a restricted continuation problem that is initializable with some atlas algorithm.

We note, for example, that there are $\frac{1}{2}(s+1)(s+2)$ ways of assigning s of the integers in $\{1,\ldots,s+2\}$ to $\mathbb{I}$. The extended continuation problem may thus be restricted in $\frac{1}{2}(s+1)(s+2)$ different ways in order to yield initializable, restricted continuation problems with, at most, a single solution point. As suggested in Sect. 2.1, $2s$ of these initializable, restricted continuation problems involve assigning either the integer 1 or the integer 2, but not both, to the index set $\mathbb{I}$. In this case, and provided that a solution exists, either the period or the mean-squared deviation attains a value on the solution point equal to the value obtained on the initial solution guess, since, by the discussion in Sect. 2.2.4, $\mu_{\mathbb{I}} = \mu_{\mathbb{I}}^* = \{\Psi_i(u_0)\}_{i \in \mathbb{I}}$.

By a similar consideration, there are $s(s-1)$ restricted continuation problems that are initializable with a 1-dimensional atlas algorithm for which either the period or the mean-squared deviation, but not both, remain fixed along the corresponding 1-dimensional solution manifold and equal to the value obtained on the initial solution guess.

2.3.2   Construction and evaluation

Rather than resulting in an unnecessary increase in complexity, the extended continuation problem formulation actually leads to a strikingly simple, and generalizable, implementation both for the construction and for the evaluation of a restriction $G(u, \mu_{\mathbb{J}})$. Specifically, we note that, at the stage of construction, it is necessary to provide only the following data to a computational implementation:

- the family of zero functions Φ, which in turn is used to compute the number m of zero functions;

- the family of monitor functions Ψ, which in turn is used to compute the number r of monitor functions and continuation parameters;

- the initial solution guess u_0, which in turn is used to compute the number n of continuation variables and the initial value of the continuation parameters $\mu_0 := \Psi(u_0)$; and

- the index set $\mathbb{I}$, which in turn is used to compute the complement $\mathbb{J}$.

In the development of a general-purpose computational tool for continuation analysis, and as will be described in greater detail in the next three chapters, this construction can be broken down into several steps. The responsibility for modifying each of the objects Φ, Ψ, u_0, and $\mathbb{I}$ may be assigned to different algorithms called in some succession.

Independently of the specific implementation of the process of construction, its outcome is a single input-output algorithm, parameterized by data produced during construction. Specifically, suppose that access to the functions Φ and Ψ is stored in the data variable Θ such that

$$\Theta(u) = \begin{pmatrix} \Phi(u) \\ \Psi(u) \end{pmatrix}. \tag{2.38}$$

Moreover, let

$$\lambda_0 := \begin{pmatrix} 0 \\ \mu_0 \end{pmatrix} \in \mathbb{R}^{m+r}. \tag{2.39}$$

Given input values for the continuation variables u and the subset $\mu_{\mathbb{J}}$ of the continuation parameters, evaluation of the left-hand side of the restricted continuation problem corresponding to the index set $\mathbb{I}$ is then equivalent to the algorithm $\Xi : \mathbb{R}^{n+|\mathbb{J}|} \to \mathbb{R}^{m+r}$ encoded in the three assignments

1. $\lambda \Leftarrow \lambda_0$,

2. $\lambda[m + \mathbb{J}] \Leftarrow \mu_{\mathbb{J}}$,

3. $\Xi(u, \mu_{\mathbb{J}}) \Leftarrow \Theta(u) - \lambda$,

where $\lambda[m + \mathbb{J}]$ denotes the set $\{\lambda_{m+i}\}_{i \in \mathbb{J}}$.

A similar algorithm applies to the evaluation of the Jacobian of the left-hand side of this restricted continuation problem with respect to u and $\mu_{\mathbb{J}}$. Specifically, suppose that access to the Jacobians of the functions Φ and Ψ is stored in the data variable $\partial_u \Theta$ such that

$$\partial_u \Theta(u) = \left(\begin{array}{c} \partial_u \Phi(u) \\ \partial_u \Psi(u) \end{array} \right). \tag{2.40}$$

Moreover, let

$$\Lambda_0 := \left(\begin{array}{c} 0 \\ -I_{\mathbb{J}} \end{array} \right) \in \mathbb{R}^{(m+r) \times |\mathbb{J}|}, \tag{2.41}$$

where $I_{\mathbb{J}}$ is the submatrix of the $r \times r$ identity matrix containing the columns indexed by $\mathbb{J}$. Then, the desired Jacobian is obtained from the algorithm $\partial \Xi : \mathbb{R}^{n+|\mathbb{J}|} \to \mathbb{R}^{(m+r) \times (n+|\mathbb{J}|)}$ encoded in the single assignment

1. $\partial \Xi(u, \mu_{\mathbb{J}}) \Leftarrow \left(\begin{array}{cc} \partial_u \Theta(u) & \Lambda_0 \end{array} \right)$.

Notably, the distinction between different restricted continuation problems is contained in the choice of elements of the index set $\mathbb{I}$. A decision as to its content can thus be made at run-time, since the implementations of Ξ or $\partial \Xi$ are entirely independent of the content of this set. The algorithms Ξ and $\partial \Xi$ are also independent of the meaning of the variables in u and $\mu_{\mathbb{J}}$, and of the content of Θ and $\partial_u \Theta$, which are defined elsewhere. As a consequence, it is possible to implement a general-purpose computational algorithm devoted entirely to the construction of an extended continuation problem and its restriction to a specific problem. Such an implementation supports all tasks common to any continuation problem independently of the definition of particular problems or problem classes.

2.4 Conclusions

As suggested in the preface to this book, there is an emphasis in its treatment on the integration of rigorous formalism, algorithm development, and software engineering design. This chapter is a case in point. The discussion has established the mathematical paradigm of extended and restricted continuation problems as advantageous also from the point of view of developing general-purpose algorithms that adhere to principles of good software design. The proposed notion of encapsulation—of binding monitor functions, zero functions, continuation variables, and continuation parameters into a coherent package—hints at modularity, interface development, and hierarchical integration. In particular, as we shall see in the next several chapters, adherence to encapsulation as a fundamental paradigm supports very naturally the idea of task embedding—a shared responsibility for the definition of a continuation problem across several encapsulated units.

In this context, it may be appropriate to point out a natural distinction between computational algorithms that serve different purposes and that are authored and deployed by different groups of individuals. At the *one-of-a-kind level*, one finds applied mathematicians, scientists, and engineers who seek to quantify the solution manifold to a particular, narrowly defined set of zero functions and, commonly, a predetermined and small set of reduced continuation problems associated with the imposition of additional constraints. An individual engaged in such an enterprise is comfortable with the manual construction of a computational implementation of these reduced continuation problems, balancing the cost of software development against the time required to generate desired results. The notions of encapsulation and extended continuation problems are not in opposition to the needs of such an individual. Unless such a general-purpose framework is already available for use, however, these notions may offer little benefit at this stage.

Further along the spectrum, at the *repeated-use level*, one finds researchers and practitioners who repeatedly encounter continuation problems of a similar nature and with a variety of constraints that cannot be captured by a small set of predetermined reduced continuation problems. To such an individual, the cost-benefit analysis may begin to lean in the direction of *toolbox development*, the construction of a general-purpose algorithmic implementation for a class of continuation problems with generalizable characteristics. As will be made abundantly clear throughout the text, the notion of encapsulation is inherent in the building of such general-purpose toolboxes.

At the far end of the spectrum, at the *core-facility level*, one finds applied mathematicians and programmers who seek to enable efficient toolbox development and deployment that supports and facilitates innovation and creativity. To such individuals, there appears to be great benefit in the upfront costs of core algorithm development, with the expectation that significant benefit will be reaped in greatly reduced implementation effort by toolbox developers and users. The extended continuation problem paradigm is an example of a general-purpose mathematical framework upon which such core facilities can be built.

In particular, as expanded upon further in Chap. 5, a key aim of the MATLAB-based *Computational Continuation Core* (abbreviated COCO throughout this text), developed by the authors, is to allow for the cooperative construction of an overall zero problem and associated monitor functions, including the definition of the vector of continuation variables and the vector of continuation parameters, by independently developed toolboxes. The extended continuation problem formulation supports such a cooperative construction by requiring that individual toolbox implementations not depend on the dimensionality of available atlas algorithms, but that they implement rather a continuation problem with the largest possible (and meaningful) dimensional deficit. In this paradigm, a toolbox developer is no longer concerned with the technicalities of a specific continuation problem, but instead provides a set of useful equations and monitor functions that only later are incorporated into a user-defined continuation problem. The actual restriction to a manifold of specific dimensionality and with particular constraints is then imposed a posteriori and completely independent of the continuation problem defined by the toolbox. In the next chapter, we formulate the basic ingredients of the corresponding COCO framework, laying the groundwork for toolbox development in Chap. 4.

Exercises

2.1. Consider the general discretization of the problem of locating periodic solutions $x(t)$ of the dynamical system in Eq. (2.1) obtained by

1. approximating x by a continuous function of $t \in [0, \mathcal{T}]$ for some unknown $\mathcal{T}$, expressed on each of N equal-sized intervals as a polynomial of degree m, and parameterized by the unknown values at $m + 1$ equidistant and maximally distributed points; and

2. imposing that this piecewise-polynomial approximant satisfy Eq. (2.1) at m suitably chosen collocation nodes in each subinterval and attain identical values at the two end points $t = 0$ and $t = \mathcal{T}$.

Show that this formulation results in $n + Nmn$ equations in the $N(m + 1)n + 1$ unknown values parameterizing the piecewise-polynomial approximant and the unknown value of $\mathcal{T}$. Write an explicit set of discrete equations for small integer choices of N, m, and n.

2.2. Show that if $x(t)$ is a periodic solution of period T of the dynamical system in Eq. (2.1), then the function $t \mapsto x(t + \tau)$ is also a periodic solution of period T for each value of τ. Explain why the corresponding state-space trajectories are geometrically indistinguishable. On what property of the vector field f does your proof depend?

2.3. Let x^* represent a known approximation to a periodic solution of the dynamical system in Eq. (2.1) and suppose that the actual solution is approximated by a polynomial of degree m as given in Eq. (2.2). What is the geometric interpretation of the phase condition

$$\left(x^*(t) - \sum_{k=0}^{m} a_k t^k \right)^T \cdot f\left(x^*(t), \lambda \right) = 0?$$

2.4. Choose a basic text on numerical methods and review the discussion of numerical quadrature. Give several examples of discretizations of the integral condition

$$\int_0^{\mathcal{T}} \left(x^*(t) - \sum_{k=0}^{m} a_k t^k \right)^T \cdot f\left(x^*(t), \lambda \right) dt = 0$$

of the form

$$\sum_{j=1}^{q} w_j \left(x^*(\beta_j \mathcal{T}) - \sum_{k=0}^{m} a_k \left(\beta_j \mathcal{T} \right)^k \right)^T \cdot f\left(x^*(\beta_j \mathcal{T}), \lambda \right) = 0$$

in terms of particular numerical choices for the quadrature weights $\{w_j\}_{j=1}^{q}$ and quadrature nodes $\{\beta_j \mathcal{T}\}_{j=1}^{q}$ on the interval $[0, \mathcal{T}]$ for different integers q.

2.5. Find an expression for the time-average

$$\frac{1}{\mathcal{T}} \int_0^{\mathcal{T}} p(s) \, ds$$

of the polynomial

$$p(t) := \sum_{k=0}^{m} a_k t^k, \, t \in [0, \mathcal{T}],$$

in terms of the coefficients $\{a_k\}_{k=0}^{m}$ and the period $\mathcal{T}$.

2.6. Consider the system of $n(m+1)+1$ equations in $n(m+1)+1$ unknowns obtained by seeking a polynomial function of degree m on the interval $[0,\mathcal{T}]$ that satisfies Eqs. (2.3)–(2.4) and one of the phase conditions in Eqs. (2.5)–(2.7) for some known values of λ. Explain why we get an overdetermined problem if we impose an additional condition, say a particular value for the approximate period $\mathcal{T}$, on the $n(m+1)+1$ unknowns. Explain why including one of the components of the parameter vector λ among the unknowns resolves the overdeterminacy. Show that there are $2s$ different closed systems of equations that result from this approach in the case of imposing a constraint on $\mathcal{T}$ or the mean-squared value in Eq. (2.8).

2.7. Choose a basic text on real analysis and review the derivation of the implicit function theorem. What is the geometric interpretation of the condition on the rank of the Jacobian $\partial_u \Phi(u^*)$?

2.8. Choose a basic text on real analysis and review the derivation of the inverse function theorem. Apply your insight to proving that if $\Phi(u^*)=0$ and $\partial_u \Phi(u^*)$ is invertible for $\Phi:\mathbb{R}^n \to \mathbb{R}^n$, i.e., in the special case of 0-dimensional deficit, then u^* is a locally unique solution to the corresponding zero problem.

2.9. Show that the rank deficit of the matrix

$$\begin{pmatrix} \partial_u \Phi(u^*) & 0 \\ \partial_u \Psi(u^*) & -I \end{pmatrix}$$

equals the rank deficit of the matrix $\partial_u \Phi(u^*)$.

2.10. Show that if $t \in \mathbb{R}^n$ is a nullvector of $\partial_u \Phi(u^*)$, then

$$\begin{pmatrix} t \\ \partial_u \Psi(u^*) \cdot t \end{pmatrix}$$

is a nullvector of

$$\begin{pmatrix} \partial_u \Phi(u^*) & 0 \\ \partial_u \Psi(u^*) & -I \end{pmatrix}.$$

2.11. Show that the dimensional deficit of the extended continuation problem equals that of the zero problem. Explain this observation in terms of the relationship between the corresponding solution manifolds.

2.12. Find the dimensions of the domain and range of the function

$$G(u,\mu_{\mathbb{J}}) := F(u,\mu)|_{\mu_{\mathbb{I}}=\mu_{\mathbb{I}}^*}$$

given in Eq. (2.11).

2.13. Show that the nullity of $\partial_{(u,\mu_{\mathbb{J}})}G(u,\mu_{\mathbb{J}})$ for the function G defined in Eq. (2.11) is at least $n-m-|\mathbb{I}|$.

2.14. Suppose that the Jacobian $\partial_{(u,\mu_{\mathbb{J}})}G(u,\mu_{\mathbb{J}})$ for the function G defined in Eq. (2.11) has full rank at $(u^*,\mu_{\mathbb{J}}^*)$. Show that u^* is a regular point of the reduced continuation problem

$$\begin{pmatrix} \Phi(u) \\ \Psi_{\mathbb{I}}(u)-\mu_{\mathbb{I}}^* \end{pmatrix} = \begin{pmatrix} 0 \\ 0 \end{pmatrix}.$$

2.15. Suppose that $\mu_{\mathbb{I}}^* = \Psi_{\mathbb{I}}(u_0)$ for some u_0. Suppose, moreover, that there exists a regular solution $(u^*, \mu_{\mathbb{J}}^*)$ to a corresponding restricted continuation problem and that $u^* \approx u_0$. Show that $\mu_{\mathbb{I}}^* = \Psi_{\mathbb{I}}(u^*)$.

2.16. Show that all points $u \in \mathbb{R}^2$ for which $\Phi(u) = 0$, where

$$\Phi(u) := u_1^2 + (u_2 - 1)^2 - 1,$$

are regular.

2.17. Show that all points $(u, \mu_{\mathbb{J}}) \in \mathbb{R}^2 \times \mathbb{R}$ for which $G(u, \mu_{\mathbb{J}}) = 0$, where

$$G(u, \mu_{\mathbb{J}}) := \begin{pmatrix} u_1^2 + (u_2 - 1)^2 - 1 \\ u_1^2 + u_2^2 - \mu_1 \end{pmatrix},$$

are regular.

2.18. Show that all points $u \in \mathbb{R}^3$ for which $\Phi(u) = 0$, where

$$\Phi(u) := u_1^2 + (u_2 - 1)^2 - 1,$$

are regular.

2.19. Show that all points $u \in \mathbb{R}^3$ for which

$$\begin{pmatrix} u_1^2 + (u_2 - 1)^2 - 1 \\ u_3 - u_1 + 1 \end{pmatrix} = \begin{pmatrix} 0 \\ 0 \end{pmatrix}$$

are regular.

2.20. Show that all points $u \in \mathbb{R}^3$ for which

$$\begin{pmatrix} u_1^2 + (u_2 - 1)^2 - 1 \\ \sqrt{u_1^2 + u_2^2} - \sqrt{2} \end{pmatrix} = \begin{pmatrix} 0 \\ 0 \end{pmatrix}$$

are regular.

2.21. Consider the extended continuation problem given by Eqs. (2.36)–(2.37). Show that this corresponds to $\frac{1}{2}(s+1)(s+2)$ different initializable restricted continuation problems.

2.22. Consider the extended continuation problem given by Eqs. (2.36)–(2.37). Show that this corresponds to $s(s-1)$ restricted continuation problems that are initializable with a 1-dimensional atlas algorithm such that either the first or second monitor function, but not both, remains fixed along the solution manifold.

2.23. Explain how you would determine the number of zero functions from a definition of Φ. Illustrate your analysis using one of the examples in Sect. 2.2.5.

2.24. Explain how you would determine the number of continuation parameters from a definition of Ψ. Illustrate your analysis using one of the examples in Sect. 2.2.5.

2.25. Explain how you would compute the initial value of the vector of continuation parameters from an initial solution guess u_0 for the vector of continuation variables. Illustrate your analysis using one of the examples in Sect. 2.2.5 and actual numerical values for the initial solution guess u_0.

2.26. Explain the meaning of each of the steps of the abstract algorithm Ξ corresponding to the evaluation of the left-hand side of a restricted continuation problem. Illustrate your analysis using one of the examples in Sect. 2.2.5 and actual numerical values for the initial solution guess u_0, the components of the vector of continuation variables u, and the subset $\mu_\mathbb{J}$ of continuation parameters.

2.27. Explain the meaning of the single step of the abstract algorithm $\partial\Xi$ corresponding to the evaluation of the Jacobian of the left-hand side of a restricted continuation problem with respect to u and $\mu_\mathbb{J}$. Illustrate your analysis using one of the examples in Sect. 2.2.5 and actual numerical values for the initial solution guess u_0, the components of the vector of continuation variables u, and the subset $\mu_\mathbb{J}$ of continuation parameters.

Chapter 3

Construction

From the discussion in Chap. 2, we recall that the construction of a restricted continuation problem requires, at the very least, the definition of

- the family of zero functions $\Phi : \mathbb{R}^n \to \mathbb{R}^m$, which in turn is used to compute the number m of zero functions;

- the family of monitor functions $\Psi : \mathbb{R}^n \to \mathbb{R}^r$, which in turn is used to compute the number r of monitor functions and continuation parameters;

- the initial solution guess $u_0 \in \mathbb{R}^n$, which in turn is used to compute the number n of continuation variables and the initial value of the vector of continuation parameters $\mu_0 := \Psi(u_0)$; and

- the index set $\mathbb{I}$, which in turn is used to compute the complement $\mathbb{J} = \{1, \ldots, r\} \setminus \mathbb{I}$.

It is useful to think of the vector of continuation variables u as a *parameterization* of the domain of the families of zero and monitor functions. Notably, while the dimension m of the range of Φ and the dimension r of the range of Ψ may be determined by inspection from the corresponding functional expressions, this is not possible for the dimension n of the domain. Instead, the latter may be obtained a posteriori from the number of components of the initial solution guess u_0.

It is the objective of this chapter to illustrate a paradigm of staged construction of a restricted continuation problem that provides for initializing the vector of continuation variables; for defining the components of Φ and Ψ; for assigning initial content to the index set $\mathbb{I}$; and for modifying this content at run-time using COCO. In Chap. 4, we demonstrate the encoding of a rudimentary COCO-compatible toolbox that automates several aspects of this construction in the context of a particular class of continuation problems.

3.1 Problem decomposition

Let $\mathcal{E}$ denote the collection of all extended continuation problems. Each element $E \in \mathcal{E}$ then corresponds to a (possibly empty) family of m zero functions Φ, a (possibly empty) family of r monitor functions Ψ, and a solution guess u_0 with n components, where $n \geq m$.

For notational convenience we include with $\mathcal{E}$ the *empty extended continuation problem* $\emptyset$, obtained in the absence of zero functions, monitor functions, and continuation variables. Two elements of $\mathcal{E}$ are said to be *equivalent* if one may be obtained from the other by reordering the family of zero functions, the family of monitor functions, and the components of the vector of continuation variables together with the corresponding initial solution guess.

3.1.1   Embeddings

Now let E denote the extended continuation problem constructed from the family of zero functions $\Phi : \mathbb{R}^n \to \mathbb{R}^m$, the family of monitor functions $\Psi : \mathbb{R}^n \to \mathbb{R}^r$, and the initial solution guess $u_0 \in \mathbb{R}^n$. Similarly, let $\hat{E}$ denote the extended continuation problem constructed from the family of zero functions $\hat{\Phi} : \mathbb{R}^{\hat{n}} \to \mathbb{R}^{\hat{m}}$, the family of monitor functions $\hat{\Psi} : \mathbb{R}^{\hat{n}} \to \mathbb{R}^{\hat{r}}$, and the initial solution guess $\hat{u}_0 \in \mathbb{R}^{\hat{n}}$.

We say that $\hat{E}$ is *embedded* in E and write $\hat{E} \subseteq E$ if there exist three index sets $\mathbb{L}_\Phi \subseteq \{1, \ldots, m\}$, $\mathbb{L}_\Psi \subseteq \{1, \ldots, r\}$, and $\mathbb{K} \subseteq \{1, \ldots, n\}$, such that $|\mathbb{L}_\Phi| = \hat{m}$, $|\mathbb{L}_\Psi| = \hat{r}$, $|\mathbb{K}| = \hat{n}$, and

$$\Phi_{\mathbb{L}_\Phi}(u) = \hat{\Phi}(u_\mathbb{K}), \tag{3.1}$$

$$\Psi_{\mathbb{L}_\Psi}(u) = \hat{\Psi}(u_\mathbb{K}), \tag{3.2}$$

$$(u_0)_\mathbb{K} = \hat{u}_0. \tag{3.3}$$

In this case, we say that the subsets $\Phi_{\mathbb{L}_\Phi}$ and $\Psi_{\mathbb{L}_\Psi}$ of zero functions and monitor functions are *independent* of the components of $u_{\{1,\ldots,n\}\setminus\mathbb{K}}$.

By definition, $\emptyset \subseteq E$ for every extended continuation problem $E \in \mathcal{E}$.

Example 3.1 Consider, for example, the composite zero problem $\Phi(u) = 0$, where

$$\Phi : u \mapsto \begin{pmatrix} f_1(v_1, v_s) \\ f_2(v_s, v_2) \end{pmatrix}, \tag{3.4}$$

$f_1 : \mathbb{R}^{n_1} \times \mathbb{R}^{n_s} \to \mathbb{R}^{m_1}$, $f_2 : \mathbb{R}^{n_s} \times \mathbb{R}^{n_2} \to \mathbb{R}^{m_2}$, $u = (v_1, v_s, v_2) \in \mathbb{R}^{n_1} \times \mathbb{R}^{n_s} \times \mathbb{R}^{n_2}$, and $u_0 := (v_{1,0}, v_{s,0}, v_{2,0})$ denotes an initial solution guess. Here, the subset v_s of continuation variables parameterizes the shared domain of the individual zero problems

$$f_1(v_1, v_s) = 0 \tag{3.5}$$

and

$$f_2(v_s, v_2) = 0. \tag{3.6}$$

In an application, such a shared variable domain may represent a connection between input ports in one system and output ports in another. The composite zero problem then corresponds to a purposeful coupling of two independent subproblems.

Now let

$$\mathbb{L}_1 := \{1, \ldots, m_1\} \tag{3.7}$$

and

$$\mathbb{K}_1 := \{1, \ldots, n_1 + n_s\}, \tag{3.8}$$

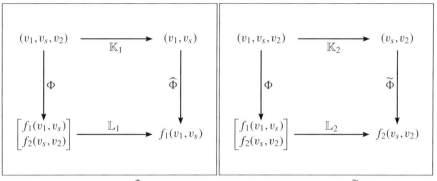

(a) Embedding of $\widehat{E}$ in E.   (b) Embedding of $\widetilde{E}$ in E.

Figure 3.1. *Let $E = (\Phi, \emptyset, u_0)$ denote the extended continuation problem with the combined zero problem $\Phi((v_1, v_s, v_2)) = (f_1(v_1, v_s), f_2(v_s, v_2))^T$ and the initial solution guess $u_0 = (v_{1,0}, v_{s,0}, v_{2,0})$. Furthermore, let $\widehat{E} = (\widehat{\Phi}, \emptyset, \widehat{u}_0)$, $\widehat{\Phi}((v_1, v_s)) = f_1(v_1, v_s)$, $\widehat{u}_0 = (v_{1,0}, v_{s,0})$ and $\widetilde{E} = (\widetilde{\Phi}, \emptyset, \widetilde{u}_0)$, $\widetilde{\Phi}((v_s, v_2)) = f_2(v_s, v_2)$, $\widetilde{u}_0 = (v_{s,0}, v_{2,0})$ denote extended continuation problems constructed from the respective individual zero problems together with the index sets $\mathbb{K}_{1/2}$ and $\mathbb{L}_{1/2}$ as defined in Example 3.1. Then, the commutative diagrams (a) and (b) imply the embeddings $\widehat{E} \subseteq E$ and $\widetilde{E} \subseteq E$. The embedding $\widehat{E} \subseteq E$ is canonical.*

and define the function $\hat{\Phi} : \mathbb{R}^{n_1 + n_s} \to \mathbb{R}^{m_1}$, such that

$$\hat{\Phi} : \hat{u} \mapsto f_1\left(\hat{u}_{\{1,\ldots,n_1\}}, \hat{u}_{\{n_1+1,\ldots,n_1+n_s\}}\right). \tag{3.9}$$

It follows that

$$\Phi_{\mathbb{L}_1}(u) = \hat{\Phi}\left(u_{\mathbb{K}_1}\right), \tag{3.10}$$

i.e., that the extended continuation problem constructed from $\hat{\Phi}$ and $\hat{u}_0 := (u_0)_{\mathbb{K}_1}$ is embedded in the extended continuation problem constructed from Φ and u_0. We illustrate this relationship in Fig. 3.1(a). In this case, f_1 is clearly independent of v_2.

Similarly, let

$$\mathbb{L}_2 := \{m_1 + 1, \ldots, m_1 + m_2\} \tag{3.11}$$

and

$$\mathbb{K}_2 := \{n_1 + 1, \ldots, n_1 + n_s + n_2\}, \tag{3.12}$$

and define the function $\tilde{\Phi} : \mathbb{R}^{n_s + n_2} \to \mathbb{R}^{m_2}$, such that

$$\tilde{\Phi} : \tilde{u} \mapsto f_2\left(\tilde{u}_{\{1,\ldots,n_s\}}, \tilde{u}_{\{n_s+1,\ldots,n_s+n_2\}}\right). \tag{3.13}$$

It follows again that

$$\Phi_{\mathbb{L}_2}(u) = \tilde{\Phi}\left(u_{\mathbb{K}_2}\right), \tag{3.14}$$

i.e., that the extended continuation problem constructed from $\tilde{\Phi}$ and $\tilde{u}_0 := (u_0)_{\mathbb{K}_2}$ is embedded in the extended continuation problem constructed from Φ and u_0. We illustrate this relationship in Fig. 3.1(b). In this case f_2 is clearly independent of v_1. ∎

We say that the extended continuation problem $\hat{E}$ is embedded *canonically* in the extended continuation problem E and write $\hat{E} \subseteq E$ if Eqs. (3.1)–(3.3) hold with $\mathbb{L}_\Phi = \{1,\ldots,\hat{m}\}$, $\mathbb{L}_\Psi = \{1,\ldots,\hat{r}\}$, and $\mathbb{K} = \{1,\ldots,\hat{n}\}$. In Example 3.1, the extended continuation problem constructed from $\hat{\Phi}$ and $\hat{u}_0$ was embedded canonically in the extended continuation problem constructed from Φ and u_0. Given the convention that $\{1,\ldots,0\} = \emptyset$, it follows that $\emptyset \subseteq E$ for every restricted continuation problem $E \in \mathcal{E}$.

An embedding $\tilde{E} \subseteq E$ can always be made canonical by replacing E with an equivalent element of $\mathcal{E}$ obtained by reordering the elements of Φ, Ψ, and u, and by modifying u_0 appropriately. As an example, the extended continuation problem constructed from $\tilde{\Phi}$ and $\tilde{u}_0$ in Example 3.1 is embedded canonically in the extended continuation problem obtained by switching f_1 and f_2 in Eq. (3.4), and by letting $u = (v_s, v_2, v_1)$ and $u_0 = (v_{s,0}, v_{2,0}, v_{1,0})$.

3.1.2 Chains of partial realizations

It is straightforward to show that $\mathcal{E}$ is a *partially ordered set* under the relation $\subseteq$. Consider, as a situation of particular interest, a *totally ordered subset* (or *chain*) of extended continuation problems

$$\emptyset = E_0 \subseteq E_1 \subseteq \cdots \subseteq E_N = E. \tag{3.15}$$

We refer to each of the extended continuation problems contained in such a chain as a *partial realization* of the greatest element E.

Example 3.2 Consider the extended continuation problem E in $u \in \mathbb{R}^4$ obtained from the family of zero functions

$$\Phi : u \mapsto \left(\begin{array}{c} f_1(u_1, u_3) \\ f_2(u_2, u_4) \end{array} \right), \tag{3.16}$$

the family of monitor functions

$$\Psi : u \mapsto \left(\begin{array}{c} g_1(u_1, u_4) \\ g_2(u_3) \end{array} \right), \tag{3.17}$$

and the initial solution guess $u_0 := (u_{1,0}, u_{2,0}, u_{3,0}, u_{4,0})$.

Consider also the extended continuation problem $\hat{E}$ in $\hat{u} \in \mathbb{R}^4$ obtained from the family of zero functions

$$\hat{\Phi} : \hat{u} \mapsto \left(\begin{array}{c} f_1(\hat{u}_1, \hat{u}_3) \\ f_2(\hat{u}_4, \hat{u}_2) \end{array} \right), \tag{3.18}$$

the family of monitor functions

$$\hat{\Psi} : \hat{u} \mapsto \left(\begin{array}{c} g_1(\hat{u}_1, \hat{u}_2) \\ g_2(\hat{u}_3) \end{array} \right), \tag{3.19}$$

and the initial solution guess $\hat{u}_0 := (\hat{u}_{1,0}, \hat{u}_{2,0}, \hat{u}_{3,0}, \hat{u}_{4,0}) := (u_{1,0}, u_{4,0}, u_{3,0}, u_{2,0})$.

It is straightforward to show that $\hat{E}$ is equivalent to E. Moreover,

$$\emptyset = E_0 \subseteq E_1 \subseteq E_2 \subseteq E_3 = \hat{E}, \tag{3.20}$$

where

- E_1 is the extended continuation problem in $w \in \mathbb{R}^2$ obtained from the empty family of zero functions, the family of monitor functions

$$\Psi^{(1)} : w \mapsto g_1(w_1, w_2), \tag{3.21}$$

  and the initial solution guess $w_0 := (\hat{u}_{1,0}, \hat{u}_{2,0})$, and

- E_2 is the extended continuation problem in $w \in \mathbb{R}^4$ obtained from the family of zero functions

$$\Phi^{(2)} : w \mapsto \begin{pmatrix} f_1(w_1, w_3) \\ f_2(w_4, w_2) \end{pmatrix}, \tag{3.22}$$

  the monitor function

$$\Psi^{(2)} : w \mapsto g_1(w_1, w_2), \tag{3.23}$$

  and the initial solution guess $w_0 := \hat{u}_0$.

Here, E_1 and E_2 constitute partial realizations of $\hat{E}$.

Alternatively, consider the extended continuation problem $\tilde{E}$ in $\tilde{u} \in \mathbb{R}^4$ obtained from the family of zero functions

$$\tilde{\Phi} : \tilde{u} \mapsto \begin{pmatrix} f_1(\tilde{u}_1, \tilde{u}_2) \\ f_2(\tilde{u}_3, \tilde{u}_4) \end{pmatrix}, \tag{3.24}$$

the family of monitor functions

$$\tilde{\Psi} : \tilde{u} \mapsto \begin{pmatrix} g_2(\tilde{u}_2) \\ g_1(\tilde{u}_1, \tilde{u}_4) \end{pmatrix}, \tag{3.25}$$

and the initial solution guess $\tilde{u}_0 := (\tilde{u}_{1,0}, \tilde{u}_{2,0}, \tilde{u}_{3,0}, \tilde{u}_{4,0}) := (u_{1,0}, u_{3,0}, u_{2,0}, u_{4,0})$.

It is again straightforward to show that $\tilde{E}$ is equivalent to E. Moreover,

$$\emptyset = E_0 \subsetneqq E_1 \subsetneqq E_2 = \tilde{E}, \tag{3.26}$$

where the partial realization E_1 is the extended continuation problem in $w \in \mathbb{R}^2$ obtained from the zero function

$$\Phi^{(1)} : w \mapsto f_1(w_1, w_2), \tag{3.27}$$

the monitor function

$$\Psi^{(1)} : w \mapsto g_2(w_2), \tag{3.28}$$

and the initial solution guess $w_0 := (\tilde{u}_{1,0}, \tilde{u}_{2,0})$. ∎

A chain

$$\emptyset = E_0 \subsetneqq E_1 \subsetneqq \cdots \subsetneqq E_N = E \tag{3.29}$$

is said to be a *nontrivial chain decomposition* of the extended continuation problem E provided that $N \geq 2$ and $E_i \neq E_N$ for some $i \in \{1, \ldots, N-1\}$. Now suppose that E is an extended continuation problem with $m + r \geq 2$. As suggested by Example 3.2, there then exists at least one nontrivial chain decomposition of some suitably constructed extended continuation problem $\hat{E}$ that is equivalent to E.

3.2   Staged construction

3.2.1   A unique parameterization

Let E again denote the extended continuation problem in $u \in \mathbb{R}^n$ obtained from the family of zero functions $\Phi : \mathbb{R}^n \to \mathbb{R}^m$, the family of monitor functions $\Psi : \mathbb{R}^n \to \mathbb{R}^r$, and the initial solution guess $u_0 \in \mathbb{R}^n$, and consider a chain

$$\emptyset = E_0 \subsetneqq E_1 \subsetneqq \cdots \subsetneqq E_N = E. \tag{3.30}$$

By definition, there exist three nondecreasing sequences of integers

$$0 = m_0 \leq m_1 \leq \cdots \leq m_N = m, \tag{3.31}$$

$$0 = r_0 \leq r_1 \leq \cdots \leq r_N = r, \tag{3.32}$$

$$0 = n_0 \leq n_1 \leq \cdots \leq n_N = n, \tag{3.33}$$

such that the partial realization E_i is an extended continuation problem in $\mathbb{R}^{n_i}$ obtained from the family of zero functions

$$\Phi^{(i)} : w \mapsto \Phi_{\{1,\ldots,m_i\}}(w, \cdot), \tag{3.34}$$

the family of monitor functions

$$\Psi^{(i)} : w \mapsto \Psi_{\{1,\ldots,r_i\}}(w, \cdot), \tag{3.35}$$

and the subset $(u_0)_{\{1,\ldots,n_i\}}$ of the initial solution guess. Here, the $\cdot$ in Eqs. (3.34)–(3.35) refers to an arbitrary numerical value, since the right-hand sides are independent of this argument.

Let $\mathbb{K}_i \subseteq \{1,\ldots,n_i\}$ denote an index set of cardinality $|\mathbb{K}_i| \geq n_i - n_{i-1}$ for $i \geq 1$, such that

$$\Phi_{\{m_{i-1}+1,\ldots,m_i\}}(u) = \phi^{(i)}\left(u_{\mathbb{K}_i}\right), \tag{3.36}$$

$$\Psi_{\{r_{i-1}+1,\ldots,r_i\}}(u) = \psi^{(i)}\left(u_{\mathbb{K}_i}\right) \tag{3.37}$$

for some functions $\phi^{(i)} : \mathbb{R}^{|\mathbb{K}_i|} \to \mathbb{R}^{m_i - m_{i-1}}$ and $\psi^{(i)} : \mathbb{R}^{|\mathbb{K}_i|} \to \mathbb{R}^{r_i - r_{i-1}}$, and such that the index set consisting of the last $n_i - n_{i-1}$ elements of $\mathbb{K}_i$ equals

$$\mathbb{K}_i^n := \{n_{i-1}+1,\ldots,n_i\}. \tag{3.38}$$

We refer to $\phi^{(i)}$ and $\psi^{(i)}$ as *representations* of $\Phi_{\{m_{i-1}+1,\ldots,m_i\}}$ and $\Psi_{\{r_{i-1}+1,\ldots,r_i\}}$, respectively, and to $\mathbb{K}_i$ as the corresponding *dependency index set*. As a special case, the functions $\Phi^{(i)}_{\{m_{i-1}+1,\ldots,m_i\}}$ and $\Psi^{(i)}_{\{r_{i-1}+1,\ldots,r_i\}}$ are representations corresponding to the trivial dependency index set $\mathbb{K}_i = \{1,\ldots,n_i\}$.

For each chain of partial realizations of an extended continuation problem E, we now arrive at a paradigm for *staged construction* of E that provides, at the ith stage of construction,

- representations $\phi^{(i)}$ and $\psi^{(i)}$ of $\Phi_{\{m_{i-1}+1,\ldots,m_i\}}$ and $\Psi_{\{r_{i-1}+1,\ldots,r_i\}}$,

- the index set $\mathbb{K}_i^o$ obtained by omitting the elements in $\mathbb{K}_i^n$ from the dependency index set $\mathbb{K}_i$, and

- the subset $(u_0)_{\mathbb{K}_i^n}$ of the initial solution guess.

We note, in particular, that $\mathbb{K}_1^o = \emptyset$ and thus that $\mathbb{K}_1 = \{1, \ldots, n_1\}$. Moreover, at each stage, $\mathbb{K}_i^n$ may be constructed from the dimension of the subset of the initial solution guess.

Example 3.3 We consider again from Example 3.2 the extended continuation problem E in $u \in \mathbb{R}^4$ obtained from the family of zero functions

$$\Phi : u \mapsto \begin{pmatrix} f_1(u_1, u_2) \\ f_2(u_3, u_4) \end{pmatrix}, \tag{3.39}$$

the family of monitor functions

$$\Psi : u \mapsto \begin{pmatrix} g_2(u_2) \\ g_1(u_1, u_4) \end{pmatrix}, \tag{3.40}$$

and the initial solution guess $u_0 := (u_{1,0}, u_{2,0}, u_{3,0}, u_{4,0})$. Recall the nontrivial chain decomposition

$$\emptyset = E_0 \subsetneqq E_1 \subsetneqq E_2 = E, \tag{3.41}$$

where E_1 is the restricted continuation problem in $w \in \mathbb{R}^2$ obtained from the zero function

$$\Phi^{(1)} : w \mapsto f_1(w_1, w_2), \tag{3.42}$$

the monitor function

$$\Psi^{(1)} : w \mapsto g_2(w_2), \tag{3.43}$$

and the initial solution guess $w_0 := (u_{1,0}, u_{2,0})$.

Following the paradigm of staged construction, we may now provide, in the first stage, the representations

$$\phi^{(1)} : (x, y) \mapsto f_1(x, y) \tag{3.44}$$

and

$$\psi^{(1)} : (x, y) \mapsto g_2(y), \tag{3.45}$$

the subset $\mathbb{K}_1^o = \emptyset$, and the subset $(u_{1,0}, u_{2,0})$ of the initial solution guess, consistent with the trivial dependency index set $\mathbb{K}_1 = \{1, 2\}$.

We then complete the construction of the extended continuation problem E by providing, in the second stage, the representations

$$\phi^{(2)} : (x, y, z) \mapsto f_2(y, z) \tag{3.46}$$

and

$$\psi^{(2)} : (x, y, z) \mapsto g_2(x, z), \tag{3.47}$$

the subset $\mathbb{K}_2^o = \{1\}$, and the subset $(u_{3,0}, u_{4,0})$ of the initial solution guess, consistent with the dependency index set $\mathbb{K}_2 = \{1, 3, 4\}$. ∎

Let E again denote an extended continuation problem in $u \in \mathbb{R}^n$ obtained from the family of zero functions $\Phi : \mathbb{R}^n \to \mathbb{R}^m$, the family of monitor functions $\Psi : \mathbb{R}^n \to \mathbb{R}^r$, and the initial solution guess u_0. Suppose that $E \subsetneqq \hat{E}$ for some other extended continuation problem $\hat{E}$. By the paradigm of staged construction, it follows that there exist functions ϕ and ψ, an index set $\mathbb{K}$, and a vector of numbers $v \in \mathbb{R}^k$, where $|\mathbb{K}| \leq n$ and the domains of ϕ and ψ both have dimension $|\mathbb{K}| + k$, such that $\hat{E}$ is the extended continuation problem in $\hat{u} \in \mathbb{R}^{n+k}$ obtained from the family of zero functions

$$\hat{\Phi} : \hat{u} \mapsto \begin{pmatrix} \Phi\left(\hat{u}_{\{1,\ldots,n\}}\right) \\ \phi\left(\hat{u}_{\mathbb{K}}, \hat{u}_{\{n+1,\ldots,n+k\}}\right) \end{pmatrix}, \tag{3.48}$$

the family of monitor functions

$$\hat{\Psi} : \hat{u} \mapsto \begin{pmatrix} \Psi\left(\hat{u}_{\{1,\ldots,n\}}\right) \\ \psi\left(\hat{u}_{\mathbb{K}}, \hat{u}_{\{n+1,\ldots,n+k\}}\right) \end{pmatrix}, \tag{3.49}$$

and the initial solution guess

$$\hat{u}_0 := (u_0, v). \tag{3.50}$$

Conversely, suppose that we are given functions ϕ and ψ, an index set $\mathbb{K}$, and a vector of numbers $v \in \mathbb{R}^k$, where the domains of ϕ and ψ both have dimension $|\mathbb{K}| + k$. It follows that for every extended continuation problem E in $u \in \mathbb{R}^n$ obtained from the family of zero functions $\Phi : \mathbb{R}^n \to \mathbb{R}^m$, the family of monitor functions $\Psi : \mathbb{R}^n \to \mathbb{R}^r$, and the initial solution guess u_0, such that $n \geq |\mathbb{K}|$, there exists an extended continuation problem

$$\hat{E} := [\phi, \psi, \mathbb{K}, v](E) \tag{3.51}$$

in $\hat{u} \in \mathbb{R}^{n+k}$ obtained from the family of zero functions in Eq. (3.48), the family of monitor functions in Eq. (3.49), and the initial solution guess in Eq. (3.50). As an example, the staged construction in Example 3.3 is given by the composition

$$E = \left(\left[\phi^{(2)}, \psi^{(2)}, \{1\}, \left(u_{3,0}, u_{4,0}\right) \right] \circ \left[\phi^{(1)}, \psi^{(1)}, \emptyset, \left(u_{1,0}, u_{2,0}\right) \right] \right) (\emptyset). \tag{3.52}$$

3.2.2   Active and inactive constraints

As suggested in Sect. 2.2, the introduction of continuation parameters μ associated with individual monitor functions enables tracking of the values of these functions along the solution manifold to the corresponding zero problem. In addition, the continuation parameters provide a mechanism for the imposition of additional nonlinear constraints on the vector of continuation variables u.

Recall the notation $\mu_{\mathbb{I}}$ and $\mu_{\mathbb{J}}$ for the subsets $\{\mu_i\}_{i \in \mathbb{I}}$ and $\{\mu_i\}_{i \in \mathbb{J}}$ ordered according to the integers in $\mathbb{I}$ and $\mathbb{J}$, respectively, and similarly for $\Psi_{\mathbb{I}}$ and $\Psi_{\mathbb{J}}$. Elements of $\mu_{\mathbb{I}}$ are referred to as *inactive continuation parameters*, as these remain constant during continuation along the solution manifold to the corresponding restricted continuation problem. Elements of $\mu_{\mathbb{J}}$ are referred to as *active continuation parameters*, as these are allowed to vary during continuation along the solution manifold to the corresponding restricted continuation problem. The set of inactive continuation parameters corresponds to a collection of *active constraints*

$$\Psi_{\mathbb{I}}(u) - \mu_{\mathbb{I}} = 0, \tag{3.53}$$

whereas the set of active continuation parameters corresponds to a collection of *inactive constraints*

$$\Psi_{\mathbb{J}}(u) - \mu_{\mathbb{J}} = 0. \tag{3.54}$$

The latter simply allow for the monitoring of the value of $\Psi_{\mathbb{J}}(u)$ through the corresponding value of $\mu_{\mathbb{J}}$.

It is natural to accompany the paradigm of staged construction described in the previous section with the designation of elements of μ as initially active or inactive continuation parameters at the time of definition of the corresponding monitor functions. As suggested in the previous chapter, however, great flexibility is afforded by enabling changes in the content of the index set $\mathbb{I}$ (and, consequently, its complement $\mathbb{J}$) even after this initial construction.

To this end, following the initial staged construction of an extended continuation problem, we allow for two additional steps of construction of the index set $\mathbb{I}$:

- a sequence of additional pairwise exchanges between elements of $\mathbb{I}$ and its complement $\mathbb{J}$; followed by

- a reallocation of a subset of elements of $\mathbb{I}$ to its complement $\mathbb{J}$.

Let $\mathbb{I}_0$ denote the content of the index set $\mathbb{I}$ after the initial staged construction. It follows that the final restricted continuation problem is initializable with a d-dimensional atlas algorithm only if $n - m \geq d \geq n - m - |\mathbb{I}_0|$. In this case, $d - n + m + |\mathbb{I}_0|$ elements must be reallocated from $\mathbb{I}$ to $\mathbb{J}$ in the final step of construction. We refer to this reallocation as a process of *deactivating* or *releasing* constraints and of *activating* the corresponding continuation parameters.

Example 3.4 The three-step construction of the index set $\mathbb{I}$ follows the philosophy of defining an initially overconstrained problem and subsequently deactivating sufficiently many constraints to obtain a solution manifold of dimension equal to that of the atlas algorithm.

As an example, consider the decomposition

$$u = (x, p) \in \mathbb{R}^m \times \mathbb{R}^{n-m} \tag{3.55}$$

into problem variables x and problem parameters p, and let $\Psi(u) = p$. In a design context, the problem parameters are considered inputs and the problem variables represent outputs of the design process. It is, therefore, natural to set $\mathbb{I}_0 = \{1, \dots, n - m\}$, such that $|\mathbb{I}_0| = n - m$. In this case, the final restricted continuation problem is initializable with a d-dimensional atlas algorithm of any dimension ($\leq n - m$) provided that exactly d elements of $\mathbb{I}$ are reallocated to $\mathbb{J}$ in the final step of construction. ∎

In the next section, we illustrate the COCO syntax for performing the staged construction of an extended continuation problem, including the designation of initially inactive or active continuation parameters, the subsequent deactivation of constraints, and the task of continuation along the corresponding solution manifold. We will return to the task of triggering pairwise exchanges between $\mathbb{I}$ and its complement $\mathbb{J}$ in Chap. 16.

3.3 The core interface

In COCO, the information used to encode a restricted continuation problem and to explore its solution manifold is stored in a *continuation problem structure*. To initialize an empty continuation problem structure `prob`, we use the `coco_prob` core utility, as in the following command.

```
prob = coco_prob();
```

In this section, we describe several COCO core utilities that provide a low-level interface for appending to the content of this structure.

3.3.1   Adding zero functions

In the simplest case, to make a (possibly vector-valued) function f available for the COCO constructor and atlas algorithms, f must be encoded in a MATLAB-compatible function with the following COCO-specific function syntax.

```
function [data y] = fname(prob, data, u)
```

Here, the function `fname` accepts three input arguments and returns two output arguments. We consider generalized versions of this function syntax in Chap. 16.

The first input argument, `prob`, contains a copy of the corresponding continuation problem structure. The content of this variable may be accessed in the function body, but changes made to the copy do not survive execution of the function. This argument is used within the function body only in very advanced applications.

As the second input argument, `data`, is identical to the first output argument, its content may be accessed and permanently modified within the function body for reference in a subsequent function call or by other COCO-related algorithms. This *function data structure* $\mathfrak{D}_f$ may include information used in the function body to guide its execution or may provide a record of past calls to the function.

The third input argument, `u`, contains a copy of a subset of the elements of the vector of continuation variables indexed by the corresponding *dependency index set* $\mathbb{K}_f$. Components of the `u` argument are extracted within the function body and used to evaluate the function value(s) that are returned in the second output argument, `y`.

Example 3.5 As an example of this syntax, consider again the zero function

$$\Phi : u \mapsto u_1^2 + (u_2 - 1)^2 - 1, u \in \mathbb{R}^2, \tag{3.56}$$

from Eq. (2.13). As discussed in Sect. 2.2.5, in the absence of additional monitor functions, the corresponding restricted continuation problem $G(u) = 0$ is initializable with a 1-dimensional atlas algorithm on a neighborhood of the point $u^* = (1,1)$.

The following definition implements a representation of the function Φ in a COCO-compatible syntax.

```
function [data y] = circ(prob, data, u)
  y = u(1)^2+(u(2)-1)^2-1;
end
```

In this case, although both `prob` and `data` must be included in the list of arguments, neither is used in the function body. Instead, the first and second elements of the `u` argument are extracted and combined as per the above formula. Notably, there is no information in the encoding of Φ as to the content of $\mathbb{K}_\Phi$ nor of the overall dimension n of the domain of the zero problem. The encoding requires only that $2 \leq |\mathbb{K}_\Phi| \leq n$. ∎

Recall from Sect. 3.2 the notation $[\phi, \psi, \mathbb{K}, v]$ for the operator whose action on an extended continuation problem E results in an extended continuation problem $\hat{E} := [\phi, \psi, \mathbb{K}, v](E)$ obtained from the family of zero functions, the family of monitor functions, and the initial solution guess given in Eqs. (3.48)–(3.50). The COCO core utility `coco_add_func` provides basic functionality in support of this operational definition such that

- at each stage of construction, one can add either a family of zero functions or a family of monitor functions, i.e., either ϕ or ψ, but not both, is empty, and

- in the case that ψ is nonempty, the corresponding continuation parameters can be assigned as all initially active or all initially inactive.

Suppose, for example, that the continuation problem structure `prob` encodes an extended continuation problem E in $\mathbb{R}^n$ obtained from the family of zero functions $\Phi : \mathbb{R}^n \to \mathbb{R}^m$, the family of monitor functions $\Psi : \mathbb{R}^n \to \mathbb{R}^r$, and the initial solution guess u_0. Consider the action on E of the operator $[f, \emptyset, \mathbb{K}_f^o, u_{f,0}]$, where $u_{f,0} \in \mathbb{R}^k$, $\mathbb{K}_f^o \subseteq \{1, \ldots, n\}$, and the domain of f has dimension $|\mathbb{K}_f^o| + k$. Let

- the variable `fid` contain a *function identifier*, i.e., a string reference to f,

- the variable `fhan` contain a *function handle* to an encoding of f,

- the variable `data` contain the function data structure $\mathfrak{D}_f$,

- the variable `uidx` contain the index set $\mathbb{K}_f^o$, and

- the variable `u0` contain the vector $u_{f,0}$.

Then, the command

```
coco_add_func(prob, fid, fhan, data, 'zero', 'uidx', uidx, 'u0', u0);
```

returns a continuation problem structure that encodes the extended continuation problem

$$\hat{E} := \left[f, \emptyset, \mathbb{K}_f^o, u_{f,0} \right](E) \tag{3.57}$$

in $\hat{u} \in \mathbb{R}^{n+k}$ obtained from the family of zero functions

$$\hat{\Phi} : \hat{u} \mapsto \left(\begin{array}{c} \Phi\left(\hat{u}_{\{1,\ldots,n\}}\right) \\ f\left(\hat{u}_{\mathbb{K}_f^o}, \hat{u}_{\{n+1,\ldots,n+k\}}\right) \end{array} \right), \tag{3.58}$$

the family of monitor functions

$$\hat{\Psi} : \hat{u} \mapsto \Psi\left(\hat{u}_{\{1,\ldots,n\}}\right), \tag{3.59}$$

and the initial solution guess

$$\hat{u}_0 := \left(u_0, u_{f,0}\right). \tag{3.60}$$

Here, the input argument `uidx` and the preceding flag `'uidx'` may be omitted if $\mathbb{K}_f^o = \emptyset$. Similarly, the input argument `u0` and the preceding flag `'u0'` may be omitted if $u_{f,0}$ is an empty vector, i.e., $k = 0$.

Example 3.6 Consider again the zero function Φ and the COCO-compatible encoding from Example 3.5. Suppose that the continuation problem structure `prob` encodes the empty extended continuation problem. The *function type* `'zero'` in the command

```
prob = coco_add_func(prob, 'fun1', @circ, [], 'zero', ...
    'u0', [0.9; 1.1]);
```

instructs `coco_add_func` to encode the function with function handle `@circ` and function identifier `'fun1'` as a zero function in the extended continuation problem stored in `prob`.

Here, the empty brackets in the third input argument represent the initial content of the function data structure $\mathfrak{D}_\Phi$. The absence of the `'uidx'` flag implies that $\mathbb{K}_\Phi^o = \emptyset$. The `'u0'` flag indicates that the subsequent argument contains the corresponding components of the initial solution guess. Since two numerical values are given, it follows that $\mathbb{K}_\Phi^n = \{1,2\}$. It follows that $\mathbb{K}_\Phi = \{1,2\}$, consistent with the requirement that $|\mathbb{K}_\Phi| \geq 2$.

Notably, there is still no information available as to the overall dimension n of the domain of the zero problem other than the observation that $n \geq 2$. At this stage of construction, however, the continuation problem structure encodes a well-defined extended continuation problem in $u \in \mathbb{R}^2$ with a compatible initial solution guess $u_0 = (0.9, 1.1)$. ∎

The continuation problem constructed in Example 3.6 has a dimensional deficit of 1. A local covering of the corresponding solution manifold near the initial solution guess u_0 may be obtained using the default 1-dimensional atlas algorithm developed for this book and included with COCO. To this end, we invoke the COCO *entry-point function* as in the command

```
coco(prob, 'run1', [], 1);
```

Here, the first argument contains the continuation problem structure `prob` returned in Example 3.6. The *run identifier* `'run1'` in the second argument provides a reference to the folder containing solution files stored during continuation. The empty brackets in the third argument indicate that the continuation problem structure contains a restricted continuation problem that is initializable with the default atlas algorithm of the dimensionality indicated by the numeral `1` in the following argument.

Example 3.7 The extract below shows the result of executing the commands described above.

```
>> prob = coco_prob();
>> prob = coco_add_func(prob, 'fun1', @circ, [], 'zero', ...
        'u0', [0.9; 1.1]);
>> coco(prob, 'run1', [], 1);
```

STEP		DAMPING		NORMS		COMPUTATION TIMES		
IT	SIT	GAMMA	||d||	||f||	||U||	F(x)	DF(x)	SOLVE
0				1.80e-01	1.42e+00	0.0	0.0	0.0
1	1	1.00e+00	9.94e-02	9.88e-03	1.49e+00	0.0	0.0	0.0
2	1	1.00e+00	4.91e-03	2.42e-05	1.49e+00	0.0	0.0	0.0
3	1	1.00e+00	1.21e-05	1.46e-10	1.49e+00	0.0	0.0	0.0
4	1	1.00e+00	7.29e-11	2.22e-16	1.49e+00	0.0	0.0	0.0

STEP	TIME	||U||	LABEL	TYPE
0	00:00:00	1.4903e+00	1	EP
10	00:00:00	8.1799e-01	2	

```
   20   00:00:01    2.2002e-02      3
   30   00:00:01    7.7764e-01      4
   40   00:00:02    1.4482e+00      5
   50   00:00:02    1.8782e+00      6  EP

STEP       TIME        ||U||   LABEL  TYPE
   0   00:00:02    1.4903e+00      7  EP
  10   00:00:03    1.9042e+00      8
  20   00:00:03    1.9901e+00      9
  30   00:00:03    1.7455e+00     10
  40   00:00:04    1.2111e+00     11
  50   00:00:04    4.7558e-01     12  EP
```

Here, the initial screen output beginning with the lines

```
     STEP   DAMPING                    NORMS              COMPUTATION TIMES
   IT SIT     GAMMA     ||d||     ||f||     ||U||     F(x)   DF(x)   SOLVE
   0                            1.80e-01  1.42e+00    0.0    0.0    0.0
   1   1   1.00e+00  9.94e-02  9.88e-03  1.49e+00    0.0    0.0    0.0
   2   1   1.00e+00  4.91e-03  2.42e-05  1.49e+00    0.0    0.0    0.0
```

provides diagnostic information during the process of locating an initial solution point u^* on the solution manifold. For example, the ||f|| and ||U|| columns contain the values of $\|\Phi(u)\|$ and $\|u\|$, respectively, after each iterate of the default COCO root-finding algorithm. Similarly, the screen output beginning with the lines

```
STEP       TIME        ||U||   LABEL  TYPE
   0   00:00:00    1.4903e+00      1  EP
  10   00:00:00    8.1799e-01      2
  20   00:00:01    2.2002e-02      3
```

provides diagnostic information during the process of continuation along the solution manifold. In this case, the ||U|| column contains the value of $\|u\|$ for each labeled point on the solution manifold. ∎

Each integer entry in the LABEL column of the screen output produced during continuation in Example 3.7 references a MATLAB-compatible data file with the name sol#.mat, where # denotes the corresponding integer. Each such *solution file* encodes a *chart structure* with numerical content representative of the corresponding point on the solution manifold, including the values of the continuation variables and any active continuation parameters. In addition, each solution file contains a (possibly empty) *data array* that can be used to store auxiliary information about the corresponding continuation problem.

The COCO *solution extractor* coco_read_solution may be invoked to extract the data array and chart structure from a given solution file. As an example, the command

```
[data chart] = coco_read_solution('fun1', 'run1', 6);
```

extracts content from the sol6.mat file stored in the folder associated with the 'run1' run identifier and the 'fun1' function identifier. On the computer used to generate the numerical results quoted in this book, the chart output argument contains a field denoted by x with the content [-0.4619; 0.1131] representing the numerical values of the components of u at the corresponding point on the solution manifold. Since, by default, no content is assigned to the solution data array, the data output argument is here empty.

3.3.2   Adding monitor functions

Monitor functions may be added, using the `coco_add_func` utility, to a continuation problem structure provided that they are encoded according to the COCO-specific function syntax shown in Sect. 3.3.1. For embedded monitor functions, the `'zero'` entry in the call to `coco_add_func` is replaced with either of the two strings `'active'` or `'inactive'`, depending on whether the corresponding continuation parameters are initially designated as active or inactive, followed by a cell array of string labels referring to individual continuation parameters.

Suppose, for example, that the continuation problem structure `prob` encodes a restricted continuation problem E in $\mathbb{R}^n$ obtained from the family of zero functions $\Phi : \mathbb{R}^n \to \mathbb{R}^m$, the family of monitor functions $\Psi : \mathbb{R}^n \to \mathbb{R}^r$, the initial solution guess u_0, and the index set $\mathbb{I}$. Consider the action on E of the operator $[\emptyset, g, \mathbb{K}_g^o, u_{g,0}]$, where $u_{g,0} \in \mathbb{R}^k$, $\mathbb{K}_g^o \subseteq \{1, \ldots, n\}$, the domain of g has dimension $|\mathbb{K}_g^o| + k$, and the range of g has dimension l. Let

- the variable `fid` contain a *function identifier*, i.e., a string reference to g,

- the variable `fhan` contain a *function handle* to an encoding of g,

- the variable `data` contain the function data structure $\mathfrak{D}_g$,

- the variable `pnames` contain a string label or a cell array of l string labels,

- the variable `uidx` contain the index set $\mathbb{K}_g^o$, and

- the variable `u0` contain the vector $u_{g,0}$.

Then, the command

```
coco_add_func(prob, fid, fhan, data, 'inactive', pnames, ...
   'uidx', uidx, 'u0', u0);
```

returns a continuation problem structure that encodes the restricted continuation problem

$$\hat{E} := \left[\emptyset, g, \mathbb{K}_g^o, u_{g,0}\right](E) \tag{3.61}$$

in $\hat{u} \in \mathbb{R}^{n+k}$ obtained from the family of zero functions

$$\hat{\Phi} : \hat{u} \mapsto \Phi\left(\hat{u}_{\{1,\ldots,n\}}\right), \tag{3.62}$$

the family of monitor functions

$$\hat{\Psi} : \hat{u} \mapsto \begin{pmatrix} \Psi\left(\hat{u}_{\{1,\ldots,n\}}\right) \\ g\left(\hat{u}_{\mathbb{K}_g^o}, \hat{u}_{\{n+1,\ldots,n+k\}}\right) \end{pmatrix}, \tag{3.63}$$

the initial solution guess

$$\hat{u}_0 := \left(u_0, u_{g,0}\right), \tag{3.64}$$

and the index set $\hat{\mathbb{I}}$ obtained by appending the set $\{|\mathbb{I}| + 1, \ldots, |\mathbb{I}| + l\}$ to $\mathbb{I}$. As before, the input argument `uidx` and the preceding flag `'uidx'` may be omitted if $\mathbb{K}_g^o = \emptyset$. Similarly, the input argument `u0` and the preceding flag `'u0'` may be omitted if $u_{g,0}$ is an empty vector, i.e., $k = 0$.

Example 3.8 We proceed to append the single monitor function

$$\Psi : u \mapsto u_1^2 + u_2^2 \tag{3.65}$$

from Eq. (2.16) to the extended continuation problem constructed in Examples 3.5 and 3.6 and to associate the corresponding element of μ with $\mathbb{I}$. The following definition implements the monitor function Ψ in Eq. (3.65) in a COCO-compatible syntax.

```
function [data y] = dist(prob, data, u)
  y = u(1)^2+u(2)^2;
end
```

The command

```
prob = coco_add_func(prob, 'fun2', @dist, [], 'inactive', 'p', ...
  'uidx', [1; 2]);
```

then augments the previously created continuation problem structure `prob`. Here, the function type `'inactive'` instructs `coco_add_func` to append the function with function handle `@dist` and function identifier `'fun2'` as a monitor function to the existing extended continuation problem and to associate the corresponding element of μ, here labeled by the string `'p'`, with $\mathbb{I}$, such that $\mathbb{I} = \{1\}$.

The empty brackets in the fourth argument represent the initial content of the function data structure $\mathfrak{D}_\Psi$. We recall that the dimension of the domain of the extended continuation problem encoded in `prob` prior to this function call equals 2. The array following the `'uidx'` flag and the absence of the `'u0'` flag then imply that $\mathbb{K}_\Psi^o = \{1, 2\}$ and $\mathbb{K}_\Psi^n = \emptyset$, and, consequently, that $\mathbb{K}_\Psi = \{1, 2\}$. ■

As discussed in Sect. 2.2.4, the restricted continuation problem constructed in Example 3.8 is initializable with a 0-dimensional atlas algorithm, since $n = 2$, $m = 1$, and $|\mathbb{I}| = 1$. The command

```
coco(prob, 'run2', [], 0);
```

locates the locally unique solution near u_0 using the default 0-dimensional atlas algorithm developed for this book and shipped with COCO.

Example 3.9 We continue the session from Example 3.7. The extract below shows the result of executing the commands described above.

```
>> prob = coco_add_func(prob, 'fun2', @dist, [], 'inactive', 'p', ...
      'uidx', [1; 2]);
>> coco(prob, 'run2', [], 0);
```

STEP	DAMPING		NORMS		COMPUTATION TIMES		
IT SIT	GAMMA	||d||	||f||	||U||	F(x)	DF(x)	SOLVE
0			1.80e-01	1.42e+00	0.0	0.0	0.0
1 1	1.00e+00	1.42e-01	2.86e-02	1.43e+00	0.0	0.0	0.0
2 1	1.00e+00	1.00e-02	1.41e-04	1.42e+00	0.0	0.0	0.0
3 1	1.00e+00	5.00e-05	3.54e-09	1.42e+00	0.0	0.0	0.0
4 1	1.00e+00	1.25e-09	9.16e-16	1.42e+00	0.0	0.0	0.0

STEP	TIME	||U||	LABEL	TYPE
0	00:00:00	1.4213e+00	1	EP

Here, the initial screen output beginning with the lines

```
     STEP   DAMPING              NORMS              COMPUTATION TIMES
   IT SIT     GAMMA      ||d||     ||f||    ||U||   F(x)  DF(x)  SOLVE
    0                            1.80e-01 1.42e+00   0.0   0.0    0.0
    1   1  1.00e+00  1.42e-01  2.86e-02 1.43e+00   0.0   0.0    0.0
    2   1  1.00e+00  1.00e-02  1.41e-04 1.42e+00   0.0   0.0    0.0
```

again provides diagnostic information during the initialization of u^*. The output

```
   STEP     TIME       ||U||  LABEL TYPE
    0   00:00:00   1.4213e+00     1  EP
```

contains diagnostic information for the single step of the 0-dimensional atlas algorithm. ∎

As discussed in Sect. 3.2.2, the initial assignment of integer indices to the index set $\mathbb{I}$ may
be overruled at run-time by a user-initiated reallocation of elements between $\mathbb{I}$ and $\mathbb{J}$. In the
case considered in Example 3.9, it is possible to obtain a restricted continuation problem of
dimensional deficit 1 by releasing the constraint on the continuation parameter associated
with the monitor function Ψ. Specifically, in the command

```
coco(prob, 'run3', [], 1, 'p', [0.1, 5]);
```

the second-to-last argument instructs the coco entry-point function to reallocate the element of the index set $\mathbb{I}$ corresponding to the component of μ labeled by 'p' to the complement $\mathbb{J}$. Finally, the last argument provides upper and lower bounds on the value of the
corresponding monitor function during continuation. As before, the numeral 1 preceding
the parameter label 'p' indicates the dimension of the corresponding atlas algorithm.

Example 3.10 We continue the session from Example 3.9. The extract below shows the
result of executing the command described above.

```
>> coco(prob, 'run3', [], 1, 'p', [0.1 5]);

     STEP   DAMPING              NORMS              COMPUTATION TIMES
   IT SIT     GAMMA      ||d||     ||f||    ||U||   F(x)  DF(x)  SOLVE
    0                            1.80e-01 2.47e+00   0.0   0.0    0.0
    1   1  1.00e+00  1.42e-01  2.86e-02 2.47e+00   0.0   0.0    0.0
    2   1  1.00e+00  1.00e-02  1.41e-04 2.47e+00   0.0   0.0    0.0
    3   1  1.00e+00  5.00e-05  3.54e-09 2.47e+00   0.0   0.0    0.0
    4   1  1.00e+00  1.25e-09  9.16e-16 2.47e+00   0.0   0.0    0.0

   STEP     TIME       ||U||  LABEL TYPE           p
    0   00:00:00   2.4699e+00     1  EP     2.0200e+00
   10   00:00:00   3.9193e-01     2         1.3530e-01
   11   00:00:00   3.3166e-01     3  EP     1.0000e-01

   STEP     TIME       ||U||  LABEL TYPE           p
    0   00:00:00   2.4699e+00     4  EP     2.0200e+00
   10   00:00:01   4.3413e+00     5         3.8700e+00
   20   00:00:01   4.4657e+00     6         3.9936e+00
   30   00:00:01   4.1921e+00     7         3.7218e+00
   40   00:00:02   2.1535e+00     8         1.7108e+00
   49   00:00:02   3.3166e-01     9  EP     1.0000e-01
```

Here, the initial screen output beginning with the lines

```
    STEP    DAMPING              NORMS                 COMPUTATION TIMES
    IT SIT   GAMMA      ||d||      ||f||      ||U||    F(x)   DF(x)  SOLVE
    0                            1.80e-01   2.47e+00   0.0    0.0    0.0
    1   1  1.00e+00   1.42e-01   2.86e-02   2.47e+00   0.0    0.0    0.0
```

provides diagnostic information during the initialization of u^* given the value $\mu_{\mathbb{I}} = \mu_{\mathbb{I}}^* = \Psi(u_0)$. For example, the $||f||$ and $||U||$ columns contain the values of $\left\| G\left(u, \mu_{\mathbb{J}}\right) \right\|$ and $\left\| \left(u, \mu_{\mathbb{J}}\right) \right\|$, respectively, after each iterate of the COCO root-finding algorithm. Similarly, the screen output beginning with the lines

```
    STEP       TIME        ||U||   LABEL  TYPE          p
     0     00:00:00    2.4699e+00      1  EP     2.0200e+00
    10     00:00:00    3.9193e-01      2         1.3530e-01
```

provides diagnostic information during the continuation. In this case, the $||U||$ column contains the value of $\left\| \left(u, \mu_{\mathbb{J}}\right) \right\|$ for each labeled solution point. ■

As before, each integer entry in the LABEL column of the screen output in Example 3.10 references a solution file with the name `sol#.mat`, where # denotes the corresponding integer. As an example, on the computer used to produce the numerical results in this book, the `chart` output argument from the command

```
[data chart] = coco_read_solution('fun1', 'run3', 7)
```

contains a field denoted by x whose value equals the array [-0.5088; 1.8609] containing the numerical values of the components of u indexed by $\mathbb{K}_\Phi$. In contrast, the command

```
[data chart] = coco_read_solution('', 'run3', 7)
```

results in `chart.x` equal to the array [-0.5088; 1.8609; 3.7218] containing the numerical values of the unknowns u and $\mu_{\mathbb{J}}$ of the corresponding restricted continuation problem.

The diagnostic information shown in the screen output in Example 3.10 represents only a subset of the numerical data obtained at selected points during the continuation run. A more complete representation of the data generated during continuation can be found by spooling the output of the `coco` entry-point function to a variable, as in the command

```
bd = coco(prob, 'run3', [], 1, 'p', [0.1, 5]);
```

or by using the `coco_bd_read` utility to extract this data from disk following the execution of the `coco` entry-point function, as in the command

```
bd = coco_bd_read('run3');
```

In this case, the bd variable is a 63×8 cell array, the first several rows of which are shown in the extract below.

```
'PT'    'StepSize'  'TIME'      '||U||'    'SLAB'  'LAB'  'TYPE'  'p'
[11]    [ 0.0634]  [0.5484]   [0.3317]    [  3]   [  3]  'EP'   [0.1000]
[10]    [ 0.0760]  [0.4415]   [0.3919]    [  2]   [  2]  'RO'   [0.1353]
[ 9]    [ 0.0944]  [0.4126]   [0.4812]    []      []     ' '    [0.1940]
[ 8]    [ 0.1214]  [0.3830]   [0.5950]    []      []     ' '    [0.2772]
```

```
[ 7]  [  0.1606]   [0.3547]   [0.7441]        []      []   ''        [0.3965]
[ 6]  [  0.2143]   [0.3231]   [0.9421]        []      []   ''        [0.5665]
[ 5]  [  0.2793]   [0.2837]   [1.1992]        []      []   ''        [0.7993]
```

3.3.3 Staged construction

In Example 3.6, a single call to `coco_add_func` was used to construct the zero problem considered therein and in the subsequent examples. As a result, the vector of continuation variables was already fully formed with the addition of the single zero function. Numerical values for the components of the corresponding initial guess were provided with the initial call to `coco_add_func`.

Consider, instead, the extended continuation problem E in $u \in \mathbb{R}^3$ obtained from the family of zero functions

$$\Phi : u \mapsto \begin{pmatrix} u_1^2 + (u_2 - 1)^2 - 1 \\ u_2 + u_3 \end{pmatrix}, \tag{3.66}$$

the monitor function

$$\Psi : u \mapsto u_1^2 - u_3^2, \tag{3.67}$$

and the initial solution guess $u_0 := (0.9, 1.1, -1.1)$. We seek a staged construction of E in terms of the nontrivial chain decomposition

$$\emptyset = E_0 \subsetneqq E_1 \subsetneqq E_2 \subsetneqq E_3 = E, \tag{3.68}$$

where

- E_1 is an extended continuation problem in $w \in \mathbb{R}^2$ obtained from the zero function

  $$\hat{\Phi}^{(1)} : w \mapsto w_1^2 + (w_2 - 1)^2 - 1, \tag{3.69}$$

  the empty family of monitor functions, and the initial solution guess $w_0 := (0.9, 1.1)$, and

- E_2 is an extended continuation problem in $w \in \mathbb{R}^3$ obtained from the family of zero functions

  $$\hat{\Phi}^{(2)} : w \mapsto \begin{pmatrix} w_1^2 + (w_2 - 1)^2 - 1 \\ w_2 + w_3 \end{pmatrix}, \tag{3.70}$$

  the empty family of monitor functions, and the initial solution guess $w_0 = (0.9, 1.1, -1.1)$.

It is straightforward to show that

$$E_1 = \left[f_1, \emptyset, \emptyset, (0.9, 1.1) \right] (E_0), \tag{3.71}$$

where a COCO-compatible encoding of the function $f_1 : (x, y) \mapsto x^2 + (y - 1)^2 - 1$ is shown below.

```
function [data y] = circ(prob, data, u)
  y = u(1)^2+(u(2)-1)^2-1;
end
```

Suppose that the variable `prob` contains an empty continuation problem structure. The assignment

```
prob = coco_add_func(prob,'fun1', @circ, [], 'zero', ...
   'u0', [0.9; 1.1]);
```

then assigns the extended continuation problem E_1 to `prob`. Here, $\mathfrak{D}_{f_1}$ is empty, $\mathbb{K}_1^o = \emptyset$, $\mathbb{K}_1^n = \{1,2\}$, and thus $\mathbb{K}_{f_1} = \{1,2\}$.

We proceed to construct the partial realization E_2 by appending to the information already stored in the problem structure `prob`. It is again straightforward to show that

$$E_2 = \big[f_2, \emptyset, \{2\}, (-1.1) \big] (E_1), \tag{3.72}$$

where a COCO-compatible encoding of the function $f_2 : (x,y) \mapsto x + y$ is shown below.

```
function [data y] = plan(prob, data, u)
   y = u(1)+u(2);
end
```

The assignment

```
prob = coco_add_func(prob, 'fun2', @plan, [], 'zero', ...
   'uidx', 2, 'u0', -1.1);
```

then encodes the extended continuation problem E_2 in the problem structure `prob`. Here, the function data structure $\mathfrak{D}_{f_2}$ is again empty. Furthermore, the integer following the `'uidx'` flag indicates that $\mathbb{K}_{f_2}^o = \{2\}$. Finally, the numerical value following the `'u0'` flag is used to grow the vector of continuation variables by one element, such that $\mathbb{K}_2^n = \{3\}$ and $\mathbb{K}_{f_2} = \{2,3\}$, and to assign the initial numerical value -1.1 to this component.

Finally, we append the monitor function Ψ to the information stored in the problem structure `prob` in order to complete the staged construction of the extended continuation problem E. In this case,

$$E_3 = \big[\emptyset, f_3, \{1,3\}, \emptyset \big] (E_2), \tag{3.73}$$

where a COCO-compatible encoding of the function $f_3 : (x,y) \mapsto x^2 - y^2$ is shown below.

```
function [data y] = hype(prob, data, u)
   y = u(1)^2-u(2)^2;
end
```

The extended continuation problem $E_3 = E$ then results from the assignment

```
prob = coco_add_func(prob, 'fun3', @hype, [], 'inactive', 'p', ...
   'uidx', [1; 3]);
```

This appends the function f_3 to the monitor functions stored in the problem structure `prob` and the corresponding integer index to the index set $\mathbb{I}$. The function data structure $\mathfrak{D}_{f_3}$ is again assigned the empty set. Finally, the `'uidx'` flag is used to indicate that $\mathbb{K}_{f_3}^o = \{1,3\}$, i.e., that the domain of f_3 is parameterized by two components of the vector of continuation variables, in that order, whose integer indices already appear in E_2. By the absence of the `'u0'` flag, it follows that $\mathbb{K}_{f_3}^n = \emptyset$ and, thus, that $\mathbb{K}_{f_3} = \{1,3\}$. Since $n = 3$,

$m = 2$, and $|\mathbb{I}| = 1$, it follows that the corresponding restricted continuation problem is initializable with a 0-dimensional atlas algorithm.

Example 3.11 The extract below shows the result of executing the sequence of commands described above.

```
>> prob = coco_prob();
>> prob = coco_add_func(prob, 'fun1', @circ, [], 'zero', ...
   'u0', [0.9; 1.1]);
>> prob = coco_add_func(prob, 'fun2', @plan, [], 'zero', ...
   'uidx', 2, 'u0', -1.1);
>> prob = coco_add_func(prob, 'fun3', @hype, [], 'inactive', 'p', ...
   'uidx', [1; 3]);
>> coco(prob, 'run4', [], 0);
```

STEP		DAMPING		NORMS		COMPUTATION TIMES		
IT	SIT	GAMMA	||d||	||f||	||U||	F(x)	DF(x)	SOLVE
0				1.80e-01	1.80e+00	0.0	0.0	0.0
1	1	1.00e+00	1.40e-01	1.43e-02	1.94e+00	0.0	0.0	0.0
2	1	1.00e+00	8.65e-03	6.19e-05	1.93e+00	0.0	0.0	0.0
3	1	1.00e+00	3.26e-05	1.05e-09	1.93e+00	0.0	0.0	0.0
4	1	1.00e+00	4.67e-10	3.51e-16	1.93e+00	0.0	0.0	0.0

STEP	TIME	||U||	LABEL	TYPE
0	00:00:00	1.9268e+00	1	EP

In this case, the command

```
[data chart] = coco_read_solution('fun2', 'run4', 1);
```

returns the components of the vector of continuation variables indexed by the dependency index set $\mathbb{K}_{f_2}$ in the x field of the chart output argument.   ∎

3.3.4 Special-purpose wrappers

In Sects. 3.3.1–3.3.3, construction of a continuation problem structure relied directly on the general-purpose, core utility coco_add_func. It is sometimes convenient to invoke an indirect calling syntax that is expanded by *special-purpose wrappers* to one or several syntactically complete calls to coco_add_func. Given an extended continuation problem in $\mathbb{R}^n$, suppose that the variable pidx contains an index subset of $\{1,\ldots,n\}$. Denote by p the corresponding ordered subset of elements of the vector of continuation variables. Moreover, let pnames be a cell array of string labels with $|p|$ elements. The command

```
prob = coco_add_pars(prob, 'pars', pidx, pnames);
```

then adds the monitor function $f : u \mapsto p$ with function identifier 'pars' to the extended continuation problem, assigns the labels contained in pnames to the corresponding elements of μ, and appends the corresponding indices of μ to the index set $\mathbb{I}$. In the expanded call to coco_add_func within the implementation of the special-purpose wrapper coco_add_pars, the 'uidx' flag is used to assign pidx to the index set $\mathbb{K}_f^o$.

Example 3.12 We continue the session from Example 3.11 and use an alternative syntax for the special-purpose wrapper coco_add_pars in order to designate the corresponding continuation parameters as initially active. As seen in the following extract, the inclusion of the corresponding string label in the call to the `coco` entry-point function results in the printing of the value of the continuation parameter to screen during continuation.

```
>> prob = coco_add_pars(prob, 'pars', 3, {'u3'}, 'active');
>> coco(prob, 'run5', [], 1, {'p' 'u3'}, [-0.4 0.4]);
```

		STEP	DAMPING		NORMS			COMPUTATION TIMES	
	IT	SIT	GAMMA	||d||	||f||	||U||	F(x)	DF(x)	SOLVE
	0				1.80e-01	2.14e+00	0.0	0.0	0.0
	1	1	1.00e+00	1.59e-01	1.43e-02	2.30e+00	0.0	0.0	0.0
	2	1	1.00e+00	9.60e-03	6.19e-05	2.29e+00	0.0	0.0	0.0
	3	1	1.00e+00	3.50e-05	1.05e-09	2.29e+00	0.0	0.0	0.0
	4	1	1.00e+00	4.83e-10	3.51e-16	2.29e+00	0.0	0.0	0.0

STEP	TIME	||U||	LABEL	TYPE	p	u3
0	00:00:00	2.2898e+00	1	EP	-4.0000e-01	-1.1708e+00
7	00:00:00	1.6292e+00	2	EP	4.0000e-01	-7.2361e-01

Notably, a run-time redesignation of the continuation parameter labeled by 'p' from $\mathbb{I}$ to $\mathbb{J}$ is here necessary in order to render a restricted continuation problem that is initializable with a 1-dimensional atlas algorithm. ∎

As a second example, given an extended continuation problem in $\mathbb{R}^n$, suppose that `vidx` and `widx` denote two index subsets of $\{1,\ldots,n\}$ of equal cardinality. Denote by v and w the corresponding ordered subsets of elements of the vector of continuation variables. The command

```
prob = coco_add_glue(prob, 'glue', vidx, widx);
```

then adds the zero function $f : u \mapsto v - w$ with function identifier 'glue' to the extended continuation problem. In the expanded call to `coco_add_func` within the implementation of the special-purpose wrapper `coco_add_glue`, the 'uidx' flag is again used to assign `vidx` and `widx`, in that order, to the index set $\mathbb{K}_f^o$.

Example 3.13 We illustrate the use of the special-purpose wrapper `coco_add_glue` by considering the problem of continuing period-n orbits of the discrete *Hénon map*:

$$\begin{pmatrix} x \\ y \end{pmatrix} \longmapsto \begin{pmatrix} y + 1 - ax^2 \\ bx \end{pmatrix}. \tag{3.74}$$

To this end, consider the function $f : \mathbb{R}^6 \mapsto \mathbb{R}^2$, where

$$f : z \mapsto \begin{pmatrix} z_5 \\ z_6 \end{pmatrix} - \begin{pmatrix} z_4 + 1 - z_1 z_3^2 \\ z_2 z_3 \end{pmatrix}, \tag{3.75}$$

encoded in the COCO-compatible function below.

```
function [data y] = henon(prob, data, u)
  y = [u(5)-u(4)-1+u(1)*u(3)^2; u(6)-u(2)*u(3)];
end
```

A period-n orbit is then a solution to the extended continuation problem E in $u \in \mathbb{R}^{2n+2}$ obtained from the family of zero functions

$$\Phi : u \mapsto \begin{pmatrix} f(u_1,u_2,u_3,u_4,u_5,u_6) \\ f(u_1,u_2,u_5,u_6,u_7,u_8) \\ \vdots \\ f(u_1,u_2,u_{2n-1},u_{2n},u_{2n+1},u_{2n+2}) \\ f(u_1,u_2,u_{2n+1},u_{2n+2},u_3,u_4) \end{pmatrix}, \tag{3.76}$$

the empty family of monitor functions, and some initial solution guess $u_0 := (u_{1,0},\ldots, u_{2n+2,0})$.

Now consider the nontrivial chain decomposition

$$\emptyset = E_0 \subsetneqq E_1 \subsetneqq \cdots \subsetneqq E_{n-1} \subsetneqq E_n = E, \tag{3.77}$$

where the partial realization E_i, for $1 \leq i < n$, is an extended continuation problem in $w \in \mathbb{R}^{2i+4}$ obtained from the family of zero functions

$$\hat{\Phi}^{(i)} : w \mapsto \begin{pmatrix} f(w_1,w_2,w_3,w_4,w_5,w_6) \\ f(w_1,w_2,w_5,w_6,w_7,w_8) \\ \vdots \\ f(w_1,w_2,w_{2i+1},w_{2i+2},w_{2i+3},w_{2i+4}) \end{pmatrix}, \tag{3.78}$$

the empty family of monitor functions, and the initial solution guess

$$w_0 := \left(u_{1,0},\ldots,u_{2i+4,0}\right). \tag{3.79}$$

It is straightforward to show that

$$E_1 = \left[f,\emptyset,\emptyset,\left(u_{1,0},u_{2,0},u_{3,0},u_{4,0},u_{5,0},u_{6,0}\right)\right](E_0). \tag{3.80}$$

Similarly, for $1 < i < n$,

$$E_i = \left[f,\emptyset,\{1,2,2i+1,2i+2\},\left(u_{2i+3,0},u_{2i+4,0}\right)\right](E_{i-1}). \tag{3.81}$$

Finally, we obtain the full extended continuation problem by the construction

$$E_n = \left[f,\emptyset,\{1,2,2n+1,2n+2,3,4\},\emptyset\right](E_{n-1}). \tag{3.82}$$

The functional encoding below implements this staged construction for an arbitrary n.

```
function prob = period(u0, n)

prob = coco_prob();
prob = coco_add_func(prob, 'henon_1', @henon, [], 'zero', ...
  'u0', u0(1:6));
for i=2:n-1
  prob = coco_add_func(prob, sprintf('henon_%d', i), @henon, [], ...
    'zero', 'uidx', [1; 2; 2*i+1; 2*i+2], 'u0', u0(2*i+3:2*i+4));
end
prob = coco_add_func(prob, sprintf('henon_%d', n), @henon, [], ...
  'zero', 'uidx', [1; 2; 2*n+1; 2*n+2; 3; 4]);

end
```

Alternatively, consider the extended continuation problem E in $u \in \mathbb{R}^{2n+4}$ obtained from the family of zero functions

$$\Phi : u \mapsto \begin{pmatrix} f(u_1, u_2, u_3, u_4, u_5, u_6) \\ f(u_1, u_2, u_5, u_6, u_7, u_8) \\ \vdots \\ f(u_1, u_2, u_{2n-1}, u_{2n}, u_{2n+1}, u_{2n+2}) \\ f(u_1, u_2, u_{2n+1}, u_{2n+2}, u_{2n+3}, u_{2n+4}) \\ u_3 - u_{2n+3} \\ u_4 - u_{2n+4} \end{pmatrix}, \tag{3.83}$$

the empty family of monitor functions, and some initial solution guess $u_0 := (u_{1,0}, \ldots, u_{2n+2,0}, u_{3,0}, u_{4,0})$. It again holds that

$$\emptyset = E_0 \subsetneqq E_1 \subsetneqq \cdots \subsetneqq E_n \subsetneqq E_{n+1} = E, \tag{3.84}$$

where the partial realization E_i, for $1 \leq i \leq n$, is an extended continuation problem in $w \in \mathbb{R}^{2i+4}$ obtained from the family of zero functions

$$\hat{\Phi}^{(i)} : w \mapsto \begin{pmatrix} f(w_1, w_2, w_3, w_4, w_5, w_6) \\ f(w_1, w_2, w_5, w_6, w_7, w_8) \\ \vdots \\ f(w_1, w_2, w_{2i+1}, w_{2i+2}, w_{2i+3}, w_{2i+4}) \end{pmatrix}, \tag{3.85}$$

the empty family of monitor functions, and the initial solution guess

$$w_0 := \left(u_{1,0}, \ldots, u_{2i+4,0} \right). \tag{3.86}$$

As before

$$E_1 = \left[f, \emptyset, \emptyset, \left(u_{1,0}, u_{2,0}, u_{3,0}, u_{4,0}, u_{5,0}, u_{6,0} \right) \right] (E_0) \tag{3.87}$$

and, this time for $1 < i \leq n$,

$$E_i = \left[f, \emptyset, \{1, 2, 2i+1, 2i+2\}, \left(u_{2i+3,0}, u_{2i+4,0} \right) \right] (E_{i-1}). \tag{3.88}$$

The final extended continuation problem is now obtained through the construction

$$E_{n+1} = \left[g, \emptyset, \{3, 4, 2n+3, 2n+4\}, \emptyset \right] (E_n), \tag{3.89}$$

where $g : (x, y) \mapsto x - y$. The modified functional encoding below implements this staged construction for an arbitrary n.

```
function prob = period(u0, n)

prob = coco_prob();
prob = coco_add_func(prob, 'henon_1', @henon, [], 'zero', ...
  'u0', u0(1:6));
for i=2:n
  prob = coco_add_func(prob, sprintf('henon_%d', i), @henon, [], ...
    'zero', 'uidx', [1; 2; 2*i+1; 2*i+2], 'u0', u0(2*i+3:2*i+4));
end
prob = coco_add_glue(prob, 'glue', [3; 4], [2*n+3; 2*n+4]);

end
```

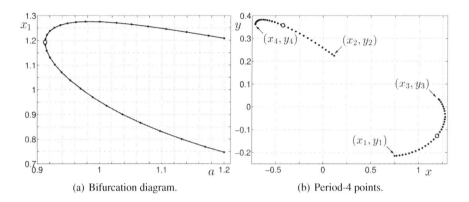

(a) Bifurcation diagram.   (b) Period-4 points.

Figure 3.2. *A family of period-4 orbits of the Hénon map obtained in Example* 3.13. *The family passes through a period-halving point marked with an open circle, where the period-4 family intersects a period-2 family. A typical bifurcation diagram is shown in* (a). *Here, we graph a projection of the solution manifold onto the hyperplane with coordinates given by the parameter a and the x component of the first point on the orbit. The full state-space orbits are shown in* (b).

The extract below shows the result of applying the `period` function to an initial approximation of a period-4 orbit found for $a = 1$ and $b = 0.3$.

```
>> u0 = [1; 0.3; 1.275; -0.031; -0.656; 0.382; 0.952; ...
     -0.197; -0.103; 0.286; 1.275; -0.031];
>> prob = period(u0, 4);
>> prob = coco_add_pars(prob, 'pars', [1 2], {'a' 'b'});
>> coco(prob, 'henon', [], 1, 'a', [0.8 1.2]);
```

STEP	DAMPING		NORMS			COMPUTATION TIMES		
IT SIT	GAMMA	||d||	||f||		||U||	F(x)	DF(x)	SOLVE
0			1.10e-03		2.64e+00	0.0	0.0	0.0
1 1	1.00e+00	9.69e-04	2.08e-07		2.64e+00	0.0	0.0	0.0
2 1	1.00e+00	6.06e-07	1.81e-13		2.64e+00	0.0	0.0	0.0

STEP	TIME	||U||	LABEL	TYPE	a
0	00:00:00	2.6375e+00	1	EP	1.0000e+00
10	00:00:00	2.5780e+00	2		9.1341e-01
20	00:00:01	2.4952e+00	3		1.0578e+00
24	00:00:01	2.5191e+00	4	EP	1.2000e+00

STEP	TIME	||U||	LABEL	TYPE	a
0	00:00:01	2.6375e+00	5	EP	1.0000e+00
7	00:00:02	2.6837e+00	6	EP	1.2000e+00

A corresponding *bifurcation diagram* is shown in Fig. 3.2(a). The state-space representation in Fig. 3.2(b) shows that the points along the period-4 orbit coalesce onto a period-2 orbit at a period-halving point (represented by the open circle in Fig. 3.2(a)).   ∎

3.3.5   Function data

Although neither of the zero or monitor functions introduced above made use of function data structures, one can easily imagine cases where this would be useful. Consider, for example, the COCO-compatible encoding shown below.

```
function [data y] = finitediff(prob, data, u)

dep = u(data.dep_idx);
par = u(data.par_idx);

ff = dep(data.f_idx);
gg = dep(data.g_idx);

f = [par(1)-(par(2)+1)*ff+ff.^2.*gg; par(2)*ff-ff.^2.*gg];

y = data.A*dep+data.B*f;

end
```

This represents a fourth-order finite-difference approximation on a uniform mesh of the left-hand sides of the coupled differential equations

$$f'' + p_1 - (p_2 + 1)f + f^2 g = 0 \qquad (3.90)$$

and

$$g'' + p_2 f - f^2 g = 0 \qquad (3.91)$$

describing steady-state solutions of a 1-dimensional *Brusselator model* on the interval $[0, 1]$. Here, given some integer N, the function data structure data is assumed to contain the fields

- dep_idx with value equal to the index array $\begin{pmatrix} 1 & \cdots & 2N+4 \end{pmatrix}$;

- par_idx with value equal to the index array $\begin{pmatrix} 2N+5 & 2N+6 \end{pmatrix}$;

- f_idx with value equal to the index array $\begin{pmatrix} 1 & 3 & \cdots & 2N+3 \end{pmatrix}$;

- g_idx with value equal to the index array $\begin{pmatrix} 2 & 4 & \cdots & 2N+4 \end{pmatrix}$;

- A with value equal to the $2N \times (2N+4)$ matrix of the form

$$\begin{pmatrix} 1 & 0 & -2 & 0 & 1 & & \\ & \ddots & \ddots & \ddots & \ddots & \ddots & \\ & & 1 & 0 & -2 & 0 & 1 \end{pmatrix}; \qquad (3.92)$$

- and B with value equal to the $2N \times (2N+4)$ matrix of the form

$$\frac{1}{12(N+1)^2} \begin{pmatrix} 1 & 0 & 10 & 0 & 1 & & \\ & \ddots & \ddots & \ddots & \ddots & \ddots & \\ & & 1 & 0 & 10 & 0 & 1 \end{pmatrix}. \qquad (3.93)$$

The computational benefit of assigning content to this function data structure at the point that the function is added to the extended continuation problem, rather than each time the function is called, should be quite evident, especially in the case when N is large.

Example 3.14 The following extract illustrates the a priori construction of the function data structure `data` and the initial solution guess `u0` and their inclusion in the subsequent call to `coco_add_func`.

```
>> N = 40;
>> data.dep_idx = (1:2*N+4)';
>> data.par_idx = [2*N+5; 2*N+6];
>> data.f_idx   = (1:2:2*N+3)';
>> data.g_idx   = (2:2:2*N+4)';
>> oneN   = ones(2*N,1);
>> zeroN  = zeros(2*N,1);
>> data.A = spdiags([oneN zeroN -2*oneN zeroN oneN], ...
      [0 1 2 3 4], 2*N, 2*N+4);
>> data.B = 1/(12*(N+1)^2)*spdiags([oneN zeroN 10*oneN zeroN oneN], ...
      [0 1 2 3 4], 2*N, 2*N+4);
>> x0   = ones(1,N+2);
>> y0   = ones(1,N+2);
>> p0   = [1; 1];
>> dep0 = [x0; y0];
>> u0   = [dep0(:); p0];
>> prob = coco_prob();
>> prob = coco_add_func(prob, 'finitediff', @finitediff, data, ...
      'zero', 'u0', u0);
```

At this stage of construction, the restricted continuation problem has a dimensional deficit of 6. We may reduce the dimensionality of the solution manifold by appending additional constraints on the vector of continuation variables. To this end, and as shown below, we rely on the `coco_add_pars` special-purpose wrapper to introduce inactive continuation parameters tracking the values of the components of u corresponding to the boundary values $f(0)$, $g(0)$, $f(1)$, and $g(1)$ and the problem parameters p_1 and p_2.

```
>> prob = coco_add_pars(prob, 'pars', [1:2 2*N+3:2*N+6]', ...
      {'f0' 'g0' 'f1' 'g1' 'p1' 'p2'});
```

The restricted continuation problem is now initializable with a 1-dimensional atlas algorithm provided that one of these constraints is removed as shown below.

```
>> coco(prob, 'brusselator', [], 1, 'g0', [0 10]);
```

	STEP	DAMPING		NORMS		COMPUTATION TIMES		
IT	SIT	GAMMA	||d||	||f||	||U||	F(x)	DF(x)	SOLVE
0				0.00e+00	9.33e+00	0.0	0.0	0.0
1	1	1.00e+00	0.00e+00	0.00e+00	9.33e+00	0.0	0.0	0.0

STEP	TIME	||U||	LABEL	TYPE	g0
0	00:00:00	9.3274e+00	1	EP	1.0000e+00
10	00:00:01	7.7315e+00	2	EP	0.0000e+00

STEP	TIME	||U||	LABEL	TYPE	g0
0	00:00:01	9.3274e+00	3	EP	1.0000e+00
10	00:00:01	1.2226e+01	4		2.0748e+00
20	00:00:02	1.6446e+01	5		3.3705e+00
30	00:00:03	2.1010e+01	6		4.6673e+00
40	00:00:04	2.5736e+01	7		5.9650e+00
50	00:00:05	3.0549e+01	8	EP	7.2636e+00

We graph the corresponding profiles $f(x)$ and $g(x)$ in Fig. 3.3. ∎

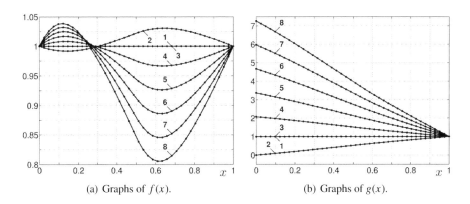

(a) Graphs of $f(x)$.   (b) Graphs of $g(x)$.

Figure 3.3. *A set of profiles representative of the family of solutions to the Brusselator boundary-value problem in Example 3.14. The boundary conditions are $f(0) = f(1) = g(1) = 1$ and $g(0) = g_0$, where g_0 is the continuation parameter. We start with the exact solution $f(x) \equiv g(x) \equiv 1$ for $g_0 = 1$ and obtain solutions with nontrivial shape as the parameter g_0 varies. The labels correspond to the session output included in the text.*

The content of the `data` input argument of the `finitediff` function in Example 3.14 represents a parameterized state of the computational algorithm implemented in the function body. After the function has been added to the extended continuation problem using the `coco_add_func` utility, this function data is stored in the continuation problem structure and available for read and write access by the corresponding function each time it is called. As explained further in the next chapter, read access to this function data also may be obtained by other functions using the `coco_get_func_data` utility.

As the `data` variable is returned unchanged by the `finitediff` function implemented in the example, and as no other function is able to directly modify this content, the parameterized state of this algorithm remains unchanged during continuation. More generally, modifications to the parameterized state of a computational algorithm implemented in a COCO-compatible function might occur as a result of changes to the function data encoded within the function body or as a result of modifications made to the function data of other functions. Support for such changes to function data across the boundary of a function definition is provided by use of the `coco_func_data` class. By storing function data among the properties of an instance of this class, we allow for changes to function data of one function to automatically propagate into the corresponding changes to the function data of another function. This concept of communication will be described further in Sect. 8.2.

3.4 Conclusions

Individual zero and monitor functions of an extended continuation problem typically depend only on a subset of the components of the vector of continuation variables. The paradigm of staged construction presented in this chapter takes advantage of this sparse dependency structure in order to realize a succession of partial realizations, each of which constitutes an extended continuation problem in its own right. The algorithmic syntax of the `coco_add_func` core utility, and of its off-spring `coco_add_pars` and `coco_add_glue`, provides a consistent framework for implementing this paradigm.

The discussion in this chapter has emphasized the great flexibility that exists in associating elements of the vector of continuation parameters with the index sets $\mathbb{I}$ and $\mathbb{J}$. Since the maximal number of elements of $\mathbb{I}$ is determined at the time of initial construction, care should be taken to provide the greatest degree of run-time flexibility for subsequent restrictions. To this end, we sometimes choose to let $\mathbb{J} = \emptyset$ initially in order to provide for maximal diversity in the number of distinct restricted continuation problems that can be constructed at run-time.

The examples in this chapter have illustrated a subset of the argument syntaxes of the `coco` entry-point function, the `coco_add_func` core utility, and the special-purpose wrappers `coco_add_pars` and `coco_add_glue`. As the purpose of this text is to instruct in the principles of constructing continuation problems within the extended continuation problem paradigm, we defer to the COCO reference manual in instances where further information is sought.

The continuation of period-n points of the Hénon map implemented in Example 3.13 demonstrated the combined use of `coco_add_func`, `coco_add_pars`, and `coco_add_glue` in order to grow successively the corresponding extended continuation problem. Notably, the `period` function shown in the example hard-coded the function handle to the encoding in `henon` as well as the integer indices for different components of the index sets $\mathbb{K}$ used at every stage of construction. A moment's reflection suggests that there may be value in reorganizing the content of `period` and the list of input arguments in order to provide a generalized implementation that could support a similar task for other iterated dynamical systems.

More broadly, it should come as no surprise that significant benefit accrues by collecting facilities for building continuation problems that belong to a particular problem context in general-purpose toolboxes and in providing these toolbox facilities as an interface to the core functions. Such facilities include *toolbox constructors* that embed multiple calls to the `coco_add_func` core utility, to its special-purpose wrappers `coco_add_pars` and `coco_add_glue`, or to other toolbox constructors in order to provide a more versatile and yet more efficient problem construction for particular problem classes. We turn to the task of constructing such COCO-compatible toolboxes in the next chapter. The full extent of the concept of task embedding and its implications for toolbox construction are then explored in Chap. 5. This pinnacle of the design fundamentals then sets the stage for the template toolboxes considered in Part II of this text.

Exercises

3.1. Explain why it is not possible to determine the dimension of the domain of the families of zero functions and monitor functions from their functional expressions.

3.2. Let the symbol $\equiv$ denote equivalence between two extended continuation problems. Show that $\equiv$ is an *equivalence relation* on $\mathcal{E}$ and describe the corresponding equivalence classes.

3.3. Suppose that E and $\hat{E}$ are equivalent extended continuation problems. Show that $E \subseteq \hat{E}$, but that $E \not\subseteq \hat{E}$ if and only if $E = \hat{E}$.

3.4. Show that $\emptyset \subseteq E$ for every extended continuation problem $E \in \mathcal{E}$. Show that $\emptyset \not\subseteq E$ for every extended continuation problem $E \in \mathcal{E}$ provided that the notation $\{1,\dots,0\}$ is understood to mean the empty set.

3.5. Suppose that $E \subseteq \hat{E}$ for some extended continuation problems $E, \hat{E} \in \mathcal{E}$. Show that there exists an extended continuation problem $\tilde{E}$, such that $\hat{E}$ is equivalent to $\tilde{E}$ and $E \not\subseteq \tilde{E}$.

3.6. Show that $\mathcal{E}$ is partially ordered under the relation $\not\subseteq$. What can you say about $\subseteq$?

3.7. Consider an extended continuation problem E obtained from a family of m zero functions, a family of r monitor functions, and an initial solution guess with n components. Now associate with E an $(m+r) \times n$ *incidence matrix* whose entries equal 0 if the zero/monitor function represented by the row index is independent of the continuation variable represented by the column index, and 1 otherwise. Translate the notions of (i) equivalence, (ii) embedding, (iii) nontrivial chain decomposition, and (iv) staged construction in terms of equivalent notions on the collection of all incidence matrices.

3.8. Repeat the analysis in Examples 3.2 and 3.3 in terms of the corresponding incidence matrices as introduced in the previous exercise.

3.9. Let $\mathbb{I}_0$ denote the content of the index set $\mathbb{I}$ after the initial staged construction of an extended continuation problem. Show that the final restricted continuation problem is initializable with a d-dimensional atlas algorithm for $d \geq 0$ only if $n - m \geq d \geq n - m - |\mathbb{I}_0|$. Explain why, in this case, $d - n + m + |\mathbb{I}_0|$ elements must be reallocated from $\mathbb{I}$ to its complement $\mathbb{J}$ prior to continuation.

3.10. Can continuation proceed when $|\mathbb{I}_0| > n - m$? Are there implications to the allowable number of monitor functions?

3.11. Consider the zero problem $\Phi(u) = 0$ corresponding to the zero function given in Example 3.5. Show that its dimensional deficit is 1 and compute the initial residual $\|\Phi(u)\|$ at the point $(0.9, 1.1)$. Does this result agree with the screen output from the `coco` entry-point function in Example 3.7?

3.12. Repeat the continuation of the 1-dimensional solution manifold of the zero problem $\Phi(u) = 0$ as shown in Example 3.7 using the default 1-dimensional atlas algorithm. Extract the solution stored in `sol11.mat` and compute the corresponding residual $\|\Phi(u)\|$.

3.13. Write the extended continuation problem corresponding to the zero function Φ and the monitor function Ψ defined in Examples 3.5 and 3.8. Write the restricted continuation problem that results from assigning the single element of μ to the index set $\mathbb{I}$. Show that the dimensional deficit of the restricted continuation problem is 0. Explain why it is necessary to reassign the index of this element of μ to $\mathbb{J}$ to enable continuation of the 1-dimensional solution manifold of the original extended continuation problem.

3.14. Repeat the continuation of the 1-dimensional solution manifold of the extended continuation problem corresponding to the zero function Φ and the monitor function Ψ as shown in Example 3.10 using the default 1-dimensional atlas algorithm. Extract the solution stored in `sol7.mat` and compute the corresponding residual norm $\|(\Phi(u), \Psi(u) - \mu)\|$.

3.15. Use the command

```
coco(prob, 'run', [], 0, 'p');
```

to locate the locally unique solution to the restricted continuation problem corresponding to the zero function Φ and the monitor function Ψ defined in Examples 3.5 and 3.8, together with the assignment of the index of the single element of μ to the index set $\mathbb{I}$. In this case, the inclusion of the fifth argument does not correspond to a deactivation of a constraint, since the dimensionality of the atlas algorithm equals 0 as per the fourth argument. What then is the result of adding this fifth argument?

3.16. Use repeated calls to `coco_add_func` to construct the restricted continuation problem in $u \in \mathbb{R}^2$ obtained from the zero function

$$\Phi(u) := u_1^2 + (u_2 - 1)^2 - 1,$$

the family of monitor functions

$$\Psi(u) := \begin{pmatrix} u_1^2 + u_2^2 \\ u_2 \\ u_1 - u_2 \end{pmatrix},$$

the initial solution guess $u_0 = (1, 1)$, and the index set $\mathbb{I} = \{2\}$.

3.17. Apply the `coco` entry-point function to locate the locally unique solution in the vicinity of the initial guess $\begin{pmatrix} 1 & 1 & 2 & 1 & 0 \end{pmatrix}$ to the restricted continuation problem constructed in the previous exercise.

3.18. Apply the `coco` entry-point function to continue the 1-dimensional solution manifold of the extended continuation problem obtained by deactivating the single constraint imposed in the restricted continuation problem considered in the previous exercise.

3.19. Repeat the construction of the extended continuation problem considered in the previous exercises by using a single COCO-compatible function for all components of Ψ as per the construction below

```
function [data y] = mon_func(prob, data, u)
  y  = [u(1)^2+u(2)^2; u(2); u(1)-u(2)];
end
```

Assign all of the corresponding components of μ to the index set $\mathbb{I}$ by using the `'inactive'` argument in the `coco_add_func` command. Perform the continuation tasks in the previous two exercises by releasing the appropriate continuation parameters in the call to the `coco` entry-point function.

3.20. Apply the `coco_add_func` utility and `coco` entry-point function to construct the restricted continuation problems corresponding to the zero function given in Eq. (2.24) and the monitor function given in Eq. (2.27) for the different assignments of the elements of μ to the index set $\mathbb{I}$ and its complement $\mathbb{J}$ as discussed in Sect. 2.2.5. What happens when you try to continue the 2-dimensional solution manifold of the extended continuation problem?

3.21. Consider all possible nontrivial chain decompositions of extended continuation problems equivalent to the extended continuation problem E introduced in Sect. 3.3.3.

Implement the corresponding staged constructions using the `coco_add_func` utility. Can this be accomplished with the encodings of `circ`, `plan`, and `hype` given in the text?

3.22. Consider the function $f : \mathbb{R}^6 \mapsto \mathbb{R}^2$, where

$$f : z \mapsto \left(\begin{array}{c} z_3 \\ z_4 \end{array} \right) - \left(\begin{array}{c} z_2 + 1 - z_5 z_1^2 \\ z_6 z_1 \end{array} \right),$$

and suppose that a COCO-compatible encoding is available in `mod_henon`. As in Example 3.13, a period-n orbit of the Hénon map is then a solution to the extended continuation problem E in $u \in \mathbb{R}^{2n+2}$ obtained from the family of zero functions

$$\Phi : u \mapsto \left(\begin{array}{c} f(u_3, u_4, u_5, u_6, u_1, u_2) \\ f(u_5, u_6, u_7, u_8, u_1, u_2) \\ \vdots \\ f(u_{2n-1}, u_{2n}, u_{2n+1}, u_{2n+2}, u_1, u_2) \\ f(u_{2n+1}, u_{2n+2}, u_3, u_4, u_1, u_2) \end{array} \right),$$

the empty family of monitor functions, and an initial solution guess $u_0 := (u_{1,0}, \ldots, u_{2n+2,0})$. Propose a suitable nontrivial chain decomposition of E and implement a functional encoding for the corresponding staged construction for arbitrary n in terms of `mod_henon`.

3.23. Generalize the `period` function in Example 3.13 to support the continuation of period-n orbits in an arbitrary finite-dimensional discrete dynamical system. Validate your code against the Hénon map shown in the example and use it to analyze a period-3 orbit in the logistic map

$$x \mapsto \lambda x (1 - x).$$

Suggest a method for finding an initial solution guess using simulation.

3.24. The coupled differential equations

$$x'' - \frac{p_4^2}{p_1} \left((p_6 + 1)x - x^2 y - p_3 \frac{e^{\frac{p_4}{\sqrt{p_5}} z} + e^{\frac{p_4}{\sqrt{p_5}}(1-z)}}{1 + e^{\frac{p_4}{\sqrt{p_5}}}} \right) = 0$$

and

$$y'' - \frac{p_4^2}{p_2} \left(x^2 y - p_6 x \right) = 0$$

for the unknown functions $x(z)$ and $y(z)$ are introduced in Chap. 4 of [Govaerts, W.J.F., *Numerical Methods for Bifurcations of Dynamical Equilibria*, SIAM, Philadelphia, 2000] as a representation in a 1-dimensional medium of a Belusov–Zhabotinsky-type chemical oscillator. Modify the zero function encoded in `finitediff` in Sect. 3.3.5 to correspond to a finite-difference discretization of these coupled differential equations for z on a uniform mesh on the interval $[0, 1]$ with $N + 1$ subintervals.

3.25. Consider solutions to the coupled differential equations shown in the previous exercise that satisfy the additional boundary conditions $x(0) = x(1) = p_3$ and $y(0) = y(1) = p_6/p_3$. Show that when $p_4 = 0$, and for fixed values of the remaining parameters, a unique solution to the discretized zero problem that also satisfies the boundary conditions has all odd elements of u equal to p_3 and all even elements of u equal to p_6/p_3. Use the `coco_add_func` utility to construct the corresponding extended continuation problem and perform continuation under variations in p_4 and with $N = 42$, $p_1 = 0.0016$, $p_2 = 0.008$, $p_3 = 2$, $p_5 = 0.5$, and $p_6 = 4.6$ as in Chap. 4 of [Govaerts, W.J.F., *Numerical Methods for Bifurcations of Dynamical Equilibria*, SIAM, Philadelphia, 2000].

Chapter 4

Toolbox Development

The `coco_add_pars` and `coco_add_glue` utilities described in Chap. 3 are examples of special-purpose wrappers that each embed a single call to `coco_add_func`. In this and the next chapter, we consider more sophisticated versions of special-purpose wrappers that embed multiple calls to `coco_add_func` or to other special-purpose wrappers. Suitably designed, these provide for the construction of extended continuation problems associated with particular classes of problems.

Consider, for example, a class of continuation problems $\mathcal{A}$ for which there is a natural decomposition of the vector of continuation variables:

$$u := (x, p) \in \mathbb{R}^m \times \mathbb{R}^{n-m}, \tag{4.1}$$

where x denotes *problem variables*, p denotes *problem parameters*, and $n > m \geq 1$. Let $\Phi : \mathbb{R}^n \to \mathbb{R}^m$ denote the corresponding family of zero functions and consider the family of $n - m$ monitor functions $\Psi : \mathbb{R}^n \to \mathbb{R}^{n-m}$, where $\Psi : u \mapsto p$. Given a regular point x^* of the zero problem $\Phi(x, p^*) = 0$ for some p^*, an initializable restricted continuation problem on a neighborhood of (x^*, p^*) now follows with the assignment $\mathbb{I} = \{1, \ldots, n - m\}$. The corresponding locally unique solution is given by $u^* = (x^*, p^*)$. Continuation with a 1-dimensional atlas algorithm, for example, then requires the release of one of the constraints on the problem parameters.

We illustrate below the development of a collection of toolbox functions particular to the construction of elements of the class $\mathcal{A}$ of initializable restricted continuation problems, as well as to the characterization of their solutions. In the process, we seek to allow for a syntactically more relaxed definition of zero functions, including the use of MATLAB anonymous functions, as well as a flexible calling syntax that supports the paradigm of staged construction.

In the absence of any distinguishing information about the origin of the family of zero functions, we refer to $\mathcal{A}$ as a class of *algebraic continuation problems* and to the toolbox developed below by the *toolbox name* `'alg'`. By convention, we use the toolbox name as the first part of every associated toolbox function.

4.1 Toolbox constructors

Let E denote the extended continuation problem in $u \in \mathbb{R}^n$ obtained from the family of zero functions Φ, the family of monitor functions Ψ, and an initial solution guess $u_0 := (x_0, p_0)$. Consider the nontrivial chain decomposition

$$\emptyset = E_0 \subsetneqq E_1 \subsetneqq E_2 = E, \tag{4.2}$$

where E_1 is the extended continuation problem in $w \in \mathbb{R}^n$ obtained from the family of zero functions Φ, the empty family of monitor functions, and the initial solution guess $w_0 := u_0$. We obtain E by the two-stage construction

$$E_1 := \big[\Phi, \emptyset, \emptyset, (x_0, p_0)\big](E_0) \tag{4.3}$$

and

$$E_2 := [\emptyset, \psi, \{m+1, \ldots, n\}, \emptyset](E_1), \tag{4.4}$$

where the representation ψ is the identity. This staged construction is clearly uniquely defined by the function that encodes Φ and by the initial values x_0 and p_0, since the integers n and m can be calculated from the dimensions of x_0 and p_0. An element of the class of algebraic continuation problems $\mathcal{A}$ is obtained by appending the assignment $\mathbb{I} = \{1, \ldots, n-m\}$ to E.

4.1.1 A basic toolbox constructor

We encode the staged construction of an element of $\mathcal{A}$ in the `alg_construct_eqn` function shown below.

```
function prob = alg_construct_eqn(fhan, x0, pnames, p0)

xdim = numel(x0);
pdim = numel(p0);

prob = coco_prob();
prob = coco_add_func(prob, 'alg', fhan, [], 'zero', ...
  'u0', [x0(:); p0(:)]);
prob = coco_add_pars(prob, 'pars', xdim+(1:pdim), pnames);

end
```

This *toolbox constructor* encodes multiple calls to the `coco_add_func` core utility. In the first call, the `'alg'` function identifier is associated with the encoding of a zero function with function handle `fhan`, with an initially empty function data structure, and with the index set $\mathbb{K} = \{1, \ldots, n\}$.

Similarly, the call to the `coco_add_pars` special-purpose wrapper associates the `'pars'` function identifier with a monitor function encoded as the identity, with an empty function data structure, and with the index set $\mathbb{K} = \{m+1, \ldots, n\}$. The call further assigns the integer indices of the corresponding $n-m$ continuation parameters to $\mathbb{I}$, thereby imposing active constraints on the components of the vector of continuation variables indexed by $\mathbb{K}$. Finally, the strings contained in `pnames` are introduced as labels for the corresponding continuation parameters.

It is tacitly assumed in the body of the `alg_construct_eqn` constructor that the encoding of Φ is compatible with COCO, i.e., that its input and output argument syntax is that described in the previous chapter. To allow for a more general formalism, in which Φ is encoded in a function with x and p as input arguments and $\Phi(x, p)$ as output argument, consider the COCO-compatible encoding of the `alg_F` *zero-function wrapper* shown below.

```
function [data y] = alg_F(prob, data, u)

x = u(data.x_idx);
p = u(data.p_idx);

y = data.fhan(x, p);

end
```

Here, fields of the function data structure `data` are used to extract the elements x and p from the u input argument and subsequently to evaluate Φ. Support for the wrapper implementation is encoded in the modified version of the `alg_construct_eqn` constructor shown below.

```
function prob = alg_construct_eqn(fhan, x0, pnames, p0)

xdim       = numel(x0);
pdim       = numel(p0);
data.fhan  = fhan;
data.x_idx = (1:xdim)';
data.p_idx = xdim+(1:pdim)';

prob = coco_prob();
prob = coco_add_func(prob, 'alg', @alg_F, data, 'zero', ...
  'u0', [x0(:); p0(:)]);
prob = coco_add_pars(prob, 'pars', data.p_idx, pnames);

end
```

Example 4.1 Consider the following definition of the function `circle`.

```
function y = circle(x, p)
  y = x^2+(p-1)^2-1;
end
```

Here, the arguments x and p are both scalar quantities. Since the function returns a single scalar output, the corresponding zero problem has a nominal dimensional deficit of 1. The extract from the command window below shows the result of invoking the `alg_construct_eqn` constructor with $x_0 = 1$, $p_0 = 1.1$, and parameter label `'y'`.

```
>> prob = alg_construct_eqn(@circle, 1, 'y', 1.1);
>> coco(prob, 'run', [], 1, 'y', [0.1 5]);
```

STEP		DAMPING		NORMS			COMPUTATION TIMES		
IT	SIT	GAMMA	||d||	||f||	||U||	F(x)	DF(x)	SOLVE	
0				1.00e-02	1.85e+00	0.0	0.0	0.0	
1	1	1.00e+00	5.00e-03	2.50e-05	1.85e+00	0.0	0.0	0.0	
2	1	1.00e+00	1.26e-05	1.58e-10	1.85e+00	0.0	0.0	0.0	
3	1	1.00e+00	7.93e-11	2.22e-16	1.85e+00	0.0	0.0	0.0	

STEP	TIME	||U||	LABEL	TYPE	y
0	00:00:00	1.8466e+00	1	EP	1.1000e+00
10	00:00:01	6.5524e-01	2		1.9555e-01
13	00:00:01	4.5826e-01	3	EP	1.0000e-01

```
STEP      TIME       ||U||   LABEL  TYPE              y
   0   00:00:01  1.8466e+00     4   EP      1.1000e+00
  10   00:00:01  2.7007e+00     5           1.8798e+00
  20   00:00:02  2.8181e+00     6           1.9903e+00
  30   00:00:02  2.5022e+00     7           1.6947e+00
  40   00:00:03  1.4239e+00     8           7.3993e-01
  49   00:00:03  4.5826e-01     9   EP      1.0000e-01
```

As before, the ||U|| column of data in the output beginning with the lines

```
    STEP  DAMPING                NORMS                 COMPUTATION TIMES
   IT SIT   GAMMA    ||d||     ||f||      ||U||      F(x)   DF(x)  SOLVE
   0                        1.00e-02  1.85e+00     0.0    0.0    0.0
    1   1  1.00e+00  5.00e-03  2.50e-05  1.85e+00     0.0    0.0    0.0
    2   1  1.00e+00  1.26e-05  1.58e-10  1.85e+00     0.0    0.0    0.0
```

and the ||U|| column in the output beginning with the lines

```
STEP      TIME       ||U||   LABEL  TYPE              y
   0   00:00:00  1.8466e+00     1   EP      1.1000e+00
  10   00:00:01  6.5524e-01     2           1.9555e-01
```

contain the value of $\|(x, p, p)\|_2$ during initialization and for each labeled point located on the solution manifold, respectively.

The introduction of the alg_F zero-function wrapper enables the use of MATLAB anonymous functions as an occasionally convenient alternative to full functional encodings. As an example, the call

```
>> prob = alg_construct_eqn(@(x,p) x^2+(p-1)^2-1, 1, 'y', 1.1);
```

to the alg_construct_eqn constructor obviates the need for the separate encoding of the function circle. ∎

4.1.2 Linearizations

The default COCO atlas algorithms rely on predictor-corrector steps that make use of the linearization of the extended continuation problem. Unless explicit derivatives are provided, such linearizations are approximated using built-in finite-difference schemes. To allow for the possibility that one or both of the linearizations $\partial_x \Phi$ and $\partial_p \Phi$ have been explicitly encoded, we modify the alg_construct_eqn constructor as shown below.

```
function prob = alg_construct_eqn(fhan, dfdxhan, dfdphan, x0, ...
  pnames, p0)

xdim          = numel(x0);
pdim          = numel(p0);
data.fhan     = fhan;
data.dfdxhan  = dfdxhan;
data.dfdphan  = dfdphan;
data.x_idx    = (1:xdim)';
data.p_idx    = xdim+(1:pdim)';

prob = coco_prob();
prob = coco_add_func(prob, 'alg', @alg_F, @alg_DFDU, data, 'zero', ...
```

```
  'u0', [x0(:); p0(:)]);
prob = coco_add_pars(prob, 'pars', data.p_idx, pnames);

end
```

Here, the `@alg_DFDU` argument in the call to `coco_add_func` references the following implementation of the Jacobian of the `alg_F` wrapper with respect to the continuation variables.

```
function [data J] = alg_DFDU(prob, data, u)

x = u(data.x_idx);
p = u(data.p_idx);

if isempty(data.dfdxhan),
  J1 = coco_ezDFDX('f(x,p)', data.fhan, x, p);
else
  J1 = data.dfdxhan(x, p);
end
if isempty(data.dfdphan)
  J2 = coco_ezDFDP('f(x,p)', data.fhan, x, p);
else
  J2 = data.dfdphan(x, p);
end
J = sparse([J1 J2]);

end
```

In this case, in the event that the `dfdxhan` and/or `dfdphan` fields are empty, finite-difference approximations of either or both of the corresponding linearizations $\partial_x \Phi$ and $\partial_p \Phi$ are produced by the COCO utilities `coco_ezDFDX` and `coco_ezDFDP`, respectively.

Example 4.2 The extract below shows the explicit inclusion of $\partial_x \Phi$ and $\partial_p \Phi$ in the call to the revised `alg_construct_eqn` constructor.

```
>> funs = {@(x,p) x^2+(p-1)^2-1, @(x,p) 2*x, @(x,p) 2*(p-1)};
>> prob = alg_construct_eqn(funs{:}, 1, 'y', 1.1);
>> coco(prob, 'run', [], 1, 'y', [0.1 5]);
```

	STEP	DAMPING			NORMS		COMPUTATION	TIMES	
IT	SIT	GAMMA	||d||	||f||	||U||	F(x)	DF(x)	SOLVE	
0				1.00e-02	1.85e+00	0.0	0.0	0.0	
1	1	1.00e+00	5.00e-03	2.50e-05	1.85e+00	0.0	0.0	0.0	
2	1	1.00e+00	1.26e-05	1.58e-10	1.85e+00	0.0	0.0	0.0	
3	1	1.00e+00	7.93e-11	2.22e-16	1.85e+00	0.0	0.0	0.0	

STEP	TIME	||U||	LABEL	TYPE	y
0	00:00:00	1.8466e+00	1	EP	1.1000e+00
10	00:00:00	6.5524e-01	2		1.9555e-01
13	00:00:00	4.5826e-01	3	EP	1.0000e-01

STEP	TIME	||U||	LABEL	TYPE	y
0	00:00:00	1.8466e+00	4	EP	1.1000e+00
10	00:00:01	2.7007e+00	5		1.8798e+00
20	00:00:01	2.8181e+00	6		1.9903e+00
30	00:00:02	2.5022e+00	7		1.6947e+00
40	00:00:02	1.4239e+00	8		7.3993e-01
49	00:00:03	4.5826e-01	9	EP	1.0000e-01

For more complicated problems, it is interesting to explore the differences, in terms of computational accuracy and cost, between the use of explicit derivatives and the built-in finite-difference approximations. ∎

4.1.3 Signals and slots

The extended continuation problem constructed by `alg_construct_eqn` makes no distinction between the x and p components of the vector of continuation variables. There is, consequently, no obvious way to identify the corresponding numerical values stored in the chart structure of a solution file. During continuation, on the other hand, the distinct nature of these components is retained in the continuation problem structure, specifically in the content of the function data structure associated with the `alg_F` zero-function wrapper. This distinction may be made available also after continuation by including the content of the function data structure in the data array stored with each solution file.

In COCO, the ability to append content to the data array of a solution file is a special case of a general *signal* and *slot* mechanism. This mechanism gives access to the runtime content of the continuation problem structure, at selected points during execution, to suitably designed utility functions. Signals may be declared and stored in the continuation problem structure by core algorithms or by COCO-compatible toolboxes using the COCO utility `coco_add_signal`. Each signal is associated with a (possibly empty) collection of *slot functions*. The execution, in some sequence, of all slot functions associated with a given signal is then triggered in response to the emission of this signal by the COCO core utility `coco_emit`.

A COCO-compatible slot function must be encoded with function syntax as shown below.

```
function [data ...] = slot_fun(prob, data, ...)
```

As before, the `prob` input argument contains a copy of the continuation problem structure used by COCO to store all properties of the extended continuation problem and any associated algorithms during execution. The content of this variable may be accessed in the function body, but changes made to the copy do not survive execution of the function. In contrast, the `data` input argument may be accessed and permanently modified within the function body for reference in a subsequent function call or by other COCO algorithms. This *slot function data structure* parameterizes the state of the slot function and may be used to guide its execution. The number of required output arguments (in addition to `data`) and the minimum required number and meaning of additional input arguments are uniquely determined by the definition of the triggering signal. In order to accommodate the possibility of different numbers of input arguments for individual slot functions associated with a given signal, we always append `varargin` to the list of input arguments.

The COCO utility `coco_add_slot` is used to append information regarding a COCO-compatible slot function to the continuation problem structure. The command

```
prob = coco_add_slot(prob, fid, fhan, data, signal);
```

assigns the function identifier `fid` as a reference to the encoding with function handle `fhan`. Although each slot function associated with a declared signal must be given a unique function identifier, the same identifier may be used for slot functions associated with different signals. The `data` variable in the fourth argument provides the initial content of the corresponding slot function data structure. Finally, the `signal` input argument identifies the name of the signal that triggers execution of this slot function.

We recall now the objective of storing, with each solution file, information sufficient for separating the vector of continuation variables into its x and p components. For simplicity, we append the entire function data structure associated with the `alg_F` zero-function wrapper to the solution data array. To this end, consider the modified encoding of the `alg_construct_eqn` constructor shown below.

```
function prob = alg_construct_eqn(fhan, dfdxhan, dfdphan, x0, ...
  pnames, p0)

xdim          = numel(x0);
pdim          = numel(p0);
data.fhan     = fhan;
data.dfdxhan  = dfdxhan;
data.dfdphan  = dfdphan;
data.x_idx    = (1:xdim)';
data.p_idx    = xdim+(1:pdim)';

prob = coco_prob();
prob = coco_add_func(prob, 'alg', @alg_F, @alg_DFDU, data, 'zero', ...
  'u0', [x0(:); p0(:)]);
prob = coco_add_pars(prob, 'pars', data.p_idx, pnames);
prob = coco_add_slot(prob, 'alg', @coco_save_data, data, 'save_full');

end
```

Here, we associate the COCO slot function `coco_save_data` with the `'save_full'` core signal and assign a copy of the `alg_F` function data structure `data` to the corresponding slot function data structure.

Example 4.3 We repeat, with the modified `alg_construct_eqn` toolbox constructor, the computation from Example 4.2.

```
>> funs = {@(x,p) x^2+(p-1)^2-1, @(x,p) 2*x, @(x,p) 2*(p-1)};
>> prob = alg_construct_eqn(funs{:}, 1, 'y', 1.1);
>> coco(prob, 'run', [], 1, 'y', [0.1 5]);
```

		STEP DAMPING		NORMS			COMPUTATION TIMES		
IT	SIT	GAMMA	||d||	||f||	||U||	F(x)	DF(x)	SOLVE
0				1.00e-02	1.85e+00	0.0	0.0	0.0
1	1	1.00e+00	5.00e-03	2.50e-05	1.85e+00	0.0	0.0	0.0
2	1	1.00e+00	1.26e-05	1.58e-10	1.85e+00	0.0	0.0	0.0
3	1	1.00e+00	7.93e-11	2.22e-16	1.85e+00	0.0	0.0	0.0

STEP	TIME	||U||	LABEL	TYPE	y
0	00:00:00	1.8466e+00	1	EP	1.1000e+00
10	00:00:00	6.5524e-01	2		1.9555e-01
13	00:00:00	4.5826e-01	3	EP	1.0000e-01

STEP	TIME	||U||	LABEL	TYPE	y
0	00:00:00	1.8466e+00	4	EP	1.1000e+00
10	00:00:01	2.7007e+00	5		1.8798e+00
20	00:00:01	2.8181e+00	6		1.9903e+00
30	00:00:02	2.5022e+00	7		1.6947e+00

```
40  00:00:02   1.4239e+00     8          7.3993e-01
49  00:00:03   4.5826e-01     9   EP     1.0000e-01
```

The call below to the `coco_read_solution` utility extracts the data array associated with the `sol3.mat` solution file stored in the run with run identifier `'run'`.

```
>> data = coco_read_solution('', 'run', 3)

data =

    'alg'     [1x1 struct]
```

As is evident from the output, the `{1,1}` element of the `data` array contains the function identifier `'alg'` associated with the slot function `coco_save_data`. We extract the content of the `{1,2}` element in the following modified call to `coco_read_solution`.

```
>> data = coco_read_solution('alg', 'run', 3)

data =

      fhan: @(x,p)x^2+(p-1)^2-1
    dfdxhan: @(x,p)2*x
    dfdphan: @(x,p)2*(p-1)
     x_idx: 1
     p_idx: 2
```

The output is identical to the `coco_save_data` slot function data structure, as ensured by the encoding of `alg_construct_eqn`. ■

The functionality illustrated in Example 4.3 supports the encoding of the *toolbox extractor* `alg_read_solution` shown below.

```
function [x p] = alg_read_solution(run, lab)

[data chart] = coco_read_solution('alg', run, lab);
x = chart.x(data.x_idx);
p = chart.x(data.p_idx);

end
```

This relies on the COCO core extractor `coco_read_solution` for extracting the data array and chart structure associated with the `'alg'` function identifier from a solution file stored in a previous run. As before, the file is identified by the label `lab` and the run identifier `run`. The `x_idx` and `p_idx` fields of the `data` structure are then used to assign numerical values to the output arguments x and p, respectively.

4.1.4 Generalized toolbox constructors

To provide flexibility in the calling syntax, we consider next a modified version of the `alg_construct_eqn` constructor, as shown in the encoding below.

```
function prob = alg_construct_eqn(data, sol)

prob = coco_prob();
prob = coco_add_func(prob, 'alg', @alg_F, @alg_DFDU, data, 'zero', ...
  'u0', sol.u);
```

```
if ~isempty(data.pnames)
  prob = coco_add_pars(prob, 'pars', data.p_idx, data.pnames);
end
prob = coco_add_slot(prob, 'alg', @coco_save_data, data, 'save_full');

end
```

Here, the content of the function data structure passed to `alg_F` and `alg_DFDU` is assumed to be provided to the constructor in the first input argument. Similarly, the initial solution guess is assumed to be contained in the `u` field of the second input argument. We proceed to encode a special-purpose wrapper for `alg_construct_eqn` that is designed to translate a list of input arguments into sufficient content for the function data structure and initial solution guess.

In accordance with the discussion in the previous section, we continue to require that a call to such a *generalized toolbox constructor* include the initial solution guesses for x_0 and p_0 as well as a function handle to the encoding of Φ. We consider the specification of function handles to encodings of the Jacobians $\partial_x \Phi$ and $\partial_p \Phi$, as well as that of the cell array of parameter labels, as optional. In the absence of function handles to the Jacobians, the finite-difference approximations encoded in `coco_ezDFDX` and `coco_ezDFDP` should then be invoked by `alg_DFDU` as described earlier. In the absence of a cell array of parameter labels, the call to `coco_add_pars` should be skipped, as shown above in the modified encoding of the `alg_construct_eqn` toolbox constructor.

To support the implementation of a generalized toolbox constructor with a combination of required and optional arguments, we rely on the COCO core class `coco_stream`. This class offers a simple interface for sequential reading of the elements of an arbitrary cell array. We may think of an object of type `coco_stream` as a handle to an input stream that provides sequential read access to a sequence of *tokens* and whose internal state includes a reference to the position of the next unprocessed token.

The class `coco_stream` includes the following member functions:

- `coco_stream(arr)` : construct a stream object from the cell array `arr`. As a special case, `arr` may contain a single object of type `coco_stream`, in which case the newly constructed stream handle will be identical to that passed as the argument, and changes to the internal state of the latter will result in changes to the internal state of the former.

- `str.peek(['cell'])` : return the next unprocessed token in the `str` input stream. A subsequent call to `peek` or `get`, described below, will return the same token. When reading beyond the list of available tokens, `peek` will return an empty array. The optional argument `'cell'` requests embedding of the token in a cell array if it is not already a cell array.

- `str.get(['cell'])` : read the next unprocessed token in the `str` input stream and mark it as processed. A subsequent call to `peek` or `get` will return the following unprocessed token, if present. When reading beyond the list of available tokens, `get` will return an empty array. The optional argument `'cell'` requests embedding of the token in a cell array if it is not already a cell array.

- `str.skip` : mark the next unprocessed token in the `str` input stream as processed. A subsequent call to `peek` or `get` will return the following unprocessed token, if present.

- `numel(str)` : compute the number of unprocessed tokens.

- `isempty(str)` : check that all tokens in the `str` input stream have been processed.

An example use of a `coco_stream` object is then shown below.

```
function prob = constructor(varargin)
  str = coco_stream(varargin{:});
  % parse arguments using str.get, str.peek, and str.skip
end
```

Here, `varargin` represents an array of input arguments of undetermined length.

As an example, consider the `alg_isol2eqn` constructor shown below together with the subfunction `is_empty_or_func`. This relies on the `coco_stream` member functions to provide support for the combination of required and optional input arguments described above.

```
function prob = alg_isol2eqn(varargin)

str = coco_stream(varargin{:});
data.fhan = str.get;
data.dfdxhan = [];
data.dfdphan = [];
if is_empty_or_func(str.peek)
  data.dfdxhan = str.get;
  if is_empty_or_func(str.peek)
    data.dfdphan = str.get;
  end
end
x0 = str.get;
data.pnames = {};
if iscellstr(str.peek('cell'))
  data.pnames = str.get('cell');
end
p0 = str.get;

alg_arg_check(data, x0, p0);
data  = alg_init_data(data, x0, p0);
sol.u = [x0(:); p0(:)];
prob  = alg_construct_eqn(data, sol);

end
```

```
function flag = is_empty_or_func(x)
  flag = isempty(x) || isa(x, 'function_handle');
end
```

It is a useful exercise to confirm that the constructor definition adheres to the argument syntax

```
varargin = alg, ...
```

where

```
alg = @f, [(@dfdx | '[]'), [(@dfdp | '[]')]], x0, [pname | pnames], p0
```

Here, the ellipsis denotes any number of additional elements; square brackets denote optional arguments or, when enclosed in apostrophes, an empty array; and the symbol | indicates a choice between two alternatives. It follows that the optional argument following the first function handle either is absent or is a function handle, an empty array, two function handles, an empty array and a function handle, a function handle and an empty array, or two empty arrays. Similarly, the second optional argument is either absent, a single string, or a cell array of strings.

The arrangement of required and optional arguments ensures that the `alg_isol2eqn` constructor knows when to stop parsing the `varargin` input argument, since various conditional statements can be used to determine the presence or absence of each optional argument. The constructor assumes that the first element of `varargin` is the function handle to the problem-specific encoding of Φ. The constructor next allows for optional specifications in `varargin` of the function handles to the Jacobians $\partial_x \Phi$ and, provided that the former is defined, $\partial_p \Phi$, respectively. In the absence of one or both of these optional arguments, the constructor assigns an empty array to the `dfdxhan` and/or `dfdphan` fields of the `data` structure, thus ensuring that the corresponding linearizations are approximated numerically. The constructor assigns the next element of `varargin` to x_0. It next allows for the optional specification of the cell array of parameter labels and assigns an empty cell array to `pnames` if no such array is provided. The constructor finally assigns the next element of `varargin` to p_0.

Following the initial argument parsing, the command

```
alg_arg_check(data, x0, p0);
```

in the `alg_isol2eqn` constructor performs a basic error check of the variables `data`, `x0`, and `p0` as encoded below.

```
function alg_arg_check(data, x0, p0)

assert(isa(data.fhan, 'function_handle'), ...
  'alg: input for ''f'' is not a function handle');
assert(isnumeric(x0), 'alg: input for ''x0'' is not numeric');
assert(isnumeric(p0), 'alg: input for ''p0'' is not numeric');
assert(numel(p0)==numel(data.pnames) || isempty(data.pnames), ...
  'alg: incompatible number of elements for ''p0'' and ''pnames''');

end
```

Here, error messages are produced in the case that the content of `data.fhan` is not a function handle, in the case that either `x0` or `p0` is a nonnumeric variable, or in the case that the number of components of `p0` differs from the number of elements of `pnames` provided that the latter is not empty. Notably, the desired function types of `data.dfdxhan` and `data.dfdphan` and of the entries of `data.pnames`, if nonempty, are automatically enforced by the `alg_isol2eqn` constructor.

Provided that the variables `data`, `x0`, and `p0` have been correctly assigned, the call to the `alg_init_data` function encoded below finalizes the content of the function data structure `data` prior to the subsequent call to the `alg_construct_eqn` constructor.

```
function data = alg_init_data(data, x0, p0)

xdim     = numel(x0);
pdim     = numel(p0);
```

```
data.x_idx = (1:xdim)';
data.p_idx = xdim+(1:pdim)';

end
```

Example 4.4 The extract below shows the use of the `alg_isol2eqn` constructor.

```
>> prob = alg_isol2eqn(@(x,p) x^2+(p-1)^2-1, 0.9, 'y', 1.1);
>> coco(prob, 'run1', [], 1, 'y', [0.1 5]);
```

		STEP	DAMPING		NORMS			COMPUTATION TIMES		
IT	SIT	GAMMA	||d||	||f||		||U||	F(x)	DF(x)	SOLVE	
0				1.80e-01	1.80e+00		0.0	0.0	0.0	
1	1	1.00e+00	1.00e-01	1.00e-02	1.85e+00		0.0	0.0	0.0	
2	1	1.00e+00	5.00e-03	2.50e-05	1.85e+00		0.0	0.0	0.0	
3	1	1.00e+00	1.26e-05	1.58e-10	1.85e+00		0.0	0.0	0.0	
4	1	1.00e+00	7.93e-11	2.22e-16	1.85e+00		0.0	0.0	0.0	

STEP	TIME	||U||	LABEL	TYPE	y
0	00:00:00	1.8466e+00	1	EP	1.1000e+00
10	00:00:00	6.5524e-01	2		1.9555e-01
13	00:00:00	4.5826e-01	3	EP	1.0000e-01

STEP	TIME	||U||	LABEL	TYPE	y
0	00:00:00	1.8466e+00	4	EP	1.1000e+00
10	00:00:01	2.7007e+00	5		1.8798e+00
20	00:00:01	2.8181e+00	6		1.9903e+00
30	00:00:02	2.5022e+00	7		1.6947e+00
40	00:00:02	1.4239e+00	8		7.3993e-01
49	00:00:03	4.5826e-01	9	EP	1.0000e-01

Here, the omission of additional function handles implies that the built-in finite-difference approximations are used for the Jacobians of the zero function. ∎

The `alg_isol2eqn` constructor processes information regarding an initial solution guess and the associated zero problem in support of the encapsulated call to the basic toolbox constructor `alg_construct_eqn`. Its encoding guarantees that all required elements of the function data structure `data` are populated with content of the required type for the construction of an initializable restricted continuation problem. We envision next an alternative special-purpose wrapper for `alg_construct_eqn` that is designed to enable continuation starting from a previously stored solution and for which a compatible function data structure is inherited from a previous run.

To this end, we first provide the following modified encoding of the toolbox extractor `alg_read_solution`.

```
function [sol data] = alg_read_solution(run, lab)

[data chart] = coco_read_solution('alg', run, lab);
sol.x = chart.x(data.x_idx);
sol.p = chart.x(data.p_idx);
sol.u = [sol.x; sol.p];

end
```

The generalized toolbox constructor `alg_sol2eqn` encoded below now achieves the desired objective of constructing an initializable restricted continuation problem using information from a previously computed solution point.

```
function prob = alg_sol2eqn(run, lab)

[sol data] = alg_read_solution(run, lab);
prob       = alg_construct_eqn(data, sol);

end
```

Example 4.5 We continue the session from Example 4.4. The extract below shows the use of the `alg_sol2eqn` constructor.

```
>> prob = alg_sol2eqn('run1', 9);
>> coco(prob, 'run2', [], 1, 'y', [0.1 1]);
```

STEP	DAMPING			NORMS		COMPUTATION TIMES		
IT SIT	GAMMA	||d||	||f||	||U||		F(x)	DF(x)	SOLVE
0			4.73e-13	4.58e-01		0.0	0.0	0.0
1 1	1.00e+00	5.43e-13	0.00e+00	4.58e-01		0.0	0.0	0.0

STEP	TIME	||U||	LABEL	TYPE	y
0	00:00:00	4.5826e-01	1	EP	1.0000e-01
10	00:00:00	1.4373e+00	2		7.5093e-01
13	00:00:00	1.7321e+00	3	EP	1.0000e+00

Here, the extended continuation problem and the initial content of the index set $\mathbb{I}$ are recreated from the chart structure and data array associated with the `'alg'` function identifier that were stored in the `sol9.mat` solution file as a result of the previous call to the `coco` entry-point function with run identifier `'run1'`. ∎

4.2 Embeddability

Let E_1 be an extended continuation problem in $u \in \mathbb{R}^{n_1}$ given by the family of zero functions $\Phi^{(1)} : \mathbb{R}^{n_1} \to \mathbb{R}^{m_1}$, the family of monitor functions $\Psi^{(1)} : \mathbb{R}^{n_1} \to \mathbb{R}^{r_1}$, and the initial solution guess $u_{1,0}$. Similarly, let E_2 be an extended continuation problem in $u \in \mathbb{R}^{n_2}$ given by the family of zero functions $\Phi^{(2)} : \mathbb{R}^{n_2} \to \mathbb{R}^{m_2}$, the family of monitor functions $\Psi^{(2)} : \mathbb{R}^{n_2} \to \mathbb{R}^{r_2}$, and the initial solution guess $u_{2,0}$. We define the *canonical sum* of E_1 and E_2 as the extended continuation problem $E_1 \oplus E_2$ in $u \in \mathbb{R}^{n_1+n_2}$ given by the family of zero functions

$$\Phi : u \mapsto \left(\begin{array}{c} \Phi^{(1)}\left(u_{\{1,\dots,n_1\}}\right) \\ \Phi^{(2)}\left(u_{\{n_1+1,\dots,n_1+n_2\}}\right) \end{array} \right), \tag{4.5}$$

the family of monitor functions

$$\Psi : u \mapsto \left(\begin{array}{c} \Psi^{(1)}\left(u_{\{1,\dots,n_1\}}\right) \\ \Psi^{(2)}\left(u_{\{n_1+1,\dots,n_2\}}\right) \end{array} \right), \tag{4.6}$$

and the initial solution guess $u_0 := \left(u_{1,0}, u_{2,0}\right)$.

Given an extended continuation problem E_2, we now define the operator $\oplus_{E_2} : \mathcal{E} \to \mathcal{E}$ such that

$$\oplus_{E_2} : E_1 \mapsto E_1 \oplus E_2. \tag{4.7}$$

It follows that $\oplus_{E_2}(\emptyset) = E_2$ and $E_1 \not\sqsubseteq \oplus_{E_2}(E_1)$. Clearly, the solutions of the extended continuation problem $\oplus_{E_2}(E_1)$ are obtained by combining solutions of the two uncoupled extended continuation problems E_1 and E_2.

Example 4.6 Consider, again, the extended continuation problem

$$E := [\emptyset, \psi, \{m+1, \ldots, n\}, \emptyset] \circ \left[\Phi, \emptyset, \emptyset, (x_0, p_0)\right](\emptyset), \tag{4.8}$$

where $(x_0, p_0) \in \mathbb{R}^m \times \mathbb{R}^{n-m}$, $\Phi : \mathbb{R}^m \times \mathbb{R}^{n-m} \to \mathbb{R}^m$, and the representation ψ is the identity. It is straightforward to show that

$$\oplus_E = \left[\emptyset, \psi, (\mathbb{K}_\Phi)_{\{m+1,\ldots,n\}}, \emptyset\right] \circ \left[\Phi, \emptyset, \emptyset, (x_0, p_0)\right]. \tag{4.9}$$

Indeed, in the special case that the right-hand-side composition is applied to the empty extended continuation problem, it follows, as before, that $\mathbb{K}_\Phi = \{1, \ldots, n\}$ and, thus, that

$$(\mathbb{K}_\Phi)_{\{m+1,\ldots,n\}} = \{m+1, \ldots, n\}. \quad \blacksquare \tag{4.10}$$

In the absence of any distinguishing information about the family of zero functions Φ or the origin of the continuation variables x and p, we refer to each element of the class of algebraic continuation problems $\mathcal{A}$ as an *equation object*. The toolbox constructors `alg_construct_eqn`, `alg_isol2eqn`, and `alg_sol2eqn` were designed to instantiate such an equation object by means of the two-stage construction in Eq. (4.8). In the final encoding of the `alg_construct_eqn` constructor, we chose to make the second stage optional in the event that the `pnames` field of the `data` structure was empty.

In this section, we modify the toolbox constructors to correspond to the action of the operator $\oplus_E$ in Eq. (4.9) on an arbitrary extended continuation problem. As the modified constructors may be invoked at any stage of a staged construction, we say that they are *embeddable*.

4.2.1   An embeddable basic constructor

We proceed by considering the changes to the basic `alg_construct_eqn` toolbox constructor that are required to accommodate its application to an arbitrary extended continuation problem. Clearly, the function must accept a nonempty continuation problem structure as an input argument. In addition, it is necessary to associate unique function identifiers with each nested call to `coco_add_func` and with each slot function associated with the `'save_full'` signal. Finally, we must replace the hard-coded reference to the index set $\{m+1, \ldots, n\}$ with a relative reference to a subset of the index set $\mathbb{K}_\Phi$.

We accomplish, in the modified encoding of the basic `alg_construct_eqn` constructor shown below, all the required changes.

```
function prob = alg_construct_eqn(prob, tbid, data, sol)

prob = coco_add_func(prob, tbid, @alg_F, @alg_DFDU, data, 'zero', ...
    'u0', sol.u);
if ~isempty(data.pnames)
    fid  = coco_get_id(tbid, 'pars');
    uidx = coco_get_func_data(prob, tbid, 'uidx');
    prob = coco_add_pars(prob, fid, uidx(data.p_idx), data.pnames);
end
```

```
prob = coco_add_slot(prob, tbid, @coco_save_data, data, 'save_full');

end
```

Here, the `prob` input argument is assumed to contain an existing (possibly empty) continuation problem structure. Moreover, instead of hard-coding the function identifiers, the modified `alg_construct_eqn` constructor makes appropriate use of the *toolbox instance identifier* `tbid` provided in the second input argument. Here, the COCO utility `coco_get_id` returns

- a concatenation of its string arguments using the '.' concatenation token provided that the first argument is nonempty, or

- the second string argument if the first argument is empty.

Here, the command

```
fid = coco_get_id(tbid, 'pars');
```

thus creates a unique function identifier `fid` consisting of a concatenation of the string contained in `tbid` with the string `.pars`. Finally, the command

```
uidx = coco_get_func_data(prob, tbid, 'uidx');
```

uses the `coco_get_func_data` utility to extract the index set $\mathbb{K}_\Phi$ associated with the function identifier `tbid`. The subset of $\mathbb{K}_\Phi$ indexed by the `p_idx` field of the `data` structure is then passed to `coco_add_pars`.

4.2.2 Embeddable generalized constructors

We proceed to modify the `alg_isol2eqn` toolbox constructor to make it embeddable. Here, we note again the need to include an additional two input arguments representing an existing continuation problem structure and a unique *object instance identifier*, respectively. An encoding of `alg_isol2eqn` that adheres to this generalized argument syntax is shown below.

```
function prob = alg_isol2eqn(prob, oid, varargin)

tbid = coco_get_id(oid, 'alg');
str  = coco_stream(varargin{:});
data.fhan = str.get;
data.dfdxhan = [];
data.dfdphan = [];
if is_empty_or_func(str.peek)
  data.dfdxhan = str.get;
  if is_empty_or_func(str.peek)
    data.dfdphan = str.get;
  end
end
x0 = str.get;
data.pnames = {};
if iscellstr(str.peek('cell'))
  data.pnames = str.get('cell');
end
```

```
p0 = str.get;

alg_arg_check(tbid, data, x0, p0);
data  = alg_init_data(data, x0, p0);
sol.u = [x0(:); p0(:)];
prob  = alg_construct_eqn(prob, tbid, data, sol);

end
```

We note, in particular, the construction of the unique toolbox instance identifier `tbid` through the following concatenation of the object instance identifier `oid` and the toolbox name.

```
tbid = coco_get_id(oid, 'alg');
```

The variable `tbid` is then used in the call to the modified `alg_arg_check` function shown below in order to allow for an instance-specific error message in the event of incompatible input arguments.

```
function alg_arg_check(tbid, data, x0, p0)

assert(isa(data.fhan, 'function_handle'), ...
  '%s: input for ''f'' is not a function handle', tbid);
assert(isnumeric(x0), '%s: input for ''x0'' is not numeric', tbid);
assert(isnumeric(p0), '%s: input for ''p0'' is not numeric', tbid);
assert(numel(p0)==numel(data.pnames) || isempty(data.pnames), ...
  '%s: incompatible number of elements for ''p0'' and ''pnames''', ...
  tbid);

end
```

Example 4.7 The extract below shows the result of invoking the modified encoding of the `alg_isol2eqn` constructor twice with the same anonymous function, but with different initial solution guesses.

```
>> prob = coco_prob();
>> fhan = @(x,p) x^2+(p-1)^2-1;
>> prob = alg_isol2eqn(prob, 'eqn1', fhan, 0.9, 'y1', 1.1);
>> prob = alg_isol2eqn(prob, 'eqn2', fhan, 0.1, 'y2', 1.9);
>> coco(prob, 'run1', [], 1, {'y1' 'y2'}, [0.5 2]);
```

	STEP	DAMPING		NORMS		COMPUTATION	TIMES	
IT	SIT	GAMMA	||d||	||f||	||U||	F(x)	DF(x)	SOLVE
0				2.55e-01	2.62e+00	0.0	0.0	0.0
1	2	5.00e-01	9.06e-01	1.43e-01	2.69e+00	0.0	0.0	0.0
2	1	1.00e+00	1.12e-01	1.07e-02	2.69e+00	0.0	0.0	0.0
3	1	1.00e+00	1.17e-02	1.36e-04	2.69e+00	0.0	0.0	0.0
4	1	1.00e+00	1.56e-04	2.45e-08	2.69e+00	0.0	0.0	0.0
5	1	1.00e+00	2.81e-08	8.88e-16	2.69e+00	0.0	0.0	0.0

STEP	TIME	||U||	LABEL	TYPE	y1	y2
0	00:00:00	2.6851e+00	1	EP	1.1000e+00	1.9000e+00
6	00:00:00	2.2472e+00	2	EP	5.0000e-01	1.9000e+00

STEP	TIME	||U||	LABEL	TYPE	y1	y2
0	00:00:00	2.6851e+00	3	EP	1.1000e+00	1.9000e+00
10	00:00:01	3.3307e+00	4		1.8798e+00	1.9000e+00

```
20  00:00:01   3.4266e+00      5           1.9903e+00   1.9000e+00
30  00:00:02   3.1719e+00      6           1.6947e+00   1.9000e+00
40  00:00:02   2.4140e+00      7           7.3993e-01   1.9000e+00
43  00:00:02   2.2472e+00      8  EP       5.0000e-01   1.9000e+00
```

In this case, the composite restricted continuation problem has a dimensional deficit of 0, so it is necessary to release only one of the two continuation parameters in order to allow for continuation using a 1-dimensional atlas algorithm. The inclusion of the second parameter label in the call to the `coco` entry-point function here results in the printing of the corresponding value to screen during continuation. ∎

Consider, next, the modified encoding of the toolbox extractor `alg_read_solution` shown below.

```
function [sol data] = alg_read_solution(oid, run, lab)

tbid        = coco_get_id(oid, 'alg');
[data chart] = coco_read_solution(tbid, run, lab);
sol.x = chart.x(data.x_idx);
sol.p = chart.x(data.p_idx);
sol.u = [sol.x; sol.p];

end
```

This uses the `tbid` toolbox instance identifier to extract the instance-specific part of the data array and chart structure from a stored solution file. For example, the command

```
[sol data] = alg_read_solution('eqn2', 'run1', 6);
```

extracts information pertaining to the instance of the `alg_F` zero function associated with the `'eqn2'` object instance identifier from the file `sol6.mat` associated with the run identifier `'run1'`.

The encoding below now shows the corresponding modifications to the generalized constructor `alg_sol2eqn`.

```
function prob = alg_sol2eqn(prob, oid, varargin)

tbid = coco_get_id(oid, 'alg');
str  = coco_stream(varargin{:});
run  = str.get;
if ischar(str.peek)
  soid = str.get;
else
  soid = oid;
end
lab = str.get;

[sol data] = alg_read_solution(soid, run, lab);
prob       = alg_construct_eqn(prob, tbid, data, sol);

end
```

Here, we again use the `coco_stream` class to enable a more flexible calling syntax in support of commencing continuation from a previously stored solution point with a possible change in instance identifier.

Example 4.8 We continue the session from Example 4.7. The extract below shows the result of invoking the revised `alg_sol2eqn` constructor twice using a solution from the earlier call to the `coco` entry-point function.

```
>> prob = coco_prob();
>> prob = alg_sol2eqn(prob, 'eqn3', 'run1', 'eqn1', 4);
>> prob = alg_sol2eqn(prob, 'eqn4', 'run1', 'eqn2', 4);
>> coco(prob, 'run2', [], 1, {'y2' 'y1'}, [0.5 2]);

     STEP   DAMPING                NORMS             COMPUTATION TIMES
   IT SIT    GAMMA    ||d||      ||f||     ||U||    F(x)   DF(x)   SOLVE
    0                           0.00e+00  3.34e+00   0.0    0.0    0.0
    1   1  1.00e+00  0.00e+00   0.00e+00  3.34e+00   0.0    0.0    0.0

   STEP      TIME      ||U||  LABEL  TYPE          y2            y1
      0   00:00:00  3.3421e+00     1   EP     1.9000e+00    1.8798e+00
     10   00:00:00  2.7961e+00     2          1.2491e+00    1.8798e+00
     17   00:00:00  2.2382e+00     3   EP     5.0000e-01    1.8798e+00

   STEP      TIME      ||U||  LABEL  TYPE          y2            y1
      0   00:00:01  3.3421e+00     4   EP     1.9000e+00    1.8798e+00
     10   00:00:01  3.4133e+00     5          1.9818e+00    1.8798e+00
     20   00:00:02  3.1190e+00     6          1.6398e+00    1.8798e+00
     30   00:00:02  2.3363e+00     7          6.4269e-01    1.8798e+00
     32   00:00:02  2.2382e+00     8   EP     5.0000e-01    1.8798e+00
```

Here, for example, the solution data array and chart structure associated with the `'eqn1'` object instance identifier are used to initialize an equation object with object instance identifier `'eqn3'`. ∎

4.3   Object-oriented design

It is no accident that the discussion in this chapter relies heavily on the terminology of object-oriented programming. As we shall see below, it is natural to conceptualize the `'alg'` toolbox as a class with *properties* as well as *methods* for initializing a class instance, and for accessing and potentially modifying the content of instance properties.

4.3.1   Function objects

A basic example of the object-oriented paradigm may already be found in a representation of zero and monitor functions in COCO as instances of a *problem function class*. As described in Chap. 3, each problem function object is characterized by the following instance properties:

- a function identifier `fid`;

- a function handle `fhan` to the COCO-compatible encoding of the corresponding zero or monitor function;

- a function handle `dfduhan`, if available, to the COCO-compatible encoding of the Jacobian of the corresponding zero or monitor function;

- a function type, say `'zero'` or `'inactive'`;

- a string array `pnames`, in the case of a monitor function, of labels for the corresponding continuation parameters;

- an index set `uidx` identifying an ordered subset of the vector of continuation variables;

- a subarray `u0` of the initial solution guess corresponding to the index set `uidx`; and

- a function data structure `data`.

In addition, each problem function object is associated with built-in instance methods used to invoke the function with function handle `fhan` and, if available, the function with function handle `dfduhan`. In each case, the input arguments consist of the continuation problem structure, the function data structure `data`, and a subarray `u` of the vector of continuation variables corresponding to the index set `uidx`. As explained in the previous chapter, each such call may access and modify the content of the function data structure in addition to evaluating the zero or monitor function.

The utility `coco_add_func` is a basic constructor for the problem function class, since it initializes content to the instance properties of a problem function object. Each successive call to `coco_add_func`, when invoked on the same continuation problem structure `prob`, appends an additional problem function object to `prob`.

The utility `coco_get_func_data` gives access to a subset of the instance properties of a problem function object. An example of the calling syntax for `coco_get_func_data` appeared in the encoding of `alg_construct_eqn` on page 82, where the function was used to extract the index set `uidx`.

Example 4.9 We continue the session from Example 3.14. The function data structure associated with the `'finitediff'` function identifier may then be obtained using the `coco_get_func_data` utility as shown below.

```
>> coco_get_func_data(prob, 'finitediff', 'data')

ans =

    dep_idx: [84x1 double]
    par_idx: [2x1 double]
      f_idx: [42x1 double]
      g_idx: [42x1 double]
          A: [80x84 double]
          B: [80x84 double]
```

Notably, multiple properties of the problem function object associated with the function identifier `fid` may be extracted simultaneously by stacking the corresponding flags, as in the call

```
[data u0 uidx] = coco_get_func_data(prob, fid, 'data, 'u0', 'uidx');
```

∎

The *slot function class* provides a representation of slot functions in COCO as class instances. In this case, each slot function object is characterized by instance properties given by

- a function identifier `fid`;

- a function handle `fhan` to the COCO-compatible encoding of the corresponding slot function; and

- a function data structure `data`.

In addition, each slot function object is associated with a single, built-in instance method used to invoke the function with function handle `fhan`, with input arguments consisting of the continuation problem structure, the function data structure `data`, and any additional input arguments as defined by the corresponding signal. As explained in this chapter, each such call may access and modify the content of the function data structure, in addition to extracting additional information from the continuation problem structure for further processing.

The utility `coco_add_slot` is a basic constructor for the slot function class, since it initializes content to the instance properties of a slot function object. Each successive call to `coco_add_slot`, when invoked on the same continuation problem structure `prob`, appends an additional slot function object to `prob`.

4.3.2   Toolbox objects

The simplest class encoding of the `'alg'` toolbox consists solely of the basic toolbox constructor `alg_construct_eqn` shown on page 70 and repeated below for immediate reference.

```
function prob = alg_construct_eqn(fhan, x0, pnames, p0)

xdim = numel(x0);
pdim = numel(p0);

prob = coco_prob();
prob = coco_add_func(prob, 'alg', fhan, [], 'zero', ...
  'u0', [x0(:); p0(:)]);
prob = coco_add_pars(prob, 'pars', xdim+(1:pdim), pnames);

end
```

Each toolbox instance is here characterized by instance properties given by

- a problem function object with function identifier `'alg'`; and

- a problem function object with function identifier `'pars'`.

In this case, no additional methods are provided for accessing or modifying the content of these instance properties.

Consider, instead, the class encoding from Sect. 4.1.2 of the `'alg'` toolbox in terms of the constructor `alg_construct_eqn`, repeated below, together with the two wrappers `alg_F` and `alg_DFDU`.

```
function prob = alg_construct_eqn(fhan, dfdxhan, dfdphan, x0, ...
  pnames, p0)

xdim        = numel(x0);
pdim        = numel(p0);
```

```
data.fhan    = fhan;
data.dfdxhan = dfdxhan;
data.dfdphan = dfdphan;
data.x_idx   = (1:xdim)';
data.p_idx   = xdim+(1:pdim)';

prob = coco_prob();
prob = coco_add_func(prob, 'alg', @alg_F, @alg_DFDU, data, 'zero', ...
  'u0', [x0(:); p0(:)]);
prob = coco_add_pars(prob, 'pars', data.p_idx, pnames);

end
```

Each toolbox instance is again characterized by

- a problem function object with function identifier `'alg'`; and

- a problem function object with function identifier `'pars'`.

Notably, here, the instance methods `alg_F` and `alg_DFDU` provide an interface between the COCO core and the user-defined zero function and its Jacobians (when available), thereby hiding the latter from the implementation of the `'alg'` toolbox. To support this encapsulation, the function data structure `data` contains information about the user-defined zero function and its Jacobians as well as about the decomposition of the vector of continuation variables into problem variables and problem parameters.

Consider, next, the class encoding from Sect. 4.1.3 of the `'alg'` toolbox in terms of the constructor `alg_construct_eqn`, repeated below, together with the wrappers `alg_F` and `alg_DFDU` and the toolbox utility function `alg_read_solution`.

```
function prob = alg_construct_eqn(fhan, dfdxhan, dfdphan, x0, ...
  pnames, p0)

xdim         = numel(x0);
pdim         = numel(p0);
data.fhan    = fhan;
data.dfdxhan = dfdxhan;
data.dfdphan = dfdphan;
data.x_idx   = (1:xdim)';
data.p_idx   = xdim+(1:pdim)';

prob = coco_prob();
prob = coco_add_func(prob, 'alg', @alg_F, @alg_DFDU, data, 'zero', ...
  'u0', [x0(:); p0(:)]);
prob = coco_add_pars(prob, 'pars', data.p_idx, pnames);
prob = coco_add_slot(prob, 'alg', @coco_save_data, data, 'save_full');

end
```

In this case, each toolbox instance is characterized by

- a problem function object with function identifier `'alg'`;

- a problem function object with function identifier `'pars'`; and

- a slot function object with function identifier `'alg'`;

We recall the use of the `coco_save_data` utility in order to append the function data structure `data` to the data array stored with each solution file during continuation.

Consider, further, the class encoding from Sect. 4.1.4 of the `'alg'` toolbox in terms of the constructor `alg_construct_eqn`, repeated below, together with the wrappers `alg_F` and `alg_DFDU`; the generalized toolbox constructors `alg_isol2eqn` and `alg_sol2eqn`; and the toolbox utility functions `alg_arg_check`, `alg_init_data`, and `alg_read_solution`.

```
function prob = alg_construct_eqn(data, sol)

prob = coco_prob();
prob = coco_add_func(prob, 'alg', @alg_F, @alg_DFDU, data, 'zero', ...
  'u0', sol.u);
if ~isempty(data.pnames)
  prob = coco_add_pars(prob, 'pars', data.p_idx, data.pnames);
end
prob = coco_add_slot(prob, 'alg', @coco_save_data, data, 'save_full');

end
```

In this case, each toolbox instance is characterized by

- a problem function object with function identifier `'alg'`;

- an optional problem function object with function identifier `'pars'`; and

- a slot function object with function identifier `'alg'`.

In particular, the problem function object with identifier `'pars'` is omitted when the `pnames` field of the function data structure `data` is empty.

Consider, finally, the class encoding from Sect. 4.2 of the `'alg'` toolbox in terms of the embeddable basic constructor `alg_construct_eqn`, repeated below, together with the wrappers `alg_F` and `alg_DFDU`; the generalized toolbox constructors `alg_isol2eqn` and `alg_sol2eqn`; and the toolbox utility functions `alg_arg_check`, `alg_init_data`, and `alg_read_solution`.

```
function prob = alg_construct_eqn(prob, tbid, data, sol)

prob = coco_add_func(prob, tbid, @alg_F, @alg_DFDU, data, 'zero', ...
  'u0', sol.u);
if ~isempty(data.pnames)
  fid  = coco_get_id(tbid, 'pars');
  uidx = coco_get_func_data(prob, tbid, 'uidx');
  prob = coco_add_pars(prob, fid, uidx(data.p_idx), data.pnames);
end
prob = coco_add_slot(prob, tbid, @coco_save_data, data, 'save_full');

end
```

This implementation of `alg_construct_eqn` is the first that adheres to a stricter interpretation of an `'alg'` *class constructor* as a function that instantiates an equation object without first creating an instance of the continuation problem structure `prob`.

4.3.3   Toolbox identifiers and data

As seen in the class encoding from Sect. 4.2 of the `'alg'` toolbox, each toolbox instance is associated with a unique toolbox instance identifier `tbid` and a unique data structure `data`. Rather than allocate separate memory for these variables, however, their content and utility are distributed across the problem function objects appended to the continuation problem structure. The toolbox instance identifier, for example, is used directly as a function identifier for the family of zero functions Φ. Similarly, the data structure is assigned *as a whole* to the function data structure associated with the encoding of Φ. Notably, the `'alg'` class object exists only in terms of the problem and slot function objects instantiated during construction and appended to the continuation problem structure `prob`.

Of the various properties of the problem and slot function objects that occur, in turn, as instance properties of an `'alg'` toolbox object, the function data structures are unique in that they may be modified by the associated zero, monitor, or slot function during execution. At each moment, their content constitutes a parameterization of the state of the function object. Since the latter is a fundamental element of a toolbox instance, it is natural to consider the totality of the function data structures assigned to all constituent function objects as a parameterization of the state of the toolbox instance. We refer to this collection of function data structures as *toolbox data*.

As demonstrated above, read access to the function data structure associated with a particular function object is available outside this function using the `coco_get_func_data` utility. In the simplest case, write access is still confined to each associated function and only to its own function data structure, even if the latter initially is a copy of a function data structure assigned to another function object. As suggested earlier, the `coco_func_data` class, to be introduced in Sect. 8.2, offers a simple mechanism for enabling individual functions to modify function data shared with other functions from the same toolbox, as well as for communication across toolbox boundaries.

4.4 Conclusions

From Chap. 2, we recall the fundamental importance of the notion of encapsulation and the formalism of extended continuation problems to the development in this text. Together, they support the implementation of a general-purpose computational algorithm devoted entirely to the abstract construction of extended continuation problems and their restriction to specific problems, with no regard for the detailed encodings of the zero or monitor functions or the dimensionality of the solution manifold. When combined with additional utility functions for performing the tasks associated with actual continuation—event handling, adaptation, and so on—such an algorithm constitutes an essential design element of the *core layer* of the COCO package.

The previous chapter introduced the theory behind, and use of, the core interface for the construction of user-specific continuation problems. In this chapter, we turned instead to the design of toolbox constructors that encode multiple calls to the core utilities in support of more broadly defined classes of continuation problems. The incremental strategy for constructing a COCO-compatible toolbox illustrated in this chapter suggests a programming paradigm that builds functionality successively and purposefully. An additional benefit of the incremental paradigm is the conceptual framework that it presents for reverse engineering existing toolboxes and for debugging toolboxes under development. We shall continue to adhere to this mode of presentation in the remainder of this text, albeit typically with fewer intermediate steps. In particular, we shall associate basic error checking functions with toolbox constructors in order to support a minimum of feedback to the user in the case of syntax errors. In a full implementation, it would be advisable to expend the required time to provide a careful set of detailed error messages.

We note that there may well occur to the reader a number of improvements to the class encodings proposed in this chapter. For example, none of the encodings of the 'alg' toolbox in previous sections supports the inclusion of user-layer function data. Indeed, the alg_F and alg_DFDU wrappers to the zero function and its Jacobians presupposed an encoding of the latter with only two input arguments, namely, x and p. It is worthwhile to consider the modifications to alg_F and alg_DFDU and to the basic and generalized constructors that would be necessary to accommodate such user-layer function data. One could also imagine providing for user-defined options to propagate through the problem construction. An example of such handling of optional arguments will be considered in the next chapter.

The paradigm of staged construction lent itself to a sequential representation of an extended continuation problem in terms of the families of zero and monitor functions, the index set, the initial solution guess, and the function data structures introduced at each successive stage of construction. We will expand significantly on this paradigm in the next chapter, in the context of building toolboxes that take advantage of the concept of task embedding.

Exercises

4.1. Suppose that $(x, p) \in \mathbb{R}^m \times \mathbb{R}^{n-m}$ and $\Phi : \mathbb{R}^n \to \mathbb{R}^m$. Show that the two-stage construction

$$E_1 := \big[\Phi, \emptyset, \emptyset, (x_0, p_0)\big](\emptyset)$$

and

$$E_2 := [\emptyset, \psi, \{m+1, \ldots, n\}, \emptyset](E_1),$$

where the representation ψ is the identity, is equivalent to the single-stage construction

$$E_2 = \big[\Phi, (x, p) \mapsto p, \emptyset, (x_0, p_0)\big](\emptyset).$$

4.2. Let E_1 and E_2 be two extended continuation problems. Show that E_1 and E_2 are both embedded in $E_1 \oplus E_2$ with disjoint index sets $\mathbb{L}_\Phi$, $\mathbb{L}_\psi$, and $\mathbb{K}$ in the notation of Sect. 3.1.1.

4.3. Show that $E_1 \nsubseteqq (E_1 \oplus E_2)$ and $E_2 \nsubseteqq (E_2 \oplus E_1)$.

4.4. Show that $E = E \oplus \emptyset = \emptyset \oplus E$ for every extended continuation problem E.

4.5. Show that the $\oplus$ operation is associative.

4.6. Let E_1 and $\hat{E}_1$ be two equivalent continuation problems. Similarly, let E_2 and $\hat{E}_2$ be two equivalent continuation problems. Show that $E_1 \oplus E_2$ is equivalent to $\hat{E}_1 \oplus \hat{E}_2$. Use this observation to define the canonical sum of two equivalence classes of extended continuation problems. Show that this operation is commutative.

4.7. Consider an extended continuation problem E obtained from a family of m zero functions, a family of r monitor functions, and an initial solution guess with n components. Following Exercise 3.7 from Chap. 3, introduce a corresponding $(m + r) \times n$ *incidence matrix* whose entries equal 0 if the zero/monitor function represented by the row index is independent of the continuation variable represented by the column index, and 1 otherwise. What is the relationship between the incidence matrices of the extended continuation problems E_1, E_2, and $E_1 \oplus E_2$?

4.8. Let $\tilde{E}$ denote an extended continuation problem in $\tilde{u} \in \mathbb{R}^{\tilde{n}}$ obtained from the family of monitor functions $\tilde{\Phi}$, the family of monitor functions $\tilde{\Psi}$, and the initial solution guess $\tilde{u}_0$. Show that

$$\hat{E} := \left(\left[\emptyset, \psi, (\mathbb{K}_\Phi)_{\{m+1,\ldots,n\}}, \emptyset \right] \circ \left[\Phi, \emptyset, \emptyset, (x_0, p_0) \right] \right) (\tilde{E})$$

is an extended continuation problem in $\hat{u} := (\tilde{u}, x, p) \in \mathbb{R}^{\tilde{n}+n}$ obtained from the family of zero functions

$$\hat{\Phi} : \hat{u} \mapsto \begin{pmatrix} \tilde{\Phi}(\tilde{u}) \\ \Phi(x, p) \end{pmatrix},$$

the family of monitor functions

$$\hat{\Psi} : \hat{u} \mapsto \begin{pmatrix} \Psi(\tilde{u}) \\ p \end{pmatrix},$$

and the initial solution guess

$$\hat{u}_0 := (\tilde{u}_0, x_0, p_0).$$

4.9. Use the result of the previous exercise to prove that

$$\oplus_E = \left[\emptyset, \psi, (\mathbb{K}_\Phi)_{\{m+1,\ldots,n\}}, \emptyset \right] \circ \left[\Phi, \emptyset, \emptyset, (x_0, p_0) \right]$$

if

$$E := \left[\emptyset, \psi, \{m+1,\ldots,n\}, \emptyset \right] \circ \left[\Phi, \emptyset, \emptyset, (x_0, p_0) \right] (\emptyset).$$

4.10. Consider the operator $\oplus_E$ defined in the previous exercise. Show that

$$\oplus_E (\emptyset) = E$$

and

$$\tilde{E} \subsetneqq \oplus_E (\tilde{E})$$

for an arbitrary extended continuation problem $\tilde{E}$.

4.11. Consider an extended continuation problem E obtained from the family of zero functions Φ, the family of monitor functions Ψ, and the initial solution guess u_0. Show that

$$\oplus_E = [\Phi, \Psi, \emptyset, u_0].$$

4.12. What is the dimensional deficit of the extended continuation problem obtained by applying the staged construction

$$[\emptyset, \psi, \{m+1,\ldots,n\}, \emptyset] \circ \left[\Phi, \emptyset, \emptyset, (x_0, p_0) \right],$$

where ψ is the identity, to the empty extended continuation problem?

4.13. Suggest a reason why one would invoke either of the generalized constructors with an empty or absent `pnames` array. Give an example where this would be useful.

4.14. Calls to the 'alg' toolbox constructors may be followed by calls to the special-purpose wrapper coco_add_glue in order to append gluing conditions between elements of the vector of continuation variables corresponding to different instances of the toolbox. Would you recommend applying such gluing conditions to problem parameters introduced in two separate instances? Would it make a difference if you had provided string labels for the corresponding continuation parameters?

4.15. Consider the final encoding of the 'alg' toolbox. Modify the alg_F zero-function wrapper so that it adds a ctr field to the function data structure and initializes this to 0 if such a field is absent, and increments this field by 1 if the field exists. Is the ctr field included in the data array stored with each solution file?

4.16. Consider the final encoding of the 'alg' toolbox. Provide a full suite of sample calls to the alg_isol2eqn toolbox constructor to verify all the conditional syntax tests in alg_arg_check.

4.17. Consider the final encoding of the 'alg' toolbox. What is the argument syntax for the alg_sol2eqn toolbox constructor?

4.18. Compute the dimensional deficits of the restricted continuation problems implemented in the examples in this chapter and verify that these are initializable by the appropriate atlas algorithm, given the run-time designation of elements from $\mathbb{I}$ to $\mathbb{J}$ indicated in the examples.

4.19. Is the final encoding of the 'alg' toolbox compatible with the case when $m = n$? If yes, demonstrate this with an example. If no, how would you modify the constructor to allow for this situation?

4.20. Does the final encoding of the 'alg' toolbox require that the number of components of the output of alg_F equal the number of problem variables? How would you modify the constructor to enforce or relax this requirement?

4.21. Identify the instance properties of the 'alg' toolbox objects created in Example 4.7. In particular, describe the function identifiers of any problem and slot function objects instantiated by the 'alg' constructors.

Chapter 5

Task Embedding

The `'alg'` toolbox developed in the previous chapter collects functionality particular to the construction of members of the class of algebraic continuation problems $\mathcal{A}$ and to the characterization of their solutions. As defined in the introduction to Chap. 4, each element in $\mathcal{A}$ is structurally anonymous in the sense of an absence of distinguishing features to identify the origin of the family of zero functions. Moreover, only a trivial family of monitor functions is included with each element of $\mathcal{A}$ in order to highlight the unique nature of the problem parameters and enable the assignment of the corresponding continuation parameter indices to $\mathbb{I}$.

As discussed in Sect. 4.3, it is natural to consider the `'alg'` toolbox as an object class encoded in terms of several class constructors, instance properties, interface methods, and utility functions. Each instance of the `'alg'` toolbox then represents an equation object whose properties, in turn, include a problem function object associated with the family of zero functions; an optional problem function object associated with the family of monitor functions; and a slot function object associated with the `coco_save_data` slot function.

Examples 4.7 and 4.8 illustrate the repeated command-line use of the `'alg'` toolbox constructors for appending multiple equation objects to a continuation problem structure. It is no stretch of the imagination to consider incorporating multiple calls to the `'alg'` toolbox constructors within another toolbox constructor. We explore such a nested object-oriented paradigm in this chapter. As an example, we consider extended continuation problems corresponding to the simultaneous continuation of solutions of a family of distinct algebraic continuation problems with shared problem parameters.

5.1 Tree decompositions

5.1.1 Canonical sums

Let E_1 be an extended continuation problem in $u \in \mathbb{R}^{n_1}$ given by the family of zero functions $\Phi^{(1)} : \mathbb{R}^{n_1} \to \mathbb{R}^{m_1}$, the family of monitor functions $\Psi^{(1)} : \mathbb{R}^{n_1} \to \mathbb{R}^{r_1}$, and the initial solution guess $u_{1,0}$. Similarly, let E_2 be an extended continuation problem in $u \in \mathbb{R}^{n_2}$ given by the family of zero functions $\Phi^{(2)} : \mathbb{R}^{n_2} \to \mathbb{R}^{m_2}$, the family of monitor functions

$\Psi^{(2)} : \mathbb{R}^{n_2} \to \mathbb{R}^{r_2}$, and the initial solution guess $u_{2,0}$. We recall, from Sect. 4.2, the definition of the canonical sum of E_1 and E_2 as the extended continuation problem $E_1 \oplus E_2$ in $u \in \mathbb{R}^{n_1+n_2}$ given by the family of zero functions

$$\Phi : u \mapsto \begin{pmatrix} \Phi^{(1)}\left(u_{\{1,\dots,n_1\}}\right) \\ \Phi^{(2)}\left(u_{\{n_1+1,\dots,n_1+n_2\}}\right) \end{pmatrix}, \tag{5.1}$$

the family of monitor functions

$$\Psi : u \mapsto \begin{pmatrix} \Psi^{(1)}\left(u_{\{1,\dots,n_1\}}\right) \\ \Psi^{(2)}\left(u_{\{n_1+1,\dots,n_1+n_2\}}\right) \end{pmatrix}, \tag{5.2}$$

and the initial solution guess $u_0 := \left(u_{1,0}, u_{2,0}\right)$.

It is straightforward to show that the $\oplus$ operation is associative. We thus write $E_1 \oplus E_2 \oplus E_3$ instead of $E_1 \oplus (E_2 \oplus E_3)$ or $(E_1 \oplus E_2) \oplus E_3$. We use the notation

$$\bigoplus_{i=1}^{N} E_i \tag{5.3}$$

to denote the canonical sum $E_1 \oplus \cdots \oplus E_N$ of the sequence of extended continuation problems $\{E_i\}_{i=1}^{N}$.

Given an extended continuation problem E obtained from the family of zero functions Φ, the family of monitor functions Ψ, and the initial solution guess u_0, we further recall the definition of the operator $\oplus_E : \mathcal{E} \to \mathcal{E}$ such that

$$\oplus_E : \tilde{E} \mapsto \tilde{E} \oplus E. \tag{5.4}$$

The application of $\oplus_E$ at any stage of construction of an extended continuation problem thus results in a new partial realization obtained by appending Φ, Ψ, and u_0 to the existing partial realization. It follows, for example, that

$$\oplus_{\bigoplus_{i=1}^{N} E_i} = \oplus_{E_N} \circ \cdots \circ \oplus_{E_1}. \tag{5.5}$$

5.1.2 Continuation problem objects

Let E denote an arbitrary extended continuation problem and suppose that there exists a sequence of extended continuation problems $\{E_i\}_{i=1}^{N}$ such that

$$\bigoplus_{i=1}^{N} E_i \subseteqq E. \tag{5.6}$$

By the paradigm of staged construction, it follows that there exists an operator $\mathcal{C}$ such that

$$\oplus_E = \mathcal{C} \circ \oplus_{E_N} \circ \cdots \circ \oplus_{E_1}. \tag{5.7}$$

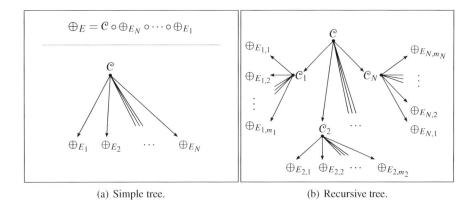

(a) Simple tree.   (b) Recursive tree.

Figure 5.1. *The construction operator $\oplus_E$ appends E to an extended continuation problem. Following the decomposition in Eq. (5.7), it may be represented as a rooted tree with root at the closer $\mathcal{C}$, as illustrated in (a). Since the operation $\oplus$ is associative, a valid sequence of application is generated by a depth-first, bottom-up node traversal starting at $\mathcal{C}$. This concept can be extended to recursive rooted trees (b), where again a valid sequence of application of the operators may be obtained by performing a depth search starting at the root.*

We say that $\mathcal{C}$ *closes* the extended continuation problem E, given the partial realization $E_1 \oplus \cdots \oplus E_N$.

We proceed to associate the relationship in Eq. (5.7) with a rooted tree structure, as shown in Fig. 5.1(a), in which each node represents one of the operators on the right-hand side of Eq. (5.7). The application of $\oplus_E$ to an extended continuation problem is then obtained by the successive application of the operators in the tree structure ordered according to a depth-first, bottom-up node traversal. Here, edges represent the possible dependence of the zero functions and/or monitor functions added at each stage of construction on the continuation variables introduced at a prior stage of construction. The tree representation thus implies that the application of $\oplus_E$ is obtained from

- the application of $\oplus_{E_1}$, followed by

- the application of $\oplus_{E_2}$ with no dependence on the continuation variables introduced in the previous step, followed by

- the application of $\oplus_{E_3}$ with no dependence on the continuation variables introduced in the previous steps, and so on until

- the application of $\oplus_{E_N}$ with no dependence on the continuation variables introduced in the previous steps, followed by

- the application of $\mathcal{C}$ with possible dependence on the continuation variables introduced at each of the previous steps.

Now suppose that there exist sequences of extended continuation problems $E_{i,j}$ such that

$$\bigoplus_{j=1}^{m_i} E_{i,j} \subseteq E_i. \tag{5.8}$$

As before, it follows that there exist operators $\mathcal{C}_i$ such that

$$\oplus_{E_i} = \mathcal{C}_i \circ \oplus_{E_{i,m_i}} \circ \cdots \circ \oplus_{E_{i,1}}. \tag{5.9}$$

We may extend the tree structure representation of E by replacing each of the terminal nodes in Fig. 5.1(a) by subtrees as shown in Fig. 5.1(b). It is straightforward to confirm that the application of $\oplus_E$ to an extended continuation problem is again obtained by the successive application of the operators in the modified tree structure ordered according to a depth-first, bottom-up node traversal.

Following the object-oriented paradigm, we refer to each subtree of a rooted tree representation of an operator $\oplus_E$ as a *continuation problem constructor* and to the corresponding extended continuation problem as a *continuation problem object*. The tree representation thus supports the interpretation of a nested definition of composite continuation problem objects as instances of toolbox classes whose properties include instances of other toolbox classes.

5.2   A composite toolbox

Consider now a class of continuation problems for which there is a natural decomposition

$$u = (x_1, \ldots, x_N, p) \in \mathbb{R}^{m_1} \times \cdots \times \mathbb{R}^{m_N} \times \mathbb{R}^{n-m}, \tag{5.10}$$

where $m := m_1 + \cdots + m_N$, $n > m$, and $m_1, \ldots, m_N > 0$. Here, $x_1, \ldots, x_N$ denote *problem variables* and p denotes *problem parameters*. Let

$$\Phi : u \mapsto \begin{pmatrix} f_1(x_1, p) \\ \vdots \\ f_N(x_N, p) \end{pmatrix}, \tag{5.11}$$

where $f_i : \mathbb{R}^{m_i} \times \mathbb{R}^{n-m} \to \mathbb{R}^{m_i}$, denote the corresponding family of zero functions and consider the family of $n - m$ monitor functions $\Psi : \mathbb{R}^n \to \mathbb{R}^{n-m}$, where $\Psi : u \mapsto p$. Given a regular point x_i^* of $\Phi^{(i)}(x, p^*)$ for each i and for some p^*, an initializable restricted continuation problem on a neighborhood of $u^* = \left(x_1^*, \ldots, x_N^*, p^*\right)$ now follows with the assignment $\mathbb{I} = \{1, \ldots, n - m\}$. The corresponding locally unique solution is then given by $u^* = \left(x_1^*, \ldots, x_N^*, p^*\right)$. Continuation with a 1-dimensional atlas algorithm, for example, requires the release of one of the constraints on the problem parameters.

Let E denote the extended continuation problem obtained from the family of zero functions Φ, the family of monitor functions Ψ, and some initial solution guess $u_0 := \left(x_{1,0}, x_{2,0}, \ldots, x_{N,0}, p_0\right)$. Moreover, let $\tilde{E}$ denote the extended continuation problem in $\tilde{u} = (x_1, p_1, \ldots, x_n, p_N) \in \mathbb{R}^{m+N(n-m)}$ obtained from the family of zero functions $\tilde{\Phi} : \mathbb{R}^{m+N(n-m)} \to \mathbb{R}^{m+(N-1)(n-m)}$ given by

$$\tilde{\Phi} : \tilde{u} \mapsto \begin{pmatrix} f_1(x_1, p_1) \\ \vdots \\ f_N(x_N, p_N) \\ p_1 - p_2 \\ \vdots \\ p_1 - p_N \end{pmatrix}, \tag{5.12}$$

the family of monitor functions $\tilde{\Psi} : \mathbb{R}^{m+N(n-m)} \to \mathbb{R}^{n-m}$ given by $\tilde{\Psi} : \tilde{u} \mapsto p_1$, and the initial solution guess $\tilde{u}_0 := (x_{1,0}, p_0, \ldots, x_{N,0}, p_0)$. By construction, solutions to the restricted continuation problem obtained from E, together with some index set $\mathbb{I}$, correspond in a one-to-one manner to solutions to the restricted continuation problem obtained from $\tilde{E}$, together with the same index set $\mathbb{I}$.

We denote the class of initializable restricted continuation problems obtained from $\tilde{E}$ together with $\mathbb{I} = \{1, \ldots, n-m\}$ by $\mathcal{CA}$. The discussion below illustrates the encoding of a collection of toolbox functions particular to the construction of elements of $\mathcal{CA}$ as well as to the characterization of their solutions. In the absence of any distinguishing information about the origin of the family of zero functions, we refer to $\mathcal{CA}$ as a class of *composite algebraic continuation problems*, to individual problems from this class as *systems*, and to the toolbox developed below by the toolbox name 'compalg'. As in the case of the 'alg' toolbox, we use the toolbox name as the first part of every associated toolbox function.

5.2.1 A toolbox closer

Let $\tilde{E}_i$ denote the extended continuation problem in $w \in \mathbb{R}^{m_i+n-m}$ obtained from the family of zero functions f_i, the empty family of monitor functions, and the initial solution guess $w := (x_{i,0}, p_0)$. It follows that

$$\bigoplus_{i=1}^{N} \tilde{E}_i \subsetneqq \tilde{E} \tag{5.13}$$

and

$$\oplus_{\tilde{E}} = \mathcal{C} \circ \oplus_{\tilde{E}_N} \circ \cdots \circ \oplus_{\tilde{E}_1}. \tag{5.14}$$

Here, $\mathcal{C} = [\phi, \psi, \mathbb{K}, v]$, where the representation $\phi : \mathbb{R}^{N(n-m)} \to \mathbb{R}^{(N-1)(n-m)}$ is given by

$$\phi : (w_1, \ldots, w_N) \mapsto \begin{pmatrix} w_1 - w_2 \\ \vdots \\ w_1 - w_N \end{pmatrix}, \tag{5.15}$$

the representation $\psi : \mathbb{R}^{N(n-m)} \to \mathbb{R}^{n-m}$ is given by

$$\psi : (w_1, \ldots, w_N) \mapsto w_1, \tag{5.16}$$

the index set $\mathbb{K}$ equals

$$\left\{ \left(\mathbb{K}_{f_1}\right)_{\{m_1+1, m_1+n-m\}}, \ldots, \left(\mathbb{K}_{f_N}\right)_{\{m_N+1, m_N+n-m\}} \right\}, \tag{5.17}$$

and v is empty. The staged construction in Eq. (5.14) is thus uniquely defined by the functions that encode $f_1, \ldots, f_N$ and by the initial values $x_{1,0}, \ldots, x_{N,0}$ and p_0, since the integers N, $m_1, \ldots, m_N$, and n can be calculated from the number of functions in the sequence $f_1, \ldots, f_N$ and the dimensions of $x_{1,0}, \ldots, x_{N,0}$ and p_0, respectively.

We proceed to encode the closing operator $\mathcal{C}$, while making the addition of $\tilde{\Psi}$ to the partial realization $\tilde{E}_1 \oplus \cdots \oplus \tilde{E}_N$ optional and contingent upon the inclusion of explicit parameter labels for the corresponding continuation parameters. To this end, we recognize each of the extended continuation problems $\tilde{E}_i$ as an *equation object*, albeit without the second stage of construction. It follows that the application of each of the $\oplus_{\tilde{E}_i}$ constructors corresponds to a unique call to the appropriate `'alg'` toolbox constructor, without the inclusion of a cell array of parameter names. Notably, the ith such call stores the index set $\{m_i + 1, \ldots, m_i + n - m\}$ in the `p_idx` field of the toolbox instance data structure for later reference. The implementation below of the `compalg_close_sys` toolbox closer encodes the closing of the extended continuation problem $\tilde{E}$ by the operator $\mathcal{C}$, as well as the addition of a slot function associated with the `'save_full'` core signal to the continuation problem structure.

```
function prob = compalg_close_sys(prob, tbid, data)

sidx = cell(1, data.neqs);
for i=1:data.neqs
  stbid   = coco_get_id(tbid, sprintf('eqn%d.alg', i));
  [fdata uidx] = coco_get_func_data(prob, stbid, 'data', 'uidx');
  sidx{i} = uidx(fdata.p_idx);
end
for i=2:data.neqs
  fid  = coco_get_id(tbid, sprintf('shared%d', i-1));
  prob = coco_add_glue(prob, fid, sidx{1}, sidx{i});
end
if ~isempty(data.pnames)
  fid  = coco_get_id(tbid, 'pars');
  prob = coco_add_pars(prob, fid, sidx{1}, data.pnames);
end
prob = coco_add_slot(prob, tbid, @coco_save_data, data, 'save_full');

end
```

Here, the `neqs` field of the `data` input argument is assumed to contain the integer N. Similarly, the `pnames` field either is empty or contains a cell array of parameter labels. Finally, the $\tilde{E}_i$ equation object is assumed to be associated with an `'alg'` toolbox instance identifier `stbid` obtained by concatenating the `'compalg'` toolbox instance identifier `tbid` with the string `'eqn#.alg'`, where # represents the integer i.

We note here the use of the `coco_get_func_data` utility to extract the function data structure $\mathfrak{D}_{f_i}$ and the index set $\mathbb{K}_{f_i}$ associated with the $\tilde{E}_i$ equation object, and to store these in the `fdata` and `uidx` variables, respectively. The command

```
sidx{i} = uidx(fdata.p_idx);
```

then assigns the index set $\left(\mathbb{K}_{f_i} \right)_{\{m_i+1, \ldots, m_i+n-m\}}$ to the ith element of the `s_idx` cell array.

5.2.2   Toolbox constructors

We proceed to implement a toolbox constructor `compalg_isol2sys` that corresponds to the action of the operator $\oplus_{\tilde{E}}$, while supporting a combination of required and optional arguments. A candidate encoding, together with the subfunction `is_empty_or_func`, is shown below.

```
function prob = compalg_isol2sys(prob, oid, varargin)

tbid = coco_get_id(oid, 'compalg');
str  = coco_stream(varargin{:});
fhans = str.get('cell');
data.neqs = numel(fhans);
dfdxhans = cell(1, data.neqs);
dfdphans = cell(1, data.neqs);
if is_empty_or_func(str.peek('cell'))
  dfdxhans = str.get('cell');
  if is_empty_or_func(str.peek('cell'))
    dfdphans = str.get('cell');
  end
end
x0 = str.get('cell');
data.pnames = {};
if iscellstr(str.peek('cell'))
  data.pnames = str.get('cell');
end
p0 = str.get;

compalg_arg_check(tbid, data, dfdxhans, dfdphans, x0, p0);
for i=1:data.neqs
  toid = coco_get_id(tbid, sprintf('eqn%d', i));
  prob = alg_isol2eqn(prob, toid, fhans{i}, dfdxhans{i}, ...
    dfdphans{i}, x0{i}, p0);
end
prob = compalg_close_sys(prob, tbid, data);

end

function flag = is_empty_or_func(x)

flag = all(cellfun('isempty', x) | ...
  cellfun('isclass', x, 'function_handle'));

end
```

Here, it is assumed that the `varargin` argument consists of at least three elements. The first of these (and also the first element in `varargin`) is either a single function handle corresponding to the function encoding $f^{(1)}$ or a cell array of function handles encoding $f^{(1)}, \ldots, f^{(N)}$. Similarly, the second of the required elements of the `varargin` argument is a single numerical vector corresponding to $x_{1,0}$ in the case of a single function handle or a cell array of numerical vectors corresponding to $x_{1,0}, \ldots, x_{N,0}$ in the case of a cell array of function handles. Finally, the third of the required elements (and also the last element in `varargin`) is assumed to contain a single numerical vector corresponding to p_0.

In between the first two required elements specified above, the constructor supports the provision of optional input arguments specifying single or, when appropriate, cell arrays of function handles to each of the Jacobians $\partial_x f^{(1)}, \ldots, \partial_x f^{(N)}$ and, if the former are given, $\partial_p f^{(1)}, \ldots, \partial_p f^{(N)}$, respectively. In the absence of one or both such optional arguments, the constructor assigns an empty cell array to `dfdxhans` and/or `dfdphans`, thus implying that the corresponding linearizations should be approximated numerically. Similarly, in between the last two required elements, the constructor supports the optional provision of a single string or a cell array of strings used to label the continuation parameters corresponding to the monitor function $\tilde{\Psi}$.

Basic error checking of the content of the `data`, `dfdxhans`, `dfdphans`, `x0`, `p0`, and `pnames` variables is provided by the function `compalg_arg_check` encoded below and invoked following the initial argument parsing in `compalg_isol2sys`.

```
function compalg_arg_check(tbid, data, dfdxhans, dfdphans, x0, p0)

assert(data.neqs~=0, '%s: insufficient number of equations', tbid);
assert(data.neqs==numel(dfdxhans), ...
  '%s: incompatible number of inputs in ''dfdxhans''', tbid);
assert(data.neqs==numel(dfdphans), ...
  '%s: incompatible number of inputs in ''dfdphans''', tbid);
assert(data.neqs==numel(x0), ...
  '%s: incompatible number of inputs in ''x0''', tbid);
assert(numel(p0)==numel(data.pnames) || isempty(data.pnames), ...
  '%s: incompatible number of elements for ''p0'' and ''pnames''', ...
  tbid);

end
```

Finally, the repeated calls to the `alg_isol2eqn` toolbox constructor shown in the extract below construct the equation objects associated with the extended continuation problems $\tilde{E}_1, \ldots, \tilde{E}_N$, respectively.

```
for i=1:data.neqs
  toid = coco_get_id(tbid, sprintf('eqn%d', i));
  prob = alg_isol2eqn(prob, toid, fhans{i}, dfdxhans{i}, ...
    dfdphans{i}, x0{i}, p0);
end
```

Here, a unique object instance identifier `toid` is included in each call to the `alg_isol2eqn` toolbox constructor. This is obtained by appending the strings `'eqn1'`, `'eqn2'`, and so on, to the `'compalg'` toolbox instance identifier `tbid`.

Example 5.1 The extract below shows the result of invoking the `compalg_isol2sys` constructor in order to perform continuation on the composite zero problem

$$x_1^2 + (y-1)^2 - 1 = 0, \qquad x_2^2 + (y-1)^2 - 1 = 0 \tag{5.18}$$

with initial solution guess $x_1 = 1$, $x_2 = -1$, $y = 1$.

```
>> funs = {@(x,p) x^2+(p-1)^2-1, @(x,p) x^2+(p-1)^2-1};
>> prob = coco_prob();
>> prob = compalg_isol2sys(prob, 'sys', funs, {1 -1}, 'y', 1);
>> coco(prob, 'run', [], 1, 'y', [0.5 1.5]);
```

STEP	DAMPING		NORMS			COMPUTATION TIMES		
IT SIT	GAMMA	||d||	||f||	||U||	F(x)	DF(x)	SOLVE
0			0.00e+00	2.24e+00	0.0	0.0	0.0
1 1	1.00e+00	0.00e+00	0.00e+00	2.24e+00	0.0	0.0	0.0

STEP	TIME	||U||	LABEL	TYPE	y
0	00:00:00	2.2361e+00	1	EP	1.0000e+00
6	00:00:00	1.5000e+00	2	EP	5.0000e-01

```
STEP      TIME         ||U||  LABEL  TYPE                   y
   0  00:00:00  2.2361e+00      3  EP          1.0000e+00
   6  00:00:00  2.8723e+00      4  EP          1.5000e+00
```

■

As with the `alg` toolbox, we provide below an alternative toolbox constructor that allows for reinitializing the extended continuation problem from the solution data array and chart structure associated with a solution file stored during a previous run.

```
function prob = compalg_sol2sys(prob, oid, varargin)

ttbid = coco_get_id(oid, 'compalg');
str = coco_stream(varargin{:});
run = str.get;
if ischar(str.peek)
  stbid = coco_get_id(str.get, 'compalg');
else
  stbid = ttbid;
end
lab = str.get;

data = coco_read_solution(stbid, run, lab);
for i=1:data.neqs
  soid = coco_get_id(stbid, sprintf('eqn%d', i));
  toid = coco_get_id(ttbid, sprintf('eqn%d', i));
  prob = alg_sol2eqn(prob, toid, run, soid, lab);
end
prob = compalg_close_sys(prob, ttbid, data);

end
```

The `compalg_sol2sys` constructor begins by extracting the data structure associated with a `compalg` toolbox instance identifier from the data array stored previously in a solution file. It proceeds to apply the `alg_sol2eqn` constructor repeatedly to reinitialize the parts of the extended continuation problem pertaining to each `alg` toolbox instance and then finally calls `compalg_close_sys` to close the extended continuation problem $\tilde{E}$.

Finally, the toolbox extractor `compalg_read_solution`, shown below, extracts the `compalg` instance-specific toolbox data and then invokes the `alg_read_solution` extractor to obtain the components of the chart structure associated with each `alg` toolbox instance.

```
function [sol data] = compalg_read_solution(oid, run, lab)

tbid = coco_get_id(oid, 'compalg');
data = coco_read_solution(tbid, run, lab);

sol = struct('x', [], 'p', []);
for i=1:data.neqs
  soid    = coco_get_id(tbid, sprintf('eqn%d', i));
  algsol  = alg_read_solution(soid, run, lab);
  sol.x{i} = algsol.x;
end
sol.p = algsol.p;

end
```

Example 5.2 We continue the session in Example 5.1. A sample output from the toolbox extractor `compalg_read_solution` is shown in the extract below.

```
>> [sol data] = compalg_read_solution('sys', 'run', 4)

sol =

    x: {[0.8660]   [-0.8660]}
    p: 1.5000


data =

    neqs: 2
    pnames: {'y'}
```

■

5.2.3   Recursive argument parsing

Recall the casting of the `varargin` input argument to each of the `alg_isol2eqn` and `alg_sol2eqn` toolbox constructors as a stream object of type `coco_stream`, as shown in the command below.

```
str  = coco_stream(varargin{:});
```

Here, the `str` variable contains a handle to a class object that provides sequential read access to a sequence of *tokens*, corresponding to the elements of the cell array `varargin{:}`, and whose internal state includes the position of the next unprocessed token.

As explained in Chap. 4 on page 77, in the special case that the `varargin` input argument is already a `coco_stream` object, the command above simply copies the object into the `str` variable. As a result, any changes to the internal state of the input stream survive the execution of the constructor. In particular, the calling function is able to access the next unprocessed element of the input stream for further parsing. Full use of this functionality then supports a recursive traversal of a `varargin` input argument by allowing portions of this argument to be parsed by embedded calls to other functions.

Example 5.3 The `alg_isol2eqn` and `alg_sol2eqn` toolbox constructors support recursive parsing by construction. It follows that they need not be invoked prior to the call to the `coco` entry-point function as was done in all previous examples. The extract below demonstrates the alternative *inline syntax* for both of these constructors.

```
>> fun = @(x,p) x^2+(p-1)^2-1;
>> alg_args = {fun, 0.9, 'y', 1};
>> coco('run1', @alg_isol2eqn, alg_args{:}, 1, 'y', [0.5 1.5]);
```

	STEP	DAMPING		NORMS		COMPUTATION TIMES		
IT	SIT	GAMMA	||d||	||f||	||U||	F(x)	DF(x)	SOLVE
0				1.90e-01	1.68e+00	0.0	0.0	0.0
1	1	1.00e+00	1.06e-01	1.11e-02	1.74e+00	0.0	0.0	0.0
2	1	1.00e+00	5.54e-03	3.07e-05	1.73e+00	0.0	0.0	0.0
3	1	1.00e+00	1.53e-05	2.36e-10	1.73e+00	0.0	0.0	0.0
4	1	1.00e+00	1.18e-10	0.00e+00	1.73e+00	0.0	0.0	0.0

```
STEP      TIME            ||U||  LABEL  TYPE                y
   0   00:00:00      1.7321e+00      1  EP         1.0000e+00
   5   00:00:00      1.1180e+00      2  EP         5.0000e-01

STEP      TIME            ||U||  LABEL  TYPE                y
   0   00:00:00      1.7321e+00      3  EP         1.0000e+00
   5   00:00:00      2.2913e+00      4  EP         1.5000e+00
>> coco('run2', @alg_sol2eqn, 'run1', 4, 1, 'y', [0.1 1.9]);

    STEP    DAMPING               NORMS               COMPUTATION TIMES
    IT SIT    GAMMA      ||d||     ||f||     ||U||   F(x)  DF(x)  SOLVE
    0                           0.00e+00  2.29e+00   0.0   0.0    0.0
    1   1  1.00e+00  0.00e+00  0.00e+00  2.29e+00   0.0   0.0    0.0

STEP      TIME            ||U||  LABEL  TYPE                y
   0   00:00:00      2.2913e+00      1  EP         1.5000e+00
  10   00:00:00      1.0832e+00      2             4.7419e-01
  17   00:00:00      4.5826e-01      3  EP         1.0000e-01

STEP      TIME            ||U||  LABEL  TYPE                y
   0   00:00:00      2.2913e+00      4  EP         1.5000e+00
   7   00:00:01      2.7221e+00      5  EP         1.9000e+00
```

Here, each constructor receives a `coco_stream` object in lieu of the `varargin` argument and proceeds to modify the internal state of this object so that further parsing may be performed by the appropriate core functions. Notably, the inline syntax implies that the object instance identifier used in the call to `alg_isol2eqn` and `alg_sol2eqn` is the empty string. Finally, in each case, the omission of a continuation problem structure as the first argument to the `coco` entry-point function indicates that such a structure should be initialized by the `coco` function and then appended to by the appropriate `'alg'` constructor. ∎

We consider next an alternative encoding for the `compalg_isol2sys` constructor that assumes an input format reminiscent of multiple successive calls to the `alg_isol2eqn` toolbox constructor. Specifically, we seek to redefine `compalg_isol2sys` so that its `varargin` argument adheres to the syntax

```
varargin = compalg, ...
```

where

```
compalg = {alg}, [pname | pnames | 'end-alg']
```

and

```
alg = @f, [(@dfdx | '[]'), [(@dfdp | '[]')]], x0, p0
```

Here, the braces { } represent one or more repetitions. In the absence of parameter labels, the optional stop token `'end-alg'` may be used to identify the conclusion of argument sequences of the form `alg`.

   We proceed to implement an embeddable `compalg_isol2sys` constructor that parses its `varargin` argument by invoking `alg_isol2eqn` repeatedly until all relevant arguments have been processed. For example, for a `varargin` argument of the form

```
alg, alg, 'end-alg'...
```

the first call to `alg_isol2eqn` should process the first `alg` group, whereas the second call should process only the second `alg` group. The `compalg_isol2sys` should then parse the toolbox-specific stop token and return control to a calling function.

A candidate implementation of the `compalg_isol2sys` constructor, which adheres to the proposed syntax, is shown below.

```
function prob = compalg_isol2sys(prob, oid, varargin)

tbid = coco_get_id(oid, 'compalg');
str  = coco_stream(varargin{:});
data.neqs = 0;
while isa(str.peek, 'function_handle')
  data.neqs = data.neqs+1;
  toid      = coco_get_id(tbid, sprintf('eqn%d', data.neqs));
  prob      = alg_isol2eqn(prob, toid, str);
end
data.pnames = {};
if strcmpi(str.peek, 'end-alg')
  str.skip;
elseif iscellstr(str.peek('cell'))
  data.pnames = str.get('cell');
end

compalg_arg_check(prob, tbid, data);
prob = compalg_close_sys(prob, tbid, data);

end
```

Here, as long as the next unprocessed token in the `str` input stream is a function handle (which is the argument type of the first element of the `alg` argument syntax), the constructor passes the stream object `str` to the `alg_isol2eqn` constructor for further processing.

The `compalg_check` error checking function is now encoded as follows.

```
function compalg_arg_check(prob, tbid, data)

assert(data.neqs~=0, '%s: insufficient number of equations', tbid);
pnum = [];
for i=1:data.neqs
  fid   = coco_get_id(tbid,sprintf('eqn%d.alg', i));
  fdata = coco_get_func_data(prob, fid, 'data');
  assert(isempty(fdata.pnames), ...
    '%s: parameter labels must not be passed to alg', tbid);
  assert(isempty(pnum) || pnum==numel(fdata.p_idx), '%s: %s', ...
    tbid, 'number of parameters must be equal for all equations');
  pnum = numel(fdata.p_idx);
end
assert(pnum==numel(data.pnames) || isempty(data.pnames), ...
  '%s: incompatible number of elements for ''pnames''', ...
  tbid);

end
```

This simply confirms that at least one instance of the `'alg'` toolbox has been constructed, that string labels occur at most once in the `varargin` input argument, and that the number of problem parameters assigned to each `'alg'` toolbox instance is identical across all instances and equal to the number of string labels, if defined.

Finally, an encoding of the `compalg_sol2sys` constructor that supports recursive parsing is shown below.

```
function prob = compalg_sol2sys(prob, oid, varargin)

ttbid = coco_get_id(oid, 'compalg');
str = coco_stream(varargin{:});
run = str.get;
if ischar(str.peek)
  stbid = coco_get_id(str.get, 'compalg');
else
  stbid = ttbid;
end
lab = str.get;

data = coco_read_solution(stbid, run, lab);
for i=1:data.neqs
  soid = coco_get_id(stbid, sprintf('eqn%d', i));
  toid = coco_get_id(ttbid, sprintf('eqn%d', i));
  prob = alg_sol2eqn(prob, toid, run, soid, lab);
end
prob = compalg_close_sys(prob, ttbid, data);

end
```

Example 5.4 We demonstrate the use of the modified `compalg_isol2sys` toolbox constructor in the extract below.

```
>> fun = @(x,p) x^2+(p-1)^2-1;
>> alg1_args = {fun, 1.1, 1};
>> alg2_args = {fun, 0.9, 1};
>> compalg_args = [alg1_args, alg2_args, 'y'];
>> coco('run3', @compalg_isol2sys, compalg_args{:}, 1, 'y', [0.5 1.5]);
```

	STEP	DAMPING		NORMS		COMPUTATION TIMES		
IT	SIT	GAMMA	||d||	||f||	||U||	F(x)	DF(x)	SOLVE
0				2.83e-01	2.24e+00	0.0	0.0	0.0
1	1	1.00e+00	1.42e-01	1.44e-02	2.24e+00	0.0	0.0	0.0
2	1	1.00e+00	7.16e-03	3.69e-05	2.24e+00	0.0	0.0	0.0
3	1	1.00e+00	1.85e-05	2.58e-10	2.24e+00	0.0	0.0	0.0
4	1	1.00e+00	1.29e-10	0.00e+00	2.24e+00	0.0	0.0	0.0

STEP	TIME	||U||	LABEL	TYPE	y
0	00:00:00	2.2361e+00	1	EP	1.0000e+00
6	00:00:00	1.5000e+00	2	EP	5.0000e-01

STEP	TIME	||U||	LABEL	TYPE	y
0	00:00:00	2.2361e+00	3	EP	1.0000e+00
6	00:00:00	2.8723e+00	4	EP	1.5000e+00

```
>> coco('run4', @compalg_sol2sys, 'run3', 2, 1, 'y', [0.5 1.5]);
```

	STEP	DAMPING		NORMS		COMPUTATION TIMES		
IT	SIT	GAMMA	||d||	||f||	||U||	F(x)	DF(x)	SOLVE
0				1.38e-14	1.50e+00	0.0	0.0	0.0
1	1	1.00e+00	7.98e-15	1.57e-16	1.50e+00	0.0	0.0	0.0

STEP	TIME	||U||	LABEL	TYPE	y
0	00:00:00	1.5000e+00	1	EP	5.0000e-01
10	00:00:00	2.7560e+00	2		1.4052e+00
12	00:00:00	2.8723e+00	3	EP	1.5000e+00

We note the absence of the optional array of string labels in the `varargin` arguments in each of the embedded calls to `alg_isol2eqn`, as enforced by `compalg_arg_check`. ∎

5.3   Toolbox settings

5.3.1   Bifurcation diagrams

We recall the use of the signal and slot mechanism in COCO for providing access to the run-time content of the continuation problem structure, in response to signals emitted at selected points during execution, to suitably designed slot functions. In Chap. 4, the coco_save_data slot function was associated with the 'save_full' signal in order to append the corresponding slot function data structure to the data array included with each solution file.

We consider next the inclusion with the 'alg' toolbox of a slot function associated with the 'bddat' core signal. This signal is emitted by the core after each successful location of a solution point during continuation. Slot functions associated with this signal typically augment the default output from the coco entry-point function with a small amount of data representative of each solution point. Such data can be useful, for example, when plotting bifurcation diagrams by characterizing families of solution points in terms of the variations in this data against variations in an active continuation parameter.

Each 'bddat'-compatible slot function must be encoded in a MATLAB function with the following COCO-specific function syntax.

```
function [data res] = bd_slot(prob, data, varargin)
```

In the default use of the 'bddat' signal, the varargin input argument is either of the form {'init'} or {'data', chart}. In the former case, the slot function assigns a cell array of strings to the second output argument res corresponding to additional column headers for the first row of the output from the coco entry-point function. In the latter case, the output argument res is assigned a cell array for each of the subsequent rows of the output array. The content of this array may, for example, be assigned by extracting numerical content from the x field of the chart structure using, wherever appropriate, information contained in the slot function data structure data.

A 'bddat'-compatible slot function appropriate for inclusion with an equation object is now shown below.

```
function [data res] = alg_bddat(prob, data, command, varargin)

res = {};
switch command
  case 'init'
    res   = '||x||';
  case 'data'
    chart = varargin{1};
    uidx  = coco_get_func_data(prob, data.tbid, 'uidx');
    u     = chart.x(uidx);
    res   = norm(u(data.x_idx), 2);
end

end
```

Here, the x_idx field of the slot function data structure is assumed to contain the index set $\{1,\dots,m\}$, where we recall the decomposition of the vector of continuation variables $u := (x, p) \in \mathbb{R}^m \times \mathbb{R}^{n-m}$. Moreover, the tbid field of the slot function data structure is assumed to contain the toolbox instance identifier of the associated instance of the `'alg'` toolbox. It follows that the norm $\|x\|$ is added to each row of the output cell array.

We provide next for the optional inclusion, in an equation object, of a slot function object associated with `alg_bddat`. This is accomplished by the modified encoding of the `alg_construct_eqn` constructor shown below.

```
function prob = alg_construct_eqn(prob, tbid, data, sol)

prob = coco_add_func(prob, tbid, @alg_F, @alg_DFDU, data, 'zero', ...
  'u0', sol.u);
if ~isempty(data.pnames)
  fid  = coco_get_id(tbid, 'pars');
  uidx = coco_get_func_data(prob, tbid, 'uidx');
  prob = coco_add_pars(prob, fid, uidx(data.p_idx), data.pnames);
end
prob = coco_add_slot(prob, tbid, @coco_save_data, data, 'save_full');
if data.alg.norm
  data.tbid = tbid;
  prob = coco_add_slot(prob, tbid, @alg_bddat, data, 'bddat');
end

end
```

Here, the toolbox instance identifier `tbid` is assigned to the `tbid` field of the `data` variable and the resultant structure is assigned to the slot function data structure of the `alg_bddat` slot function. Notably, the slot function object associated with the `'bddat'` signal is omitted from the toolbox instance in the event that the `alg.norm` field of the toolbox data structure equals the boolean value `false`.

Questions now arise as to how a decision regarding the inclusion or omission of the additional output column can be made at run-time, how this decision may trickle down to individual equation objects embedded within a larger continuation problem object, and how collective settings may be imposed on all instances of a particular toolbox within a (sub)tree structure.

5.3.2 The settings tree

A mechanism for accomplishing the tasks listed above is provided within COCO through the `coco_set` and `coco_get` utilities.

The `coco_set` utility may be used to assign optional settings to individual nodes of a continuation problem object. Each setting is associated with a unique *path identifier*, a unique *property name*, and a corresponding *value*. A call of the form

```
prob = coco_set(prob, pid, 'prop1', val1, 'prop2', val2, ...);
```

appends the values `val1`, `...` of the properties `'prop1'`, `...` associated with the path identifier `pid` to the `prob` continuation problem structure. As an example, the command

```
prob = coco_set(prob, 'eqn.alg', 'norm', true);
```

assigns the boolean value `true` to the `'norm'` property associated with the `'eqn.alg'` path identifier, here identical to the identifier of an instance of the `'alg'` toolbox.

For properties assigned to the continuation problem structure using `coco_set`, the `coco_get` function can be used to extract the corresponding values. In particular, the command

```
coco_get(prob, pid, 'property')
```

returns the value of the property `'property'` associated with the path identifier `pid` in the `prob` continuation problem structure provided that a value has been assigned to this property, and an empty array otherwise. Thus, given the previous assignment using `coco_set`, the command

```
coco_get(prob, 'eqn.alg', 'norm');
```

extracts the value of the `'norm'` property associated with the `'eqn.alg'` instance of the `'alg'` toolbox from the `prob` continuation problem structure.

Now consider the modified encoding of the `alg_isol2eqn` toolbox constructor shown below.

```
function prob = alg_isol2eqn(prob, oid, varargin)

tbid = coco_get_id(oid, 'alg');
str  = coco_stream(varargin{:});
data.fhan = str.get;
data.dfdxhan = [];
data.dfdphan = [];
if is_empty_or_func(str.peek)
  data.dfdxhan = str.get;
  if is_empty_or_func(str.peek)
    data.dfdphan = str.get;
  end
end
x0 = str.get;
data.pnames = {};
if iscellstr(str.peek('cell'))
  data.pnames = str.get('cell');
end
p0 = str.get;

alg_arg_check(tbid, data, x0, p0);
data = alg_get_settings(prob, tbid, data);
data = alg_init_data(data, x0, p0);
sol.u = [x0(:); p0(:)];
prob = alg_construct_eqn(prob, tbid, data, sol);

end
```

This differs from previous encodings only in the call to the `alg_get_settings` function shown below.

```
function data = alg_get_settings(prob, tbid, data)

data.alg.norm = coco_get(prob, tbid, 'norm');
if isempty(data.alg.norm)
  data.alg.norm = false;
end
assert(islogical(data.alg.norm), ...
  '%s: input for ''norm'' option is not boolean', tbid);

end
```

Here, the `coco_get` utility is used to extract the value of the `'norm'` property associated with the `tbid` toolbox instance of the `'alg'` toolbox from the `prob` continuation problem structure. As explained above, in the event that this has not been previously assigned, `coco_get` returns an empty array. The conditional statement thus provides default content to the `alg.norm` field of the `data` structure.

Example 5.5 The extract below shows the use of `coco_set` to assign a value to the `'norm'` property of the `'alg'` instance of the `'alg'` toolbox.

```
>> fun = @(x,p) x^2+(p-1)^2-1;
>> alg_args = {fun, 1.1, 'mu', 1};
>> prob = coco_prob();
>> prob = coco_set(prob, 'alg', 'norm', true);
>> bd1 = coco(prob, 'run1', @alg_isol2eqn, alg_args{:}, 1, ...
   'mu', [0.5 1.5]);
```

	STEP	DAMPING		NORMS			COMPUTATION TIMES		
IT	SIT	GAMMA	||d||	||f||	||U||	F(x)	DF(x)	SOLVE
0				2.10e-01	1.79e+00	0.0	0.0	0.0
1	1	1.00e+00	9.55e-02	9.11e-03	1.73e+00	0.0	0.0	0.0
2	1	1.00e+00	4.54e-03	2.06e-05	1.73e+00	0.0	0.0	0.0
3	1	1.00e+00	1.03e-05	1.06e-10	1.73e+00	0.0	0.0	0.0
4	1	1.00e+00	5.29e-11	0.00e+00	1.73e+00	0.0	0.0	0.0

STEP	TIME	||U||	LABEL	TYPE	mu
0	00:00:00	1.7321e+00	1	EP	1.0000e+00
5	00:00:00	1.1180e+00	2	EP	5.0000e-01

STEP	TIME	||U||	LABEL	TYPE	mu
0	00:00:00	1.7321e+00	3	EP	1.0000e+00
5	00:00:00	2.2913e+00	4	EP	1.5000e+00

Inspection of the `bd1` output array shows the inclusion of an additional column containing the norm of the problem variable. ∎

Now suppose that the `'alg'` toolbox constructor is invoked to construct a subtree of a larger continuation problem object, for example, as part of a call to the `compalg_isol2sys` toolbox constructor. In this case, each equation object is associated with a unique toolbox instance identifier, and `coco_set` may be used to assign (nondefault) values to the `'norm'` property for each individual object.

Together with the `coco_merge` utility, the `coco_set`/`coco_get` pair offers an alternative mechanism for providing default property values for all instances of a toolbox appearing within a particular subtree. Consider, for example, a continuation problem object associated with an instance of the `'compalg'` toolbox with toolbox instance identifier `'compalg'`. We assign the default value of `true` to the `'norm'` property of any equation object that appears within this continuation problem object using the command

```
prob = coco_set(prob, 'compalg.alg', 'norm', true);
```

Now consider the modified encoding of the `alg_get_settings` function shown below.

```
function data = alg_get_settings(prob, tbid, data)

defaults.norm = false;
data.alg = coco_merge(defaults, coco_get(prob, tbid));
assert(islogical(data.alg.norm), ...
```

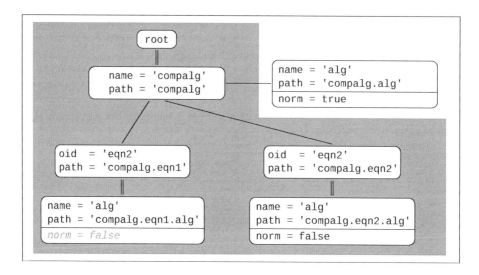

Figure 5.2. *The settings tree utility implemented in* COCO *facilitates the definition of toolbox options for individual or groups of toolbox instances. In its basic form it mimics the tree structure generated during the construction of an extended continuation problem. For example, the subtree inside the gray shaded area is a representation of the toolbox tree obtained in Example 5.6 if one interprets boxes connected by a double line as a single node. As one can see, each leaf has a unique path identifier, which allows the definition of toolbox options for every single instance individually. In the example, the* 'norm' *property of the toolbox instance with path* 'compalg.eqn2.alg' *has been set to* false *explicitly, while no such assignment has been made for the toolbox with path* 'compalg.eqn1.alg'. *In this case, the toolbox default is used. In a slight deviation from the toolbox tree concept, the settings tree supports the addition of leafs that do not correspond to an actual object instance, but rather define options for groups of toolboxes. In the example, the assignment of a value to the* 'norm' *property for the path* 'compalg.alg' *adds the leaf to the settings tree that is outside the shaded area. This leaf defines options for all instances of a toolbox with name* 'alg' *that reside on a lower level in the hierarchy than the node with path identifier* 'compalg'. *Note that explicit assignments on lower levels have precedence over assignments on higher levels. In the example under consideration, extracting the property* 'norm' *from* 'compalg.eqn1.alg' *will, therefore, result in* true, *while the same operation for* 'compalg.eqn2.alg' *will result in* false.

```
    '%s: input for ''norm'' option is not boolean', tbid);

end
```

Here, the `coco_get` utility returns a structure containing all properties associated with the toolbox instance identifier `tbid` and, in their absence, default values assigned using `coco_set` to any parent nodes (see Fig. 5.2). The `coco_merge` utility is used here to merge this structure into the `defaults` structure, overwriting the latter, where possible, with content from the former.

Example 5.6 The extract below shows the use of coco_set to assign a default value to the 'norm' property of any instance of the 'alg' toolbox inside the top-level instance of the 'compalg' toolbox and to override this default value in the case of the 'compalg.eqn2. alg' toolbox instance.

```
>> fun = @(x,p) x^2+(p-1)^2-1;
>> alg1_args = {fun, 1.1, 1};
>> alg2_args = {fun, 0.9, 1};
>> compalg_args = [alg1_args, alg2_args, 'mu'];
>> prob = coco_prob();
>> prob = coco_set(prob, 'compalg.alg', 'norm', true);
>> prob = coco_set(prob, 'compalg.eqn2.alg', 'norm', false);
>> bd2 = coco(prob, 'run2', @compalg_isol2sys, compalg_args{:}, ...
      1, 'mu', [0.5 1.5]);
```

		STEP DAMPING		NORMS			COMPUTATION TIMES		
IT	SIT	GAMMA	||d||	||f||	||U||	F(x)	DF(x)	SOLVE	
0				2.83e-01	2.24e+00	0.0	0.0	0.0	
1	1	1.00e+00	1.42e-01	1.44e-02	2.24e+00	0.0	0.0	0.0	
2	1	1.00e+00	7.16e-03	3.69e-05	2.24e+00	0.0	0.0	0.0	
3	1	1.00e+00	1.85e-05	2.58e-10	2.24e+00	0.0	0.0	0.0	
4	1	1.00e+00	1.29e-10	0.00e+00	2.24e+00	0.0	0.0	0.0	

STEP	TIME	||U||	LABEL	TYPE	mu
0	00:00:00	2.2361e+00	1	EP	1.0000e+00
6	00:00:00	1.5000e+00	2	EP	5.0000e-01

STEP	TIME	||U||	LABEL	TYPE	mu
0	00:00:00	2.2361e+00	3	EP	1.0000e+00
6	00:00:00	2.8723e+00	4	EP	1.5000e+00

Inspection of the bd2 output array shows the inclusion of an additional column containing the norm of the problem variable corresponding to the first equation object. We show a graphical representation of the corresponding toolbox settings tree in Fig. 5.2. ∎

Finally, we show the inclusion of a call to alg_get_settings also in the alg_sol2eqn constructor.

```
function prob = alg_sol2eqn(prob, oid, varargin)

tbid = coco_get_id(oid, 'alg');
str  = coco_stream(varargin{:});
run  = str.get;
if ischar(str.peek)
  soid = str.get;
else
  soid = oid;
end
lab = str.get;

[sol data] = alg_read_solution(soid, run, lab);
data       = alg_get_settings(prob, tbid, data);
prob       = alg_construct_eqn(prob, tbid, data, sol);

end
```

5.4   Conclusions

This chapter emphasizes several principles of task embedding. The development further highlights the need to accommodate object and toolbox instance identifiers that distinguish between different instances of a single toolbox. This principle was used by the `'compalg'` toolbox constructors in order to differentiate between successive embedded calls to the `'alg'` toolbox constructors. In addition, the chapter emphasizes the formulation of grammars that support recursive parsing of input arguments, including the use, whenever appropriate or required, of toolbox-specific stop tokens. This principle allows for alternative implementations of the `'compalg'` toolbox constructors that rely directly on the argument parsing functionality inherent in the embedded calls to the generalized `'alg'` toolbox constructors.

We recall, from the previous chapter, the use of toolbox data associated with the `'alg'` toolbox in support of embedded calls to the user-defined zero function within the zero-function wrapper `alg_F`. In contrast, it is not the case that the `'compalg'` toolbox data subsumes that of the `'alg'` toolbox instances, nor that the `'compalg'` toolbox encodes a zero function that hides the distinct instances of `alg_F` from the COCO core. Instead, in accordance with the embedding paradigm, the `'compalg'` toolbox relies on repeated calls to the `alg_construct_eqn` constructor to append the relevant information regarding the user-defined zero function and the corresponding toolbox data to the extended continuation problem. In addition, the `'compalg'` toolbox constructs its own distinct toolbox data. This contains only that information necessary to append the gluing conditions to the extended continuation problem, to introduce inactive continuation parameters for the problem parameters, and to enable a reconstruction of the continuation problem from a given solution point.

The discussion in Sect. 5.3 suggests the assignment of optional toolbox settings to a subfield of the toolbox data structure as a mechanism for propagating user-defined choices for the execution of individual toolboxes. In the final encoding of `alg_get_settings`, for example, any current settings for the optional `'norm'` property of the `data.alg` structure were overwritten by the default value of `false` or by a user-specified choice, with precedence given to the latter. In Chap. 7, we consider an alternative encoding of a toolbox-specific settings function that gives precedence to current settings over select default values.

We make extensive and repeated use of all of the concepts discussed above in the next part of this text. In particular, as shown in Chap. 9, the `'compalg'` toolbox serves as a useful template for a composite toolbox for the continuation of multisegment boundary-value problems, e.g., with application to periodic trajectories in hybrid dynamical systems.

Exercises

5.1. Let $E_1, \ldots, E_N$ denote N extended continuation problems. Show that

$$\oplus_{\oplus_{i=1}^N E_i} = \oplus_{E_N} \circ \cdots \circ \oplus_{E_1}.$$

5.2. Consider several rooted tree structures and write down expressions for all continuation problem constructors found in each tree.

5.3. Verify Eq. (5.14), where ϕ, ψ, and $\mathbb{K}$ are given in Eqs. (5.15)–(5.17).

5.4. Consider the encoding of the `compalg_isol2sys` and `compalg_sol2sys` toolbox constructors on pages 101 and 103. Verify that the construction of the embedded equation objects $\tilde{E}_i$ omits the inclusion of the optional monitor functions associated with the problem parameters.

5.5. Use the `compalg_sol2sys` toolbox constructor on page 103 to commence continuation along the solution manifold to the zero problem

$$x_1^2 + (y-1)^2 - 1 = 0, \qquad x_2^2 + (y-1)^2 - 1 = 0$$

by reinitializing the corresponding restricted continuation problem from a solution file generated by the `coco` entry-point function in Example 5.1.

5.6. Consider the encoding of the `compalg_isol2sys` and `compalg_sol2sys` toolbox constructors on pages 106 and 107. Does the construction of the embedded equation objects $\tilde{E}_i$ omit the inclusion of the optional monitor functions associated with the problem parameters? Can you modify the constructors to enforce or relax this requirement?

5.7. Is it necessary to include the optional stop token `'end-alg'` in the `compalg` argument syntax? What are the potential consequences of omitting this argument when embedding the `'compalg'` toolbox within another toolbox?

5.8. How would you modify the `'alg'` generalized toolbox constructors to support the inclusion of an optional stop token in the `alg` argument syntax? Is such a stop token necessary? Is it useful?

5.9. Consider the two different versions of the `'compalg'` toolbox described in this chapter. Provide a full suite of sample calls to each of the `compalg_isol2sys` toolbox constructors to verify all the conditional tests in `compalg_arg_check`.

5.10. Consider the first encoding of the `'compalg'` toolbox described in this chapter. What is the argument syntax for the `compalg_isol2sys` and `compalg_sol2sys` toolbox constructors?

5.11. Consider the second encoding of the `'compalg'` toolbox described in this chapter. What is the argument syntax for the `compalg_sol2sys` toolbox constructor?

5.12. In the exercises section at the end of the previous chapter, we considered modifications to the `'alg'` toolbox to accommodate the case when $m = n$ in the case of a single zero problem, corresponding to the absence of problem parameters. Revisit the observations you made there in the context of the composite toolbox constructed in this chapter and modify it as appropriate.

5.13. Write an alternative to the `'compalg'` toolbox that relies on a single instance of the `'alg'` toolbox, albeit suitably modified to accommodate cell arrays of function handles and initial values for the corresponding problem variables as inputs. Consider the differences in the number of variables and equations implemented by this toolbox as compared to a corresponding call to the `'compalg'` toolbox and comment on the implications to computational efficiency.

5.14. Continuation of stationary points of reaction-diffusion equations of the form $u_t = F(u,\mu)$, $u : \Omega \subseteq \mathbb{R}^n \to \mathbb{R}$, with boundary condition $F_{\mathrm{bc}}(u|_{\partial\Omega}) = 0$, leads to systems of nonlinear algebraic equations after discretization. The *linear stability* of such an approximate stationary point can be determined from the sequence of eigenvalues λ_k

of the Jacobian of the discretization of F. Write an $'alg'$-compatible slot function, associated with the $'bddat'$ core signal, that includes the *stability indicator* $stab = (0 < \max_k(real(\lambda_k)))$ in the bifurcation data. Use a boolean $'eigs'$ property of the $'alg'$ toolbox to flag the optional inclusion of the stability indicator.

5.15. Implement a new generalized constructor for the $'compalg'$ toolbox that supports recursive parsing by either $alg\_isol2eqn$ or $alg\_sol2eqn$ for each embedded equation object. What is the modified $compalg$ argument grammar?

5.16. Let $f : \mathbb{R}^m \times \mathbb{R} \times \mathbb{R}^{n-m-1} \to \mathbb{R}^m$ be continuously differentiable. For each $\mu \in \mathbb{R}^{n-m-1}$, consider the problem of finding a solution $x(\alpha; \mu)$ of the equation

$$f(x(\alpha; \mu), \alpha, \mu) = 0$$

for all α on some interval $[a, b]$. Formulate a family of zero functions similar to those in Eq. (5.12) for computing a piecewise-linear interpolant of $x(\alpha; \mu)$ in terms of its nodal values $x_j := x(\alpha_j; \mu)$ on the mesh

$$a = \alpha_0 < \alpha_1 < \cdots < \alpha_{N-1} < \alpha_N = b$$

for some integer N.

5.17. Consider the extended continuation problem in

$$u := (x_0, \ldots, x_N, \alpha_0, \ldots, \alpha_N, \mu)$$

obtained from the family of zero functions in the previous exercise, the family of monitor functions $\Psi : u \mapsto (\alpha_0, \ldots, \alpha_N, \mu)$, and the initial solution guess $u_0 := (x_0^*, \ldots, x_N^*, \alpha_0^*, \ldots, \alpha_N^*, \mu^*)$. Modify the $'compalg'$ toolbox to encode the initializable restricted continuation problem obtained by setting $\mathbb{J} = \emptyset$. Implement a generalized constructor that enables continuation to commence from a sequence of solution points of the equation

$$f(x, \alpha, \mu) = 0$$

obtained using continuation applied to a single equation object constructed with the $'alg'$ toolbox for $\mu = \mu^*$.

5.18. Consider the 2-dimensional manifold in $\mathbb{R}^3$ implicitly defined by

$$p_1 - x\left(p_2 - x^2\right) = 0$$

for $p_1, p_2 \in [-5, 5]$. Apply the modified $'compalg'$ toolbox from the previous exercise to perform continuation of interpolating curve segments along this manifold, with respect to different parameters.

5.19. Consider the modified interpolation problem of finding an interpolating curve $(x(s), \alpha(s))$ such that $f(x(s), \alpha(s), \mu) = 0$, $f : \mathbb{R}^m \times \mathbb{R} \times \mathbb{R}^{n-m-1} \to \mathbb{R}^n$ on the interval $\alpha \in [a, b]$. Derive a family of zero functions similar to Eq. (5.12) for computing a piecewise linear interpolating curve, where the implicit parameterization in s is defined using the conditions $\alpha(0) = a$, $\alpha(1) = b$ and $\|(x(s_{i+1}), \alpha(s_{i+1})) - (x(s_i), \alpha(s_i))\|_2^2 = h^2$ for $i = 1, \ldots, N - 1$. What is the dimensional deficit of the corresponding zero problem? What is the geometric meaning of s and h? Repeat Exercise 5.17 for the family of zero functions obtained in Exercise 5.18.

5.20. Consider again the family of zero functions in Eq. (5.12). Modify the `'compalg'`
toolbox closer to allow for a user-defined selection of those components of p_i that
are included in the gluing conditions and the introduction of the N constraints

$$\|(x_i, \tilde{p}_i) - (x_{i+1}, \tilde{p}_{i+1})\|_2^2 = h^2, i = 1, \ldots, N-1,$$

and

$$\|(x_N, \tilde{p}_N) - (x_1, \tilde{p}_1)\|_2^2 = h^2$$

in terms of the remaining components $\tilde{p}_i$ for some fixed scalar h. What is the nominal dimensional deficit of this problem? Show that this problem has a free phase
similar to the problem of continuing periodic solutions associated with a given vector field in Sect. 2.1. Test your toolbox with geometrically closed families of solution points (e.g., as obtained by intersecting the solution manifold with a suitably
chosen hyperplane) for which you can specify an explicit global phase condition.

5.21. Investigate the possibility of applying the algorithm from Exercise 5.20 to the case
of a closed interpolating curve, that is, an interpolating curve with $(x(1), \alpha(1)) = (x(0), \alpha(0))$.

Part II

Toolbox Templates

Chapter 6

Discretization

The class of algebraic continuation problems we considered in Chap. 4 is characterized by our ignorance of the origin of the family of zero functions. Accordingly, its encoding in a COCO-compatible toolbox requires no problem-specific choices, other than the optional inclusion of parameter labels. The absence of distinguishing information about the inherent meaning of the continuation variables, meanwhile, inspires only the use of $\|x\|$—the norm of the vector of problem variables—as a possible solution measure.

In this chapter, we consider the infinite-dimensional problem associated with finding a solution of an ordinary differential equation. Here, a process of *discretization* results in a finite-dimensional family of zero functions, expressed in terms of a finite number of continuation variables. Collectively, these variables parameterize members of a class of functional approximants of the original solution. In contrast to the case of algebraic continuation problems, the *collocation zero problem* formulated in this chapter depends on decisions made in the process of discretization. These include nonunique and nontrivial choices for

- the class of functional approximants,

- the means to parameterize members from this class, and

- the associated family of zero functions.

From the origin of the collocation zero problem, there follow various natural solution measures that approximate functional properties of the exact solution of the original differential equation.

By the nature of the process of discretization, some fidelity to the original solution is lost in the finite-dimensional approximation. Increased accuracy may require a *refined discretization* of the infinite-dimensional continuation problem. This is usually accomplished by enlarging the class of functional approximants and the number of continuation variables used to parameterize these functions. Such a refined discretization often leads to a relatively high-dimensional collocation zero problem for which significant computational time may be spent computing the zero functions and their Jacobians. In order to reduce the cost of computation, we spend considerable effort in this chapter discussing methods of *vectorization*, whereby iterative constructs are replaced with array operations. The notation and terminology introduced here is then used repeatedly in subsequent chapters.

6.1   The collocation zero problem

Consider a solution $y(t) \in \mathbb{R}^n$ on the interval $[0, T]$, for some yet-to-be-determined scalar T, to the ordinary differential equation

$$\frac{dy}{dt} = f(y, p), \tag{6.1}$$

where the *vector field* $f : \mathbb{R}^n \times \mathbb{R}^q \rightarrow \mathbb{R}^n$ is parameterized by a vector of *problem parameters* $p \in \mathbb{R}^q$. We seek below to approximate the unknown function y on the interval $[0, T]$ in terms of a continuous function of t, expressed on each of N equal-sized intervals as a polynomial of degree m and parameterized by the unknown values at $m + 1$ base points.

For convenience, consider the linear transformation

$$t = t(\tau) := T\tau, \tau \in [0, 1], \tag{6.2}$$

such that $\upsilon(\tau) := y(t(\tau))$ satisfies the equation

$$\frac{d\upsilon}{d\tau} = Tf(\upsilon, p) \tag{6.3}$$

on the interval $[0, 1]$. Given the positive integer N, we now consider the uniform partition

$$0 = \tau_1 < \cdots < \tau_j := \frac{j-1}{N} < \cdots < \tau_{N+1} = 1 \tag{6.4}$$

(cf. Fig. 6.1(a)) and the linear transformation

$$\tau = \tau^{(j)}(\sigma) := \tau_j + \frac{(1+\sigma)}{2}(\tau_{j+1} - \tau_j), \sigma \in [-1, 1], \tag{6.5}$$

on the interval $[\tau_j, \tau_{j+1}]$ for $j \in \{1, \ldots, N\}$. It follows that $\upsilon^{(j)}(\sigma) := \upsilon(\tau^{(j)}(\sigma))$ satisfies the equation

$$\frac{d\upsilon^{(j)}}{d\sigma} = \frac{T}{2N}f\left(\upsilon^{(j)}, p\right) \tag{6.6}$$

on the interval $[-1, 1]$. Notably,

$$\tau^{(j)}(1) = \tau_{j+1} = \tau^{(j+1)}(-1) \tag{6.7}$$

for every $1 \leq j < N$. Continuity of the original solution across the interval $[0, T]$ then implies that

$$\upsilon^{(j)}(1) = \upsilon^{(j+1)}(-1) \tag{6.8}$$

for $1 \leq j < N$.

For $t \in [T\tau_j, T\tau_{j+1}]$, where $1 \leq j \leq N$, we now obtain

$$y(t) = \upsilon^{(j)}\left(2\frac{\frac{t}{T} - \tau_j}{\tau_{j+1} - \tau_j} - 1\right). \tag{6.9}$$

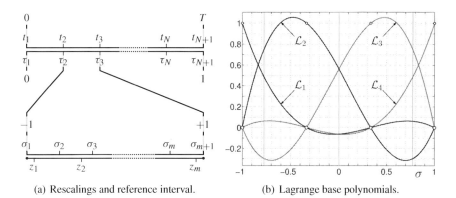

(a) Rescalings and reference interval.   (b) Lagrange base polynomials.

Figure 6.1. *To construct a collocation zero problem, we first rescale the variable $t \in [0,T]$ to $\tau \in [0,1]$. In a second step, we construct a uniform partition $0 = \tau_1 < \tau_2 < \cdots < \tau_{N+1} = 1$ of size N of the unit interval. In a third step, we map each subinterval $[\tau_j, \tau_{j+1}]$ onto the reference interval $\sigma \in [-1,1]$. In a fourth step, we introduce a partition $-1 = \sigma_1 < \sigma_2 < \cdots < \sigma_{m+1} = 1$ of size m of the reference interval and represent the sought-after solution using Lagrange polynomial interpolation. In a last step, we require that the interpolating polynomial satisfy the ordinary differential equation $dy/dt = f(y,p)$ at m collocation nodes z_l, $l = 1, \ldots, m$. Panel (a) shows the hierarchy of scalings and partitions and panel (b) the family of Lagrange polynomials of degree 3 on the reference interval $[-1,1]$. Values at the base points are shown as circles, and the locations of the three corresponding Gauss collocation nodes are indicated by vertical gray lines.*

As per the stated objective, we proceed to approximate $v^{(j)}$ by a polynomial of degree m. To this end, consider further the partition

$$-1 = \sigma_1 < \cdots < \sigma_i < \cdots < \sigma_{m+1} = 1 \tag{6.10}$$

and define the $m+1$ *Lagrange polynomials*

$$\mathcal{L}_i(\sigma) := \prod_{k=1, k \neq i}^{m+1} \frac{\sigma - \sigma_k}{\sigma_i - \sigma_k}, i = 1, \ldots, m+1, \tag{6.11}$$

of degree m (cf. Fig. 6.1(b)) such that

$$\mathcal{L}'_i(\sigma) := \sum_{k=1, k \neq i}^{m+1} \frac{1}{\sigma_i - \sigma_k} \prod_{l=1, l \neq i, k}^{m+1} \frac{\sigma - \sigma_l}{\sigma_i - \sigma_l}. \tag{6.12}$$

In particular, $\mathcal{L}_i(\sigma_i) = 1$ and $\mathcal{L}_i(\sigma_l) = 0$ for $l \neq i$, which imply that the polynomial

$$g_j(\sigma) := \sum_{i=1}^{m+1} \mathcal{L}_i(\sigma) v_{(m+1)(j-1)+i} \tag{6.13}$$

interpolates the values $v_{(m+1)(j-1)+i} \in \mathbb{R}^n$ attained at $\sigma = \sigma_i$ for $i \in \{1, \ldots, m+1\}$.

Finally, introduce the *collocation nodes* z_l, $l = 1, \ldots m$, on the interval $[-1, 1]$ (cf. Fig. 6.1(a)). The *collocation zero problem* is then given by

- the imposition of *collocation conditions* corresponding to Eq. (6.6) on the polynomial interpolants g_j at the collocation nodes z_l, i.e.,

$$0 = \frac{dg_j}{d\sigma}(z_l) - \frac{T}{2N} f\left(g_j(z_l), p\right), \tag{6.14}$$

where

$$g_j(z_l) = \sum_{i=1}^{m+1} \mathcal{L}_i(z_l) \upsilon_{(m+1)(j-1)+i} \tag{6.15}$$

and

$$\frac{dg_j}{d\sigma}(z_l) = \sum_{i=1}^{m+1} \mathcal{L}_i'(z_l) \upsilon_{(m+1)(j-1)+i} \tag{6.16}$$

for $j = 1, \ldots, N$ and $l = 1, \ldots, m$; and

- the imposition of *continuity conditions* corresponding to Eq. (6.8) on the concatenation of the polynomial interpolants g_j at the interior end points, i.e.,

$$0 = g_{j+1}(\sigma_1) - g_j(\sigma_{m+1}) = \upsilon_{(m+1)j+1} - \upsilon_{(m+1)(j-1)+m+1} \tag{6.17}$$

for $j = 1, \ldots, N-1$.

As seen from the definition, the collocation zero problem sets up a total of $Nmn + (N-1)n$ equations for the, at most, $1 + Nn(m+1)$ unknowns T and $\left\{\upsilon_{(m+1)(j-1)+i}\right\}_{j=1, i=1}^{j=N, i=m+1}$. For known values of the problem parameters, the nominal dimensional deficit of the collocation zero problem is therefore $n + 1$. The number of unknowns is reduced if, for example, the interval length T is known.

6.2 Vectorization

We have reason, in the following, to take advantage of an efficient computational implementation of the collocation zero problem, in which loops in the computational flow over various indices are replaced by array and matrix operations.

6.2.1 Arrays and indexing

An *n-dimensional array* on a set $\mathbb{A}$ is a finite sequence of *array elements* $\{a_i\}_{i=1}^m$ in $\mathbb{A}$ for some integer $m \geq 1$ together with a factorization of m into $n \geq 1$ positive integers $\{m_i\}_{i=1}^n$. The integers m_i, $i = 1, \ldots, n$, are the *periods* of the array, the sequence $\{m_i\}_{i=1}^n$ is the *shape* of the array, and m is the *length* of the array. Each index $1 \leq i \leq n$ represents an array *dimension*. In particular, if $m_i = 1$ for some integer index $1 \leq i \leq n$, then we refer to the corresponding dimension as a *singleton*. If, for a given n-dimensional array, the set

$$\left\{1 \leq i \leq n-1 \mid m_j = 1 \ \forall j \in [i+1, n]\right\} \tag{6.18}$$

is nonempty with smallest element k, then we refer to the last $n - k$ dimensions as *trailing singletons*. Every n-dimensional array is isomorphic to the array obtained by eliminating all trailing singletons.

It follows from the definition that an n-dimensional array in $\mathbb{A}$ with nth period m_n for $n > 1$ is isomorphic to a 1-dimensional array of m_n, $(n - 1)$-dimensional arrays in $\mathbb{A}$ with $(n - 1)$th period m_{n-1}. We use the *array delimiters* $[\![$ and $]\!]$ to visualize the nested structure implied by this recursive definition of a multidimensional array. For example, by the definition, a 1-dimensional array in $\mathbb{A}$ is simply a sequence $\{a_i\}_{i=1}^{m}$ in $\mathbb{A}$ with a single period m. We represent such an array by a linear arrangement of the elements of the sequence enclosed by the array delimiters as in

$$[\![\quad a_1 \quad \cdots \quad a_m \quad]\!] . \tag{6.19}$$

Since a 2-dimensional array of periods m_1 and m_2 and with elements $\{a_i\}_{i=1}^{m_1 m_2}$ is simply a 1-dimensional array of m_2, 1-dimensional arrays with m_1 elements each, it may be visualized as follows:

$$[\![\quad [\![\quad a_1 \quad \cdots \quad a_{m_1} \quad]\!] \quad \cdots \quad [\![\quad a_{(m_2-1)m_1+1} \quad \cdots \quad a_{m_2 m_1} \quad]\!] \quad]\!] . \tag{6.20}$$

It should be clear from this example how to generalize the nested use of $[\![$ and $]\!]$ to represent an arbitrary n-dimensional array.

For 1- or 2-dimensional arrays of real or complex numbers, we typically omit the $[\![$ and $]\!]$ delimiters and, instead, organize the sequence of array elements in a matrix. In particular, a 1-dimensional array is represented by a column matrix. For $a_i \in \mathbb{R}$, the array shown in Eq. (6.19) is represented by the $m \times 1$ column matrix

$$\begin{pmatrix} a_1 \\ \vdots \\ a_m \end{pmatrix} . \tag{6.21}$$

Similarly, a 2-dimensional array may be represented by a rectangular matrix in which the elements of each nested 1-dimensional array are organized along individual columns. Thus, for $a_i \in \mathbb{R}$, the array shown in Eq. (6.20) is represented by the $m_1 \times m_2$ matrix

$$\begin{pmatrix} a_1 & \cdots & a_{(m_2-1)m_1+1} \\ \vdots & \ddots & \vdots \\ a_{m_1} & \cdots & a_{m_2 m_1} \end{pmatrix} . \tag{6.22}$$

The elements of an n-dimensional array A may be indexed by n-*tuples* of integers $(i_1, \ldots, i_n)$, where $1 \le i_j \le m_j$ for each $j = 1, \ldots, n$. Here, the n-tuple $(i_1, \ldots, i_n)$ refers to the element indexed by $(i_1, \ldots, i_{n-1})$ in the i_nth $(n - 1)$-dimensional array in the recursive definition of A. Thus, in the array

$$[\![\quad [\![\quad a_1 \quad \cdots \quad a_{m_1} \quad]\!] \quad \cdots \quad [\![\quad a_{(m_2-1)m_1+1} \quad \cdots \quad a_{m_2 m_1} \quad]\!] \quad]\!] , \tag{6.23}$$

the element indexed by the 2-tuple (i_1, i_2) can be found at the intersection of the i_1th row and i_2th column of the matrix representation in Eq. (6.22).

For given periods $\{m_i\}_{i=1}^{n}$, we may translate between the *linear index* i of an element a_i in the sequence $\{a_i\}_{i=1}^{m}$ and the corresponding n-tuple by the following algorithms, where r/s equals the integer quotient of the integers r and s.

Given the index i :

$$k \Leftarrow n$$

$$l \Leftarrow m_1 m_2 \cdots m_n$$

while $k > 0$ do

$$l \Leftarrow l/m_k$$

$$q \Leftarrow (i-1)/l$$

$$i_k \Leftarrow q+1$$

$$i \Leftarrow i - ql$$

$$k \Leftarrow k-1$$

end

Given the n-tuple $(i_1, \ldots, i_n)$:

$$k \Leftarrow n$$

$$l \Leftarrow m_1 m_2 \cdots m_n$$

$$i \Leftarrow 1$$

while $k > 0$ do

$$l \Leftarrow l/m_k$$

$$i \Leftarrow i + (i_k - 1)l$$

$$k \Leftarrow k-1$$

end

It is a useful exercise to consider the influence of singleton dimensions on these algorithms.
An n-dimensional array

$$\left(\{a_i\}_{i=1}^m , \{m_i\}_{i=1}^n \right) \tag{6.24}$$

may be *reshaped* by replacing the sequence $\{m_i\}_{i=1}^n$ with a different sequence $\{m_i'\}_{i=1}^{n'}$ provided that

$$\prod_{i=1}^n m_i = \prod_{i=1}^{n'} m_i' = m. \tag{6.25}$$

Clearly, this has no effect on the sequence $\{a_i\}_{i=1}^m$ of array elements, but does on the association between linear indices and n-tuples introduced above, and consequently on the arrangement of the array delimiters $[\![$ and $]\!]$ in a visual representation of the array. As a special case, denote by $\mathfrak{red}$ the map from an n-dimensional array to an $(n-1)$-dimensional array obtained by substituting the product $m_{n-1}m_n$ for the $(n-1)$th period and omitting the nth period of the original array, i.e.,

$$\mathfrak{red} : \left(\{a_i\}_{i=1}^m , \{m_i\}_{i=1}^n \right) \longmapsto \left(\{a_i\}_{i=1}^m , \{m_1, \ldots, m_{n-2}, m_{n-1}m_n\} \right). \tag{6.26}$$

For an n-dimensional array with $n-k$ trailing singletons, the application of $n-k$ iterates of $\mathfrak{red}$ to the original array yields the isomorphic k-dimensional array obtained by ignoring the trailing singleton dimensions.

The $\mathfrak{vec}$ operator given by

$$\mathfrak{vec} : \left(\{a_i\}_{i=1}^m , \{m_i\}_{i=1}^n \right) \longmapsto \left(\{a_i\}_{i=1}^m , \{m\} \right) \tag{6.27}$$

is equivalent to $n-1$ successive applications of $\mathfrak{red}$ to the original array. In the case of real or complex array entries, the $\mathfrak{vec}$ operator may be thought of as mapping a multidimensional array to a column matrix. Similarly, if m is divisible by c, applying the operator

$$\mathfrak{vec}_c : \left(\{a_i\}_{i=1}^m , \{m_i\}_{i=1}^n \right) \longmapsto \left(\{a_i\}_{i=1}^m , \{c, m/c\} \right) \tag{6.28}$$

to an array of real or complex elements corresponds to converting a multidimensional array to a $c \times m/c$ rectangular matrix.

For a given n-dimensional array A with shape $\{m_i\}_{i=1}^n$, we let the functional notation $A(i_1,\ldots,i_n)$ represent the element of the array indexed by $(i_1,\ldots,i_n)$. Now suppose that $\{s_i\}_{i=1}^k$ is a sequence of $k \geq n$ sequences of integers such that

$$1 \leq s_{i,j} \leq m_i,\, j = 1,\ldots,|s_i|, \tag{6.29}$$

for $1 \leq i \leq k$, where we set $m_i = 1$ for $i > n$. We then define

$$A(s_1,\ldots,s_k) \tag{6.30}$$

as the k-dimensional array with shape $\{|s_i|\}_{i=1}^k$ whose element indexed by $(i_1,\ldots,i_k)$ equals

$$A\left(s_{1,i_1},\ldots,s_{n,i_n}\right). \tag{6.31}$$

As a special case, we let a colon ":" in lieu of the ith sequence s_i represent the sequence $\{1,\ldots,m_i\}$. In particular, the array obtained by replacing all n arguments of A with a colon is the original array A. Moreover, if A is the 1-dimensional array

$$[\![\ a_1\ \cdots\ a_{m_1}\ ]\!], \tag{6.32}$$

then $A\left(:,\{1\}_{i=1}^{m_2}\right)$ is the 2-dimensional array consisting of m_2 copies of A, i.e., the $m_1 \times m_2$ matrix

$$\begin{pmatrix} a_1 & \cdots & a_1 \\ \vdots & \vdots & \vdots \\ a_{m_1} & \cdots & a_{m_1} \end{pmatrix}. \tag{6.33}$$

We may further generalize the functional notation to the case of a sequence $\{s_i\}_{i=1}^k$ of $k < n$ sequences of integers such that

$$1 \leq s_{i,j} \leq m_i,\, j = 1,\ldots,|s_i|, \tag{6.34}$$

for $1 \leq i \leq k-1$ and

$$1 \leq s_{k,j} \leq m_k \cdots m_n,\, j = 1,\ldots,|s_k|. \tag{6.35}$$

In this case, we define

$$A(s_1,\ldots,s_k) \tag{6.36}$$

as the k-dimensional array with shape $\{|s_i|\}_{i=1}^k$ whose element indexed by $(i_1,\ldots,i_k)$ equals

$$\mathfrak{red}^{n-k}(A)\left(s_{1,i_1},\ldots,s_{k,i_k}\right). \tag{6.37}$$

As a special case, the array obtained by replacing each of the k arguments of A with a colon is $\mathfrak{red}^{n-k}(A)$. It follows that

$$A(:) = \mathfrak{vec}(A). \tag{6.38}$$

Example 6.1 In MATLAB, multidimensional arrays of real or complex numbers may be constructed either by associating individual n-tuples with specific values or by assigning a 1-dimensional array of numbers to a variable and then reshaping it as appropriate. For example, the assignment

```
>> A = (1:12)'

A =

     1
     2
     3
     4
     5
     6
     7
     8
     9
    10
    11
    12
```

stores in the variable A the 1-dimensional array consisting of the first 12 integers. Following the proposed custom, MATLAB represents this array visually as a column matrix. The command

```
>> A = reshape(A, [2 3 2])

A(:,:,1) =

     1     3     5
     2     4     6


A(:,:,2) =

     7     9    11
     8    10    12
```

then reshapes A to a 3-dimensional array with periods 2, 3, and 2. The extract shows how MATLAB uses the functional notation A(:,:,1) and A(:,:,2) to represent the 2-dimensional arrays obtained by fixing the last index to 1 and 2, respectively. The shape of the array may be accessed using the MATLAB size function as shown below.

```
>> size(A)

ans =

     2     3     2
```

The generalized functional notation and its consequence to the shape of the array is illustrated in the following sequence of commands.

```
>> A = A(:,:)

A =

     1     3     5     7     9    11
     2     4     6     8    10    12

>> size(A)

ans =

     2     6
```

```
>> A = A(:)

A =

        1
        2
        3
        4
        5
        6
        7
        8
        9
       10
       11
       12

>> size(A)

ans =

       12        1
```

We note, in particular, that MATLAB retains a trailing singleton for the 1-dimensional array. In contrast, for higher-dimensional arrays, trailing singletons are always omitted, as seen below.

```
>> A = reshape(A, [2 3 2 1 1])

A(:,:,1) =

        1        3        5
        2        4        6


A(:,:,2) =

        7        9       11
        8       10       12

>> size(A)

ans =

        2        3        2
```

As suggested above, the functional notation may be used to form new multidimensional arrays from the elements of a given array, for example in order to construct multidimensional tilings of a given array. As an example, the commands

```
>> A = A(:,:);
>> B = A(:,:,[1 1])

B(:,:,1) =

        1        3        5        7        9       11
        2        4        6        8       10       12
```

```
B(:,:,2) =

     1     3     5     7     9    11
     2     4     6     8    10    12
```

first reshape A to a 2-dimensional array with periods 2 and 6, and then assign to B the 3-dimensional array obtained by storing two copies of A along the third dimension. The array B may be reshaped to a 2-dimensional array with periods 2 and 12 by means of the functional notation shown below.

```
>> B = B(:,:)

B =

     1     3     5     7     9    11     1     3     5     7     9    11
     2     4     6     8    10    12     2     4     6     8    10    12
```

It follows that B consists of two copies of the 2-dimensional version of A placed side by side.

More complex tilings of an array are possible using the MATLAB function repmat. As an example, the command

```
>> repmat(A, [3 1])

ans =

     1     3     5     7     9    11
     2     4     6     8    10    12
     1     3     5     7     9    11
     2     4     6     8    10    12
     1     3     5     7     9    11
     2     4     6     8    10    12
```

creates a new 2-dimensional array represented by a rectangular matrix consisting of three copies of A stacked vertically. ∎

For 1- and 2-dimensional arrays of real or complex numbers, the horizontal and vertical tiling of the corresponding matrix representations by the MATLAB function repmat may be expressed in terms of the Kronecker product operator $\otimes$. Specifically, for two rectangular matrices A and B, we define $A \otimes B$ as the block matrix whose (i, j)th block equals $A_{ij} B$. It follows that the number of rows (columns) of $A \otimes B$ equals the product of the number of rows (columns) of A and the number of rows (columns) of B. Now let the notation $1_{k,l}$ refer to a $k \times l$ matrix whose entries all equal 1. Then, for a given $m \times n$ matrix A, the product $1_{k,l} \otimes A$ is an $mk \times nl$ matrix consisting of k rows with l copies of A in each row.

Example 6.2 In MATLAB, the Kronecker product operator $\otimes$ is represented by the kron function, although this command may not always provide the most computationally efficient or memory-efficient implementation. For example, for an arbitrary matrix A, the Kronecker product $1_{k,l} \otimes A$ may be alternatively obtained using the kron function, as in

```
>> A = [1 2 3; 4 5 6];
>> kron(ones(2,3), A)
```

```
ans =

     1     2     3     1     2     3     1     2     3
     4     5     6     4     5     6     4     5     6
     1     2     3     1     2     3     1     2     3
     4     5     6     4     5     6     4     5     6
```

or using the `repmat` command, as in

```
>> repmat(A, [2, 3])

ans =

     1     2     3     1     2     3     1     2     3
     4     5     6     4     5     6     4     5     6
     1     2     3     1     2     3     1     2     3
     4     5     6     4     5     6     4     5     6
```

A less trivial example is the Kronecker product $I_k \otimes A$, where I_k represents the $k \times k$ identity matrix. This yields a block-diagonal matrix with k copies of the $m \times n$ matrix A along the diagonal. Since the resultant matrix is highly sparse, the MATLAB `sparse` function may provide an alternative and computationally efficient implementation for large values of k, m, and n. Specifically, consider the index arrays

$$r = 1_{n,1} \otimes \mathfrak{vec}_m \left(\begin{bmatrix} 1 & \cdots & km \end{bmatrix} \right) \tag{6.39}$$

and

$$c = 1_{m,1} \otimes \mathfrak{vec}_1 \left(\begin{bmatrix} 1 & \cdots & kn \end{bmatrix} \right). \tag{6.40}$$

The (r_i, c_i) entry in the matrix $I_k \otimes A$ then equals the ith element of the array $1_{1,k} \otimes A$, and all other elements of this matrix equal 0. The following sequence of commands generates the corresponding sparse representation for $k = 3$.

```
>> k = 3;
>> [m n] = size(A);
>> A = repmat(A, [1 k]);
>> r = reshape(1:k*m, [m k]);
>> r = repmat(r, [n 1]);
>> c = repmat(1:k*n, [m 1]);
>> sparse(r, c, A)

ans =

   (1,1)        1
   (2,1)        4
   (1,2)        2
   (2,2)        5
   (1,3)        3
   (2,3)        6
   (3,4)        1
   (4,4)        4
   (3,5)        2
   (4,5)        5
   (3,6)        3
   (4,6)        6
   (5,7)        1
   (6,7)        4
   (5,8)        2
   (6,8)        5
```

```
    (5,9)        3
    (6,9)        6
```

We visualize the sparse matrix in its rectangular form using the MATLAB `full` command.

```
>> full(ans)

ans =

     1     2     3     0     0     0     0     0     0
     4     5     6     0     0     0     0     0     0
     0     0     0     1     2     3     0     0     0
     0     0     0     4     5     6     0     0     0
     0     0     0     0     0     0     1     2     3
     0     0     0     0     0     0     4     5     6
```

∎

A similar use of the `sparse` function as in Example 6.2 also applies when constructing the block-diagonal form of the Jacobians of the vector field evaluated at the collocation nodes, as seen later in this section. In this case, the *row* and *column* index arrays r and c may be constructed a priori as they depend only on the state-space dimension n and the discretization parameters N and m.

6.2.2   Vectorized algorithms

A MATLAB implementation of an algorithm is said to be *vectorized* if its encoding can be accomplished without the explicit use of `for`-loops and other iterative operations.

Recall, for example, the construction of the Lagrange polynomials

$$\mathcal{L}_j(\sigma) = \prod_{k=1, k \neq j}^{m+1} \frac{\sigma - \sigma_k}{\sigma_j - \sigma_k}, j = 1, \ldots, m+1, \tag{6.41}$$

from which it follows that

$$\mathcal{L}'_j(\sigma) = \sum_{k=1, k \neq j}^{m+1} \frac{1}{\sigma_j - \sigma_k} \prod_{l=1, l \neq j,k}^{m+1} \frac{\sigma - \sigma_l}{\sigma_j - \sigma_l}, \tag{6.42}$$

and let L and L' denote the $m \times (m+1)$ matrices whose (i, j) entries equal $\mathcal{L}_j(z_i)$ and $\mathcal{L}'_j(z_i)$, respectively. We describe next an equivalent definition of $\mathcal{L}_j(z_i)$ and $\mathcal{L}'_j(z_i)$ that supports a vectorized implementation.

Specifically, define the 3-dimensional arrays t_1 and t_2 such that

$$t_1(i,j,k) = \begin{cases} z_i - \sigma_k, & k \neq j, \\ 1, & k = j, \end{cases} \tag{6.43}$$

$$t_2(i,j,k) = \begin{cases} \sigma_j - \sigma_k, & k \neq j, \\ 1, & k = j, \end{cases} \tag{6.44}$$

for $i = 1, \ldots, m$ and $j, k = 1, \ldots, m+1$. Eq. (6.41) then implies that

$$\mathcal{L}_j(z_i) = \prod_{k=1}^{m+1} \frac{t_1(i,j,k)}{t_2(i,j,k)}. \tag{6.45}$$

Similarly, consider the 3-dimensional array t_3 and the 4-dimensional arrays t_4 and t_5 such that

$$t_3(i,j,k) = \begin{cases} (\sigma_j - \sigma_k)^{-1}, & k \neq j, \\ 0, & k = j, \end{cases} \tag{6.46}$$

$$t_4(i,j,k,l) = \begin{cases} z_i - \sigma_l, & l \neq j,k, \\ 1, & l = j,k, \end{cases} \tag{6.47}$$

$$t_5(i,j,k,l) = \begin{cases} \sigma_j - \sigma_l, & l \neq j,k, \\ 1, & l = j,k, \end{cases} \tag{6.48}$$

for $i = 1, \ldots, m$ and $j,k,l = 1, \ldots, m+1$. From Eq. (6.42), we then obtain

$$\mathcal{L}'_j(z_i) = \sum_{k=1}^{m+1} t_3(i,j,k) \prod_{l=1}^{m+1} \frac{t_4(i,j,k,l)}{t_5(i,j,k,l)}. \tag{6.49}$$

Now let the notations $t_1 \div t_2$ and $t_1 * t_2$ denote the arrays obtained by element-by-element division and multiplication, respectively, of the elements of t_1 and t_2. Moreover, let P_k and S_k denote the mappings from the space of n-dimensional arrays (with $n \geq k$) to the space of $(n-1)$-dimensional arrays obtained from

$$P_k(A)(i_1, \ldots, i_{k-1}, i_{k+1}, \ldots, i_n) := \prod_{i_k} A(i_1, \ldots, i_n) \tag{6.50}$$

and

$$S_k(A)(i_1, \ldots, i_{k-1}, i_{k+1}, \ldots, i_n) := \sum_{i_k} A(i_1, \ldots, i_n), \tag{6.51}$$

respectively. It then follows that

$$L = P_3(t_1 \div t_2) \tag{6.52}$$

and

$$L' = S_3(t_3 * P_4(t_4 \div t_5)). \tag{6.53}$$

Example 6.3 We encode the vectorized algorithms in Eqs. (6.52)–(6.53) in the `coll_L` and `coll_Lp` functions shown below.

```
function A = coll_L(ts, tz)

q = numel(ts);
p = numel(tz);

zi = repmat(reshape(tz, [p 1 1]), [1 q q]);
sj = repmat(reshape(ts, [1 q 1]), [p 1 q]);
sk = repmat(reshape(ts, [1 1 q]), [p q 1]);

t1 = zi-sk;
t2 = sj-sk;
idx = find(abs(t2)<=eps);
t1(idx) = 1;
t2(idx) = 1;

A = prod(t1./t2, 3);

end
```

```
function A = coll_Lp(ts, tz)

q = numel(ts);
p = numel(tz);

zi = repmat(reshape(tz, [p 1 1 1]), [1 q q q]);
sj = repmat(reshape(ts, [1 q 1 1]), [p 1 q q]);
sk = repmat(reshape(ts, [1 1 q 1]), [p q 1 q]);
sl = repmat(reshape(ts, [1 1 1 q]), [p q q 1]);

t3 = sj(:,:,:,1)-sk(:,:,:,1);
t4 = zi-sl;
t5 = sj-sl;

idx1 = find(abs(t5)<=eps);
idx2 = find(abs(t3)<=eps);
idx3 = find(abs(sk-sl)<=eps);
t5(union(idx1, idx3)) = 1;
t4(union(idx1, idx3)) = 1;
t3(idx2) = 1;
t3       = 1.0./t3;
t3(idx2) = 0;

A = sum(t3.*prod(t4./t5, 4), 3);

end
```

These functions return the matrices L and L', respectively, when called with the arrays $[\![\ \sigma_1\ \cdots\ \sigma_{m+1}\ ]\!]$ and $[\![\ z_1\ \cdots\ z_m\ ]\!]$ assigned to the input arguments $\texttt{ts}$ and $\texttt{tz}$, respectively. As an example, the assignment

```
A = prod(t1./t2, 3);
```

uses the MATLAB function $\texttt{prod}$ to replace the 3-dimensional array $t_1 \div t_2$ in its first argument with the 2-dimensional array $P_3 (t_1 \div t_2)$. ∎

The implementations of the vectorized algorithms in Eqs. (6.52)–(6.53) in Example 6.3 make extensive use of the MATLAB $\texttt{reshape}$ and $\texttt{repmat}$ functions for the construction of the intermediate array variables $t_1,\ldots,t_5$. Notably, although $\texttt{for}$-loops and other explicitly iterative operations are avoided at the level of the MATLAB interpreter, some cost is born in the form of additional memory allocated for the components of the temporary arrays $\texttt{zi}$, $\texttt{sj}$, $\texttt{sk}$, and $\texttt{sl}$.

6.2.3 Vectorized functions

Let $[\![\mathbb{A}]\!]$ denote the set of all finite-dimensional arrays in $\mathbb{A}$. A function $h : [\![\mathbb{A}]\!] \to [\![\mathbb{B}]\!]$ is said to be *vectorized* if the image of $h|_\mathbb{A}$ is in $\mathbb{B}$ and if

$$h : \left(\{a_i\}_{i=1}^m, \{m_i\}_{i=1}^n\right) \longmapsto \left(\{h(a_i)\}_{i=1}^m, \{m_i\}_{i=1}^n\right). \tag{6.54}$$

In particular, we say that h is the *vectorized extension* of $h|_\mathbb{A}$. Similarly, let $[\![\mathbb{A}]\!] \,\bar{\times}\, [\![\mathbb{B}]\!]$ denote the subset of all pairs $(a,b) \in [\![\mathbb{A}]\!] \times [\![\mathbb{B}]\!]$ for which the shape of a equals the shape of b. Then, $h : [\![\mathbb{A}]\!] \,\bar{\times}\, [\![\mathbb{B}]\!] \to [\![\mathbb{C}]\!]$ is said to be vectorized if the image of $h|_{\mathbb{A}\times\mathbb{B}}$ is in $\mathbb{C}$ and if

$$h : \left(\left(\{a_i\}_{i=1}^m, \{m_i\}_{i=1}^n\right), \left(\{b_i\}_{i=1}^m, \{m_i\}_{i=1}^n\right)\right) \longmapsto \left(\{h(a_i,b_i)\}_{i=1}^m, \{m_i\}_{i=1}^n\right). \tag{6.55}$$

It is straightforward to generalize this construction to arbitrary numbers of arguments.

By the above definition, the action of a vectorized function on an n-dimensional array may be reduced to its action on the individual array elements, independently of the shape of the array. As long as the shape of the original array is suitably stored, it thus suffices to encode the algorithm represented by h only for 1-dimensional arrays. A given n-dimensional array may be converted to a 1-dimensional array by the application of the vec operator, and the original shape may then be restored after application of h.

As an example, let $\mathbb{A} = \mathbb{R}^s$ for some integer s. It follows that the elements of $\mathbb{A}$ are 1-dimensional arrays of real numbers and that an n-dimensional array of elements in $\mathbb{A}$ corresponds to an $(n+1)$-dimensional array of real numbers. Now consider a function $h : \mathbb{A} \to \mathbb{A}$ and its vectorized extension, also denoted by h. As per the above discussion, it suffices to encode the algorithm represented by h only for inputs and outputs in the form of 1-dimensional arrays in $\mathbb{A}$, i.e., for rectangular matrices with s rows.

Example 6.4 Consider, for example, the `catenary` function shown below.

```
function y = catenary(x)

x1 = x(1,:);
x2 = x(2,:);

y(1,:) = x2;
y(2,:) = (1+x2.^2)./x1;

end
```

This function encodes the vectorized extension of the function

$$\begin{pmatrix} x_1 \\ x_2 \end{pmatrix} \longmapsto \begin{pmatrix} x_2 \\ \left(1+x_2^2\right)/x_1 \end{pmatrix} \tag{6.56}$$

for $x_1, x_2 \in \mathbb{R}$ corresponding to a first-order form of the Euler–Lagrange equations in Eq. (1.7). Here, the `x1` and `x2` variables contain 1-dimensional arrays of elements in $\mathbb{R}$. The function returns a 1-dimensional array of elements in $\mathbb{R}^2$, to be reshaped to the shape of the original input by the calling function.

A more complicated example is afforded by the `catenary_DFDX` function shown below.

```
function J = catenary_DFDX(x)

x1 = x(1,:);
x2 = x(2,:);

J = zeros(2,2, numel(x1));
J(1,2,:) = 1;
J(2,1,:) = -(1+x2.^2)./x1.^2;
J(2,2,:) = -2*x2./x1;

end
```

This function encodes the vectorized extension of the function

$$\begin{pmatrix} x_1 \\ x_2 \end{pmatrix} \longmapsto \begin{pmatrix} 0 & 1 \\ -\left(1+x_2^2\right)/x_1^2 & -2x_2/x_1 \end{pmatrix}, \tag{6.57}$$

corresponding to the Jacobian of the function in Eq. (6.56). Here, the input argument is converted to two 1-dimensional arrays of elements in $\mathbb{R}$ and the function returns a 1-dimensional array of elements in $\mathbb{R}^{2\times 2}$. ∎

The vectorized encoding of the Jacobian shown in Example 6.4 was designed to return a 3-dimensional array of numbers corresponding to a 1-dimensional array of rectangular (here, square) matrices. Let

$$\llbracket\ A_1\quad \ldots\quad A_n\ \rrbracket, A_i \in \mathbb{R}^{k\times l} \tag{6.58}$$

denote such an array. We define the block-diagonalization operator $\mathfrak{diag}$ as follows:

$$\mathfrak{diag}\left(\llbracket\ A_1\quad \ldots\quad A_n\ \rrbracket\right) := \begin{pmatrix} A_1 & & \\ & \ddots & \\ & & A_n \end{pmatrix}, \tag{6.59}$$

where all other entries of the $nk \times nl$ matrix on the right-hand side are set equal to zero. Similarly, let $\mathfrak{transp}$ denote the block-transposition operation

$$\mathfrak{transp}\left(\llbracket\ A_1\quad \ldots\quad A_n\ \rrbracket\right) := \begin{pmatrix} A_1 \\ \vdots \\ A_n \end{pmatrix} \in \mathbb{R}^{nk\times l}. \tag{6.60}$$

Each of these operations may be expressed in terms of suitably defined index vectors and by use of the MATLAB sparse function.
In particular, let

$$r = 1_{l,1} \otimes \mathfrak{vec}_k\left(\llbracket\ 1\quad \cdots\quad nk\ \rrbracket\right) \tag{6.61}$$

and

$$c = 1_{k,1} \otimes \mathfrak{vec}_1\left(\llbracket\ 1\quad \cdots\quad nl\ \rrbracket\right). \tag{6.62}$$

It follows that the (r_i, c_i) entry in $\mathfrak{diag}\left(\llbracket\ A_1\quad \ldots\quad A_n\ \rrbracket\right)$ equals the ith element of the 3-dimensional array of numbers corresponding to the original array $\llbracket\ A_1\quad \ldots\quad A_n\ \rrbracket$ and that all other elements of the matrix equal 0. Similarly, let

$$r = 1_{l,1} \otimes \mathfrak{vec}_k\left(\llbracket\ 1\quad \cdots\quad nk\ \rrbracket\right) \tag{6.63}$$

and

$$c = 1_{k,n} \otimes \mathfrak{vec}_1\left(\llbracket\ 1\quad \cdots\quad l\ \rrbracket\right). \tag{6.64}$$

It is again straightforward to show that the (r_i, c_i) entry in $\mathfrak{transp}\left(\llbracket\ A_1\quad \ldots\quad A_n\ \rrbracket\right)$ equals the ith numerical element of the array $\llbracket\ A_1\quad \ldots\quad A_n\ \rrbracket$.

6.3   A vectorized zero problem

We return to the vector field f and the associated Jacobians $\partial_y f$ and $\partial_p f$, and consider their vectorized extensions, also denoted by f, $\partial_y f$, and $\partial_p f$, respectively. From the above discussion, it follows that it suffices to encode each of these functions for inputs in the form of 1-dimensional arrays of elements in $\mathbb{R}^n$ with return values in the form of 1-dimensional arrays of elements in $\mathbb{R}^n$, $\mathbb{R}^{n \times n}$, and $\mathbb{R}^{n \times q}$, respectively.

Example 6.5 In the case that explicit expressions exist for the Jacobians $\partial_y f$ and $\partial_p f$, these may be encoded in the same way as was done for the catenary_DFDX function in Example 6.4. If explicit expressions are unavailable, the Jacobians may be estimated, instead, using suitable finite-difference approximations, e.g., the two-sided approximation

$$\frac{\partial f}{\partial y_i}(y, p) \approx \frac{f(y + he_i, p) - f(y - he_i, p)}{2h} \tag{6.65}$$

for some appropriate choice for $h \ll 1 + |y_i|$ and with e_i equal to the 1-dimensional array whose jth element equals 1 if $j = i$ and 0 otherwise.

Suppose, for example, that y represents a single 1-dimensional array of n real numbers and that p represents a single 1-dimensional array of q real numbers. In this case, and assuming a vectorized implementation of f, the Jacobian $\partial_y f$ evaluated at (y, p) may be approximated by the array

$$\frac{1}{2h}\left(f\left(1_{1,n} \otimes y + hI_n, 1_{1,n} \otimes p\right) - f\left(1_{1,n} \otimes y - hI_n, 1_{1,n} \otimes p\right)\right). \tag{6.66}$$

A vectorized form of this formula, generalized to an arbitrary number of components of the function f, is implemented in the function num_DFDX shown below.

```
function J = num_DFDX(F, x, p)

x       = x(:,:);
p       = p(:,:);
[m n]   = size(x);
idx     = repmat(1:n, [m 1]);
x0      = x(:,idx);
p0      = p(:,idx);

idx = repmat(1:m, [1 n]);
idx = sub2ind([m m*n], idx, 1:m*n);

h = 1.0e-8*(1.0+abs(x0(idx)));
x = x0;

x(idx)  = x0(idx)+h;
fr      = F(x, p0);
x(idx)  = x0(idx)-h;
fl      = F(x, p0);

l  = size(fr, 1);
hi = repmat(0.5./h, [l 1]);
J  = reshape(hi.*(fr-fl), [l m n]);

end
```

Here, a function handle to an encoding of f and arrays of numerical values for y and p are provided in the first, second, and third input arguments, respectively.

The implementation of the num_DFDX function assumes that the x and p input arguments may be represented as 1-dimensional arrays of equal length. It proceeds to reshape these to 2-dimensional arrays in $\mathbb{R}$ with the resulting periods of x stored in the variables m and n, respectively. The MATLAB sub2ind function is here used to compute the linear indices of the elements in the $m \times mn$ matrix whose row indices are contained in the `idx` array and whose column indices are contained in the sequence $\{1,\dots,mn\}$. Finally, the command

```
h = 1.0e-8*(1.0+abs(x0(idx)));
```

assigns to h a 1-dimensional array of scalars that adapts the step size h to the magnitude of the corresponding element of y.

The vectorized function `num_DFDP` shown below implements a similar two-sided finite-difference approximation for the Jacobian $\partial_p f(y,p)$.

```
function J = num_DFDP(F, x, p)

x   = x(:,:);
p   = p(:,:);
m   = size(p, 1);
n   = size(x, 2);
idx = repmat(1:n, [m 1]);
x0  = x(:,idx);
p0  = p(:,idx);

idx = repmat(1:m, [1 n]);
idx = sub2ind([m m*n], idx, 1:m*n);

h = 1.0e-8*(1.0+abs(p0(idx)));
p = p0;

p(idx) = p0(idx)+h;
fr     = F(x0, p);
p(idx) = p0(idx)-h;
fl     = F(x0, p);

l  = size(fr, 1);
hi = repmat(0.5./h, [l 1]);
J  = reshape(hi.*(fr-fl), [l m n]);

end
```

The commands

```
m   = size(p, 1);
```

and

```
l   = size(fr, 1);
```

are here used to determine the shape of the Jacobian matrices obtained by fixing the third dimension of the output argument.

The algorithms encoded in `num_DFDX` and `num_DFDP` are both implemented in core functions included with COCO. ■

Denote now by υ_{cn} the $n \times Nm$ rectangular matrix whose columns are the values of the polynomial interpolants g_j in Eq. (6.13) for $j = 1,\dots,N$ at the collocation nodes z_l, for $l = 1,\dots,m$. It follows that the columns of the $n \times Nm$ matrix

$$f\left(\upsilon_{cn}, 1_{1,Nm} \otimes p\right) \qquad (6.67)$$

contain the values of the vector field at the corresponding points in state space. Similarly, the Nm $n \times n$ and $n \times q$ matrices obtained by fixing the third dimension of the 3-dimensional arrays

$$\partial_y f \left(v_{cn}, 1_{1,Nm} \otimes p \right) \tag{6.68}$$

and

$$\partial_p f \left(v_{cn}, 1_{1,Nm} \otimes p \right), \tag{6.69}$$

respectively, contain the values of the corresponding Jacobians of the vector field at these points in state space.

From the discussion in the previous section, it follows that if

$$r = 1_{n,1} \otimes \mathfrak{vec}_n \left(\begin{bmatrix} 1 & \cdots & Nmn \end{bmatrix} \right) \tag{6.70}$$

and

$$c = 1_{n,1} \otimes \mathfrak{vec}_1 \left(\begin{bmatrix} 1 & \cdots & Nmn \end{bmatrix} \right), \tag{6.71}$$

then the (r_i, c_i) entry in the matrix

$$\mathfrak{diag} \left(\partial_y f \left(v_{cn}, 1_{1,Nm} \otimes p \right) \right) \tag{6.72}$$

equals the ith element of the array in Eq. (6.68) and all other elements of this matrix equal 0. Similarly, if

$$r = 1_{q,1} \otimes \mathfrak{vec}_n \left(\begin{bmatrix} 1 & \cdots & Nmn \end{bmatrix} \right) \tag{6.73}$$

and

$$c = 1_{n,Nm} \otimes \mathfrak{vec}_1 \left(\begin{bmatrix} 1 & \cdots & q \end{bmatrix} \right), \tag{6.74}$$

then the (r_i, c_i) entry in the matrix

$$\mathfrak{transp} \left(\partial_p f \left(v_{cn}, 1_{1,Nm} \otimes p \right) \right) \tag{6.75}$$

equals the ith element of the array in Eq. (6.69).

Now define the two $Nmn \times N(m+1)n$-dimensional block-diagonal matrices

$$W = I_N \otimes (L \otimes I_n) \tag{6.76}$$

and

$$W' = I_N \otimes \left(L' \otimes I_n \right), \tag{6.77}$$

where I_k again represents the $k \times k$ identity matrix and the (i, j) entries of the $m \times (m+1)$ matrices L and L' are given by $\mathcal{L}_j(z_i)$ and $\mathcal{L}'_j(z_i)$, respectively. Finally, let

$$v_{bp} = \begin{pmatrix} \vdots \\ v_{(m+1)(j-1)+1} \\ \vdots \\ v_{(m+1)j} \\ \vdots \end{pmatrix} \tag{6.78}$$

denote the $N(m+1)n$-dimensional column matrix whose entries are the components of the values of the polynomial interpolants at the corresponding base points. By construction, it follows that the matrix products $W \cdot \upsilon_{bp}$ and $W' \cdot \upsilon_{bp}$ contain the values of the polynomial interpolants and their derivatives, respectively, at the collocation nodes. In particular,

$$\upsilon_{cn} = \mathfrak{vec}_n \left(W \cdot \upsilon_{bp} \right). \tag{6.79}$$

The collocation conditions in Eq. (6.14) now correspond to the vanishing of the matrix difference

$$\frac{T}{2N} \, \mathfrak{vec} \left(f \left(\mathfrak{vec}_n \left(W \cdot \upsilon_{bp} \right), 1_{1,Nm} \otimes p \right) \right) - W' \cdot \upsilon_{bp}. \tag{6.80}$$

By analysis of this expression it is straightforward to show that the Jacobians of the collocation conditions with respect to υ_{bp}, T, and p equal

$$\frac{T}{2N} \, \mathfrak{diag} \left(\partial_y f \left(\mathfrak{vec}_n \left(W \cdot \upsilon_{bp} \right), 1_{1,Nm} \otimes p \right) \right) \cdot W - W', \tag{6.81}$$

$$\frac{1}{2N} \, \mathfrak{vec} \left(f \left(\mathfrak{vec}_n \left(W \cdot \upsilon_{bp} \right), 1_{1,Nm} \otimes p \right) \right), \tag{6.82}$$

and

$$\frac{T}{2N} \, \mathfrak{transp} \left(\partial_p f \left(\mathfrak{vec}_n \left(W \cdot \upsilon_{bp} \right), 1_{1,Nm} \otimes p \right) \right), \tag{6.83}$$

respectively.

Similarly, there exists a constant $(N-1)n \times N(m+1)n$ matrix Q whose entries take the values -1, 0, or 1 such that the continuity conditions in Eq. (6.17) correspond to the vanishing of the matrix product

$$Q \cdot \upsilon_{bp}. \tag{6.84}$$

It follows immediately that the Jacobians of the continuity conditions with respect to υ_{bp}, T, and p equal Q, 0, and 0, respectively.

6.4   Conclusions

The formulation of the collocation zero problem considered in this chapter is a special case of a large class of discretization methods. These methods seek to arrive at finite-dimensional problems whose solutions converge to the solution of the original infinite-dimensional problem, with desirable convergence properties as the discretization is further refined. We consider alternative methods of discretization, including both specializations and generalizations of the collocation formulation considered here, in later chapters.

Considerable mileage is already obtained in this part of the text from the collocation zero problem in Eqs. (6.14)–(6.17) and its vectorized version in Eqs. (6.80) and (6.84). In particular, in the next chapter we translate this formalism into a full encoding of the 'coll' toolbox, complete with a zero problem, its Jacobian, and embeddable toolbox constructors. Various embeddings of the task associated with the collocation zero problem are then considered in Chaps. 8 and 9. These include the development of toolboxes for continuing solutions to two-point boundary-value problems in a single independent variable, periodic orbits in autonomous dynamical systems, and, more generally, sets of constrained orbit segments of hybrid dynamical systems.

Exercises

6.1. Show that if $y(t)$ satisfies

$$\frac{dy}{dt} = f(y, p),$$

then

$$y(T\tau)$$

satisfies

$$\frac{dy}{d\tau} = Tf(y, p).$$

6.2. Given the positive integers N and j, show that if $y(\tau)$ satisfies

$$\frac{dy}{d\tau} = Tf(y, p),$$

then $y\left(\frac{j-1}{N} + \frac{1+\sigma}{2N}\right)$ satisfies

$$\frac{dy}{d\sigma} = \frac{T}{2N} f(y, p).$$

6.3. Write out explicit expressions for the Lagrange polynomials

$$\mathcal{L}_i(\sigma) = \prod_{k=1, k \neq i}^{m+1} \frac{\sigma - \sigma_k}{\sigma_i - \sigma_k}$$

and their derivatives

$$\mathcal{L}_i'(\sigma) := \sum_{k=1, k \neq i}^{m+1} \frac{1}{\sigma_i - \sigma_k} \prod_{l=1, l \neq i, k}^{m+1} \frac{\sigma - \sigma_l}{\sigma_i - \sigma_l},$$

where $i = 1, \ldots, m+1$ for $m = 1$, $m = 2$, and $m = 3$, and graph these for $-1 \leq \sigma \leq 1$. Use your observations of these graphs to show that the matrix whose (i, j) entry is given by $\mathcal{L}_i(\sigma_j)$ equals the identity matrix.

6.4. Suppose that $n = 2$, $N = 2$, and $m = 3$ in the collocation zero problem. Identify the unknowns associated with the discretization of a solution to the original differential equation and write out the full set of zero functions. Confirm that, in the case of known values for the problem parameters, the dimensional deficit of the collocation zero problem equals $n + 1$.

6.5. Verify that, in general, the collocation zero problem consists of $Nmn + (N-1)n$ equations in at most $Nn(m+1) + 1$ unknowns.

6.6. Consider the algorithms on page 126 for translating between the linear index of an element a_i in the sequence $\{a_i\}_{i=1}^m$ and the corresponding n-tuple for a given sequence of periods $\{m_i\}_{i=1}^n$. Are these algorithms each other's inverses? What is the effect of singleton dimensions?

6.7. The translation from an n-tuple to the corresponding linear index shown in the second algorithm on page 126 uses integer division. Propose an alternative algorithm that relies on multiplication. How would this change if array indices started with 0 instead of 1?

6.8. For a given n-dimensional array A, we associate each permutation π of the integers $\{1,\dots,n\}$ with the n-dimensional array $\pi(A)$, where

$$\pi(A) : (i_1,\dots,i_n) \mapsto A\left(\pi_{i_1},\dots,\pi_{i_n}\right).$$

What permutation of a 2-dimensional array of real or complex numbers corresponds to the transpose of the corresponding rectangular matrix? Can you combine permutations and indexing operations in order to reproduce the functionality of the MATLAB `repmat` function?

6.9. Given a $k \times l$ matrix A, consider the matrix $A^* = A^T \otimes I_n$. Show that A^* is the unique $nl \times nk$ matrix such that

$$\mathfrak{vec}_n\left(w\right) \cdot A = \mathfrak{vec}_n\left(A^* \cdot w\right)$$

for every nk-dimensional vector w. Show that the number of nonzero elements of A^* is at most nkl.

6.10. Let

$$r = 1_{n,1} \otimes \mathfrak{vec}_m\left(\begin{bmatrix} 1 & \cdots & km \end{bmatrix}\right)$$

and

$$c = 1_{m,1} \otimes \mathfrak{vec}_1\left(\begin{bmatrix} 1 & \cdots & kn \end{bmatrix}\right)$$

be two arrays. Show that the (r_i, c_i) entry in $I_k \otimes A$ equals the ith element of the array $1_{1,k} \otimes A$ and that all other elements of $I_k \otimes A$ equal 0.

6.11. In MATLAB, the Kronecker product $I_k \otimes A$ may be obtained with the command

```
kron(speye(k), A)
```

or with the explicit sequence of commands

```
r = repmat(reshape(1:k*u, [u k]), [v 1]);
c = repmat(1:k*v, [u 1]);
sparse(r,c,repmat(A, [1 k]));
```

where `u` and `v` contain the number of rows and columns, respectively, of the matrix A. Verify this and compare the computational time involved in either alternative for several large matrices `A` and large values of k.

6.12. Let

$$r = 1_{l,1} \otimes \mathfrak{vec}_k\left(\begin{bmatrix} 1 & \cdots & nk \end{bmatrix}\right)$$

and

$$c = 1_{k,1} \otimes \mathfrak{vec}_1\left(\begin{bmatrix} 1 & \cdots & nl \end{bmatrix}\right)$$

be two arrays. Show that the (r_i, c_i) entry in $\mathfrak{diag}\left(\begin{bmatrix} A_1 & \cdots & A_n \end{bmatrix}\right)$ equals the ith numerical element of $\begin{bmatrix} A_1 & \cdots & A_n \end{bmatrix}$ and all other elements of $\mathfrak{diag}\left(\begin{bmatrix} A_1 & \cdots & A_n \end{bmatrix}\right)$ equal 0.

6.13. Let
$$r = 1_{l,1} \otimes \mathfrak{vec}_k\left(\begin{bmatrix} 1 & \cdots & nk \end{bmatrix}\right)$$
and
$$c = 1_{k,n} \otimes \mathfrak{vec}_1\left(\begin{bmatrix} 1 & \cdots & l \end{bmatrix}\right)$$

be two arrays. Show that the (r_i, c_i) entry in $\mathfrak{transp}\left(\begin{bmatrix} A_1 & \cdots & A_n \end{bmatrix}\right)$ equals the ith numerical element of the array $\begin{bmatrix} A_1 & \cdots & A_n \end{bmatrix}$.

6.14. Verify Eqs. (6.52) and (6.53).

6.15. Verify that the functions `coll_L` on page 133 and `coll_Lp` on page 134 encode the definitions in Eqs. (6.52) and (6.53), respectively.

6.16. Suppose that y represents a 1-dimensional array of n real numbers and that p represents a 1-dimensional array of q real numbers. Assuming a vectorized implementation of f, show that the Jacobian $\partial_y f$ evaluated at (y, p) may be approximated by the array

$$\frac{1}{2h}\left(f\left(1_{1,n} \otimes y + h I_n, 1_{1,n} \otimes p\right) - f\left(1_{1,n} \otimes y - h I_n, 1_{1,n} \otimes p\right)\right).$$

6.17. Verify that the function `num_DFDX` on page 137 computes vectorized midpoint approximations for the Jacobian of the function `F` with respect to its first argument. What is the dimension of the output of `num_DFDX`?

6.18. Verify that the function `num_DFDP` on page 138 computes vectorized midpoint approximations for the Jacobian of the function `F` with respect to its second argument. What is the dimension of the output of `num_DFDP`?

6.19. Express the algorithm `num_DFDP` using an expression similar to Eq. (6.66).

6.20. The theory of vectorized functions has been developed with interpreters such as MATLAB in mind. For such implementations, the reduction of computation time through elimination of loops is evident. However, the theory as such applies to any programming language, for example, C, C++, and FORTRAN. An application of particular interest is the implementation of vectorized extensions. In C, this corresponds to changing code of the form

```
void func(double *y, double *x, double *p) {
    /* do something with x and p and write result to y */
}

for(int i=0;i<N;i++) func(y+i*n, x+i*n, p+i*q);
```

to code of the form

```
void func(int N, int n, int q,
    double *y, double *x, double *p) {
  for(int i=0;i<N;i++) {
    loc_x = x+i*n;
    loc_p = p+i*q;
```

```
    loc_y = y+i*n;
    /* do something with loc_x and loc_p
       and write result to loc_y */
    }
}
```

Implement both versions of the `catenary` function in Example 6.4 to test the impact of the vectorization paradigm. Make sure that the function is called as external, so that you avoid optimization by the compiler that would eliminate the distinction between the two implementations. A good way is to define the to-be-called function in a shared object, linked at run-time.

6.21. Verify the statements regarding size and content of the matrices in Eqs. (6.67)–(6.69).

6.22. Let $n = 2$, $N = 2$, and $m = 3$ and write out the matrices

$$r = 1_{n,1} \otimes \mathfrak{vec}_n \left(\begin{bmatrix} 1 & \cdots & Nmn \end{bmatrix} \right)$$

and

$$c = 1_{n,1} \otimes \mathfrak{vec}_1 \left(\begin{bmatrix} 1 & \cdots & Nmn \end{bmatrix} \right).$$

Use this to verify, in general, that the (r_i, c_i) entry in the matrix

$$\mathfrak{diag} \left(\partial_y f \left(\upsilon_{cn}, 1_{1,Nm} \otimes p \right) \right)$$

equals the ith element of the array

$$\partial_y f \left(\upsilon_{cn}, 1_{1,Nm} \otimes p \right)$$

and that all other elements of this matrix equal 0.

6.23. Let $n = 2$, $N = 2$, and $m = 3$ and write out the matrices

$$r = 1_{q,1} \otimes \mathfrak{vec}_n \left(\begin{bmatrix} 1 & \cdots & Nmn \end{bmatrix} \right)$$

and

$$c = 1_{n,Nm} \otimes \mathfrak{vec}_1 \left(\begin{bmatrix} 1 & \cdots & q \end{bmatrix} \right).$$

Use this to verify that the (r_i, c_i) entry in the matrix

$$\mathfrak{transp} \left(\partial_y f \left(\upsilon_{cn}, 1_{1,Nm} \otimes p \right) \right)$$

equals the ith element of the array

$$\partial_p f \left(\upsilon_{cn}, 1_{1,Nm} \otimes p \right).$$

6.24. Write a sequence of MATLAB commands for computing each of the arrays

$$1_{n,1} \otimes \mathfrak{vec}_n \left(\begin{bmatrix} 1 & \cdots & Nmn \end{bmatrix} \right),$$
$$1_{n,1} \otimes \mathfrak{vec}_1 \left(\begin{bmatrix} 1 & \cdots & Nmn \end{bmatrix} \right),$$
$$1_{q,1} \otimes \mathfrak{vec}_n \left(\begin{bmatrix} 1 & \cdots & Nmn \end{bmatrix} \right),$$
$$1_{n,Nm} \otimes \mathfrak{vec}_1 \left(\begin{bmatrix} 1 & \cdots & q \end{bmatrix} \right).$$

6.25. Denote by v_{cn} the $n \times Nm$ rectangular matrix whose columns are the values of the polynomial interpolants at the collocation nodes. Denote by v_{bp} the $N(m+1)n$-dimensional column matrix whose entries are the components of the values of the polynomial interpolants at the corresponding base points. Show that

$$v_{cn} = \mathfrak{vec}_n \left(W \cdot v_{bp} \right),$$

where W is given in Eq. (6.76).

6.26. Show that the Jacobians of the vector-valued function

$$\frac{T}{2N} \, \mathfrak{vec} \left(f \left(\mathfrak{vec}_n \left(W \cdot v_{bp} \right), 1_{1,Nm} \otimes p \right) \right) - W' \cdot v_{bp}$$

with respect to v_{bp}, T, and p are given by Eqs. (6.81)–(6.83).

6.27. Let $n = 2$, $N = 2$, and $m = 3$ and write out a matrix Q such that

$$Q \cdot v_{bp} = 0$$

corresponds to the continuity conditions in the collocation zero problem.

6.28. Let $v^*(\tau) \in \mathbb{R}^n$ denote a solution of the differential equation

$$\frac{dv}{d\tau} = T^* f \left(v, p^* \right)$$

on the interval $[0,1]$ and suppose that f is continuously differentiable on some neighborhood of the curve $v^*([0,1]) \times \{p^*\}$ in $\mathbb{R}^n \times \mathbb{R}^q$. It follows that there exists a unique continuously differentiable function $\phi(v, \tau, T, p)$ for $v \approx v^*(0)$, τ on some neighborhood of $[0,1]$, $T \approx T^*$, and $p \approx p^*$ such that

$$\partial_\tau \phi(v, \tau, T, p) = T f \left(\phi(v, \tau, T, p), p \right)$$

and $\phi(v, 0, T, p) = v$. Show that

$$\phi \left(v^*(0), \tau, T^*, p^* \right) = v^*(\tau).$$

6.29. Consider the continuously differentiable function $\phi(v, \tau, T, p)$ introduced in the previous exercise and let

$$\Delta_\tau^*(\tau) := \partial_\tau \phi \left(v^*(0), \tau, T^*, p^* \right).$$

Show that

$$\Delta_\tau^*(\tau) = T^* f \left(v^*(\tau), p^* \right)$$

and

$$\partial_\tau \Delta_\tau^*(\tau) = T^* \partial_v f \left(v^*(\tau), p^* \right) \cdot \Delta_\tau^*(\tau).$$

Use these results to find a basis for the nullspace of the Jacobian of the collocation zero problem with respect to v_{bp}.

6.30. Consider the continuously differentiable function $\phi(v, t, T, p)$ introduced in the previous exercises and let

$$\Delta_v^*(\tau) := \partial_v \phi\left(v^*(0), \tau, T^*, p^*\right),$$
$$\Delta_T^*(\tau) := \partial_T \phi\left(v^*(0), \tau, T^*, p^*\right),$$
$$\Delta_p^*(\tau) := \partial_p \phi\left(v^*(0), \tau, T^*, p^*\right).$$

Show that

$$\partial_\tau \Delta_v^*(\tau) = T^* \partial_v f\left(v^*(\tau), p^*\right) \cdot \Delta_v^*(t),$$
$$\partial_\tau \Delta_T^*(\tau) = T^* \partial_v f\left(v^*(\tau), p^*\right) \cdot \Delta_T^*(t) + f\left(v^*(\tau), p^*\right),$$
$$\partial_\tau \Delta_p^*(\tau) = T^* \partial_v f\left(v^*(\tau), p^*\right) \cdot \Delta_p^*(t) + T^* \partial_p f\left(v^*(\tau), p^*\right),$$

where

$$\Delta_v^*(0) = I_n, \ \Delta_T^*(0) = 0, \ \Delta_p^*(0) = 0.$$

Use these results to find a basis for the nullspace of the Jacobian of the collocation zero problem with respect to the vector of continuation variables $u := \left(v_{bp}, T, p\right)$.

Chapter 7

The Collocation Continuation Problem

The `'alg'` and `'compalg'` toolboxes described in Part I of this text are each designed to allow for the construction of initializable restricted continuation problems, as well as to support this construction when embedded within an encapsulating toolbox. In this chapter, we consider the construction of an *auxiliary toolbox* that is specifically designed for embedding within another toolbox, but not necessarily intended to support the construction of an initializable restricted continuation problem as a stand-alone toolbox.

We focus, in particular, on an implementation of the *collocation continuation problem* as a reference to a class of restricted continuation problems obtained from the discretization of a system of first-order ordinary differential equations and the corresponding solution segments. Three extended examples at the end of the chapter serve to illustrate the versatility of the resulting `'coll'` toolbox as a stand-alone tool, deferring examples of its embedding to subsequent chapters.

7.1 Problem definition

We recall from the previous chapter the vectorized form of the collocation zero problem in terms of the family of zero functions $\Phi : \mathbb{R}^{N(m+1)n+1+q} \to \mathbb{R}^{Nmn+(N-1)n}$, where

$$\Phi : \left(v_{bp}, T, p \right) \mapsto \left(\begin{array}{c} \frac{T}{2N} \, \mathfrak{vec} \left(f \left(\mathfrak{vec}_n \left(W \cdot v_{bp} \right), 1_{1,Nm} \otimes p \right) \right) - W' \cdot v_{bp} \\ Q \cdot v_{bp} \end{array} \right) \tag{7.1}$$

in terms of the known matrices $W = I_N \otimes (L \otimes I_n)$, $W' = I_N \otimes \left(L' \otimes I_n \right)$, and Q. Here, the (i, j) entries of the $m \times (m+1)$ matrices L and L' are given by

$$\prod_{k=1, k \neq j}^{m+1} \frac{z_i - \sigma_k}{\sigma_j - \sigma_k} \tag{7.2}$$

and

$$\sum_{k=1, k \neq j}^{m+1} \frac{1}{\sigma_j - \sigma_k} \prod_{l=1, l \neq j, k}^{m+1} \frac{z_i - \sigma_l}{\sigma_j - \sigma_l}, \tag{7.3}$$

respectively, in terms of the *mesh nodes* σ_j, $j = 1, \ldots, m+1$, and the *collocation nodes* z_l, $l = 1, \ldots, m$.

With

$$
v_{bp} = \begin{pmatrix} \vdots \\ v_{(m+1)(j-1)+1} \\ \vdots \\ v_{(m+1)j} \\ \vdots \end{pmatrix} \tag{7.4}
$$

we reconnect to a solution of the differential equation

$$
\frac{dy}{dt} = f(y, p) \tag{7.5}
$$

on the interval $[0, T]$ by substitution for $v^{(j)}$ in Eq. (6.9) in terms of the polynomial interpolant in Eq. (6.13) to obtain the approximation

$$
y(t) \approx \sum_{i=1}^{m+1} v_{(m+1)(j-1)+i} \prod_{k=1, k \neq i}^{m+1} \frac{\frac{2Nt}{T} - 2j + 1 - \sigma_k}{\sigma_i - \sigma_k} \tag{7.6}
$$

on the interval

$$
T \frac{j-1}{N} \leq t \leq T \frac{j}{N} \tag{7.7}
$$

for $1 \leq j \leq N$.

Example 7.1 We choose in this chapter to assume a uniform partition in σ on each subinterval such that

$$
\sigma_i := 2 \frac{i-1}{m} - 1 \tag{7.8}
$$

for $i = 1, \ldots, m+1$. The piecewise-polynomial approximant in Eq. (7.6) is then given by

$$
y(t) \approx \sum_{i=1}^{m+1} v_{(m+1)(j-1)+i} \prod_{k=1, k \neq i}^{m+1} \frac{m}{i-k} \left(\frac{Nt}{T} - j + 1 - \frac{k-1}{m} \right) \tag{7.9}
$$

on the interval

$$
T \frac{j-1}{N} \leq t \leq T \frac{j}{N} \tag{7.10}
$$

for $1 \leq j \leq N$. In particular, for $1 \leq l \leq m+1$, we obtain

$$
y \left(\frac{T}{N} \left(j - 1 + \frac{l-1}{m} \right) \right) \approx v_{(m+1)(j-1)+l}. \tag{7.11}
$$

From the continuity conditions, it follows that an approximate collection of samples of the sought solution $y(t)$ is given by the columns of the matrix $\mathfrak{vec}_n(v_{bp})$ after omitting the columns with column indices $1 + i(m+1)$ for $i = 1, \ldots, N-1$. Moreover, let

$$
t_{(m+1)(j-1)+i} := \frac{T}{N} \left(j - 1 + \frac{i-1}{m} \right) \tag{7.12}
$$

for $j = 1, \ldots, N$ and $i = 1, \ldots, m+1$ and consider the array

$$t_{bp} = \begin{pmatrix} \vdots \\ t_{(m+1)(j-1)+1} \\ \vdots \\ t_{(m+1)j} \\ \vdots \end{pmatrix}. \tag{7.13}$$

The corresponding samples in t are then given by the 1-dimensional array obtained by omitting the elements in t_{bp} indexed by the linear indices $1 + i(m+1)$ for $i = 1, \ldots, N-1$. ∎

Together with the choice of mesh nodes in Example 7.1, a well-defined zero problem follows with the selection of an appropriate set of collocation nodes. In this chapter, we rely on an *orthogonal collocation*, in which the collocation nodes z_l, $l = 1, \ldots, m$, are chosen to coincide with the m consecutive roots of the mth-order Legendre polynomial on the interval $[-1, 1]$, i.e., the *Gauss nodes* associated with Gauss–Legendre quadrature. The function coll_nodes shown below computes the polynomial roots and the corresponding Gauss–Legendre quadrature weights from the eigenvalues and eigenvectors of the matrix

$$J = - \begin{pmatrix} 0 & g_1 & & & \\ g_1 & 0 & g_2 & & \\ & g_2 & \ddots & \ddots & \\ & & \ddots & 0 & g_{m-1} \\ & & & g_{m-1} & 0 \end{pmatrix}, \tag{7.14}$$

where

$$g_i = \frac{i}{\sqrt{4i^2 - 1}}. \tag{7.15}$$

```
function [nds wts] = coll_nodes(m)

n = (1:m-1)';
g = n.*sqrt(1./(4*n.^2-1));
J = -diag(g,1)-diag(g,-1);

[w x] = eig(J);
nds   = diag(x);
wts   = 2*w(1,:).^2;

end
```

Example 7.2 Denote the 1-dimensional array of quadrature weights obtained from the coll_nodes function by w and consider the vectorized extension of some function $g : \mathbb{R}^n \to \mathbb{R}$, also denoted by g. We then obtain the vectorized quadrature approximation

$$\int_0^T g(y(t))\, dt \approx \frac{T}{2N} g\left(\mathfrak{vec}_n\left(W \cdot v_{bp}\right)\right) \cdot \left(1_{N,1} \otimes w\right) \tag{7.16}$$

in terms of the array $1_{N,1} \otimes w$ and the value of g on the array $v_{cn} = \mathfrak{vec}_n(W \cdot v_{bp})$.

Similarly, consider the vectorized extension of some function $h : \mathbb{R} \to \mathbb{R}^n$, also denoted by h. Let h_{cn} denote the $n \times Nm$ matrix whose columns equal the value of h at the collocation nodes. It then follows that

$$\int_0^T h(t)^T \cdot y(t)\, dt \approx \frac{T}{2N} \mathfrak{vec}(h_{cn})^T \cdot (I_N \otimes (\Omega \otimes I_n)) \cdot W \cdot \upsilon_{bp}, \qquad (7.17)$$

where Ω is the $m \times m$ diagonal matrix whose diagonal entries equal w. ∎

We now define a *segment object* as a restricted continuation problem in $u := (\upsilon_{bp}, T, p)$ obtained from the family of zero functions Φ, the family of monitor functions $\Psi : u \mapsto p$, and the index set $\mathbb{I} = \{1, \ldots, q\}$. It follows that a segment object has a nominal dimensional deficit of $n + 1$ on a neighborhood of a regular solution point $(\upsilon_{bp}^*, T^*, p^*)$.

7.2 Encoding

7.2.1 The collocation zero problem

We encode the zero functions corresponding to the collocation zero problem in Eq. (7.1) in the COCO-compatible function `coll_F` below.

```
function [data y] = coll_F(prob, data, u)

x = u(data.xbp_idx);
T = u(data.T_idx);
p = u(data.p_idx);

xx = reshape(data.W*x, data.x_shp);
pp = repmat(p, data.p_rep);

ode = data.fhan(xx, pp);
ode = (0.5*T/data.coll.NTST)*ode(:)-data.Wp*x;
cnt = data.Q*x;

y = [ode; cnt];

end
```

We assume here that the `u` input argument contains the array obtained by concatenating the arrays υ_{bp}, $[\![\, T \,]\!]$, and p, and that the `data` function data structure contains the fields

- `xbp_idx` with value equal to $[\![\, 1 \quad \cdots \quad N(m+1)n \,]\!]$,

- `T_idx` with value equal to $N(m+1)n + 1$,

- `p_idx` with value equal to $[\![\, N(m+1)n+2 \quad \cdots \quad N(m+1)n+1+q \,]\!]$,

- `x_shp` with value equal to $[\![\, n \quad Nm \,]\!]$,

- `p_rep` with value equal to $[\![\, 1 \quad Nm \,]\!]$,

- `W` and `Wp` with values equal to W and W',

- `coll.NTST` with value equal to N,

- `Q` with value equal to Q, and

- `fhan` with value equal to a function handle to the encoding of the vectorized extension of the vector field f.

It follows, for example, that the assignment

```
ode = data.fhan(xx, pp);
```

yields the matrix given by

$$f\left(\mathfrak{vec}_n\left(W\cdot v_{bp}\right), 1_{1,Nm}\otimes p\right). \tag{7.18}$$

The Jacobian of the collocation zero functions with respect to the components of the input array `u` is now encoded as follows.

```
function [data J] = coll_DFDU(prob, data, u)

x = u(data.xbp_idx);
T = u(data.T_idx);
p = u(data.p_idx);

xx = reshape(data.W*x, data.x_shp);
pp = repmat(p, data.p_rep);

if isempty(data.dfdxhan)
  dxode = coco_ezDFDX('f(x,p)v', data.fhan, xx, pp);
else
  dxode = data.dfdxhan(xx, pp);
end
dxode = sparse(data.dxrows, data.dxcols, dxode(:));
dxode = (0.5*T/data.coll.NTST)*dxode*data.W-data.Wp;

dTode = data.fhan(xx, pp);
dTode = (0.5/data.coll.NTST)*dTode(:);

if isempty(data.dfdphan)
  dpode = coco_ezDFDP('f(x,p)v', data.fhan, xx, pp);
else
  dpode = data.dfdphan(xx, pp);
end
dpode = sparse(data.dprows, data.dpcols, dpode(:));
dpode = (0.5*T/data.coll.NTST)*dpode;

J = [dxode dTode dpode; data.Q data.dTpcnt];

end
```

Here, in addition to the fields defined above, the `data` function data structure is assumed to contain the fields

- `dfdxhan` that is either empty or equal to a function handle to the encoding of the vectorized extension of $\partial_y f$,

- `dfdphan` that is either empty or equal to a function handle to the encoding of the vectorized extension of $\partial_p f$,

- dxrows with value equal to $1_{n,1} \otimes \mathfrak{vec}_n \left(\begin{bmatrix} 1 & \cdots & Nmn \end{bmatrix} \right)$,

- dxcols with value equal to $1_{n,1} \otimes \mathfrak{vec}_1 \left(\begin{bmatrix} 1 & \cdots & Nmn \end{bmatrix} \right)$,

- dprows with value equal to $1_{q,1} \otimes \mathfrak{vec}_n \left(\begin{bmatrix} 1 & \cdots & Nmn \end{bmatrix} \right)$,

- dpcols with value equal to $1_{n,Nm} \otimes \mathfrak{vec}_1 \left(\begin{bmatrix} 1 & \cdots & q \end{bmatrix} \right)$, and

- dTpcnt with value equal to the $n(N-1) \times (1+q)$ zero matrix.

It follows, for example, that the assignments

```
if isempty(data.dfdxhan)
  dxode = coco_ezDFDX('f(x,p)v', data.fhan, xx, pp);
else
  dxode = data.dfdxhan(xx, pp);
end
dxode = sparse(data.dxrows, data.dxcols, dxode(:));
dxode = (0.5*T/data.coll.NTST)*dxode*data.W-data.Wp;
```

yield the matrix

$$\frac{T}{2N} \, \mathfrak{diag}\left(\partial_y f\left(\mathfrak{vec}_n\left(W \cdot \upsilon_{bp}\right), 1_{1,Nm} \otimes p\right)\right) \cdot W - W'. \tag{7.19}$$

Similarly, the assignments

```
if isempty(data.dfdphan)
  dpode = coco_ezDFDP('f(x,p)v', data.fhan, xx, pp);
else
  dpode = data.dfdphan(xx, pp);
end
dpode = sparse(data.dprows, data.dpcols, dpode(:));
dpode = (0.5*T/data.coll.NTST)*dpode;
```

yield the matrix

$$\frac{T}{2N} \, \mathfrak{transp}\left(\partial_p f\left(\mathfrak{vec}_n\left(W \cdot \upsilon_{bp}\right), 1_{1,Nm} \otimes p\right)\right). \tag{7.20}$$

In the event that data.dfdxhan and/or data.dfdphan is empty, coll_DFDU calls one or both of the built-in, vectorized finite-difference approximations similar to the differentiation routines shown in Example 6.5 on page 137.

7.2.2   An embeddable constructor

The function below implements an embeddable toolbox constructor compatible with the collocation zero problem formulated above and modeled on alg_construct_eqn discussed in Chap. 4.

```
function prob = coll_construct_seg(prob, tbid, data, sol)

prob = coco_add_func(prob, tbid, @coll_F, @coll_DFDU, data, 'zero', ...
  'u0', sol.u);
if ~isempty(data.pnames)
```

```
   fid  = coco_get_id(tbid, 'pars');
   uidx = coco_get_func_data(prob, tbid, 'uidx');
   prob = coco_add_pars(prob, fid, uidx(data.p_idx), data.pnames);
 end
 prob = coco_add_slot(prob, tbid, @coco_save_data, data, 'save_full');

 end
```

7.2.3 An embeddable generalized constructor

We consider next an embeddable generalized toolbox constructor compatible with the basic `coll_construct_seg` toolbox constructor described above and with the sentence syntax

```
varargin = coll, ...
```

where

```
coll = @f, [(@dfdx | '[]'), [(@dfdp | '[]')]], t0, x0, [pnames], p0
```

The toolbox constructor `coll_isol2seg`, shown with the subfunction `is_empty_or_func` below, provides the desired functionality.

```
function prob = coll_isol2seg(prob, oid, varargin)

tbid = coco_get_id(oid, 'coll');
str  = coco_stream(varargin{:});
data.fhan = str.get;
data.dfdxhan  = [];
data.dfdphan  = [];
if is_empty_or_func(str.peek)
  data.dfdxhan = str.get;
  if is_empty_or_func(str.peek)
    data.dfdphan = str.get;
  end
end
t0 = str.get;
x0 = str.get;
data.pnames = {};
if iscellstr(str.peek('cell'))
  data.pnames = str.get('cell');
end
p0 = str.get;

coll_arg_check(tbid, data, t0, x0, p0);
data = coll_get_settings(prob, tbid, data);
data = coll_init_data(data, x0, p0);
sol  = coll_init_sol(data, t0, x0, p0);
prob = coll_construct_seg(prob, tbid, data, sol);

end

function flag = is_empty_or_func(x)
  flag = isempty(x) || isa(x, 'function_handle');
end
```

Following initial parsing, the constructor invokes each of the auxiliary toolbox utilities `coll_arg_check`, `coll_get_settings`, `coll_init_data`, and `coll_init_sol`.

Specifically, basic argument error checking modeled on the `alg_arg_check` utility is provided by the `coll_arg_check` error checking routine shown below.

```
function coll_arg_check(tbid, data, t0, x0, p0)

assert(isa(data.fhan, 'function_handle'), ...
  '%s: input for ''f'' is not a function handle', tbid);
assert(isnumeric(t0), '%s: input for ''t0'' is not numeric', tbid);
assert(isnumeric(x0), '%s: input for ''x0'' is not numeric', tbid);
assert(isnumeric(p0), '%s: input for ''p0'' is not numeric', tbid);
assert(ndims(t0)==2 && min(size(t0))==1, ...
  '%s: input for ''t0'' is not a vector', tbid);
assert(ndims(x0)==2, ...
  '%s: input for ''x0'' is not an array of vectors', tbid);
assert(size(x0, 1)==numel(t0), ...
  '%s: dimensions of ''t0'' and ''x0'' do not match', tbid);
assert(numel(p0)==numel(data.pnames) || isempty(data.pnames), ...
  '%s: incompatible number of elements for ''p0'' and ''pnames''', ...
  tbid);

end
```

For example, this function enforces the requirement that the `t0` input argument be a 1-dimensional array and that the `x0` input argument be a 2-dimensional array with as many rows as the number of elements of `t0`.

The `coll_get_settings` function that is encoded below is modeled similarly on the `alg_get_settings` function, albeit with the difference that existing values take precedence over default value.

```
function data = coll_get_settings(prob, tbid, data)

defaults.NTST = 10;
defaults.NCOL = 4;
if ~isfield(data, 'coll')
  data.coll = [];
end
data.coll = coco_merge(defaults, coco_merge(data.coll, ...
  coco_get(prob, tbid)));
NTST = data.coll.NTST;
assert(numel(NTST)==1 && isnumeric(NTST) && mod(NTST,1)==0, ...
  '%s: input for option ''NTST'' is not an integer', tbid);
NCOL = data.coll.NCOL;
assert(numel(NCOL)==1 && isnumeric(NCOL) && mod(NCOL,1)==0, ...
  '%s: input for option ''NCOL'' is not an integer', tbid);

end
```

Here, values are assigned to the NTST (N) and NCOL (m) fields of the `coll` property of the `data` structure. Indeed, if these fields have been assigned outside `coll_get_settings`, such values take precedence over the default values corresponding to $N = 10$ and $m = 4$. Additional error checking of these optional settings is performed to ensure that NTST and NCOL both contain integers.

The command

```
data = coll_init_data(data, x0, p0);
```

in `coll_isol2seg` invokes the `coll_init_data` function shown below in order to populate the fields of the toolbox data structure `data`.

```
function data = coll_init_data(data, x0, p0)

NTST = data.coll.NTST;
NCOL = data.coll.NCOL;
dim  = size(x0, 2);
pdim = numel(p0);

data.dim  = dim;
data.pdim = pdim;

bpnum  = NCOL+1;
bpdim  = dim*(NCOL+1);
xbpnum = (NCOL+1)*NTST;
xbpdim = dim*(NCOL+1)*NTST;
cndim  = dim*NCOL;
xcnnum = NCOL*NTST;
xcndim = dim*NCOL*NTST;
cntnum = NTST-1;
cntdim = dim*(NTST-1);

data.xbp_idx = (1:xbpdim)';
data.T_idx   = xbpdim+1;
data.p_idx   = xbpdim+1+(1:pdim)';
data.tbp_idx = setdiff(1:xbpnum, 1+bpnum*(1:cntnum))';
data.x_shp   = [dim xcnnum];
data.xbp_shp = [dim xbpnum];
data.p_rep   = [1 xcnnum];

tm = linspace(-1, 1, bpnum)';
t  = repmat((0.5/NTST)*(tm+1), [1 NTST]);
t  = t+repmat((0:cntnum)/NTST, [bpnum 1]);
data.tbp     = t(:)/t(end);

data.x0_idx = (1:dim)';
data.x1_idx = xbpdim-dim+(1:dim)';

[tc wts]     = coll_nodes(NCOL);
wts          = repmat(wts, [dim NTST]);
data.wts1    = wts(1,:);
data.wts2    = spdiags(wts(:), 0, xcndim, xcndim);

pmap         = coll_L(tm, tc);
dmap         = coll_Lp(tm, tc);
rows         = reshape(1:xcndim, [cndim NTST]);
rows         = repmat(rows, [bpdim 1]);
cols         = repmat(1:xbpdim, [cndim 1]);
W            = repmat(kron(pmap, eye(dim)), [1 NTST]);
Wp           = repmat(kron(dmap, eye(dim)), [1 NTST]);
data.W       = sparse(rows, cols, W);
data.Wp      = sparse(rows, cols, Wp);

data.dxrows = repmat(reshape(1:xcndim, [dim xcnnum]), [dim 1]);
data.dxcols = repmat(1:xcndim, [dim 1]);
data.dprows = repmat(reshape(1:xcndim, [dim xcnnum]), [pdim 1]);
data.dpcols = repmat(1:pdim, [dim xcnnum]);

temp         = reshape(1:xbpdim, [bpdim NTST]);
Qrows        = [1:cntdim 1:cntdim];
Qcols        = [temp(1:dim, 2:end) temp(cndim+1:end, 1:end-1)];
```

```
Qvals       = [ones(cntdim,1) -ones(cntdim,1)];
data.Q      = sparse(Qrows, Qcols, Qvals, cntdim, xbpdim);
data.dTpcnt = sparse(cntdim, 1+pdim);

end
```

The encoding contains calls to the `coll_nodes`, `coll_L`, and `coll_Lp` functions shown in the previous section and in Example 6.3 on pages 133–134, respectively, in order to generate the collocation nodes and the corresponding quadrature weights, as well as to compute the matrices L and L'. The function further appends the additional fields `wts1`, `wts2`, `x0_idx`, and `x1_idx` to the toolbox data structure. These contain the arrays $1_{N,1} \otimes w$ and $I_N \otimes (\Omega \otimes I_n)$ from Eqs. (7.16)–(7.17) and the indices in the column matrix v_{bp} corresponding to the boundary points v_1 and $v_{N(m+1)}$, respectively. While the latter enable encapsulating toolboxes to impose boundary conditions on the solution trajectory, the former may be used to approximate characteristic integral measures of the solution to the infinite-dimensional problem, for example its $\mathcal{L}_2$ norm.

Finally, the command

```
sol = coll_init_sol(data, t0, x0, p0);
```

in `coll_isol2seg` relies on the function `coll_init_sol` shown below to compute an initial guess for T and the base point values v_{bp} by linear interpolation. The corresponding base point mesh is stored by `coll_init_data` in the `tbp` field of the `data` structure.

```
function sol = coll_init_sol(data, t0, x0, p0)

t0 = t0(:);
T0 = t0(end)-t0(1);
t0 = (t0-t0(1))/T0;
x0 = interp1(t0, x0, data.tbp)';

sol.u = [x0(:); T0; p0];

end
```

The function below implements an embeddable generalized toolbox constructor modeled on the `alg_sol2eqn` toolbox constructor discussed in Chap. 4.

```
function prob = coll_sol2seg(prob, oid, varargin)

tbid = coco_get_id(oid, 'coll');
str  = coco_stream(varargin{:});
run  = str.get;
if ischar(str.peek)
  soid = str.get;
else
  soid = oid;
end
lab = str.get;

[sol data] = coll_read_solution(soid, run, lab);
data       = coll_get_settings(prob, tbid, data);
data       = coll_init_data(data, sol.x, sol.p);
sol        = coll_init_sol(data, sol.t, sol.x, sol.p);
prob       = coll_construct_seg(prob, tbid, data, sol);

end
```

The encoding supports continuation from a solution obtained in a previous run with a change in object identifiers and a possible change in the values of N and m. The toolbox extractor `coll_read_solution`, which may be used to extract the chart structure and the associated data structure from a stored solution file, is shown below.

```
function [sol data] = coll_read_solution(oid, run, lab)

tbid        = coco_get_id(oid, 'coll');
[data chart] = coco_read_solution(tbid, run, lab);

sol.t = data.tbp(data.tbp_idx)*chart.x(data.T_idx);
xbp   = reshape(chart.x(data.xbp_idx), data.xbp_shp)';
sol.x = xbp(data.tbp_idx,:);
sol.p = chart.x(data.p_idx);

end
```

Here, the `tbp_idx` field of the `data` structure is used to eliminate the duplicate values of the solution vector at each of the $N-1$ internal boundary points.

7.3 Examples

In the following several examples, we illustrate the calling syntax and the utility of the collocation toolbox. These examples serve to motivate further development of encapsulating toolboxes that rely on the auxiliary collocation toolbox for setting up the part of the zero problem associated with trajectory discretization.

7.3.1 A shooting method for boundary-value problems

Consider again the two-point boundary-value problem

$$f'' = \frac{1+f'^2}{f}, \tag{7.21}$$

where $f(0) = 1$ and $f(1) = Y$. This was obtained in Chap. 1 from the Euler–Lagrange theory applied to the problem of finding the catenary curve between two points in the plane.

   Now suppose that a numerical solution is sought for the case when $Y = 3$. In the absence of an explicit solution to the two-point boundary-value problem, other than for certain discrete values of Y, we proceed to generate the sought solution(s) using a so-called *shooting method* in three steps (cf. Fig. 7.1).

Step 1 Find a solution to the system of first-order ordinary differential equations

$$\frac{dy_1}{dt} = y_2, \tag{7.22}$$

$$\frac{dy_2}{dt} = \frac{1+y_2^2}{y_1}, \tag{7.23}$$

in the unknown functions $y_1(t)$ and $y_2(t)$ with initial conditions

$$y_1(0) = 1, \, y_2(0) = 0$$

for t on the interval $[0, T]$ for some small T.

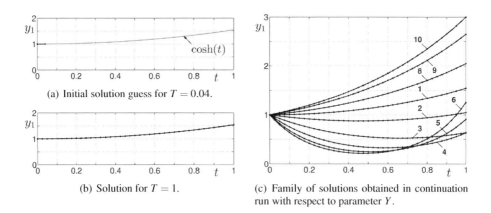

(a) Initial solution guess for $T = 0.04$.

(b) Solution for $T = 1$.

(c) Family of solutions obtained in continuation run with respect to parameter Y.

Figure 7.1. *A technique for constructing initial solutions to nonlinear boundary-value problems is continuation in the interval length T. Here, one starts with a short trajectory segment for small $T \ll 1$ and continues in T until reaching the desired value. This process is called* growing an initial orbit *and is illustrated in Sect. 7.3.1 in the context of the problem from calculus of variations, introduced in Chap. 1. Panel* (a) *shows an initial guess obtained with an Euler step at $t = 0$ with step size $h = 0.04$ and compares this approximation with the exact solution. The grown solution is shown in panel* (b). *From this solution one can, finally, start a continuation in Y. The labeled solution curves in* (c) *correspond to the labels in the session output included in the text.*

Step 2 Continue this solution in T until $T = 1$.

Step 3 Continue this solution in $y_1(T)$ for fixed T but without the constraint on $y_2(0)$ until $y_1(T) = 3$.

The method obtains its name from Step 2, in which a short solution segment is extended until a terminating condition is satisfied and additional conditions are imposed.

The vector field f is encoded in the function below in a `'coll'`-compatible syntax.

```
function y = catenary(x, p)

x1 = x(1,:);
x2 = x(2,:);

y(1,:) = x2;
y(2,:) = (1+x2.^2)./x1;

end
```

For $t \approx 0$, we may approximate the solution to the initial-value problem by the explicit Euler approximation

$$y_1(t) \approx 1, \, y_2(t) \approx t, \tag{7.24}$$

as suggested in Fig. 7.1(a). To provide this starting guess to the collocation constructor, we store a two-point representation on the interval $[0, 0.04]$ in the `t0` and `x0` variables as per the commands

```
>> t0 = [0; 0.04];
>> x0 = [1 0; 1 0.04];
```

where each row of x0 corresponds to the state at the corresponding value of t0.

We proceed to invoke `coll_isol2seg` to construct the collocation zero problem associated with the system of first-order ordinary differential equations on the interval [0, 0.04].

```
>> prob = coll_isol2seg(coco_prob(), '', @catenary, t0, x0, []);
```

Here, the empty brackets in the last argument denote the absence of a problem parameter. The nominal dimensional deficit of this zero problem is 3. For later reference, we extract the function data structure for the collocation zero problem using the `coco_get_func_data` utility.

```
>> data = coco_get_func_data(prob, 'coll', 'data');
```

In order to impose the initial condition on y_1 and to allow for controlled variations in the values of T, $y_2(0)$, and $y_1(T)$, we invoke the `coco_add_pars` special-purpose wrapper as per the command

```
>> prob = coco_add_pars(prob, 'pars', ...
       [data.x0_idx; data.x1_idx(1); data.T_idx], ...
       {'y1s' 'y2s' 'y1e' 'T'});
```

Here, the integers contained in `data.x0_idx`, `data.x1_idx(1)`, and `data.T_idx` are the indices of the elements of the vector of continuation variables associated with $y_1(0)$, $y_2(0)$, $y_1(T)$, and T, respectively, and `'y1s'`, `'y2s'`, `'y1e'`, and `'T'` label the corresponding components of μ. It follows that $|\mathbb{I}| = 4$ and, consequently, that two constraints must be released in order to make the corresponding restricted continuation problem initializable with a 1-dimensional atlas algorithm. As seen in the extract below, this is achieved by allowing for variations in $y_1(T)$ and T while keeping $y_2(0)$ constant and equal to its value on the initial guess.

```
>> coco(prob, 'run1', [], 1, {'T' 'y1e'}, [0 1]);
```

STEP	DAMPING		NORMS			COMPUTATION	TIMES	
IT SIT	GAMMA	||d||	||f||		||U||	F(x)	DF(x)	SOLVE
0			2.92e-04	7.14e+00		0.0	0.0	0.0
1 1	1.00e+00	2.66e-03	2.71e-09	7.15e+00		0.0	0.0	0.0
2 1	1.00e+00	1.10e-08	2.25e-15	7.15e+00		0.0	0.0	0.0

STEP	TIME	||U||	LABEL	TYPE	T	y1e
0	00:00:00	7.1455e+00	1	EP	4.0000e-02	1.0008e+00
2	00:00:00	7.1414e+00	2	EP	0.0000e+00	1.0000e+00

STEP	TIME	||U||	LABEL	TYPE	T	y1e
0	00:00:00	7.1455e+00	3	EP	4.0000e-02	1.0008e+00
10	00:00:00	9.1121e+00	4		8.7192e-01	1.4048e+00
12	00:00:01	9.7534e+00	5	EP	1.0000e+00	1.5431e+00

We illustrate the result of the execution of the `coco` entry-point function by extracting the final discretized trajectory for $T = 1$ from the corresponding `sol#.mat` file and graphing $y_1(t)$ using the following sequence of commands.

```
>> sol = coll_read_solution('', 'run1', 5);
>> plot(sol.t, sol.x(:,1), 'r')
```

The graph of the piecewise-polynomial approximant is shown in Fig. 7.1(b).

Finally, in order to solve for the initial condition of y_2 that will ensure the satisfaction of the right-hand boundary condition on y_1, we restart continuation from the final discretized trajectory for $T = 1$, this time releasing the constraints associated with $y_1(T)$ and $y_2(0)$ while keeping T constant. As seen in the extract below, with the default number of continuation steps, continuation locates one solution trajectory for which $T = 1$ and $y_1(T) = 3$.

```
>> prob = coll_sol2seg(coco_prob(), '', 'run1', 5);
>> data = coco_get_func_data(prob, 'coll', 'data');
>> prob = coco_add_pars(prob, 'pars', ...
     [data.x0_idx; data.x1_idx(1); data.T_idx], ...
     {'y1s' 'y2s' 'y1e' 'T'});
>> coco(prob, 'run2', [], 1, {'y1e' 'y2s'}, [0 3]);
```

STEP	DAMPING		NORMS			COMPUTATION	TIMES	
IT SIT	GAMMA	||d||	||f||	||U||	F(x)	DF(x)	SOLVE	
0			3.77e-15	9.70e+00	0.0	0.0	0.0	
1 1	1.00e+00	1.03e-14	3.42e-15	9.70e+00	0.0	0.0	0.0	

STEP	TIME	||U||	LABEL	TYPE	y1e	y2s
0	00:00:00	9.7020e+00	1	EP	1.5431e+00	-1.6150e-15
10	00:00:00	7.1558e+00	2		1.0456e+00	-5.5308e-01
20	00:00:01	7.2068e+00	3		6.3732e-01	-1.6247e+00
30	00:00:01	9.8530e+00	4		6.3093e-01	-2.8907e+00
40	00:00:02	1.3639e+01	5		9.0829e-01	-3.9308e+00
50	00:00:02	1.7716e+01	6	EP	1.2529e+00	-4.5709e+00

STEP	TIME	||U||	LABEL	TYPE	y1e	y2s
0	00:00:02	9.7020e+00	7	EP	1.5431e+00	-1.6150e-15
10	00:00:03	1.3122e+01	8		2.0471e+00	3.5727e-01
20	00:00:03	1.7618e+01	9		2.6443e+00	6.6301e-01
27	00:00:04	2.0430e+01	10	EP	3.0000e+00	8.0936e-01

The graphs of the piecewise-polynomial approximants corresponding to the labeled solution points are shown in Fig. 7.1(c).

7.3.2 A connecting orbit

Consider the dynamical system

$$\frac{dy}{dt} = f(y, p) \tag{7.25}$$

given by the vector field $f : \mathbb{R}^2 \times \mathbb{R}^2 \to \mathbb{R}^2$, where

$$f(y, p) = \begin{pmatrix} y_2 \\ p_2 y_2 - y_1(1 - y_1)(y_1 - p_1) \end{pmatrix} \tag{7.26}$$

for $0 < p_1 < 1$ and $p_2 \geq 0$. Solutions to the differential equation in Eq. (7.25) describe the shape of traveling wave fronts in the so-called *Huxley equation* used to describe nerve cell dynamics.

This system possesses three equilibria at $y^* = (0,0)$, $y^{**} = (p_1, 0)$, and $y^{***} = (1,0)$. Of these, y^* and y^{***} are saddles with unstable and stable eigendirections given by

$$v_u^* = \begin{pmatrix} \sqrt{4p_1 + p_2^2} - p_2 \\ 2p_1 \end{pmatrix}, \tag{7.27}$$

$$v_s^* = \begin{pmatrix} -\sqrt{4p_1 + p_2^2} - p_2 \\ 2p_1 \end{pmatrix}, \tag{7.28}$$

and

$$v_u^{***} = \begin{pmatrix} \sqrt{4(1-p_1) + p_2^2} - p_2 \\ 2(1-p_1) \end{pmatrix}, \tag{7.29}$$

$$v_s^{***} = \begin{pmatrix} -\sqrt{4(1-p_1) + p_2^2} - p_2 \\ 2(1-p_1) \end{pmatrix}, \tag{7.30}$$

respectively. The equilibrium at y^{**} is a center for $p_2 = 0$, an unstable focus provided that $0 < p_2 < 2\sqrt{p_1(1-p_1)}$, an unstable node for $p_2 = 2\sqrt{p_1(1-p_1)}$, and a saddle for $p_2 > 2\sqrt{p_1(1-p_1)}$. As an example, the direction field for $p_1 = 1/2$ and $p_2 = 0$ is illustrated in Fig. 7.2.

As long as

$$p_2 = \frac{1 - 2p_1}{\sqrt{2}}, \tag{7.31}$$

the system possesses a heteroclinic connecting orbit from y^* to y^{***} (cf. Fig. 7.2) for which

$$y_1(t) = \frac{1}{1 + e^{-t/\sqrt{2}}}. \tag{7.32}$$

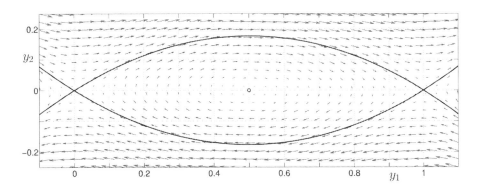

Figure 7.2. *Skeleton of the dynamics of the Huxley dynamical system given by the vector field in Eq. (7.26) and its direction field for $p_1 = 1/2$ and $p_2 = 0$. The system has three equilibria: a center at $(1/2, 0)$ and two saddles that are connected in a heteroclinic cycle. We demonstrate a continuation of the upper heteroclinic orbit in Sect. 7.3.2.*

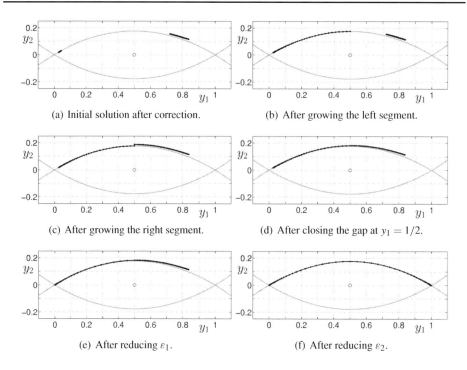

Figure 7.3. *In contrast to the example in Sect. 7.3.1, the construction of an initial solution for the continuation of a heteroclinic orbit in the Huxley dynamical system given by the vector field in Eq. (7.26) consists of a sequence of steps. We start with two short orbit segments close to each saddle, initial guesses to which can be obtained by Euler steps in the respective invariant eigenspace. The two segments after initial correction are shown in panel (a). In the subsequent step we grow both segments until they terminate on the hyperplane $y_1 = 0.5$, (b) and (c). At this point, the two end points are separated by a small gap, the so-called* Lin gap, *which we close in the next step, (d). In the last step, (e) and (f), we grow the two segments toward the saddle equilibria and obtain an approximation to a connecting orbit, which can be used as a starting point for subsequent continuation runs.*

This orbit departs y^* along the unstable eigendirection v_u^* and approaches y^{***} along the stable eigendirection v_s^{***}. In the discussion below, we seek to locate and subsequently continue a discretized approximant of this heteroclinic orbit under variations in p_1.

To this end, we consider two simultaneous collocation zero problems associated with the vector field $f(y, p)$. We proceed to generate an initial discretized approximant of the heteroclinic orbit for $p_1 = 0.5$ and $p_2 = 0$ in four steps (see Fig. 7.3):

Step 1 Find solutions $y^{(1)}(t)$ and $y^{(2)}(t)$ to the system of first-order ordinary differential equations

$$\frac{dy^{(1)}}{dt} = f\left(y^{(1)}, p\right), \tag{7.33}$$

$$\frac{dy^{(2)}}{dt} = f\left(y^{(2)}, p\right),\tag{7.34}$$

with boundary conditions

$$y^{(1)}(0) = y^* + \varepsilon_1 \frac{v_u^*}{\|v_u^*\|}\tag{7.35}$$

and

$$y^{(2)}(T_2) = y^{***} + \varepsilon_2 \frac{v_s^{***}}{\|v_s^{***}\|},\tag{7.36}$$

respectively, for some small parameter values ε_1 and ε_2 over some intervals $[0, T_1]$ and $[0, T_2]$, respectively.

Step 2 Continue this solution in $y_1^{(1)}(T_1)$ and then in $y_1^{(2)}(0)$ for fixed ε_1, ε_2, p_1, and p_2 until $y_1^{(1)}(T_1) = y_1^{(2)}(0) = 0.5$.

Step 3 Continue this solution in $y_2^{(1)}(T_1) - y_2^{(2)}(0)$ for fixed ε_1, ε_2, p_1, $y_1^{(1)}(T_1)$, and $y_1^{(2)}(0)$ until $y_2^{(1)}(T_1) = y_2^{(2)}(0)$.

Step 4 Continue this solution in ε_1 and then in ε_2 for fixed p_1, $y_2^{(1)}(T_1) - y_2^{(2)}(0)$, $y_1^{(1)}(T_1)$, and $y_1^{(2)}(0)$ until $\varepsilon_1 = \varepsilon_2 = 10^{-3}$.

Here, "shooting" is used in Step 2 in order to extend the initial solution segments until the terminal conditions are reached.

We approximate the solutions to the initial-value problems in Step 1 with $T_1 = T_2 = 1$ by a single step of the forward Euler method:

$$y^{(1)}(T_1) \approx y^* + \varepsilon_1 \frac{v_u^*}{\|v_u^*\|} + f\left(y^* + \varepsilon_1 \frac{v_u^*}{\|v_u^*\|}\right),\tag{7.37}$$

$$y^{(2)}(0) \approx y^{***} + \varepsilon_2 \frac{v_s^{***}}{\|v_s^{***}\|} - f\left(y^{***} + \varepsilon_2 \frac{v_s^{***}}{\|v_s^{***}\|}\right).\tag{7.38}$$

To provide this starting guess to the collocation constructor, we store a two-point representation for each trajectory segment on the interval $[0, 1]$ in the segs structure, as per the commands

```
>> p0    = [0.5; 0];
>> eps0  = [0.03; 0.2];
>> vu    = [sqrt(4*p0(1)+p0(2)^2)-p0(2); 2*p0(1)];
>> vu    = vu/norm(vu, 2);
>> segs(1).t0 = [0; 1];
>> x0         = eps0(1)*vu;
>> segs(1).x0 = [x0   x0+huxley(x0, p0)]';
>> segs(1).p0 = p0;
>> vs    = [-sqrt(4*(1-p0(1))+p0(2)^2)-p0(2); 2*(1-p0(1))];
>> vs    = vs/norm(vs, 2);
>> segs(2).t0 = [0; 1];
>> x0         = [1; 0]+eps0(2)*vs;
>> segs(2).x0 = [x0-huxley(x0, p0)   x0]';
>> segs(2).p0 = p0;
```

where each row of `seg.x0` corresponds to the state at the corresponding value of `seg.t0` and we have assigned arbitrarily the initial values $\varepsilon_1 = 0.03$ and $\varepsilon_2 = 0.2$.

We turn to the construction of an extended continuation problem corresponding to the algorithm described above. As in the example in the previous section, each of the steps described above may be achieved by applying a 1-dimensional atlas algorithm to a suitable restriction of this continuation problem. We shall find it convenient to overconstrain the continuation problem during the construction stage and subsequently release an appropriate number of constraints on the continuation parameters. In all but the first step, construction relies on a previously computed solution, rather than the initial guess shown above.

Consider the vectorized encoding below of the vector field f.

```
function y = huxley(x, p)

x1 = x(1,:);
x2 = x(2,:);
p1 = p(1,:);
p2 = p(2,:);

y(1,:) = x2;
y(2,:) = p2.*x2-x1.*(1-x1).*(x1-p1);

end
```

The following constructor then invokes `coll_isol2seg` twice to construct the two collocation zero problems associated with the two systems of first-order ordinary differential equations in Eqs. (7.33)–(7.34), given the initial solution guesses contained in the input argument `segs`.

```
function prob = huxley_isol2het(prob, segs, eps0)

prob = coll_isol2seg(prob, 'huxley1', @huxley, ...
  segs(1).t0, segs(1).x0, segs(2).p0);
prob = coll_isol2seg(prob, 'huxley2', @huxley, ...
  segs(2).t0, segs(2).x0, segs(2).p0);

prob = huxley_close_het(prob, eps0);

end
```

At this stage of construction, the nominal dimensional deficit of the zero problem is 10, as the problem parameters have been introduced twice into the continuation problem.

We collect the remainder of the construction of the overconstrained restricted continuation problem in the `huxley_close_het` closer function shown below.

```
function prob = huxley_close_het(prob, epsv)

[data1 uidx1] = coco_get_func_data(prob, 'huxley1.coll', ...
  'data', 'uidx');
[data2 uidx2] = coco_get_func_data(prob, 'huxley2.coll', ...
  'data', 'uidx');

prob = coco_add_glue(prob, 'shared', uidx1(data1.p_idx), ...
  uidx2(data2.p_idx));

prob = coco_add_func(prob, 'bcs', @huxley_bcs, [], 'zero', 'uidx', ...
  [uidx1(data1.x0_idx); uidx2(data2.x1_idx); uidx1(data1.p_idx)], ...
```

```
    'u0', epsv);
uidx = coco_get_func_data(prob, 'bcs', 'uidx');
data.eps_idx = [numel(uidx)-1; numel(uidx)];
prob = coco_add_slot(prob, 'bcs', @coco_save_data, data, 'save_full');

prob = coco_add_glue(prob, 'gap', uidx1(data1.x1_idx(2)), ...
    uidx2(data2.x0_idx(2)), 'gap', 'inactive');

prob = coco_add_pars(prob, 'pars', ...
    [uidx1(data1.p_idx); uidx(data.eps_idx); ...
    uidx1(data1.x1_idx(1)); uidx2(data2.x0_idx(1))], ...
    {'p1' 'p2' 'eps1' 'eps2' 'y11e' 'y21e'});

end
```

Here, in order to compensate for parameter duplication, we first extract the function data structure for each of the two collocation zero problems. Following the syntax introduced in the context of the 'compalg' toolbox, the command

```
prob = coco_add_glue(prob, 'shared', uidx1(data1.p_idx), ...
    uidx2(data2.p_idx));
```

then appends the appropriate gluing conditions to the zero problem. As this adds two more equations to the extended continuation problem, the nominal dimensional deficit is now reduced to 8.

In order to impose boundary conditions on the initial points along the trajectory segments, the command

```
prob = coco_add_func(prob, 'bcs', @huxley_bcs, [], 'zero', 'uidx', ...
    [uidx1(data1.x0_idx); uidx2(data2.x1_idx); uidx1(data1.p_idx)], ...
    'u0', epsv);
```

appends the zero function encoded below to the extended continuation problem.

```
function [data y] = huxley_bcs(prob, data, u)

x10 = u(1:2);
x20 = u(3:4);
par = u(5:6);
eps = u(7:8);

vu = [sqrt(4*par(1)+par(2)^2)-par(2); 2*par(1)];
vu = vu/norm(vu, 2);
vs = [-sqrt(4*(1-par(1))+par(2)^2)-par(2); 2*(1-par(1))];
vs = vs/norm(vs, 2);

y = [x10-eps(1)*vu; x20-([1; 0]+eps(2)*vs)];

end
```

Specifically, the 'uidx' optional argument is here used to identify the index set $\mathbb{K}^o$ corresponding to the function encoded in huxley_bcs. In addition, the variables ε_1 and ε_2 are here added to the zero problem, and their initial values are given following the 'u0' optional input argument.

The coco_get_func_data utility function is used by huxley_close_het in the command

```
uidx = coco_get_func_data(prob, 'bcs', 'uidx');
```

to assign the index set $\mathbb{K}$ corresponding to the function `huxley_bcs` to the variable `uidx`. With the addition of four boundary conditions and the two additional variables ε_1 and ε_2, the nominal dimensional deficit of the zero problem is now 6.

To accommodate the continuation task in Step 3, the assignment

```
prob = coco_add_glue(prob, 'gap', uidx1(data1.x1_idx(2)), ...
    uidx2(data2.x0_idx(2)), 'gap', 'inactive');
```

in `huxley_close_het` appends the difference $y_2^{(1)}(T_1) - y_2^{(2)}(0)$ to the family of monitor functions, assigns the corresponding index to $\mathbb{I}$, and labels the corresponding component of μ by `'gap'`. Finally, in the call

```
prob = coco_add_pars(prob, 'pars', ...
    [uidx1(data1.p_idx); uidx(data.eps_idx); ...
    uidx1(data1.x1_idx(1)); uidx2(data2.x0_idx(1))], ...
    {'p1' 'p2' 'eps1' 'eps2' 'y11e' 'y21e'});
```

the integer index sets in the third argument identify the elements of the vector of continuation variables associated with p_1, p_2, ε_1, ε_2, $y_1^{(1)}(T_1)$, and $y_1^{(2)}(0)$, respectively. The command assigns the parameter labels `'p1'`, `'p2'`, `'eps1'`, `'eps2'`, `'y11e'`, and `'y21e'` to the corresponding components of μ and assigns the corresponding integer indices to $\mathbb{I}$. At this stage of construction, $|\mathbb{I}| = 7$. It follows that two constraints must be released in order to make the corresponding restricted continuation problem initializable with a 1-dimensional atlas algorithm.

As an example that accomplishes Step 1 and the first half of Step 2 in the algorithm above, consider the following extract. Here, we allow for variations in $y_1^{(1)}(T_1)$ and $y_2^{(1)}(T_1) - y_2^{(2)}(0)$ and restrict $y_1^{(1)}(T_1)$ to the interval $[0, 0.5]$.

```
>> prob = huxley_isol2het(coco_prob(), segs, eps0);
>> coco(prob, 'run1', [], 1, {'y11e', 'gap'}, [0 0.5]);
```

	STEP	DAMPING		NORMS		COMPUTATION	TIMES	
IT	SIT	GAMMA	‖d‖	‖f‖	‖U‖	F(x)	DF(x)	SOLVE
0				8.76e-03	5.83e+00	0.0	0.0	0.0
1	1	1.00e+00	2.71e-01	1.54e-04	5.79e+00	0.0	0.0	0.0
2	1	1.00e+00	5.59e-03	6.23e-08	5.79e+00	0.0	0.0	0.0
3	1	1.00e+00	1.14e-06	9.85e-15	5.79e+00	0.0	0.0	0.0
4	1	1.00e+00	1.73e-13	1.70e-15	5.79e+00	0.0	0.0	0.0

STEP	TIME	‖U‖	LABEL	TYPE	y11e	gap
0	00:00:00	5.7880e+00	1	EP	4.1815e-02	-1.2701e-01
10	00:00:00	6.5448e+00	2	EP	0.0000e+00	-1.5179e-01

STEP	TIME	‖U‖	LABEL	TYPE	y11e	gap
0	00:00:01	5.7880e+00	3	EP	4.1815e-02	-1.2701e-01
10	00:00:01	7.5532e+00	4		4.0943e-01	1.5420e-02
12	00:00:02	7.9504e+00	5	EP	5.0000e-01	2.1219e-02

The two trajectory segments corresponding to the solution with label 1 are shown in Fig. 7.3(a). Similarly, the two trajectory segments in Fig. 7.3(b) correspond to the solution with label 5. In the subsequent steps of the algorithm, we repeat the above construction, but this time with an initial solution guess given by a solution found in the previous continuation run. In order to use the `huxley_close_het` closer, it is necessary to provide numerical values for ε_1 and ε_2. As these are not part of the collocation zero problem, they must be extracted directly from the solution stored in the corresponding `sol#.mat` file.

To enable the identification of the corresponding elements of the `chart.x` vector, the command

```
data.eps_idx = [numel(uidx)-1; numel(uidx)];
```

in `huxley_close_het` stores the elements of the index set $\mathbb{K}$ of the encoding in `huxley_bcs` corresponding to ε_1 and ε_2 in the `eps_idx` field of the `data` structure. The command

```
prob = coco_add_slot(prob, 'bcs', @coco_save_data, data, 'save_full');
```

appends the `coco_save_data` slot function to the continuation problem structure in order to store the content of this data structure in the solution data array associated with each solution file. The `huxley_sol2het` constructor shown below loads the corresponding data from the solution file associated with the label `lab` and the run identifier `run`.

```
function prob = huxley_sol2het(prob, run, lab)

prob = coll_sol2seg(prob, 'huxley1', run, lab);
prob = coll_sol2seg(prob, 'huxley2', run, lab);

[data chart] = coco_read_solution('bcs', run, lab);
epsv = chart.x(data.eps_idx);

prob = huxley_close_het(prob, epsv);

end
```

Repeated application of `huxley_sol2het` now serves to complete the remainder of the algorithm. In each case, the interval of continuation for the first element in the list of active continuation parameters has been chosen to include the value at the starting solution and to be limited at one end by the desired value of this parameter. The extract below shows the completion of the four steps of the algorithm described above.

```
>> prob = huxley_sol2het(coco_prob(), 'run1', 5);
>> coco(prob, 'run2', [], 1, {'y21e', 'gap'}, [0.5 1]);

    STEP   DAMPING              NORMS                COMPUTATION TIMES
    IT SIT     GAMMA     ||d||     ||f||     ||U||    F(x)   DF(x)   SOLVE
     0                          2.12e-15  7.97e+00    0.0    0.0     0.0
     1   1  1.00e+00  7.28e-15  1.90e-15  7.97e+00    0.0    0.0     0.0

  STEP      TIME       ||U||  LABEL  TYPE          y21e           gap
     0   00:00:00  7.9674e+00      1  EP      7.2123e-01    2.1219e-02
     5   00:00:00  7.7724e+00      2  EP      5.0000e-01   -1.0930e-02

  STEP      TIME       ||U||  LABEL  TYPE          y21e           gap
     0   00:00:00  7.9674e+00      3  EP      7.2123e-01    2.1219e-02
     8   00:00:01  8.8969e+00      4  EP      1.0000e+00    1.1358e-01
>> prob = huxley_sol2het(coco_prob(), 'run2', 2);
>> coco(prob, 'run3', [], 1, {'gap', 'p2'}, [-0.2 0]);
```

```
       STEP   DAMPING              NORMS              COMPUTATION TIMES
    IT SIT      GAMMA     ||d||     ||f||     ||U||    F(x)   DF(x)  SOLVE
    0                              1.63e-15  7.76e+00  0.0    0.0    0.0
    1    2  5.00e-01  2.87e-15  1.42e-15  7.76e+00  0.0    0.0    0.0

    STEP     TIME       ||U||  LABEL TYPE           gap          p2
     0   00:00:00  7.7563e+00     1   EP    -1.0930e-02   1.1879e-16
     6   00:00:00  9.0584e+00     2   EP    -2.0000e-01  -2.8752e-01

    STEP     TIME       ||U||  LABEL TYPE           gap          p2
     0   00:00:00  7.7563e+00     3   EP    -1.0930e-02   1.1879e-16
     1   00:00:00  7.7070e+00     4   EP     0.0000e+00   1.6332e-02
>> prob = huxley_sol2het(coco_prob(), 'run3', 4);
>> coco(prob, 'run4', [], 1, {'eps1', 'p2'}, [1e-3 eps0(1)]);

       STEP   DAMPING              NORMS              COMPUTATION TIMES
    IT SIT      GAMMA     ||d||     ||f||     ||U||    F(x)   DF(x)  SOLVE
    0                              1.88e-15  7.71e+00  0.0    0.0    0.0
    1    1  1.00e+00  2.30e-15  1.69e-15  7.71e+00  0.0    0.0    0.0

    STEP     TIME       ||U||  LABEL TYPE          eps1          p2
     0   00:00:00  7.7071e+00     1   EP     3.0000e-02   1.6332e-02
    10   00:00:00  1.0797e+01     2          1.6663e-03   1.6397e-02
    12   00:00:01  1.1407e+01     3   EP     1.0000e-03   1.6397e-02
>> prob = huxley_sol2het(coco_prob(), 'run4', 3);
>> coco(prob, 'run5', [], 1, {'eps2', 'p2'}, [1e-3 eps0(2)]);

       STEP   DAMPING              NORMS              COMPUTATION TIMES
    IT SIT      GAMMA     ||d||     ||f||     ||U||    F(x)   DF(x)  SOLVE
    0                              2.17e-15  1.14e+01  0.0    0.0    0.0
    1    1  1.00e+00  8.11e-15  1.70e-15  1.14e+01  0.0    0.0    0.0

    STEP     TIME       ||U||  LABEL TYPE          eps2          p2
     0   00:00:00  1.1408e+01     1   EP     2.0000e-01   1.6397e-02
    10   00:00:00  1.3282e+01     2          1.6885e-02   1.1037e-05
    19   00:00:01  1.5692e+01     3   EP     1.0000e-03   3.2544e-16
```

Here, the solution with label 2 from the run with run identifier 'run2' is represented by the trajectory segments in Fig. 7.3(c). The solution with label 4 from the run with run identifier 'run3' is represented by the trajectory segments in Fig. 7.3(d), where we note the absence of a gap at $y_1 = 1/2$. Finally, the solutions with label 3 in each of the runs with run identifiers 'run4' and 'run5', respectively, are represented by the trajectory segments in Figs. 7.3(e) and 7.3(f). As an example, we may illustrate the result of the last execution of the coco entry-point function by extracting the approximate heteroclinic trajectory from the corresponding sol#.mat file, using the following sequence of commands.

```
>> hold on
>> sol = coll_read_solution('huxley1', 'run5', 3);
>> plot(sol.x(:,1), sol.x(:,2), 'r')
>> sol = coll_read_solution('huxley2', 'run5', 3);
>> plot(sol.x(:,1), sol.x(:,2), 'r')
>> hold off
```

Continuation of this heteroclinic approximant under variations in p_1 and p_2 is now achieved by the sequence of commands shown below.

```
>> prob = huxley_sol2het(coco_prob(), 'run5', 3);
>> coco(prob, 'run6', [], 1, {'p1', 'p2'}, [0.25 0.75]);
```

STEP	DAMPING		NORMS			COMPUTATION TIMES		
IT SIT	GAMMA	$\|\|d\|\|$	$\|\|f\|\|$	$\|\|U\|\|$		F(x)	DF(x)	SOLVE
0			2.21e-15	1.57e+01		0.0	0.0	0.0
1 1	1.00e+00	2.10e-13	2.14e-15	1.57e+01		0.0	0.0	0.0

STEP	TIME	$\|\|U\|\|$	LABEL	TYPE	p1	p2
0	00:00:00	1.5700e+01	1	EP	5.0000e-01	3.8298e-16
4	00:00:00	1.5694e+01	2	EP	2.5000e-01	3.5355e-01

STEP	TIME	$\|\|U\|\|$	LABEL	TYPE	p1	p2
0	00:00:00	1.5700e+01	3	EP	5.0000e-01	3.8298e-16
4	00:00:00	1.5741e+01	4	EP	7.5000e-01	-3.5355e-01

7.3.3 Coupling `'coll'` and `'alg'`

As a final example, consider the dynamical system given by the vector field $f : \mathbb{R}^2 \times \mathbb{R}^2 \to \mathbb{R}^2$, where

$$f(y, p) = \begin{pmatrix} 1 - y_1^2 \\ p_1 y_1 + p_2 y_2 \end{pmatrix}, \tag{7.39}$$

with $p_2 > 0$. This system has two equilibria at $y^* = (-1, p_1/p_2)$ and $y^{**} = (1, -p_1/p_2)$. Of these, y^* is an unstable node with eigendirections given by

$$v_{u,1}^* = \begin{pmatrix} 0 \\ 1 \end{pmatrix}, v_{u,2}^* = \begin{pmatrix} 2 - p_2 \\ p_1 \end{pmatrix}, \tag{7.40}$$

and y^{**} is a saddle with stable eigendirection given by

$$v_s^{**} = \begin{pmatrix} -2 - p_2 \\ p_1 \end{pmatrix}. \tag{7.41}$$

For arbitrary values of p_1 and p_2 there exists a heteroclinic orbit connecting the two equilibria. For example, for $p_1 = p_2 = 1$ (see Fig. 7.4(a)), the heteroclinic orbit connecting the two equilibria is given by

$$y_1(t) = \tanh t, \, y_2(t) = 1 - 2e^t \cot^{-1} e^t, \tag{7.42}$$

from which it follows that

$$\lim_{t \to -\infty} \frac{y_2^* - y_2(t)}{y_1^* - y_1(t)} = -\infty; \tag{7.43}$$

i.e., the heteroclinic orbit departs from y^* along the $\begin{pmatrix} 0 & -1 \end{pmatrix}^T$ direction. In the discussion below, we seek to locate and subsequently continue a discretized approximant of this heteroclinic orbit under variations in the problem parameters. In contrast to the previous example, we include in the determination of the heteroclinic orbit the numerical task of locating the equilibria y^* and y^{**} as well as the stable eigendirection at y^{**}.

To this end, we again consider two simultaneous collocation zero problems associated with the vector field $f(y, p)$. We proceed to generate an initial discretized approximant of the heteroclinic orbit for $p_1 = p_2 = 1$ in the following four steps (see panels (b)–(e) in Fig. 7.4):

Step 1 Find simultaneous solutions to the algebraic equations

$$f\left(y^*, p\right) = 0,$$
$$f\left(y^{**}, p\right) = 0,$$
$$\partial_y f\left(y^{**}, p\right) \cdot v_s^{**} = \lambda_s^{**} v_s^{**},$$
$$v_s^{**T} \cdot v_s^{**} = 1$$

and to the system of first-order ordinary differential equations

$$\frac{dy^{(1)}}{dt} = f\left(y^{(1)}, p\right), \tag{7.44}$$

$$\frac{dy^{(2)}}{dt} = f\left(y^{(2)}, p\right) \tag{7.45}$$

in the unknown functions $y^{(1)}(t)$ and $y^{(2)}(t)$ with initial conditions

$$y^{(1)}(0) = y^* + \varepsilon_1 \begin{pmatrix} \cos\theta \\ \sin\theta \end{pmatrix} \tag{7.46}$$

and

$$y^{(2)}(T_2) = y^{**} + \varepsilon_2 \frac{v_s^{**}}{\left\| v_s^{**} \right\|} \tag{7.47}$$

for some unknown angle θ and some small parameter values ε_1 and ε_2 over some intervals $[0, T_1]$ and $[0, T_2]$, respectively.

Step 2 Continue this solution in $y_2^{(1)}(T_1)$ and then in $y_2^{(2)}(0)$ for fixed ε_1, ε_2, p_1, p_2, and $y_1^{(1)}(T_1) - y_1^{(2)}(0)$ until $y_2^{(1)}(T_1) = y_2^{(2)}(0) = 0$.

Step 3 Continue this solution in $y_1^{(1)}(T_1) - y_1^{(2)}(0)$ for fixed ε_1, ε_2, p_1, p_2, $y_2^{(1)}(T_1)$, and $y_2^{(2)}(0)$ until $y_1^{(1)}(T_1) = y_1^{(2)}(0)$.

Step 4 Continue this solution in ε_1 and then in ε_2 for fixed p_1, p_2, $y_2^{(1)}(T_1)$, $y_2^{(2)}(0)$, and $y_1^{(1)}(T_1) - y_1^{(2)}(0)$ until $\varepsilon_1 = \varepsilon_2 = 10^{-3}$.

We may again approximate the solution to the initial-value problems in Step 1 with $T_1 = T_2 = 1$ by a single step of the forward Euler method:

$$y^{(1)}(T_1) \approx y^* + \varepsilon_1 \begin{pmatrix} \cos\theta \\ \sin\theta \end{pmatrix} + f\left(y^* + \varepsilon_1 \begin{pmatrix} \cos\theta \\ \sin\theta \end{pmatrix}\right), \tag{7.48}$$

$$y^{(2)}(0) \approx y^{**} + \varepsilon_2 \frac{v_s^{**}}{\left\| v_s^{**} \right\|} - f\left(y^{**} + \varepsilon_2 \frac{v_s^{**}}{\left\| v_s^{**} \right\|}\right). \tag{7.49}$$

To provide this starting guess to the collocation constructor, we store a two-point representation for each trajectory segment on the interval $[0, 1]$ in the `segs` structure using the following commands.

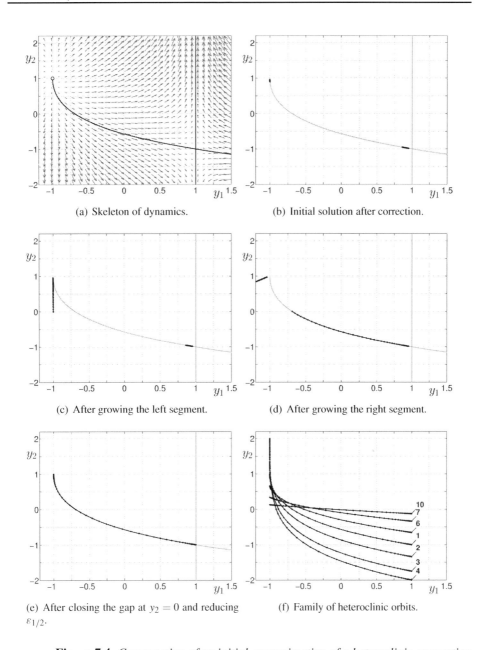

(a) Skeleton of dynamics.

(b) Initial solution after correction.

(c) After growing the left segment.

(d) After growing the right segment.

(e) After closing the gap at $y_2 = 0$ and reducing $\varepsilon_{1/2}$.

(f) Family of heteroclinic orbits.

Figure 7.4. *Computation of an initial approximation of a heteroclinic connection in the dynamical system with vector field given in Eq. (7.39), following the methodology described in Sect. 7.3.3; cf. Fig. 7.3. Here, we combine two instances of the `'coll'` toolbox with two instances of the `'alg'` toolbox, which are used to solve for the equilibria and an invariant eigenspace. A family of connecting orbits is shown in panel (f). The labels correspond to the session output of the last run included in the text.*

```
>> p0 = [1; 1];
>> eps0 = [0.05; 0.05];
>> th0 = -pi/2;
>> eqs10 = [-1; 1];
>> eqs20 = [1; -1];
>> vec0 = [-3/sqrt(10); 1/sqrt(10)];
>> lam0 = -2;
>> segs(1).t0 = [0; 1];
>> x0          = eqs10+eps0(1)*[cos(th0); sin(th0)];
>> segs(1).x0 = [x0  x0+doedel(x0, p0)]';
>> segs(1).p0 = p0;
>> segs(2).t0 = [0; 1];
>> x0          = eqs20+eps0(2)*vec0;
>> segs(2).x0 = [x0-doedel(x0, p0) x0]';
>> segs(2).p0 = p0;
>> algs(1).x0 = eqs10;
>> algs(1).p0 = p0;
>> algs(2).x0 = eqs20;
>> algs(2).p0 = p0;
```

Each row of seg.x0 corresponds to the state at the corresponding value of seg.t0, and we have again assigned arbitrarily the initial values $\varepsilon_1 = 0.05$ and $\varepsilon_2 = 0.05$. The initial solution guess $\theta = -\pi/2$ follows from Eq. (7.43). Finally, the initial value of λ^{**} has here been stored in the lam0 variable.

We turn to the construction of an extended continuation problem corresponding to the algorithm described above. As in the examples in the previous sections, each of the steps described above may be achieved by applying a 1-dimensional atlas algorithm to a suitable restriction of this continuation problem. We again find it convenient to overconstrain the continuation problem during the construction stage and subsequently to release an appropriate number of constraints on the system variables.

The vector field f and its Jacobian $\partial_y f$ are encoded in the functions shown below.

```
function y = doedel(x, p)

x1 = x(1,:);
x2 = x(2,:);
p1 = p(1,:);
p2 = p(2,:);

y(1,:) = 1-x1.^2;
y(2,:) = p1.*x1+p2.*x2;

end


function J = doedel_DFDX(x, p)

x1 = x(1,:);
p1 = p(1,:);
p2 = p(2,:);

J = zeros(2,2,numel(x1));
J(1,1,:) = -2*x1;
J(2,1,:) = p1;
J(2,2,:) = p2;

end
```

The following constructor invokes the `coll_isol2seg` constructor twice to construct the two collocation zero problems associated with the two systems of first-order ordinary differential equations in Eqs. (7.44)–(7.45) given the initial solution guesses contained in the input argument `segs`.

```
function prob = doedel_isol2het(prob, segs, algs, eps0, th0, vec0, lam0)

prob = coll_isol2seg(prob, 'doedel1', @doedel, @doedel_DFDX, ...
  segs(1).t0, segs(1).x0, segs(1).p0);
prob = coll_isol2seg(prob, 'doedel2', @doedel, @doedel_DFDX, ...
  segs(2).t0, segs(2).x0, segs(2).p0);

prob = alg_isol2eqn(prob, 'doedel3', @doedel, @doedel_DFDX, ...
  algs(1).x0, algs(1).p0);
prob = alg_isol2eqn(prob, 'doedel4', @doedel, @doedel_DFDX, ...
  algs(2).x0, algs(2).p0);

prob = doedel_close_het(prob, eps0, th0, vec0, lam0);

end
```

As the continuation problem includes the determination of the locus of the two equilibria, `doedel_isol2het` also invokes the `alg_isol2eqn` constructor twice. At this stage of construction, the nominal dimensional deficit of the zero problem is 14, as the problem parameters have been introduced four times into the continuation problem.

We collect the remainder of the construction of the overconstrained restricted continuation problem in the `doedel_close_het` closer function shown below.

```
function prob = doedel_close_het(prob, eps, th, vec, lam)

[data1 uidx1] = coco_get_func_data(prob, 'doedel1.coll', ...
  'data', 'uidx');
[data2 uidx2] = coco_get_func_data(prob, 'doedel2.coll', ...
  'data', 'uidx');
[data3 uidx3] = coco_get_func_data(prob, 'doedel3.alg', ...
  'data', 'uidx');
[data4 uidx4] = coco_get_func_data(prob, 'doedel4.alg', ...
  'data', 'uidx');

prob = coco_add_glue(prob, 'shared', ...
  [uidx1(data1.p_idx); uidx1(data1.p_idx); uidx1(data1.p_idx)], ...
  [uidx2(data2.p_idx); uidx3(data3.p_idx); uidx4(data4.p_idx)]);

prob = coco_add_func(prob, 'evs', @doedel_evs, [], 'zero', ...
  'uidx', [uidx4(data4.p_idx); uidx4(data4.x_idx)], 'u0', [vec; lam]);
uidx_evs = coco_get_func_data(prob, 'evs', 'uidx');
data_evs.vec_idx = [numel(uidx_evs)-2; numel(uidx_evs)-1];
data_evs.lam_idx = numel(uidx_evs);
prob = coco_add_slot(prob, 'evs', @coco_save_data, data_evs, ...
  'save_full');

prob = coco_add_func(prob, 'bcs', @doedel_bcs, [], 'zero', 'uidx', ...
    [uidx1(data1.x0_idx); uidx2(data2.x1_idx); ...
    uidx3(data3.x_idx); uidx4(data4.x_idx); ...
    uidx_evs(data_evs.vec_idx)], 'u0', [eps; th]);
uidx_bcs = coco_get_func_data(prob, 'bcs', 'uidx');
data_bcs.eps_idx = [numel(uidx_bcs)-2; numel(uidx_bcs)-1];
```

```
data_bcs.th_idx  = numel(uidx_bcs);
prob = coco_add_slot(prob, 'bcs', @coco_save_data, data_bcs, ...
  'save_full');

prob = coco_add_glue(prob, 'lin', uidx1(data1.x1_idx(1)), ...
  uidx2(data2.x0_idx(1)), 'gap', 'inactive');

prob = coco_add_pars(prob, 'pars', ...
  [uidx1(data1.p_idx); uidx_bcs(data_bcs.eps_idx); ...
  uidx1(data1.x1_idx(2)); uidx2(data2.x0_idx(2))], ...
    {'p1' 'p2' 'eps1' 'eps2' 'y12e' 'y22e'});

end
```

We again compensate for parameter duplication by extracting the function data structure for each of the two collocation zero problems and each of the two algebraic zero problems. The corresponding gluing conditions are added to the continuation problem by the assignment

```
prob = coco_add_glue(prob, 'shared', ...
  [uidx1(data1.p_idx); uidx1(data1.p_idx); uidx1(data1.p_idx)], ...
  [uidx2(data2.p_idx); uidx3(data3.p_idx); uidx4(data4.p_idx)]);
```

As this adds six more equations to the zero problem, the nominal dimensional deficit is now reduced to 8.

In order to impose the condition on the stable eigenvector of the y^{**} equilibrium, the assignment

```
prob = coco_add_func(prob, 'evs', @doedel_evs, [], 'zero', ...
  'uidx', [uidx4(data4.p_idx); uidx4(data4.x_idx)], 'u0', [vec; lam]);
```

appends the zero function encoded below to the extended continuation problem.

```
function [data y] = doedel_evs(prob, data, u)

par = u(1:2);
eqs = u(3:4);
vec = u(5:6);
lam = u(7);

jac = doedel_DFDX(eqs, par);

y = [jac*vec-lam*vec; vec'*vec-1];

end
```

The 'uidx' optional argument is used in the call to coco_add_func to identify the index set $\mathbb{K}^o$ for the doedel_evs function. In addition, the variables v_s^{**} and λ_s^{**} are here added to the zero problem and their initial values are given following the 'u0' optional input argument. The index set $\mathbb{K}$ corresponding to the function doedel_evs is assigned to the variable uidx_evs by the subsequent call to the coco_get_func_data utility function. With the addition of three eigenvector/eigenvalue conditions and the two additional variables v_s^{**} and λ_s^{**}, the nominal dimensional deficit of the zero problem remains equal to 8.

The assignment

```
prob = coco_add_func(prob, 'bcs', @doedel_bcs, [], 'zero', 'uidx', ...
    [uidx1(data1.x0_idx); uidx2(data2.x1_idx); ...
    uidx3(data3.x_idx); uidx4(data4.x_idx); ...
    uidx_evs(data_evs.vec_idx)], 'u0', [eps; th]);
```

in doedel_close_het appends the COCO-compatible function encoded below to the family
of zero functions in order to impose the boundary conditions given in Step 1.

```
function [data y] = doedel_bcs(prob, data, u)

x10 = u(1:2);
x20 = u(3:4);
eqs = u(5:8);
vec = u(9:10);
eps = u(11:12);
th  = u(13);

y = [x10-(eqs(1:2)+eps(1)*[cos(th); sin(th)]);...
    x20-(eqs(3:4)+eps(2)*vec)];

end
```

Again, the 'uidx' optional argument is used in the call to coco_add_func to identify the
index set $\mathbb{K}^o$ corresponding to the doedel_bcs function. In addition, the variables ε_1, ε_2,
and θ are here added to the zero problem and their initial values are given following the
'u0' optional input argument. The index set $\mathbb{K}$, associated with the function doedel_bcs,
is assigned to the variable uidx_bcs by the subsequent call to the coco_get_func_data
utility function. With the addition of four conditions on the initial points of the trajectory
segments and the three additional variables ε_1, ε_2, and θ, the nominal dimensional deficit
of the zero problem is now 7.

To accommodate the continuation task in Step 3, the assignment

```
prob = coco_add_glue(prob, 'lin', uidx1(data1.x1_idx(1)), ...
    uidx2(data2.x0_idx(1)), 'gap', 'inactive');
```

appends the difference $y_1^{(1)}(T_1) - y_1^{(2)}(0)$ to the monitor functions, assigns the corresponding index to $\mathbb{I}$, and assigns the label 'gap' to the corresponding component of μ. Finally,
in the subsequent call to the coco_add_pars function,

```
prob = coco_add_pars(prob, 'pars', ...
    [uidx1(data1.p_idx); uidx_bcs(data_bcs.eps_idx); ...
    uidx1(data1.x1_idx(2)); uidx2(data2.x0_idx(2))], ...
    {'p1' 'p2' 'eps1' 'eps2' 'y12e' 'y22e'});
```

uidx1(data1.p_idx), uidx_bcs(data_bcs.eps_idx), uidx1(data1.x1_idx(2)), and
uidx2(data2.x0_idx(2)) identify the indices of the elements of the vector of continu-
ation variables associated with p_1, p_2, ε_1, ε_2, $y_2^{(1)}(T_1)$, and $y_2^{(2)}(0)$, respectively, and the
parameter labels 'p1', 'p2, 'eps1', 'eps2', 'y12e', and 'y22e' reference the corre-
sponding components of μ. At this stage of construction $|\mathbb{I}| = 7$. It follows that a single
constraint must be released in order to make the corresponding restricted continuation problem initializable with a 1-dimensional atlas algorithm.

As an example corresponding to Step 1 and the first half of Step 2 in the algorithm above, consider the following extract. Here we allow for variations in $y_2^{(1)}(T_1)$ restricted to the interval $[0, 0.99]$.

```
>> prob = coco_prob();
>> prob = doedel_isol2het(prob, segs, algs, eps0, th0, vec0, lam0);
>> coco(prob, 'run1', [], 1, 'y12e', [0 0.99]);
```

STEP		DAMPING		NORMS			COMPUTATION	TIMES	
IT	SIT	GAMMA	||d||	||f||	||U||	F(x)	DF(x)	SOLVE	
0				3.42e-02	1.43e+01	0.0	0.0	0.0	
1	1	1.00e+00	4.99e-01	2.25e-03	1.43e+01	0.0	0.0	0.0	
2	1	1.00e+00	4.18e-02	9.14e-06	1.42e+01	0.0	0.0	0.0	
3	1	1.00e+00	1.07e-04	2.10e-10	1.42e+01	0.0	0.0	0.0	
4	1	1.00e+00	4.91e-09	4.59e-15	1.42e+01	0.0	0.0	0.0	

STEP	TIME	||U||	LABEL	TYPE	y12e
0	00:00:00	1.4250e+01	1	EP	9.0000e-01
10	00:00:00	1.4015e+01	2		3.8077e-01
13	00:00:01	1.3942e+01	3	EP	0.0000e+00

STEP	TIME	||U||	LABEL	TYPE	y12e
0	00:00:01	1.4250e+01	4	EP	9.0000e-01
10	00:00:02	1.4378e+01	5		9.8373e-01
19	00:00:03	1.4402e+01	6	EP	9.9000e-01

Here, the solution with label 1 is represented by the two trajectory segments in Fig. 7.4(b), whereas that with label 3 is represented by the two trajectory segments in Fig. 7.4(c).

As explained above, the indices of the vector of continuation variables associated with v_s^{**}, λ_s^{**}, ε_1, ε_2, and θ are stored in the appropriate fields of the data structure by repeated use of the coco_get_func_data utility function in the closer doedel_close_het. As in the previous example, a slot function is included with the continuation problem that associates the identifier 'doedel_save' with this data structure in the sol#.mat files stored during continuation. The function shown below loads the corresponding data from the sol#.mat file associated with the label lab and the run identifier run.

```
function prob = doedel_sol2het(prob, run, lab)

prob = coll_sol2seg(prob, 'doedel1', run, lab);
prob = coll_sol2seg(prob, 'doedel2', run, lab);
prob = alg_sol2eqn(prob, 'doedel3', run, lab);
prob = alg_sol2eqn(prob, 'doedel4', run, lab);

[data chart] = coco_read_solution('evs', run, lab);
vec  = chart.x(data.vec_idx);
lam  = chart.x(data.lam_idx);

[data chart] = coco_read_solution('bcs', run, lab);
eps  = chart.x(data.eps_idx);
th   = chart.x(data.th_idx);

prob = doedel_close_het(prob, eps, th, vec, lam);

end
```

Repeated application of doedel_sol2het now serves to complete the remainder of the algorithm. In each case, the interval of continuation for the first element in the list of active continuation parameters has been chosen to include the value at the starting solution and to be limited at one end by the desired value of this parameter. The extract below shows the completion of the four steps of the algorithm described above.

```
>> prob = doedel_sol2het(coco_prob(), 'run1', 3);
>> coco(prob, 'run2', [], 1, 'y22e', [-0.995 0]);
```

```
     STEP    DAMPING              NORMS               COMPUTATION TIMES
     IT SIT    GAMMA     ||d||     ||f||     ||U||    F(x)   DF(x)   SOLVE
      0                           4.70e-15  1.40e+01   0.0    0.0     0.0
      1   1  1.00e+00  2.39e-14   3.55e-15  1.40e+01   0.0    0.0     0.0


     STEP    TIME        ||U||    LABEL  TYPE          y22e
      0  00:00:00   1.3974e+01     1    EP       -9.5257e-01
      9  00:00:00   1.4171e+01     2    EP       -9.9500e-01


     STEP    TIME        ||U||    LABEL  TYPE          y22e
      0  00:00:01   1.3974e+01     3    EP       -9.5257e-01
     10  00:00:01   1.3729e+01     4             -6.4831e-01
     20  00:00:02   1.4476e+01     5    EP        0.0000e+00
>> prob = doedel_sol2het(coco_prob(), 'run2', 5);
>> coco(prob, 'run3', [], 1, 'gap', [-2 0]);


     STEP    DAMPING              NORMS               COMPUTATION TIMES
     IT SIT    GAMMA     ||d||     ||f||     ||U||    F(x)   DF(x)   SOLVE
      0                           4.65e-15  1.46e+01   0.0    0.0     0.0
      1   1  1.00e+00  1.42e-14   4.25e-15  1.46e+01   0.0    0.0     0.0


     STEP    TIME        ||U||    LABEL  TYPE           gap
      0  00:00:00   1.4595e+01     1    EP       -1.8599e+00
      3  00:00:00   1.4872e+01     2    EP       -2.0000e+00


     STEP    TIME        ||U||    LABEL  TYPE           gap
      0  00:00:00   1.4595e+01     3    EP       -1.8599e+00
     10  00:00:01   1.2940e+01     4             -7.2081e-01
     14  00:00:01   1.2476e+01     5    EP        0.0000e+00
>> prob = doedel_sol2het(coco_prob(), 'run3', 5);
>> coco(prob, 'run4', [], 1, 'eps1', [1e-3 eps0(1)]);


     STEP    DAMPING              NORMS               COMPUTATION TIMES
     IT SIT    GAMMA     ||d||     ||f||     ||U||    F(x)   DF(x)   SOLVE
      0                           3.98e-15  1.25e+01   0.0    0.0     0.0
      1   1  1.00e+00  6.94e-15   3.01e-15  1.25e+01   0.0    0.0     0.0


     STEP    TIME        ||U||    LABEL  TYPE          eps1
      0  00:00:00   1.2476e+01     1    EP        5.0000e-02
     10  00:00:00   1.4512e+01     2              1.0958e-03
     11  00:00:01   1.4563e+01     3    EP        1.0000e-03
>> prob = doedel_sol2het(coco_prob(), 'run4', 3);
>> coco(prob, 'run5', [], 1, 'eps2', [1e-3 eps0(2)]);


     STEP    DAMPING              NORMS               COMPUTATION TIMES
     IT SIT    GAMMA     ||d||     ||f||     ||U||    F(x)   DF(x)   SOLVE
      0                           4.83e-15  1.46e+01   0.0    0.0     0.0
      1   1  1.00e+00  1.77e-13   3.63e-15  1.46e+01   0.0    0.0     0.0


     STEP    TIME        ||U||    LABEL  TYPE          eps2
      0  00:00:00   1.4563e+01     1    EP        5.0000e-02
     10  00:00:01   1.5750e+01     2    EP        1.0000e-03
```

The solution with label 5 in the run with run identifier `run2` is represented by the two trajectory segments in Fig. 7.4(d). We note that the initial tangent vector of the left trajectory segment (governed by the continuation variable θ) has changed during this continuation run. The approximate connecting orbit obtained in the run with run identifier `run5`, after closing the Lin gap and reducing ε_1 and ε_2, is shown in Fig. 7.4(e). Continuation of this heteroclinic approximant under variations in p_2 is now achieved by the sequence of commands shown below.

```
>> prob = doedel_sol2het(coco_prob(), 'run5', 2);
>> coco(prob, 'run6', [], 1, 'p2', [0.5 8]);
```

STEP	DAMPING		NORMS			COMPUTATION TIMES		
IT SIT	GAMMA	||d||	||f||		||U||	F(x)	DF(x)	SOLVE
0			4.41e-15		1.58e+01	0.0	0.0	0.0
1 1	1.00e+00	1.69e-13	4.09e-15		1.58e+01	0.0	0.0	0.0

STEP	TIME	||U||	LABEL	TYPE	p2
0	00:00:00	1.5782e+01	1	EP	1.0000e+00
10	00:00:00	1.8920e+01	2		7.4459e-01
20	00:00:01	2.3141e+01	3		5.7093e-01
27	00:00:02	2.5864e+01	4	EP	5.0000e-01

STEP	TIME	||U||	LABEL	TYPE	p2
0	00:00:02	1.5782e+01	5	EP	1.0000e+00
10	00:00:03	1.3368e+01	6		1.5268e+00
20	00:00:04	1.3008e+01	7		2.9667e+00
30	00:00:05	1.5730e+01	8		5.1045e+00
40	00:00:06	1.9585e+01	9		7.3252e+00
44	00:00:06	2.0857e+01	10	EP	8.0000e+00

The family of connecting trajectories shown in Fig. 7.4(f) corresponds to a subset of the labeled solutions from this run.

7.4 Conclusions

The treatment in this chapter provides a foundation for the task of continuation of state-space trajectory segments of systems of ordinary differential equations that supports arbitrary boundary conditions, interior conditions, or integral conditions. Notably, the indices of the elements of the vector of continuation variables corresponding to the boundary points along such a trajectory segment are explicitly assigned to the data function data structure by the `coll_init_data` function. As shown in the next several chapters, this enables a straightforward embedding of the `coll` toolbox in toolboxes for continuation of the solutions to two-point boundary-value problems, multipoint boundary-value problems, and periodic boundary-value problems.

A number of generalizations of the `coll` toolbox follow immediately from the theoretical discussion at the beginning of Chap. 6. These include allowing for nonuniform partitions of the interval $[0,1]$ in the independent variable, allowing for adaptive partitions that change during continuation to accommodate bounds on some error tolerance, and providing a collocation discretization for systems of second-order ordinary differential equations. Such changes constitute, in many cases, a substantial recoding of the `coll` toolbox. The implementation paradigm described here should enable a straightforward construction even in these instances. In order to support successive debugging and validation, it is recommended that the formal mathematical treatment, including that of the vectorized structures, be performed concurrently with the realization in code. We will provide examples of adaptive implementations of the `'coll'` toolbox in Part V of this text.

Exercises

7.1. Suppose that

$$
y(t) \approx \sum_{i=1}^{m+1} \upsilon_{(m+1)(j-1)+i} \prod_{k=1,k\neq i}^{m+1} \frac{m}{i-k}\left(\frac{Nt}{T} - j + 1 - \frac{k-1}{m}\right)
$$

on the interval

$$
T\frac{j-1}{N} \leq t \leq T\frac{j}{N}
$$

for $1 \leq j \leq N$. Show that

$$
y\left(\frac{T}{N}\left(j-1+\frac{l-1}{m}\right)\right) \approx \upsilon_{(m+1)(j-1)+l}
$$

for $1 \leq l \leq m+1$.

7.2. What is the number of approximate samples of the unknown solution y to the original differential equation that are obtained from the columns of the matrix $\mathfrak{vec}_n\left(\upsilon_{bp}\right)$ after omitting the columns with column indices $1+i(m+1)$ for $i=1,\ldots,N-1$?

7.3. Review the construction of the Gauss nodes and Gauss–Legendre quadrature weights from the eigenstructure of the matrix J obtained in `coll_nodes` on page 149 using an advanced numerical methods book. Do similar methods exist for other choices of collocation nodes, say the *Lobatto* nodes?

7.4. Denote the 1-dimensional array of quadrature weights computed in `coll_nodes` on page 149 by w and consider the vectorized extension of some function $g : \mathbb{R}^n \to \mathbb{R}$, also denoted by g. Show that

$$
\int_0^T g(y(t))\, dt \approx \frac{T}{2N} g\left(\mathfrak{vec}_n\left(W \cdot \upsilon_{bp}\right)\right) \cdot \left(1_{N,1} \otimes w\right).
$$

7.5. Denote the 1-dimensional array of quadrature weights computed in `coll_nodes` on page 149 by w and consider the vectorized extension of some function $h : \mathbb{R} \to \mathbb{R}^n$, also denoted by h. Let h_{cn} denote the $n \times Nm$ matrix whose columns equal the value of h at the collocation nodes. Show that

$$
\int_0^T h(t)^T \cdot y(t)\, dt \approx \frac{T}{2N}\mathfrak{vec}(h_{cn})^T \cdot (I_N \otimes (\Omega \otimes I_n)) \cdot W \cdot \upsilon_{bp},
$$

where Ω is the $m \times m$ diagonal matrix whose diagonal entries equal w.

7.6. What is the dimensional deficit of a segment object?

7.7. Verify the construction of the Jacobian in `coll_DFDU`. Use the `coco_ezDFDX` utility with first argument `'f(o,d,x)'` to compute a corresponding finite-difference approximation.

7.8. Verify that the handling of the `'NTST'` and `'NCOL'` options in the `coll_get_settings` function is consistent with the precedence for existing settings over default values. What has precedence over existing settings?

7.9. Replace the construction of `data.W` and `data.Wp` in `coll_init_data` with two nested calls to the MATLAB `kron` function.

7.10. For the example considered in Sect. 7.3.1, verify the statement regarding the nominal dimensional deficits at various stages of construction.

7.11. Identify the step in the encoding of the example in Sect. 7.3.1 that guarantees that $y_1(0) = 1$ throughout continuation.

7.12. Replace the initial solution guess in the example in Sect. 7.3.1 by a discretization of $y_1(t) = 1 + t^2/2$, $y_2(t) = t$ and repeat the analysis. What differences do you observe?

7.13. Verify that the `'coll'` toolbox is compatible with nonzero values for T. How would you modify the encoding to handle the case when $T = 0$?

7.14. Use your observations from the previous exercise to modify the analysis in Sect. 7.3.1 so as to generate the desired solution by shooting backward in time from the right-end boundary.

7.15. Verify the statements regarding the existence and linear stability properties of equilibria for the vector field

$$f(y, p) = \begin{pmatrix} y_2 \\ p_2 y_2 - y_1(1 - y_1)(y_1 - p_1) \end{pmatrix}.$$

7.16. Verify that

$$y(t) = \left(1 + e^{-t/\sqrt{2}}\right)^{-1}$$

is a solution to the ordinary differential equation

$$y'' - \frac{1 - 2p}{\sqrt{2}} y' + y(1 - y)(y - p) = 0$$

for all p. Find the limits of y as $t \to \pm\infty$.

7.17. For the example considered in Sect. 7.3.2, verify the statement regarding the nominal dimensional deficits at various stages of construction.

7.18. Modify the analysis in Sect. 7.3.2 so as to generate the desired solution by shooting backward in time from one of the equilibria.

7.19. Show the intermediate results of the continuation steps in Sect. 7.3.2 by using the `coll_read_solution` command to extract solution data.

7.20. Verify Eq. (7.31) for the family of heteroclinic connections found using numerical continuation in Sect. 7.3.2 in the run with run identifier `'run6'`.

7.21. Why is it necessary to introduce the ε_1 and ε_2 problem parameters in the examples in Sects. 7.3.2 and 7.3.3?

7.22. Review [Doedel, E.J. and Friedman, M.J., "Numerical computation of heteroclinic orbits," *Journal of Computational and Applied Mathematics*, 26, pp. 155–170, 1989] for an alternative treatment of the problem of computing connecting orbits for the dynamical systems discussed in Sects. 7.3.2 and 7.3.3.

7.23. Verify the statements regarding the existence and linear stability properties of equilibria for the vector field

$$f(y,p) = \begin{pmatrix} 1 - y_1^2 \\ p_1 y_1 + p_2 y_2 \end{pmatrix}.$$

7.24. Verify that
$$y_1(t) = \tanh t, \; y_2(t) = 1 - 2e^t \cot^{-1} e^t$$

satisfy the system of ordinary differential equations

$$y_1' = 1 - y_1^2, \; y_2' = y_1 + y_2.$$

7.25. For the example considered in Sect. 7.3.3, verify the statement regarding the nominal dimensional deficits at various stages of construction.

7.26. Find a general solution of the dynamical system in Sect. 7.3.3, and verify the claim regarding the existence of a heteroclinic orbit for arbitrary values of p_1/p_2.

7.27. In the collocation condition in Eq. (6.80), the same continuation variables were used to represent the problem parameters p for all $N(m+1)$ mesh points. This leads to full columns in the Jacobian in Eq. (6.83). An alternative is to use separate continuation variables to represent p (a) for each collocation interval or (b) for each collocation node and then add suitable gluing conditions, as was done in the 'compalg' toolbox. Sketch the structure of the Jacobian corresponding to Eq. (6.80) for each of the cases (a) and (b) defined above. What is the size of each Jacobian?

7.28. Implement the modified collocation problems introduced in the previous exercise and compare the computation times for large values of N, for example, with the catenary problem considered in Sect. 7.3.1. Which version is computationally most efficient? Why?

7.29. Given a vector field f and a solution $y(t)$ for $t \in [0,1]$ of the differential equation

$$\frac{dy}{dt} = Tf(y,p),$$

consider the *variational equation*

$$\frac{dM}{dt} = T \partial_y f(y,p) \cdot M,$$

where $M(t)$ is an $n \times n$ matrix. Derive a collocation problem that discretizes the variational equation on $[0,1]$ separately on each of N discretization intervals. Impose the condition that the solution on the jth interval equal the identity matrix at the left boundary of the interval and let M_j denote the value at the right boundary. Describe a method for computing the sequence $\{M_j\}_{j=1}^N$. Implement your method in code.

7.30. Consider the discretization of the variational equation in the previous exercise. Suppose that you wanted to solve the variational equation for a continuous function $M(t)$, such that $M(0) = I_n$. How would you use the sequence $\{M_j\}_{j=1}^N$ to compute $M(1)$?

7.31. In Sect. 7.3.3, the equation

$$\partial_y f(y^{**}, p) \cdot v_s^{**} = \lambda_s^{**} v_s^{**}$$

was introduced for computing the eigenvector v_s^{**}. The linearization of this equation includes the second-order term

$$A = \partial_y \left(\partial_y f\left(y, p\right) \cdot v_s^{**}\right)\big)\big|_{y=y^{**}}.$$

Assume that $\|v_s^{**}\| = 1$. Show that the formula

$$\frac{1}{2h}\left(\partial_y f\left(y^{**} + h v_s^{**}, p\right) - \partial_y f\left(y^{**} - h v_s^{**}, p\right)\right)$$

is an h^2 approximation to A. Compare the computational effort when using this formula with the effort of numerical differentiation of

$$g(y) := \partial_y f(y, p) \cdot v_s^{**}$$

with respect to y. Use this result to implement the Jacobian corresponding to the function `doedel_evs` and repeat the computational analysis in Sect. 7.3.3. What differences do you observe?

Chapter 8

Single-Segment Continuation Problems

The collocation toolbox developed in the previous chapter defines a zero problem with a dimensional deficit of $n + q + 1$, where n is the range dimension of the vector field $f : \mathbb{R}^n \times \mathbb{R}^q \to \mathbb{R}^n$ and q equals the number of problem parameters p. For a given set of values of the problem parameters, the corresponding solution manifold corresponds to an $(n + 1)$-dimensional family of parameterizations of continuous piecewise-polynomial approximants of solutions $\upsilon(\tau)$, for $\tau \in [0, 1]$, to the differential equation

$$\frac{d\upsilon}{d\tau} = Tf(\upsilon, p), \tag{8.1}$$

in turn parameterized by the interval length T.

We obtain an extended continuation problem with dimensional deficit equal to the number of problem parameters by providing $n + 1$ additional conditions on the vector of continuation variables. In the object-oriented continuation paradigm, we accomplish this by embedding a *segment object*, obtained from the `'coll'` toolbox, in a parent continuation problem object. This chapter describes two candidate embeddings, each of which implements a zero problem associated with a separate problem class.

In the first case, we provide for the imposition of arbitrary *boundary conditions* on the values of the solution approximant at $\tau = 0$ and $\tau = 1$, as well as on the interval length T. We refer to the corresponding continuation problem object as a *constrained segment object*. The `'bvp'` toolbox associated with the construction of a constrained segment object finds general use for a large collection of classical boundary-value problems. It also serves as a development platform for other special-purpose toolboxes associated with the construction of more restricted classes of boundary-value problems.

In the second case, we combine a gluing condition on the end points of a segment with an *integral condition* in order to trace families of single-segment periodic orbits of the dynamical system in Eq. (8.1). We refer to the corresponding continuation problem object as a *periodic orbit object*. The implementation of the `'po'` toolbox associated with the construction of a periodic orbit object illustrates the use of a slot function for updating the toolbox data structure prior to each new continuation step. In particular, we make use of the `coco_func_data` class in order to support safe communication and content access to toolbox data within and across toolbox boundaries.

8.1   Boundary-value problems

A fully constrained two-point boundary-value continuation problem results from appending an $(n+1)$-dimensional zero problem of the form

$$f_{bc}\left(T, v_1, v_{N(m+1)}, p\right) = 0, \qquad (8.2)$$

in terms of a function $f_{bc} : \mathbb{R} \times \mathbb{R}^n \times \mathbb{R}^n \times \mathbb{R}^q \to \mathbb{R}^{n+1}$, to the collocation zero problem. In the notation of the previous chapter, v_1 and $v_{N(m+1)}$ denote the base point values of the piecewise-polynomial approximant at $\tau = 0$ and $\tau = 1$, respectively. This formulation thus supports arbitrary constraints on the end points of a segment, including conditions that couple the values at the two points, as well as conditions on the interval length T. As a special case, given known constants $T^* \in \mathbb{R}$ and $v_1^* \in \mathbb{R}^n$, the boundary-value problem obtained from the function

$$f_{bc} : \left(T, v_1, v_{N(m+1)}, p\right) \mapsto \begin{pmatrix} T - T^* \\ v_1 - v_1^* \end{pmatrix} \qquad (8.3)$$

corresponds to the imposition of a fixed interval length and known initial conditions on the solution approximants.

8.1.1   Encoding

We present a complete encoding of an embeddable `'bvp'` toolbox that conforms with the paradigm of toolbox construction and task embedding presented in Chaps. 4 and 5. Here, the construction of an initializable restricted continuation problem relies on embedding a call to the corresponding `'coll'` toolbox constructor within each of the `'bvp'` toolbox constructors and, subsequently, appending the additional zero functions particular to the boundary conditions in Eq. (8.2).

To this end, consider the family of zero functions

$$u \mapsto f_{bc}\left(T, v_1, v_{N(m+1)}, p\right). \qquad (8.4)$$

Let the corresponding index set $\mathbb{K}$ consist of the integer indices of the components of the vector of continuation variables corresponding to T, v_1, $v_{N(m+1)}$, and p, in that order. Let the corresponding function data structure $\mathfrak{D}$ contain the properties

- `T_idx` with value equal to 1,

- `x0_idx` with value equal to $[\![\ 2 \ \cdots \ n+1\ ]\!]$,

- `x1_idx` with value equal to $[\![\ n+2 \ \cdots \ 2n+1\ ]\!]$,

- `p_idx` with value equal to $[\![\ 2n+2 \ \cdots \ 2n+1+q\ ]\!]$,

- `fhan` with value equal to a function handle to an encoding of f_{bc}, and

- `dfdxhan` that is either an empty array or equal to a function handle to an encoding of the Jacobian $\partial_{(T, v_1, v_{N(m+1)}, p)} f_{bc}$.

The COCO-compatible zero-function wrappers bvp_F and bvp_DFDU, shown below, encode this zero function and its Jacobian with respect to the subset of components of the vector of continuation variables indexed by $\mathbb{K}$.

```
function [data y] = bvp_F(prob, data, u)

T  = u(data.T_idx);
x0 = u(data.x0_idx);
x1 = u(data.x1_idx);
p  = u(data.p_idx);

y  = data.fhan(T, x0, x1, p);

end

function [data J] = bvp_DFDU(prob, data, u)

T  = u(data.T_idx);
x0 = u(data.x0_idx);
x1 = u(data.x1_idx);
p  = u(data.p_idx);

J  = data.dfdxhan(T, x0, x1, p);

end
```

The bvp_isol2seg function, shown below together with the is_empty_or_func sub-function, encodes an embeddable toolbox constructor compatible with the sentence syntax

```
varargin = coll, fhan, [dfdxhan]
```

where coll was given on page 153.

```
function prob = bvp_isol2seg(prob, oid, varargin)

tbid   = coco_get_id(oid, 'bvp');
segoid = coco_get_id(tbid, 'seg');
str    = coco_stream(varargin{:});
prob   = coll_isol2seg(prob, segoid, str);
data.fhan = str.get;
data.dfdxhan = [];
if is_empty_or_func(str.peek)
  data.dfdxhan = str.get;
end

data = bvp_init_data(prob, tbid, data);
bvp_arg_check(prob, tbid, data);
prob = bvp_close_seg(prob, tbid, data);

end

function flag = is_empty_or_func(x)
  flag = isempty(x) || isa(x, 'function_handle');
end
```

The call to the embedded coll_isol2seg toolbox constructor results in the construction of a segment object with toolbox instance identifier given by appending 'bvp.seg.coll' to the oid input argument.

The function `bvp_init_data`, shown below, is used to populate the fields of the `'bvp'` toolbox data structure.

```
function data = bvp_init_data(prob, tbid, data)

segtbid = coco_get_id(tbid, 'seg.coll');
fdata   = coco_get_func_data(prob, segtbid, 'data');

data.T_idx  = 1;
data.x0_idx = 1+(1:fdata.dim)';
data.x1_idx = 1+fdata.dim+(1:fdata.dim)';
data.p_idx  = 1+2*fdata.dim +(1:fdata.pdim)';

end
```

Basic error checking is performed by the function `bvp_arg_check`, shown below.

```
function bvp_arg_check(prob, tbid, data)

assert(isa(data.fhan, 'function_handle'), ...
  '%s: input for ''f'' is not a function handle', tbid);

end
```

Finally, the `bvp_close_seg` function, shown below, is invoked in order to close the constrained segment object by appending the zero function in Eq. (8.4).

```
function prob = bvp_close_seg(prob, tbid, data)

segtbid   = coco_get_id(tbid, 'seg.coll');
[fdata uidx] = coco_get_func_data(prob, segtbid, 'data', 'uidx');
uidx = uidx([fdata.T_idx; fdata.x0_idx; fdata.x1_idx; fdata.p_idx]);
if isempty(data.dfdxhan)
  prob = coco_add_func(prob, tbid, @bvp_F, data, 'zero', 'uidx', uidx);
else
  prob = coco_add_func(prob, tbid, @bvp_F, @bvp_DFDU, data, 'zero', ...
    'uidx', uidx);
end
prob = coco_add_slot(prob, tbid, @coco_save_data, data, 'save_full');

end
```

Here, the built-in finite-difference approximation of the Jacobian of the zero function, with respect to the corresponding components of u, is computed in the event that the `dfdxhan` property is an empty array.

The `bvp_sol2seg` function, shown below, encodes an embeddable toolbox constructor compatible with the sentence syntax

```
varargin = run, [source], label
```

where `source` denotes an object instance identifier associated with a previously computed solution point.

```
function prob = bvp_sol2seg(prob, oid, varargin)

ttbid = coco_get_id(oid, 'bvp');
str   = coco_stream(varargin{:});
run   = str.get;
```

```
if ischar(str.peek)
  stbid = coco_get_id(str.get, 'bvp');
else
  stbid = ttbid;
end
lab = str.get;

data = coco_read_solution(stbid, run, lab);
toid = coco_get_id(ttbid, 'seg');
soid = coco_get_id(stbid, 'seg');
prob = coll_sol2seg(prob, toid, run, soid, lab);
data = bvp_init_data(prob, ttbid, data);
prob = bvp_close_seg(prob, ttbid, data);

end
```

Finally, the function `bvp_read_solution`, shown below, provides a candidate toolbox extractor that is compatible with the closure of the constrained segment object encoded in `bvp_close_seg`.

```
function [sol data] = bvp_read_solution(oid, run, lab)

tbid = coco_get_id(oid, 'bvp');
data = coco_read_solution(tbid, run, lab);

segoid = coco_get_id(tbid, 'seg');
sol    = coll_read_solution(segoid, run, lab);

end
```

8.1.2 Examples

Consider, again, the two-point boundary-value problem

$$f'' = \frac{1 + f'^2}{f}, \tag{8.5}$$

where $f(0) = 1$ and $f(1) = Y$. This boundary-value problem was obtained in Chap. 1 by applying Euler–Lagrange theory to the problem of finding a local minimum for the functional in Eq. (1.1). We arrive at a formulation compatible with the 'coll' toolbox by rewriting Eq. (8.5) as the following system of first-order differential equations:

$$y_1' = y_2, \tag{8.6}$$

$$y_2' = \frac{1 + y_2^2}{y_1}, \tag{8.7}$$

corresponding to the vector field

$$f(y, p) = \begin{pmatrix} y_2 \\ (1 + y_2^2)/y_1 \end{pmatrix}. \tag{8.8}$$

If we let $p = Y$, then the boundary conditions are equivalent to the vanishing of the function

$$f_{bc}\left(T, \upsilon_1, \upsilon_{N(m+1)}, p\right) := \begin{pmatrix} T - 1 \\ \upsilon_{1,1} - 1 \\ \upsilon_{N(m+1),1} - p \end{pmatrix}, \tag{8.9}$$

whose Jacobian is given by

$$\partial_{(T,v_1,v_{N(m+1)},p)} f_{bc}\left(T,v_1,v_{N(m+1)},p\right) = \begin{pmatrix} 1 & 0 & 0 & 0 & 0 & 0 \\ 0 & 1 & 0 & 0 & 0 & 0 \\ 0 & 0 & 0 & 1 & 0 & -1 \end{pmatrix}. \qquad (8.10)$$

Example 8.1 The vector field in Eq. (8.8) is encoded in the vectorized function `catn`, shown below.

```
function y = catn(x, p)

x1 = x(1,:);
x2 = x(2,:);

y(1,:) = x2;
y(2,:) = (1+x2.^2)./x1;

end
```

Similarly, the boundary condition function in Eq. (8.9) and its Jacobian in Eq. (8.10) are encoded in the functions `catn_bc` and `catn_bc_DFDX`, respectively, shown below.

```
function fbc = catn_bc(T, x0, x1, p)
  fbc = [T-1; x0(1)-1; x1(1)-p];
end
```

```
function Jbc = catn_bc_DFDX(T, x0, x1, p)

Jbc = zeros(3,6);
Jbc(1,1) = 1;
Jbc(2,2) = 1;
Jbc(3,4) = 1;
Jbc(3,6) = -1;

end
```

We provide an initial solution guess and continue the solution under variations in Y as shown in the following extract.

```
>> t0 = [0; 1];
>> x0 = [1 0; 1 0];
>> Y0 = 1;
>> coll_args = {@catn, t0, x0, 'Y', Y0};
>> bvp_args = [coll_args, {@catn_bc, @catn_bc_DFDX}];
>> prob = bvp_isol2seg(coco_prob(), '', bvp_args{:});
>> coco(prob, 'run', [], 1, 'Y', [0 3]);
```

STEP		DAMPING		NORMS		COMPUTATION	TIMES	
IT	SIT	GAMMA	||d||	||f||	||U||	F(x)	DF(x)	SOLVE
0				3.16e-01	7.28e+00	0.0	0.0	0.0
1	1	1.00e+00	2.38e+00	4.67e-02	7.03e+00	0.0	0.0	0.0
2	1	1.00e+00	2.41e-01	6.46e-04	7.05e+00	0.0	0.0	0.0
3	1	1.00e+00	2.20e-03	6.06e-08	7.05e+00	0.0	0.0	0.0
4	1	1.00e+00	1.57e-07	2.51e-15	7.05e+00	0.0	0.0	0.0

STEP	TIME	||U||	LABEL	TYPE	Y
0	00:00:00	7.0510e+00	1	EP	1.0000e+00
10	00:00:00	6.9343e+00	2		6.5549e-01

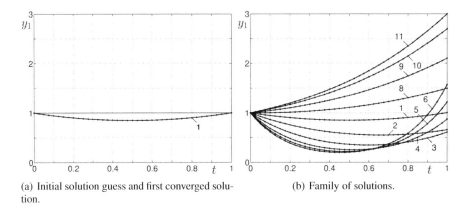

(a) Initial solution guess and first converged solution.

(b) Family of solutions.

Figure 8.1. *Solutions to the boundary-value problem given by the system of differential equations in Eqs. (8.6)–(8.7) and the given boundary conditions at $x = 0$ and $x = 1$. We start with the function $y_1 = 1$ shown in panel (a) as an initial guess. The initial correction converges to solution 1. A family of solutions obtained using numerical continuation is shown in panel (b). The labels correspond to the session output included in the text.*

STEP	TIME	$||U||$	LABEL	TYPE	Y
20	00:00:01	8.9765e+00	3		6.0841e-01
30	00:00:01	1.2650e+01	4		8.7096e-01
40	00:00:02	1.6709e+01	5		1.2150e+00
50	00:00:02	2.1043e+01	6	EP	1.5705e+00

STEP	TIME	$||U||$	LABEL	TYPE	Y
0	00:00:02	7.0510e+00	7	EP	1.0000e+00
10	00:00:03	9.5298e+00	8		1.4959e+00
20	00:00:03	1.3666e+01	9		2.1006e+00
30	00:00:04	1.8190e+01	10		2.6936e+00
36	00:00:04	2.0633e+01	11	EP	3.0000e+00

The initial solution guess and the solution with label 1 are represented by the trajectories shown in Fig. 8.1(a). A family of solutions to the boundary-value problem corresponding to the labeled solutions resulting from this computation is shown in Fig. 8.1(b). ∎

The *Bratu differential equation*

$$z''(t) + pe^{z(t)} = 0, \tag{8.11}$$

together with the boundary conditions

$$z(0) = z(1) = 0, \tag{8.12}$$

for some scalar parameter p, collectively model a steady-state temperature distribution of an exothermic reaction in a 1-dimensional medium with boundaries connected to heat baths. The rate of change of the vector-valued function

$$y := \begin{pmatrix} z \\ z' \end{pmatrix}, \tag{8.13}$$

with respect to t, is then given by the vector field

$$f(y, p) := \begin{pmatrix} z' \\ -pe^z \end{pmatrix}. \tag{8.14}$$

A two-point boundary-value problem now results from appending additional conditions of the form in Eq. (8.2) to the corresponding collocation zero problem. Here,

$$f_{bc}(T, v_1, v_{N(m+1)}, p) = \begin{pmatrix} T-1 \\ v_{1,1} \\ v_{N(m+1),1} \end{pmatrix}, \tag{8.15}$$

where the second subscript denotes the vector component. It follows that

$$\partial_{(T, v_1, v_{N(m+1)}, p)} f_{bc}(T, v_1, v_{N(m+1)}, p) = \begin{pmatrix} 1 & 0 & 0 & 0 & 0 & 0 \\ 0 & 1 & 0 & 0 & 0 & 0 \\ 0 & 0 & 0 & 1 & 0 & 0 \end{pmatrix}. \tag{8.16}$$

Example 8.2 The vectorized function `brat`, shown below, encodes the vector field in Eq. (8.14).

```
function y = brat(x, p)

x1 = x(1,:);
x2 = x(2,:);
p1 = p(1,:);

y(1,:) = x2;
y(2,:) = -p1.*exp(x1);

end
```

Similarly, the boundary condition function f_{bc} and its Jacobian with respect to T, v_1, $v_{N(m+1)}$, and p are encoded in the functions `brat_bc` and `brat_bc_DFDX`, respectively, shown below.

```
function fbc = brat_bc(T, x0, x1, p)
   fbc = [T-1; x0(1); x1(1)];
end
```

```
function Jbc = brat_bc_DFDX(T, x0, x1, p)

Jbc = zeros(3,6);
Jbc(1,1) = 1;
Jbc(2,2) = 1;
Jbc(3,4) = 1;

end
```

We provide an initial solution guess and continue the solution under variations in the single problem parameter p, as shown in the following extract.

```
>> coll_args = {@brat, [0;1], zeros(2), 'p', 0};
>> bvp_args  = [coll_args, {@brat_bc, @brat_bc_DFDX}];
>> prob = coco_prob();
```

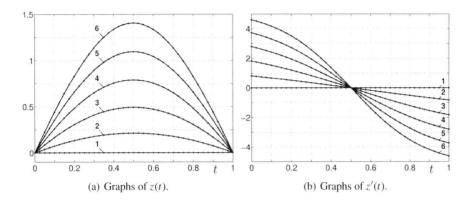

(a) Graphs of $z(t)$.   (b) Graphs of $z'(t)$.

Figure 8.2. *Solutions to the Bratu boundary-value problem given by Eq. (8.11) and the given boundary conditions at $t = 0$ and $t = 1$. We start with the exact solution $z(t) \equiv 0$ at $p = 0$ and obtain a family of solutions with increasing amplitudes. The labels correspond to the session output included in the text.*

```
>> coco(prob, 'run', @bvp_isol2seg, bvp_args{:}, 1, 'p', [0 4]);

     STEP   DAMPING                  NORMS                   COMPUTATION TIMES
     IT SIT    GAMMA    ||d||      ||f||      ||U||    F(x)   DF(x)   SOLVE
     0                           0.00e+00   1.00e+00    0.0    0.0     0.0
     1   1  1.00e+00  0.00e+00   0.00e+00   1.00e+00    0.0    0.0     0.0


     STEP      TIME         ||U||  LABEL  TYPE              p
       0   00:00:00    1.0000e+00     1   EP      0.0000e+00
      10   00:00:00    4.2659e+00     2           1.4248e+00
      20   00:00:01    9.1634e+00     3           2.6124e+00
      30   00:00:01    1.4048e+01     4           3.2646e+00
      40   00:00:02    1.8893e+01     5           3.5026e+00
      50   00:00:02    2.3716e+01     6   EP      3.4605e+00
```

In this case, the initial solution guess $\upsilon_{bp} = 0$ for $p = 0$ corresponds to the exact and unique solution of the original boundary-value problem for this parameter value. Fig. 8.2 shows graphs of $z(t)$ and its derivative for each of the labeled points located on the solution manifold. ∎

8.2 Periodic orbits

A *periodic orbit continuation problem*, corresponding to a single-segment periodic orbit, results from appending the $(n + 1)$-dimensional zero problem

$$\begin{pmatrix} \upsilon_1 - \upsilon_{N(m+1)} \\ \upsilon_{bp}^{*T} \cdot W'^T \cdot (I_N \otimes (\Omega \otimes I_n)) \cdot W \cdot \upsilon_{bp} \end{pmatrix} = 0 \qquad (8.17)$$

to the collocation zero problem. Here, υ_{bp}^* denotes a 1-dimensional reference array of length $nN(m + 1)$ and Ω is an $m \times m$ diagonal matrix with nonzero entries given by the Gauss–Legendre quadrature weights, as in Eq. (7.17). The last component of the zero problem in Eq. (8.17) is thus a discretized version of the *integral phase condition*

$$\int_0^1 \left(\frac{dv^*}{d\tau}(\tau)\right)^T \cdot v(\tau)\,d\tau = 0 \tag{8.18}$$

for some piecewise-differentiable function v^*, approximated by Gauss quadrature.

If v_{bp}^* satisfies the collocation zero problem for some T and p, then $v_{bp} = v_{bp}^*$ satisfies the periodic orbit continuation problem provided that $v_1^* = v_{N(m+1)}^*$. It is reasonable, therefore, to initialize v_{bp}^* with the initial solution guess $v_{bp,0}$ and, subsequently, to update v_{bp}^* to equal a previously obtained and suitably selected solution v_{bp} prior to each continuation step.

8.2.1   Encoding

We present a complete encoding of an embeddable 'po' toolbox that conforms with the paradigm of toolbox construction and task embedding presented in Chaps. 4 and 5. As was the case with the 'bvp' toolbox, the construction of an initializable restricted continuation problem relies on embedding a call to the corresponding 'coll' toolbox constructor within each of the 'po' toolbox constructors and, subsequently, appending the additional zero functions in Eq. (8.17).

To this end, consider the zero function

$$u \mapsto \begin{pmatrix} v_1 - v_{N(m+1)} \\ v_{bp}^{*T} \cdot W'^T \cdot (I_N \otimes (\Omega \otimes I_n)) \cdot W \cdot v_{bp} \end{pmatrix}. \tag{8.19}$$

Let the index set $\mathbb{K}$ consist of the integer indices of the components of the vector of continuation variables corresponding to v_{bp}. Let the function data structure $\mathfrak{D}$ contain the properties

- x0_idx with value equal to $[\![\ 1\ \ \cdots\ \ n\ ]\!]$,

- x1_idx with value equal to $[\![\ N(m+1)n - n + 1\ \ \cdots\ \ N(m+1)n\ ]\!]$,

- xp0 with value equal to $v_{bp}^{*T} \cdot W'^T \cdot (I_N \otimes (\Omega \otimes I_n)) \cdot W$, and

- J with value equal to

$$\begin{pmatrix} \begin{pmatrix} I_n & 0 & -I_n \end{pmatrix} \\ v_{bp}^{*T} \cdot W'^T \cdot (I_N \otimes (\Omega \otimes I_n)) \cdot W \end{pmatrix}.$$

The COCO-compatible zero-function wrappers po_F and po_DFDU, shown below, now encode the zero function and its Jacobian with respect to the subset of components of the vector of continuation variables indexed by $\mathbb{K}$.

```
function [data y] = po_F(prob, data, u)

x0 = u(data.x0_idx);
x1 = u(data.x1_idx);
```

```
y = [x0-x1; data.xp0*u];

end

function [data J] = po_DFDU(prob, data, u)
  J = data.J;
end
```

The `po_isol2orb` function, shown below, encodes an embeddable toolbox constructor compatible with the sentence syntax

```
varargin = coll
```

where `coll` was given on page 153.

```
function prob = po_isol2orb(prob, oid, varargin)

tbid    = coco_get_id(oid, 'po');
str     = coco_stream(varargin{:});
segoid  = coco_get_id(tbid, 'seg');
prob    = coll_isol2seg(prob, segoid, str);

data = struct();
data = po_init_data(prob, tbid, data);
prob = po_close_orb(prob, tbid, data);

end
```

The call to the embedded `coll_isol2seg` toolbox constructor results in the construction of a segment object with toolbox instance identifier given by appending `'po.seg.coll'` to the `oid` object instance identifier.

The function `po_init_data`, shown below, is used to populate the fields of the `'po'` toolbox data structure.

```
function data = po_init_data(prob, tbid, data)

stbid     = coco_get_id(tbid, 'seg.coll');
[fdata u0] = coco_get_func_data(prob, stbid, 'data', 'u0');

dim  = fdata.dim;
NTST = fdata.coll.NTST;
NCOL = fdata.coll.NCOL;
rows = [1:dim 1:dim];
cols = [fdata.x0_idx fdata.x1_idx];
vals = [ones(1,dim) -ones(1,dim)];
J    = sparse(rows, cols, vals, dim, dim*NTST*(NCOL+1));

data.x0_idx = fdata.x0_idx;
data.x1_idx = fdata.x1_idx;
data.intfac = fdata.Wp'*fdata.wts2*fdata.W;
data.xp0    = u0(fdata.xbp_idx)'*data.intfac;
data.J      = [J; data.xp0];

end
```

Here, the `intfac` field is introduced in order to support subsequent updates to v_{bp}^*. Finally, the `po_close_orb` function, shown below, is invoked in order to close the periodic orbit object by appending the zero function in Eq. (8.19).

```
function prob = po_close_orb(prob, tbid, data)

data.tbid = tbid;
data = coco_func_data(data);
prob = coco_add_slot(prob, tbid, @po_update, data, 'update');
segtbid      = coco_get_id(tbid, 'seg.coll');
[fdata uidx] = coco_get_func_data(prob, segtbid, 'data', 'uidx');
prob = coco_add_func(prob, tbid, @po_F, @po_DFDU, data, 'zero', ...
  'uidx', uidx(fdata.xbp_idx));
fid  = coco_get_id(tbid, 'period');
prob = coco_add_pars(prob, fid, uidx(fdata.T_idx), fid, 'active');
prob = coco_add_slot(prob, tbid, @coco_save_data, data, 'save_full');

end
```

Here, the command

```
prob = coco_add_pars(prob, fid, uidx(fdata.T_idx), fid, 'active');
```

appends a monitor function whose value equals the period T, assigns the corresponding index to $\mathbb{J}$, and assigns the string label contained in `fid` as a reference to the corresponding continuation parameter.

Following the above discussion, the discretized phase condition includes a reference array v_{bp}^* that must be initialized prior to continuation and updated prior to each new continuation step. Write access to the copy of the `'po'` toolbox data structure used by the zero function encoding the boundary conditions must therefore be provided to a separate function that performs the initialization/update. We enable such communication across function boundaries, but within the same toolbox, by converting the `'po'` toolbox data structure to an instance of the COCO utility class `coco_func_data` as shown in the command

```
data = coco_func_data(data);
```

We assign this instance as the function data structure to all of the toolbox functions that require write and read access to the current content of the toolbox data structure. The execution of the `po_update` slot function, shown below, is then triggered by the emission of the `'update'` signal by the core, once prior to continuation and then again before each new continuation step.

```
function data = po_update(prob, data, cseg, varargin)

fid          = coco_get_id(data.tbid, 'seg.coll');
[fdata uidx] = coco_get_func_data(prob, fid, 'data', 'uidx');
u            = cseg.src_chart.x;
data.xp0     = u(uidx(fdata.xbp_idx))'*data.intfac;
data.J(end,:) = data.xp0;

end
```

At the outset of continuation, the `src_chart.x` field of the `cseg` argument contains the initial solution guess u_0. In contrast, when the function is called prior to each new continuation step, this field contains a previously computed point on the solution manifold, from which continuation proceeds.

As an alternative to `po_isol2orb`, the `po_sol2orb` function, shown below, encodes an embeddable toolbox constructor compatible with the sentence syntax

```
varargin = run, [source], label
```

where `source` denotes an object instance identifier associated with a previously computed solution point.

```
function prob = po_sol2orb(prob, oid, varargin)

ttbid = coco_get_id(oid, 'po');
str   = coco_stream(varargin{:});
run   = str.get;
if ischar(str.peek)
  stbid = coco_get_id(str.get, 'po');
else
  stbid = ttbid;
end
lab = str.get;

toid = coco_get_id(ttbid, 'seg');
soid = coco_get_id(stbid, 'seg');
prob = coll_sol2seg(prob, toid, run, soid, lab);
data = coco_read_solution(stbid, run, lab);
data = po_init_data(prob, ttbid, data);
prob = po_close_orb(prob, ttbid, data);

end
```

Finally, the function `po_read_solution`, shown below, provides a candidate toolbox extractor compatible with the `po_close_orb` toolbox closer.

```
function [sol data] = po_read_solution(oid, run, lab)

tbid = coco_get_id(oid, 'po');
data = coco_read_solution(tbid, run, lab);

segoid = coco_get_id(tbid, 'seg');
sol    = coll_read_solution(segoid, run, lab);

end
```

8.2.2 Examples

The `po_isol2orb` toolbox constructor may be used to construct a periodic orbit object from a given initial solution guess, as may have been found using forward simulation or from a preliminary theoretical analysis. Consider, for example, continuation along a family of periodic orbits emanating from a generic Hopf bifurcation of an equilibrium of a dynamical system. Let p^* and y^* denote the parameter vector and equilibrium point associated with the Hopf bifurcation such that

$$f\left(y^*, p^*\right) = 0. \tag{8.20}$$

Let $\pm i\omega^*$ denote a pair of purely imaginary, conjugate eigenvalues of the Jacobian matrix $\partial_y f\left(y^*, p^*\right)$ with corresponding eigenvectors $v_R \pm iv_I$. From normal-form theory we know that the tangent space to the family of periodic orbits at y^* is the 2-dimensional subspace spanned by v_R and v_I. Indeed, suppose that the equilibrium point y^* belongs to a family of equilibria for p on some neighborhood of p^* such that there occurs a nondegenerate change in linear stability along this family as p crosses p^*. Then, for ε sufficiently small, there exists a family of periodic orbits, emanating from the equilibrium at y^* when $p = p^*$, and approximately given by

$$y(t, \varepsilon) = y^* + \varepsilon \left(v_R \cos \omega t - v_I \sin \omega t \right) + \mathcal{O}\left(\varepsilon^2 \right) \tag{8.21}$$

for $t \in [0, 2\pi/\omega]$, where

$$\omega = \omega(\varepsilon) = \omega^* + \mathcal{O}\left(\varepsilon^2 \right) \tag{8.22}$$

and

$$p = p(\varepsilon) = p^* + \mathcal{O}\left(\varepsilon^2 \right). \tag{8.23}$$

Example 8.3 As an example, consider the vector field

$$f(y, p) = \begin{pmatrix} p_1 y_1 + y_2 + p_2 y_1^2 \\ -y_1 + p_1 y_2 + y_2 y_3 \\ \left(p_1^2 - 1 \right) y_2 - y_1 - y_3 + y_1^2 \end{pmatrix} \tag{8.24}$$

encoded in the function marsden, shown below.

```
function y = marsden(x, p)

x1 = x(1,:);
x2 = x(2,:);
x3 = x(3,:);
p1 = p(1,:);
p2 = p(2,:);

y(1,:) = p1.*x1+x2+p2.*x1.^2;
y(2,:) = -x1+p1.*x2+x2.*x3;
y(3,:) = (p1.^2-1).*x2-x1-x3+x1.^2;

end
```

The equilibrium at $y^* = 0$ undergoes a nondegenerate Hopf bifurcation as p_1 crosses $p_1^* = 0$ for arbitrary values of p_2. In each case, $\omega^* = 1$, $v_R = \begin{pmatrix} 1 & 0 & -1 \end{pmatrix}^T$, and $v_I = \begin{pmatrix} 0 & 1 & 0 \end{pmatrix}^T$. Let $\varepsilon = 0.01$. Then, the commands

```
>> t0 = (0:2*pi/100:2*pi)';
>> x0 = 0.01*(cos(t0)*[1 0 -1]-sin(t0)*[0 1 0]);
>> p0 = [0; 6];
```

assign content to the t0, x0, and p0 variables for use in the call to the po_isol2orb constructor, as shown in the extract below.

```
>> prob = coco_prob();
>> prob = po_isol2orb(prob, '', @marsden, t0, x0, {'p1' 'p2'}, p0);
>> coco(prob, 'run1', [], 1, {'p1' 'po.period'}, [-1 1]);
```

| STEP | | DAMPING | | NORMS | | COMPUTATION TIMES | | |
IT	SIT	GAMMA	||d||	||f||	||U||	F(x)	DF(x)	SOLVE
0				7.56e-04	1.07e+01	0.0	0.0	0.0
1	1	1.00e+00	6.96e-03	4.26e-07	1.07e+01	0.0	0.0	0.0
2	1	1.00e+00	8.25e-06	4.55e-13	1.07e+01	0.0	0.0	0.0
3	1	1.00e+00	6.95e-11	4.68e-17	1.07e+01	0.0	0.0	0.0

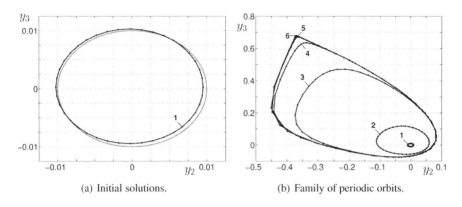

(a) Initial solutions.   (b) Family of periodic orbits.

Figure 8.3. *We compute a family of periodic orbits emanating from a Hopf bifurcation point of the dynamical system given by the vector field in Eq. (8.24). We obtain an initial solution guess from normal form analysis, shown in gray in panel* (a). *The initial correction step converges to orbit* 1. *Panel* (b) *shows the family of periodic orbits of increasing amplitudes that seem to approach a homoclinic orbit, indicated by the corner that develops in the top left part of the plot and allocates many mesh points due to slow dynamics. The labels correspond to the session output included in the text.*

```
STEP      TIME        ||U||   LABEL  TYPE         p1    po.period
   0   00:00:00   1.0727e+01       1  EP    -1.5307e-04   6.2874e+00
  10   00:00:00   1.1135e+01       2        -9.6848e-03   6.6100e+00
  20   00:00:01   1.4514e+01       3        -1.6663e-02   9.1683e+00
  30   00:00:02   1.9109e+01       4        -1.3393e-02   1.2571e+01
  40   00:00:02   2.3842e+01       5        -1.3151e-02   1.6048e+01
  50   00:00:03   2.8651e+01       6  EP    -1.3215e-02   1.9553e+01

STEP      TIME        ||U||   LABEL  TYPE         p1    po.period
   0   00:00:03   1.0727e+01       7  EP    -1.5307e-04   6.2874e+00
  10   00:00:04   1.0876e+01       8        -4.1165e-03   6.4058e+00
  20   00:00:04   1.2885e+01       9        -2.0073e-02   7.9495e+00
  30   00:00:05   1.7371e+01      10        -1.3793e-02   1.1288e+01
  40   00:00:06   2.2065e+01      11        -1.3191e-02   1.4745e+01
  50   00:00:07   2.6849e+01      12  EP    -1.3166e-02   1.8243e+01
```

A subset of the corresponding family of periodic orbits is shown in Fig. 8.3(b). ∎

We note a dramatic increase in the approximate period T under relatively small variations in $p_1 \approx -0.013$ during the execution of the `coco` entry-point function in Example 8.3. To explore this behavior, consider the state-space representation of the approximate periodic orbit corresponding to the solution label 6, shown in Fig. 8.4. This graph reveals an apparent "corner" developing along the periodic orbit, at which the discretization points accumulate. We surmise, consequently, the nearby existence of a saddle-like equilibrium with an associated homoclinic trajectory that is closely approximated by the periodic orbit for sufficiently large values of T.

Example 8.4 In the extract below, we use the `po_read_solution` extractor to load the sampled points along the approximate periodic orbit obtained with $T = T_0 \approx 19.55$ during the execution of the `coco` entry-point function in Example 8.3. By searching for a minimum

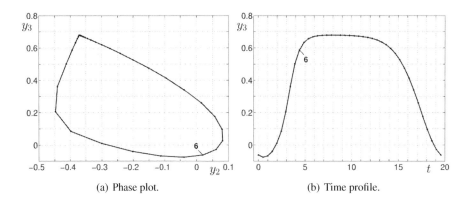

(a) Phase plot.   (b) Time profile.

Figure 8.4. *The last periodic orbit 6 obtained in the continuation run shown in Fig. 8.3. The left panel shows the phase plot and the right panel the time profile. The time profile shows a phase of slow dynamics, indicating existence of a nearby equilibrium point, followed by a fast excursion. To confirm our hypothesis, we use this orbit and extend the part that seems to be close to an equilibrium; see Fig. 8.5.*

of the norm of the vector field f evaluated at each such point, we locate the sample point that lies closest to the surmised saddle equilibrium.

```
>> sol1 = po_read_solution('', 'run1', 6);
>> f = marsden(sol1.x', repmat(sol1.p, [1 size(sol1.x, 1)]));
>> [mn idx] = min(sqrt(sum(f.*f, 1)));
```

We now modify the sampled data in order to obtain a closer approximation to the homoclinic trajectory. In particular, although we retain the sampled values of v_{bp}, we assume that a lengthy time interval, several multiples times the approximate period T_0 obtained previously, is spent in the vicinity of the saddle equilibrium (cf. Fig. 8.5(a)). In the extract below, the `idx` variable is used to separate the time sample into a subset of time steps that precede the passage near the saddle and a complementary subset of time steps that follow this passage.

```
>> scale = 25;
>> T  = sol1.t(end);
>> t0 = [sol1.t(1:idx,1) ; T*(scale-1)+sol1.t(idx+1:end,1)];
>> x0 = sol1.x;
>> p0 = sol1.p;
```

We again call the `po_isol2orb` constructor in order to construct an initializable restricted continuation problem corresponding to a periodic orbit of period equal to $25T_0$. We rediscretize the segment object with a larger value of N in order to resolve the excursion away from the equilibrium.

```
>> prob = coco_set(coco_prob(), 'coll', 'NTST', ceil(scale*10));
>> prob = po_isol2orb(prob, '', @marsden, t0, x0, {'p1' 'p2'}, p0);
```

In order to continue an approximate homoclinic trajectory under variations in the problem parameters, we finally reassign the integer index of the continuation parameter whose value

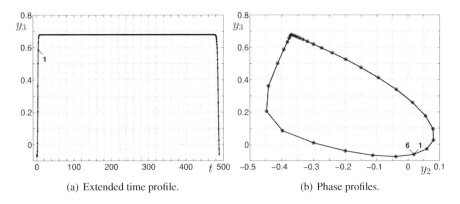

(a) Extended time profile.   (b) Phase profiles.

Figure 8.5. *Starting with orbit 6 from Fig. 8.4, we insert a long segment of constant dynamics and rescale the period such that the shape of the orbit in phase space should be unchanged if there exists a nearby homoclinic orbit. The extended time profile after the initial correction step is shown in panel* (a). *We clearly observe an elongated phase of near-constant dynamics. We overlay this new solution* 1 *(black dot) on top of the previous orbit* 6 *(gray circle) in panel* (b). *The phase plots, including the distribution of mesh points, are virtually identical, which supports the assumption that a nearby homoclinic orbit exists. We continue a family of high-period orbits in Fig. 8.6.*

equals the period T to $\mathbb{I}$. In the extract below, this is accomplished by imposing a pairwise switch between $\mathbb{I}$ and $\mathbb{J}$ of the integer indices corresponding to the parameter labels 'p2' and 'po.period' using the `coco_xchg_pars` utility.

```
>> prob = coco_xchg_pars(prob, 'p2', 'po.period');
>> coco(prob, 'run2', [], 1, {'p1' 'p2'}, [-1 1]);
```

	STEP	DAMPING		NORMS		COMPUTATION	TIMES	
IT	SIT	GAMMA	||d||	||f||	||U||	F(x)	DF(x)	SOLVE
0				4.06e-02	4.90e+02	0.0	0.0	0.0
1	1	1.00e+00	5.72e-02	6.96e-05	4.90e+02	0.0	0.0	0.5
2	1	1.00e+00	4.01e-05	1.18e-10	4.90e+02	0.0	0.1	0.8
3	1	1.00e+00	3.65e-10	1.13e-14	4.90e+02	0.0	0.1	1.0

STEP	TIME	||U||	LABEL	TYPE	p1	p2
0	00:00:01	4.8974e+02	1	EP	-1.3145e-02	6.0224e+00
10	00:00:11	4.8992e+02	2		-1.9656e-02	4.5153e+00
20	00:00:20	4.9021e+02	3		-3.0840e-02	3.2553e+00
30	00:00:29	4.9057e+02	4		-4.5004e-02	2.4202e+00
40	00:00:37	4.9098e+02	5		-6.1259e-02	1.8504e+00
50	00:00:45	4.9145e+02	6	EP	-7.8914e-02	1.4464e+00

STEP	TIME	||U||	LABEL	TYPE	p1	p2
0	00:00:45	4.8974e+02	7	EP	-1.3145e-02	6.0224e+00
10	00:00:54	4.8963e+02	8		-9.0300e-03	7.9986e+00
20	00:01:03	4.8961e+02	9		-6.2870e-03	1.0877e+01
30	00:01:13	4.8968e+02	10		-4.8412e-03	1.4072e+01
40	00:01:23	4.8983e+02	11		-3.9817e-03	1.7421e+01
50	00:01:32	4.9005e+02	12	EP	-3.3994e-03	2.0846e+01

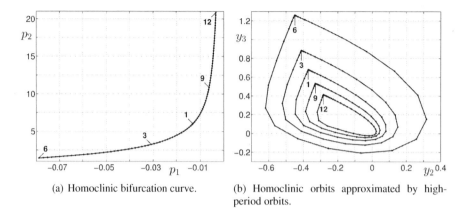

(a) Homoclinic bifurcation curve.   (b) Homoclinic orbits approximated by high-period orbits.

Figure 8.6. *Continuation of the periodic orbit with high period, illustrated in Fig. 8.5, while keeping the period constant, resulting in an approximation to a* homoclinic bifurcation curve *(a). Each point on this curve corresponds to a terminal point along a family of periodic orbits emanating from a Hopf bifurcation under variations in* p_2. *Panel* (b) *shows selected members of the family of high-period orbits. The labels correspond to the session output included in the text.*

The *two-parameter bifurcation diagram* in Fig. 8.6(a) shows the *homoclinic bifurcation curve*, corresponding here to the termination of the family of periodic orbits emanating from the original Hopf bifurcation. Fig. 8.6(b) shows example periodic orbits obtained using this approach. ■

8.3   Alternative embeddings

The 'bvp' toolbox appends a set of algebraic constraints on the system parameters p, the interval length T, and the boundary values v_1 and $v_{N(m+1)}$ to the continuation problem object constructed by the 'coll' toolbox. It does this without a change in the dimension of the domain of the continuation problem. The constraints thus reduce the dimensionality of the solution manifold from $n + q + 1$ to something more manageable. In the case of a fully constrained two-point boundary-value problem, the dimensional deficit equals the number of problem parameters. The toolbox encoding, however, also supports partially constrained boundary-value problems in which between 1 and $n + 1$ constraints are imposed on p, T, v_1, and $v_{N(m+1)}$.

The task-embedding principle described in Chap. 5 is realized in the 'bvp' toolbox by the call to coll_isol2seg in the bvp_isol2seg constructor, and similarly in the case of coll_sol2seg and bvp_sol2seg. This call is then followed by use of the toolbox utility bvp_init_data to extract arrays of the integer indices of the continuation variables corresponding to T, v_1, $v_{N(m+1)}$, and p, respectively, from the corresponding 'coll' toolbox data structure, and to assign these arrays to the 'bvp' toolbox data structure. It is clear that, other than the vital role played by bvp_init_data, the remainder of the 'bvp' toolbox is mostly about bookkeeping.

As seen in the examples, the `bvp` toolbox supports the continuation of solutions of two-point boundary-value problems in the special case of separable boundary conditions, for fixed interval length. Due to the generality of the formulation of the function f_{bc}, the toolbox also supports the continuation of initial-value and final-value problems defined for fixed interval length, as well as two-point boundary-value problems with terminal points constrained to suitable hypersurfaces in $\mathbb{R}^n$. Consider, for example, the function

$$f_{bc} : \left(T, \upsilon_1, \upsilon_{N(m+1)}, p\right) \mapsto \begin{pmatrix} \upsilon_1 - \upsilon_{N(m+1)} \\ \left(\upsilon_1 - \upsilon_1^*\right)^T \cdot f\left(\upsilon_1^*, p^*\right) \end{pmatrix} \qquad (8.25)$$

for some known constant $\upsilon_1^* \in \mathbb{R}^n$. Solutions to the corresponding continuation problem correspond to periodic orbits, albeit with the requirement that the $\tau = 0$ end point lie on the hyperplane in $\mathbb{R}^n$ through υ_1^* and perpendicular to $f\left(\upsilon_1^*, p^*\right)$.

We note the absence of support in the implementation of the `bvp` toolbox for user-level function data used to parameterize the user-supplied function f_{bc}. As a consequence, the values of υ_1^* and p^* in Eq. (8.25) must be set explicitly in the encoding of f_{bc} and cannot be changed during continuation. It follows that continuation may trace only periodic orbits that intersect the given hyperplane through υ_1^*. A solution branch must fold over and retrace the same family of periodic orbits as continuation reaches a parameter value corresponding to a tangential contact with this hyperplane.

To overcome the problem of disappearing intersections of the periodic orbit with the hyperplane during continuation, one could again introduce an update to the value of υ_1^* and the vector $f\left(\upsilon_1^*, p^*\right)$ at opportune moments during continuation. Notably, a solution υ_{bp} of the boundary-value problem, for some value υ_1^*, is by definition also a solution of the boundary-value problem obtained after the update $\upsilon_1^* \leftarrow \upsilon_1$. Under small variations in the active continuation parameters, the corresponding periodic orbit would still be expected to intersect the hyperplane through the updated value of υ_1^* and perpendicular to the updated vector $f\left(\upsilon_1^*, p^*\right)$.

As suggested above, however, such an update mechanism cannot be accomplished within the `bvp` toolbox implemented in Sect. 8.1. At this point, a choice may be made from among the following:

- terminating continuation along the solution manifold as a point of tangential contact with the hyperplane is approached and then restarting continuation with a different value for υ_1^* and the vector $f\left(\upsilon_1^*, p^*\right)$ inserted into the encoding of f_{bc};

- generalizing the `bvp` toolbox to offer support for mutable, user-level function data; or

- encoding a separate toolbox that appends the zero functions in Eq. (8.25) explicitly to a collocation continuation problem, stores υ_1^* in the toolbox data structure, and provides a toolbox-specific slot function for updating the value of υ_1^* and the vector $f\left(\upsilon_1^*, p^*\right)$ prior to each new continuation step.

Merit may be found in each of these alternatives, albeit under different circumstances and depending on one's computational priorities. While the mechanism afforded by the first alternative may appear to require the least upfront effort, it nevertheless relies on monitoring the angle between the vectors $f\left(\upsilon_1, p\right)$ and $f\left(\upsilon_1^*, p^*\right)$ during continuation, extracting υ_1

from a solution file, and modifying the encoding of f_{bc}. Doing so more than a couple of times during a continuation run quickly turns into a tedium that calls for automation.

Such automation is afforded with a minimum of change to the 'bvp' toolbox by the mechanism proposed in the second alternative listed above. To this end, consider the following modified encoding of the bvp_F and bvp_DFDU zero-function wrappers.

```
function [data y] = bvp_F(prob, data, u)

T  = u(data.T_idx);
x0 = u(data.x0_idx);
x1 = u(data.x1_idx);
p  = u(data.p_idx);

y  = data.fhan(data.bc_data, T, x0, x1, p);

end

function [data J] = bvp_DFDU(prob, data, u)

T  = u(data.T_idx);
x0 = u(data.x0_idx);
x1 = u(data.x1_idx);
p  = u(data.p_idx);

J  = data.dfdxhan(data.bc_data, T, x0, x1, p);

end
```

Here, function data used by the user-supplied function f_{bc} and its Jacobian is assumed to be stored in the bc_data field of the toolbox data structure. We assign content to this field and to an additional bc_update field by modifying the bvp_isol2seg constructor to accommodate the modified argument syntax

```
varargin = coll, fhan, [dfdxhan], [bc_data, [bc_update]]
```

as shown below.

```
function prob = bvp_isol2seg(prob, oid, varargin)

tbid   = coco_get_id(oid, 'bvp');
segoid = coco_get_id(tbid, 'seg');
str    = coco_stream(varargin{:});
prob   = coll_isol2seg(prob, segoid, str);
data.fhan = str.get;
data.dfdxhan = [];
if is_empty_or_func(str.peek)
  data.dfdxhan = str.get;
end
data.bc_data   = struct();
data.bc_update = [];
if isstruct(str.peek)
  data.bc_data = str.get;
  if isa(str.peek, 'function_handle')
    data.bc_update = str.get;
  end
end

data = bvp_init_data(prob, tbid, data);
bvp_arg_check(prob, tbid, data);
```

```
prob = bvp_close_seg(prob, tbid, data);

end
```

Here, `data.bc_update` is assumed to contain a function handle to a user-level function for updating the content of the `bc_data` field in response to the `'update'` signal, as shown below in the encoding of the `bvp_bc_update` function.

```
function data = bvp_bc_update(prob, data, cseg, varargin)

uidx = coco_get_func_data(prob, data.tbid, 'uidx');
u    = cseg.src_chart.x(uidx);
T    = u(data.T_idx);
x0   = u(data.x0_idx);
x1   = u(data.x1_idx);
p    = u(data.p_idx);
data.bc_data = data.bc_update(data.bc_data, T, x0, x1, p);

end
```

As the `'update'` signal is emitted prior to continuation as well as before each new continuation step, this function serves a dual purpose of initializing the `bc_data` field and of updating this field during continuation. The following encoding of the `bvp_close_seg` closer makes the necessary changes to the `prob` continuation problem structure.

```
function prob = bvp_close_seg(prob, tbid, data)

if ~isempty(data.bc_update)
  data.tbid = tbid;
  data = coco_func_data(data);
  prob = coco_add_slot(prob, tbid, @bvp_bc_update, data, 'update');
end
segtbid  = coco_get_id(tbid, 'seg.coll');
[fdata uidx] = coco_get_func_data(prob, segtbid, 'data', 'uidx');
uidx = uidx([fdata.T_idx; fdata.x0_idx; fdata.x1_idx; fdata.p_idx]);
if isempty(data.dfdxhan)
  prob = coco_add_func(prob, tbid, @bvp_F, data, 'zero', 'uidx', uidx);
else
  prob = coco_add_func(prob, tbid, @bvp_F, @bvp_DFDU, data, 'zero', ...
    'uidx', uidx);
end
prob = coco_add_slot(prob, tbid, @coco_save_data, data, 'save_full');

end
```

We proceed to encode the function f_{bc} in Eq. (8.25) and its Jacobian in the functions `per_bc` and `per_bc_DFDX`, shown below.

```
function fbc = per_bc(data, T, x0, x1, p)
  fbc = [x0-x1; data.f0*(x0-data.x0)];
end


function Jbc = per_bc_DFDX(data, T, x0, x1, p)
  Jbc = data.J;
end
```

Finally, the function `per_bc_update`, shown below, provides for an update to the content of v_1^*, as discussed above.

```
function data = per_bc_update(data, T, x0, x1, p)

n = numel(x0);
q = numel(p);

data.x0 = x0;
data.f0 = data.fhan(x0,p)';
data.J  = [sparse(n,1), speye(n,n), -speye(n,n), sparse(n,q);
           sparse(1,1), data.f0,    sparse(1,n), sparse(1,q)];

end
```

Example 8.5 We demonstrate the use of the modified 'bvp' toolbox in the context of continuation of periodic orbits of the *Lienard* dynamical system

$$\frac{dy}{dt} = f(y, p) \tag{8.26}$$

obtained from the vector field

$$f(y, p) = \begin{pmatrix} y_2 \\ py_2 - y_2^3 - y_1 \end{pmatrix}, \tag{8.27}$$

implemented in the vectorized function below.

```
function y = lienard(x, p)

x1 = x(1,:);
x2 = x(2,:);
p1 = p(1,:);

y(1,:) = x2;
y(2,:) = p1.*x2-x2.^3-x1;

end
```

In the following extract, we first generate an initial solution guess for $p = p_0 := 1$ using a single stage of forward integration. The results of integration are used to assemble content for function data used by `per_bc`, `per_bc_DFDX`, and `per_bc_update`.

```
>> p0 = 1;
>> x0 = [0.4; -1.2];
>> f  = @(t,x) lienard(x, p0);
>> [t0 x0] = ode45(f, [0 6.7], x0);
>> coll_args = {@lienard, t0, x0, 'p', p0};
>> data = struct();
>> data.fhan = @lienard;
>> data = per_bc_update(data, [], x0(1,:)', [], p0);
```

The result of including updates to v_1^* and the vector $f\left(v_1^*, p^*\right)$ is shown below in the output from the call to the `coco` entry-point function and in Fig. 8.7.

```
>> bvp_args = {@per_bc, @per_bc_DFDX, data, @per_bc_update};
>> prob = bvp_isol2seg(coco_prob(), '', coll_args{:}, bvp_args{:});
>> coco(prob, 'run_moving', [], 1, 'p', [-1 1]);
```

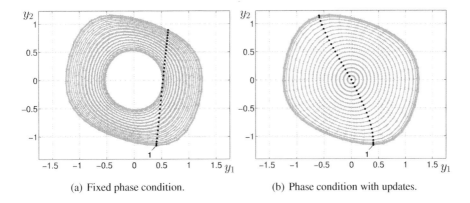

(a) Fixed phase condition.   (b) Phase condition with updates.

Figure 8.7. *Continuation of periodic orbits of the Lienard system in Example* 8.5. *Both continuation runs start at the same initial solution, marked with label* 1. *The initial point y*(0) *on each orbit is emphasized. With a fixed phase condition, all initial points must lie on a straight line* (a), *and we are unable to compute the full family due to a tangency. Updating the phase condition in each continuation step results in the full family; the initial points now lie on a curve passing through the Hopf bifurcation point at the origin within numerical accuracy* (b).

	STEP	DAMPING		NORMS		COMPUTATION	TIMES	
IT	SIT	GAMMA	||d||	||f||	||U||	F(x)	DF(x)	SOLVE
0				7.08e-02	1.09e+01	0.0	0.0	0.0
1	1	1.00e+00	2.92e-01	5.73e-03	1.09e+01	0.0	0.0	0.0
2	1	1.00e+00	9.13e-03	1.39e-05	1.09e+01	0.0	0.0	0.0
3	1	1.00e+00	1.62e-05	5.11e-11	1.09e+01	0.0	0.0	0.0
4	1	1.00e+00	7.13e-11	3.69e-15	1.09e+01	0.0	0.0	0.0

STEP	TIME	||U||	LABEL	TYPE	p
0	00:00:00	1.0900e+01	1	EP	1.0000e+00
10	00:00:00	8.0650e+00	2		3.6536e-01
20	00:00:01	6.2832e+00	3		3.5932e-08
30	00:00:02	8.0634e+00	4		3.6502e-01
38	00:00:03	1.0900e+01	5	EP	1.0000e+00

Continuation here generates a discrete sample of periodic orbits emanating from a Hopf bifurcation at $p = y = 0$, including at least a single orbit of very small amplitude. In contrast, the omission of updates to v_1^* and $f\left(v_1^*, p^*\right)$ results in a family of orbits with p bounded from below by ≈ 0.1.

```
>> bvp_args = {@per_bc, @per_bc_DFDX, data};
>> prob = bvp_isol2seg(coco_prob(), '', coll_args{:}, bvp_args{:});
>> coco(prob, 'run_fixed', [], 1, 'p', [-1 1]);
```

	STEP	DAMPING		NORMS		COMPUTATION	TIMES	
IT	SIT	GAMMA	||d||	||f||	||U||	F(x)	DF(x)	SOLVE
0				7.08e-02	1.09e+01	0.0	0.0	0.0
1	1	1.00e+00	2.92e-01	5.73e-03	1.09e+01	0.0	0.0	0.0
2	1	1.00e+00	9.13e-03	1.39e-05	1.09e+01	0.0	0.0	0.0
3	1	1.00e+00	1.62e-05	5.11e-11	1.09e+01	0.0	0.0	0.0
4	1	1.00e+00	7.13e-11	3.69e-15	1.09e+01	0.0	0.0	0.0

STEP	TIME	||U||	LABEL	TYPE	p
0	00:00:00	1.0900e+01	1	EP	1.0000e+00
10	00:00:00	8.1627e+00	2		3.8668e-01
20	00:00:01	7.4707e+00	3		2.3792e-01
30	00:00:01	9.8514e+00	4		7.6409e-01
33	00:00:02	1.0902e+01	5	EP	1.0000e+00

The artificial turning point along the solution manifold is here a consequence of a tangency of the periodic orbit with the straight line in the (y_1, y_2) plane through $v_1^* = (0.4, -1.2)$ and perpendicular to $f\left(v_1^*, p_0\right) = (-1.2, 0.128)^T$, as shown in Fig. 8.7(a).   ∎

8.4 Conclusions

The $'bvp'$ and $'po'$ toolboxes demonstrate the simplicity with which composite continuation problems may be constructed by embedded calls to the $'coll'$ toolbox. In the next chapter, we consider the natural extension to toolboxes that embed multiple instances of the $'coll'$ toolbox. This provides functionality in support of continuation of families of constrained trajectory segments, with applications to covering invariant manifolds and tracking multisegmented trajectories in hybrid dynamical systems.

As shown in the previous section, the $'bvp'$ toolbox may be modified to give greater flexibility at the user level by including parameterizations of user-level functions that may vary during continuation. The changes described above support the use of the $'bvp'$ toolbox as a general-purpose development platform for more narrowly defined classes of boundary-value problems, e.g., the continuation of periodic orbits realized in terms of the function f_{bc} in Eq. (8.25). Experimentation with functions external to the $'bvp'$ toolbox might, at some point of development, give impetus to the independent development of a stand-alone toolbox. Whether such a toolbox should embed the $'bvp'$ toolbox or simply provide a special-purpose embedding of the $'coll'$ toolbox would depend on the amount of duplication and the extent to which inheritance of $'bvp'$ functionality is desired.

A similar consideration applies to the question of whether the $'po'$ toolbox should embed calls to the $'coll'$ toolbox directly, as was done in this chapter, or only indirectly by first constructing a suitable two-point boundary-value continuation problem object in terms of the zero function

$$f_{bc} : \left(T, v_1, v_{N(m+1)}, p\right) \mapsto v_1 - v_{N(m+1)} \tag{8.28}$$

and then appending the discretized integral phase condition. As a further alternative, one might imagine developing a general-purpose $'int'$ toolbox for handling integral constraints of the form

$$\int_0^1 g\left(v\left(\tau\right)\right) d\tau = 0 \tag{8.29}$$

and then embedding calls to $'bvp'$ and $'int'$ in a modified $'po'$ toolbox. The degree to which inheritance and task decomposition should be pursued is a matter of software engineering judgment that is best developed through practice and experimentation.

As described in the context of the $'po'$ toolbox, the encapsulation of toolbox data in an instance of the `coco_func_data` class provides for selective read/write access in order to enable safe communication across function and toolbox boundaries. Care should be taken when relying on this functionality to avoid unanticipated consequences that may result from unintentional calls-by-reference to toolbox or user-level functions. To this end, safe use of an instance of the `coco_func_data` class suggests that it be used only as an argument to the `coco_add_func` function object constructor. It follows that full read-/write access to the fields of this instance is now afforded only to the class methods of the function object. In contrast, only read access is available to functions that rely on the `coco_get_func_data` utility to extract the function data structure.

Exercises

8.1. Consider the `'bvp'` toolbox developed in Sect. 8.1. Verify that the `bvp_sol2seg` constructor is able to restart continuation from a previously computed solution.

8.2. Consider the linear differential equation

$$my'' + cy' + ky = A\cos(\omega t + \theta)$$

and use the `'bvp'` toolbox to investigate the solution manifold of the boundary-value problem

$$y(0) = y(T) = \delta, \, y'(T) = 0$$

for fixed values of m, c, k, A, ω, and δ. Repeat your analysis for a range of positive and negative values of δ.

8.3. Does the `'bvp'` toolbox require $n+1$ additional zero functions? What can you say about the nominal dimensional deficit of the constrained segment object constructed by the `'bvp'` toolbox?

8.4. Consider again the linear differential equation

$$my'' + cy' + ky = A\cos(\omega t + \theta).$$

Use your observations from the previous exercise to investigate the solution manifold of the boundary-value problem

$$y(0) = y(T) = \delta$$

for fixed values of m, c, k, A, ω, and δ. What is its dimensionality? What is the relationship between this manifold and that obtained in the presence of the additional boundary condition $y'(T) = 0$? Repeat your analysis for a range of positive and negative values of δ.

8.5. Let D and a be two positive numbers. Furthermore, suppose that the function $f : \mathbb{R} \to \mathbb{R}$ is continuous and positive on the interval $(0,1)$ and that the function $J : \mathbb{R}^2 \to \mathbb{R}$ is continuous in its first argument and continuously differentiable in its second argument, with a negative partial derivative with respect to its second argument on the interval $(0,1)$. In [Lory, P., "Enlarging the domain of convergence for multiple shooting by the homotopy method," *Numerische Mathematik*, 35, pp. 231–240, 1980], it is shown that a solution $(C(x), v(x))$ to the two-point boundary-value problem

$$-DC''(x) + (v(x)C(x))' = f(x),$$
$$v'(x) + \tau J(x, C(x)) = 0$$

for $C'(0) = 0$, $C(1) = a$, and $v(0) = 0$ can be uniquely parameterized by $\tau \in [0, 1]$. In particular, when $\tau = 0$, the solution is given explicitly by

$$C_0(x) = a + \frac{1}{D} \int_x^1 \left(\int_0^s f(t)dt \right) ds, v_0(x) = 0.$$

Use the 'bvp' toolbox to perform continuation of this solution under variations in τ until $\tau = 1$ for some example choices for D, a, f, and J.

8.6. Show that the general solution to the Bratu differential equation

$$z''(t) + pe^{z(t)} = 0$$

with the initial condition

$$z(0) = 0$$

is given by

$$z(t) = \ln \left(\frac{2C \left(p + (2C - p)\cosh\left(\sqrt{2C}t\right) + 2\sqrt{C(C-p)}\sinh\left(\sqrt{2C}t\right) \right)}{\left(2C - p + p\cosh\left(\sqrt{2C}t\right)\right)^2} \right),$$

where

$$C = p + \frac{1}{2} \left(z'(0)\right)^2.$$

What is a limiting expression for $z(t)$ in the limit $p \to 0$? In this limit, what value of C guarantees that $z(1) = 0$?

8.7. Consider the solution to the Bratu differential equation with the initial condition $z(0) = 0$ given in the previous exercise. Show that a maximal value of $z(t)$ is obtained for

$$t = \frac{1}{\sqrt{2C}} \cosh^{-1}\left(\frac{2C}{p} - 1\right).$$

Show that $z(t)$ satisfies the boundary condition $z(1) = 0$ only if $z(1-t) = z(t)$. Use this fact to show that

$$z(t) = \ln \left(\frac{1 + \cosh\left(\sqrt{\frac{C}{2}}\right)}{1 + \cosh\left(\sqrt{\frac{C}{2}}(1 - 2t)\right)} \right)$$

satisfies the Bratu differential equation and the boundary conditions $z(0) = z(1) = 0$ provided that C and p are related through an algebraic condition. Use the 'alg' toolbox to continue solutions of this algebraic zero problem. What are your observations regarding the existence and uniqueness of solutions for different ranges of the problem parameter p? Can you reproduce these results with the 'bvp' toolbox?

8.8. Consider the two-point boundary-value problem in the previous exercise. Explore the accuracy of the orthogonal collocation discretization scheme by comparing the approximate solution obtained with the 'bvp' toolbox with the exact solution for different values of N and m.

8.9. Consider a collection of differential equations

$$\frac{dy^{(j)}}{dt} = f^{(j)}\left(y^{(j)}, p\right)$$

for $j = 1, \ldots, M$ in terms of the vector fields $f^{(j)} : \mathbb{R}^{n_j} \times \mathbb{R}^q \to \mathbb{R}^{n_j}$. Let

$$y = \begin{pmatrix} y^{(1)} \\ \vdots \\ y^{(M)} \end{pmatrix}$$

and show that

$$\frac{dy}{dt} = f(y, p)$$

for some suitable vector field $f : \mathbb{R}^{n_1 + \cdots + n_M} \times \mathbb{R}^q \to \mathbb{R}^{n_1 + \cdots + n_M}$. Use this observation to reformulate a system of differential equations with solutions coupled only through the boundary conditions with a two-point boundary-value problem suitable for analysis with the 'bvp' toolbox. Implement a corresponding 'bvp'-compatible constructor and compare your approach to that described in [Dercole, F. and Kuznetsov, Y.A., "SlideCont: An Auto97 driver for bifurcation analysis of Filippov systems," *ACM Transactions on Mathematical Software*, 31(1), pp. 95–119, 2005].

8.10. Modify the 'bvp' toolbox to include a discretized approximation of the integral norm

$$\|v\|_2 := \sqrt{\int_0^1 v(\tau)^T \cdot v(\tau) \, d\tau}$$

for each point on the solution manifold in the output from the coco entry-point function. Apply your modified toolbox to the Bratu problem from Sect. 8.1.2. What do you observe for solutions with $v \approx 0$? How would you modify the toolbox further to enable monitoring of the integral norm without singularities?

8.11. Implement an 'int' toolbox to enable the imposition of integral constraints of the form $F(v, p) = 0$, where

$$F(v, p) := \int_0^1 g(v(\tau), p) d\tau,$$

for some scalar-valued function g, on solutions of a differential equation

$$\frac{dv}{d\tau} = f(v, p).$$

8.12. Use the `'bvp'` toolbox to implement your own version of the MANBVP algorithm developed in [England, J.P., Krauskopf, B., and Osinga, H.M., "Computing one-dimensional global manifolds of Poincaré maps by continuation," *SIAM Journal on Applied Dynamical Systems*, 4(4), pp. 1008–1041, 2005].

8.13. Verify that the bottom component of Eq. (8.17) corresponds to the discretization of the integral phase condition in Eq. (8.18).

8.14. Suppose that $v^*(\tau)$ is a periodic function of τ on the interval $[0, 1]$. Show that the integral

$$\int_0^1 \left\| v(\tau) - v^*(\tau + \sigma) \right\|_2^2 d\tau$$

attains a local extremum when σ is chosen so that

$$\int_0^1 \left(\frac{dv^*}{d\tau}(\tau + \sigma) \right)^T \cdot \left(v(\tau) - v^*(\tau + \sigma) \right) d\tau = 0.$$

Use the periodicity of v^* to relate this observation to the integral phase condition in Eq. (8.18).

8.15. Show that the discretization of the integral phase condition in Eq. (8.18) is satisfied by $v_{bp} = v_{bp}^*$ provided that $v_1^* = v_{N(m+1)}^*$. How would this change if the Gauss–Legendre quadrature weights were omitted in the discretization? What if some other set of quadrature weights were used?

8.16. Suppose that v_{bp}^* satisfies the collocation zero problem and $v_1^* = v_{N(m+1)}^*$. Show that $W' \cdot v_{bp}^*$ lies in the nullspace of the Jacobian of the family of collocation zero functions with respect to v_{bp}. Show that $W' \cdot v_{bp}^*$ lies in the nullspace of the Jacobian of the periodicity zero function $u \mapsto v_1 - v_{N(m+1)}$ with respect to v_{bp}. Interpret these observations in light of the condition number of the Jacobian of the family of zero functions used to define a periodic orbit object.

8.17. Modify the `'po'` toolbox to rely on the `coco_add_glue` special-purpose wrapper to impose the periodicity condition on the segment end points.

8.18. Confirm that the `'coll'` toolbox applies only to autonomous vector fields. Explicit time dependence must therefore be accommodated through the introduction of the time variable or one or several one-to-one functions of the time variable as additional state variables. Consider, for example, a harmonically excited dynamical system and let $\theta \in \mathbb{S}$ equal the phase of excitation. How would you use the modified `'bvp'` toolbox from Sect. 8.3 to perform continuation of periodic orbits in such a system?

8.19. Consider the following coupled, ordinary differential equations:

$$\frac{d\alpha}{dt} = \alpha - \omega\beta - \alpha \left(\alpha^2 + \beta^2 \right),$$
$$\frac{d\beta}{dt} = \omega\alpha + \beta - \beta \left(\alpha^2 + \beta^2 \right).$$

Show that $\alpha \tan \omega t - \beta \equiv 0$ and that the unit circle in the (α, β) plane is invariant and attractive under the corresponding dynamics. Use these observations to suggest a continuous-in-time alternative to the phase θ to capture the harmonic excitation discussed in the previous exercise. How would you generalize this observation to arbitrary instances of periodic excitation? Hint: consider truncated Fourier series decompositions.

8.20. Verify that the encoding of the `po_update` function agrees with the theoretical discussion.

8.21. Implement a modified `'po'` toolbox that allows the user to choose between the integral phase condition and an additional boundary condition based on Eq. (8.25) with appropriate updates. Use this toolbox to explore the differences between the two formulations by applying each to the smooth dynamical system given by the vector field

$$f(y,\lambda) = \begin{pmatrix} (1-\lambda)y_1 - y_2 \\ y_1 + y_1^2 \end{pmatrix}$$

following the discussion in [Beyn, W.-J. and Thümmler, V., "Phase conditions, symmetries and PDE continuation," in *Numerical Continuation Methods for Dynamical Systems*, Krauskopf, B., Osinga, H.M., and Galán-Vioque, J. (Eds.), Canopus Publishing Limited, Bristol, UK, pp. 301–330, 2007].

8.22. Suppose that the linearization of a smooth dynamical system about an equilibrium point y^* has a conjugate pair of purely imaginary eigenvalues $\pm i\omega$ with corresponding eigenvectors $v_R \pm iv_I$. Show that the 2-dimensional affine space through y^* and spanned by v_R and v_I is invariant under the linearized dynamics and that all solutions lie on periodic orbits given by

$$y(t,\varepsilon) = y^* + \varepsilon(v_R \cos\omega t - v_I \sin\omega t),$$

with period $T = 2\pi/\omega$.

8.23. Explain the appearance of a "corner" in the state-space representation of the periodic orbits computed in Example 8.3 as the homoclinic bifurcation curve is approached. Use a graph of the time dependence of the components of y to relate this to a lengthy duration of quiescent behavior near an equilibrium followed by a relatively rapid excursion away from the equilibrium.

8.24. Confirm that the period T is constant in the continuation run in Example 8.4.

8.25. Consider the mean-squared value of the deviation of a solution approximant from its time-averaged value

$$\int_0^1 \left\| v(\tau) - \int_0^1 v(s)\,ds \right\|^2 d\tau.$$

Modify the `'po'` toolbox to support the optional inclusion of a monitor function whose value equals this integral. Denote the corresponding continuation parameter by `'msd'` and suppose that the corresponding index belongs to $\mathbb{J}$ by default. Repeat the continuation of the family of periodic orbits in Example 8.3 and graph the variations in the value of `'msd'` along the solution manifold. Use the `coco_xchg_pars` utility to enable the following call to the `coco` entry-point function:

```
coco(prob, '', 1, 'msd', [0 1]);
```

8.26. What functions are responsible for initialization of the `bc_data` field in Example 8.5?

8.27. Graph the sequence of points of intersection of the periodic orbits found in Example 8.5 with the hyperplane through v_1^* and perpendicular to $f\left(v_1^*, p^*\right)$ as a function of p^*, in the absence of updates to the `bc_data` field. What do you observe? Do your observations agree with the claim that the solution manifold folds back on itself and retraces the same branch in the reverse direction?

8.28. Review the theory of Hopf bifurcations in [Marsden, J.E. and McCracken, M., *The Hopf Bifurcation and Its Applications*, Springer-Verlag, New York, 1976]. Use the `'alg'`, `'po'`, and modified `'bvp'` toolboxes to investigate the examples in Sect. 4B of this classical reference, including the vector fields in Examples 8.3 and 8.5 in the present chapter. You will be able to explore the stability properties of periodic orbits and detect Hopf bifurcations along branches of equilibria after studying Chaps. 10 and 17, respectively, of the present text.

Chapter 9

Multisegment Continuation Problems

By a slight extension to the `'bvp'` toolbox, we up the ante significantly in this chapter in terms of the class of continuation problems available for analysis. We demonstrate, in particular, the embedding of multiple instances of the `'coll'` toolbox within a general-purpose, embeddable toolbox for continuation of collections of constrained trajectory segments. As in the case of the `'compalg'` toolbox, this requires the inclusion of additional gluing conditions in the extended continuation problem. We rely on this implementation as the basis for a special-purpose toolbox that supports continuation of families of multisegment periodic trajectories as occur, for example, in hybrid dynamical systems.

9.1 Boundary-value problems

Given a collection of vector fields $\left\{ f^{(j)} : \mathbb{R}^{n_j} \times \mathbb{R}^q \to \mathbb{R}^{n_j} \right\}_{j=1}^{M}$, a corresponding *multisegment boundary-value problem* is obtained by seeking a collection

$$\left\{ \upsilon^{(j)} : [0,1] \to \mathbb{R}^{n_j} \right\}_{j=1}^{M} \tag{9.1}$$

of M smooth curves such that

$$\frac{d\upsilon^{(j)}}{dt} = T_j f^{(j)} \left(\upsilon^{(j)}, p \right) \tag{9.2}$$

and

$$f_{bc} \left(T_1, \ldots, T_M, \upsilon^{(1)}(0), \ldots, \upsilon^{(M)}(0), \upsilon^{(1)}(1), \ldots, \upsilon^{(M)}(1), p \right) = 0. \tag{9.3}$$

The class of two-point boundary-value problems is clearly equivalent to the special case of a single-segment boundary-value problem with $M = 1$.

We provide below an encoding of the `'msbvp'` toolbox for constructing continuation problem objects compatible with the definition above and inspired by the modified version of the `'bvp'` toolbox shown in Sect. 8.3. In particular, we provide wrappers for f_{bc} and its Jacobian that are independent of the value of M, deferring the detailed definition of these functions to the user level. We further include support for function data associated with the user-level implementations of f_{bc} and its Jacobian, as well as for a user-level function for updating such function data. Finally, the implementation mirrors partially the 'compalg' toolbox in Chap. 5. Specifically, we rely on multiple embedded calls to either the coll_isol2seg or the coll_sol2seg toolbox constructor. These are accompanied by appropriate gluing conditions to reflect the shared nature of the problem parameters across all segment objects.

Consider the encoding of the msbvp_isol2segs toolbox constructor, shown below together with the is_empty_or_func subfunction.

```
function prob = msbvp_isol2segs(prob, oid, varargin)

tbid = coco_get_id(oid, 'msbvp');
str  = coco_stream(varargin{:});
data.nsegs = 0;
while isa(str.peek, 'function_handle')
  data.nsegs = data.nsegs+1;
  segoid = coco_get_id(tbid, sprintf('seg%d', data.nsegs));
  prob   = coll_isol2seg(prob, segoid, str);
end
data.pnames = {};
if strcmpi(str.peek, 'end-coll')
  str.skip;
else
  data.pnames = str.get('cell');
end
data.fbchan = str.get;
data.dfbcdxhan = [];
if is_empty_or_func(str.peek)
  data.dfbcdxhan = str.get;
end
data.bc_data   = struct();
data.bc_update = [];
if isstruct(str.peek)
  data.bc_data = str.get;
  if is_empty_or_func(str.peek)
    data.bc_update = str.get;
  end
end

msbvp_arg_check(prob, tbid, data);
data = msbvp_init_data(prob, tbid, data);
prob = msbvp_close_segs(prob, tbid, data);

end

function flag = is_empty_or_func(x)
  flag = isempty(x) || isa(x, 'function_handle');
end
```

The varargin argument syntax for msbvp_isol2segs is given by

```
varargin = {coll}, (pnames | 'end-coll'), @fbc, [@dfbcdx],
    [bc_data, [@bc_update]]
```

where

```
coll = @f, [(@dfdx | '[]'), [(@dfdp | '[]')]], t0, x0, p0
```

Here, the stop token `'end-coll'` is included in order to indicate termination of the sequence of arguments of the form `coll` in the absence of parameter labels. The repeated calls to the embedded `coll_isol2seg` toolbox constructor result in multiple instances of the collocation zero problem with toolbox instance identifiers given by appending the string `'msbvp.seg#.coll'`, where # denotes a unique integer for each trajectory segment, to the `oid` object instance identifier. Basic error checking is provided by `msbvp_arg_check`, shown below.

```
function msbvp_arg_check(prob, tbid, data)

assert(data.nsegs~=0, '%s: insufficient number of segments', tbid);
pnum = [];
for i=1:data.nsegs
  fid   = coco_get_id(tbid,sprintf('seg%d.coll', i));
  fdata = coco_get_func_data(prob, fid, 'data');
  assert(isempty(fdata.pnames), ...
    '%s: parameter labels must not be passed to coll', tbid);
  assert(isempty(pnum) || pnum==numel(fdata.p_idx), '%s: %s', ...
    tbid, 'number of parameters must be equal for all segments');
  pnum = numel(fdata.p_idx);
end
assert(iscellstr(data.pnames) || isempty(data.pnames), ...
  '%s: incorrect format for parameter labels', tbid);
assert(pnum==numel(data.pnames) || isempty(data.pnames), ...
  '%s: incompatible number of parameter labels', ...
  tbid);
assert(isa(data.fbchan, 'function_handle'), ...
  '%s: input for ''fbc'' is not a function handle', tbid);

end
```

We initialize content of the `'msbvp'` toolbox data structure in the `msbvp_init_data` function, shown below.

```
function data = msbvp_init_data(opts, tbid, data)

xnum = 0;
for i=1:data.nsegs
  stbid = coco_get_id(tbid, sprintf('seg%d.coll', i));
  fdata = coco_get_func_data(opts, stbid, 'data');
  xnum  = xnum+numel(fdata.x0_idx);
end

data.T_idx  = (1:data.nsegs)';
data.x0_idx = data.nsegs+(1:xnum)';
data.x1_idx = data.nsegs+xnum +(1:xnum)';
data.p_idx  = data.nsegs+2*xnum+(1:fdata.pdim)';

end
```

Here, the `nsegs` field is assumed to contain the integer *M* and the `xnum` variable keeps a running tally of the accumulated state-space dimension. The `T_idx`, `x0_idx`, `x1_idx`, and `p_idx` fields of the toolbox data structure are used in the encodings of the `msbvp_F` and `msbvp_DFDU` wrappers for f_{bc} and its Jacobian, as shown below.

```
function [data y] = msbvp_F(prob, data, u)

T   = u(data.T_idx);
```

```
x0 = u(data.x0_idx);
x1 = u(data.x1_idx);
p  = u(data.p_idx);

y  = data.fbchan(data.bc_data, T, x0, x1, p);

end

function [data J] = msbvp_DFDU(prob, data, u)

T  = u(data.T_idx);
x0 = u(data.x0_idx);
x1 = u(data.x1_idx);
p  = u(data.p_idx);

J  = data.dfbcdxhan(data.bc_data, T, x0, x1, p);

end
```

In this case, the u input argument is assumed to contain the continuation variables corresponding to $T_1, \ldots, T_M, v_1^{(1)}, \ldots, v_1^{(M)}, v_{N_1(m_1+1)}^{(1)}, \ldots, v_{N_M(m_M+1)}^{(M)}$, and p, in that order. Here, the notation $v_i^{(j)}$ refers to the corresponding element of the v_{bp} array for the jth segment. The implementation thus allows for segment-specific choices for the number of discretization intervals N and the degree of the polynomial approximants m.

The function msbvp_bc_update provides a wrapper for a user-defined function for updating the bc_data field of the toolbox data structure, as shown below.

```
function data = msbvp_bc_update(prob, data, cseg, varargin)

uidx = coco_get_func_data(prob, data.tbid, 'uidx');
u    = cseg.src_chart.x(uidx);
T    = u(data.T_idx);
x0   = u(data.x0_idx);
x1   = u(data.x1_idx);
p    = u(data.p_idx);
data.bc_data = data.bc_update(data.bc_data, T, x0, x1, p);

end
```

The msbvp_close_segs closer, shown below, now appends the boundary conditions to the collection of collocation zero problems constructed by the embedded calls to the 'coll' constructors and provides for the necessary gluing conditions and inactive continuation parameters.

```
function prob = msbvp_close_segs(prob, tbid, data)

if ~isempty(data.bc_update)
  data.tbid = tbid;
  data = coco_func_data(data);
  prob = coco_add_slot(prob, tbid, @msbvp_bc_update, data, 'update');
end
T_idx  = zeros(data.nsegs,1);
x0_idx = [];
x1_idx = [];
s_idx  = cell(1, data.nsegs);
for i=1:data.nsegs
  fid       = coco_get_id(tbid,sprintf('seg%d.coll', i));
  [fdata uidx] = coco_get_func_data(prob, fid, 'data', 'uidx');
```

```
    T_idx(i) = uidx(fdata.T_idx);
    x0_idx   = [x0_idx; uidx(fdata.x0_idx)];
    x1_idx   = [x1_idx; uidx(fdata.x1_idx)];
    s_idx{i} = uidx(fdata.p_idx);
  end
  uidx = [T_idx; x0_idx; x1_idx; s_idx{1}];
  if isempty(data.dfbcdxhan)
    prob = coco_add_func(prob, tbid, @msbvp_F, data, ...
      'zero', 'uidx', uidx);
  else
    prob = coco_add_func(prob, tbid, @msbvp_F, @msbvp_DFDU, data, ...
      'zero', 'uidx', uidx);
  end
  for i=2:data.nsegs
    fid  = coco_get_id(tbid, sprintf('shared%d', i-1));
    prob = coco_add_glue(prob, fid, s_idx{1}, s_idx{i});
  end
  if ~isempty(data.pnames)
    fid  = coco_get_id(tbid, 'pars');
    prob = coco_add_pars(prob, fid, s_idx{1}, data.pnames);
  end
  prob = coco_add_slot(prob, tbid, @coco_save_data, data, 'save_full');

end
```

The `msbvp_sol2segs` toolbox constructor, shown below, enables continuation to start from a previously computed solution point.

```
function prob = msbvp_sol2segs(prob, oid, varargin)

ttbid = coco_get_id(oid, 'msbvp');
str = coco_stream(varargin{:});
run = str.get;
if ischar(str.peek)
  stbid = coco_get_id(str.get, 'msbvp');
else
  stbid = ttbid;
end
lab = str.get;

data = coco_read_solution(ttbid, run, lab);
for i=1:data.nsegs
  toid = coco_get_id(ttbid, sprintf('seg%d', i));
  soid = coco_get_id(stbid, sprintf('seg%d', i));
  prob = coll_sol2seg(prob, toid, run, soid, lab);
end
data = msbvp_init_data(prob, ttbid, data);
prob = msbvp_close_segs(prob, ttbid, data);

end
```

Finally, a basic toolbox extractor is implemented in `msbvp_read_solution`, shown below.

```
function [sol data] = msbvp_read_solution(oid, run, lab)

tbid = coco_get_id(oid, 'msbvp');
data = coco_read_solution(tbid, run, lab);

sol = cell(1, data.nsegs);
for i=1:data.nsegs
  segoid = coco_get_id(tbid, sprintf('seg%d', i));
```

```
    sol{i} = coll_read_solution(segoid, run, lab);
  end

end
```

9.2   Quasi-periodic invariant tori

In this section, we consider an example application of the `msbvp` toolbox to the continuation of quasi-periodic invariant tori of a smooth dynamical system. As we shall see below, the formulation takes maximal advantage of the multisegment boundary-value problem formulation. It relies on a large number of segment objects representing solution curves on the torus and all-to-all boundary conditions.

In order to provide a context for the theoretical treatment, consider the *Langford* dynamical system given by the vector field

$$f(y,p) := \begin{pmatrix} (y_3 - 0.7)\,y_1 - \omega y_2 \\ \omega y_1 + (y_3 - 0.7)\,y_2 \\ 0.6 + y_3 - \frac{1}{3} y_3^3 - \left(y_1^2 + y_2^2\right)(1 + \rho y_3) + \varepsilon y_3 y_1^3 \end{pmatrix}, \qquad (9.4)$$

where $p = (\omega, \rho, \varepsilon)$. This may be obtained from a center-manifold reduction of a model from hydrodynamics in the vicinity of a critical bifurcation. We encode this vector field and its Jacobians with respect to y and p in the vectorized functions shown below.

```
function y = lang(x, p)

x1  = x(1,:);
x2  = x(2,:);
x3  = x(3,:);
om  = p(1,:);
ro  = p(2,:);
eps = p(3,:);

y(1,:) = (x3-0.7).*x1-om.*x2;
y(2,:) = om.*x1+(x3-0.7).*x2;
y(3,:) = 0.6+x3-x3.^3/3-(x1.^2+x2.^2).*(1+ro.*x3)+eps.*x3.*x1.^3;

end


function J = lang_DFDX(x, p)

x1  = x(1,:);
x2  = x(2,:);
x3  = x(3,:);
om  = p(1,:);
ro  = p(2,:);
eps = p(3,:);

J = zeros(3,3,numel(x1));
J(1,1,:) = (x3-0.7);
J(1,2,:) = -om;
J(1,3,:) = x1;
J(2,1,:) = om;
J(2,2,:) = (x3-0.7);
J(2,3,:) = x2;
J(3,1,:) = -2*x1.*(1+ro.*x3)+3*eps.*x3.*x1.^2;
```

```
J(3,2,:) = -2*x2.*(1+ro.*x3);
J(3,3,:) = 1-x3.^2-ro.*(x1.^2+x2.^2)+eps.*x1.^3;

end


function J = lang_DFDP(x, p)

x1 = x(1,:);
x2 = x(2,:);
x3 = x(3,:);

J = zeros(3,4,numel(x1));
J(1,1,:) = -x2;
J(2,1,:) = x1;
J(3,2,:) = -x3.*(x1.^2+x2.^2);
J(3,3,:) = x3.*x1.^3;

end
```

The introduction in the analysis below of a fourth problem parameter is here anticipated in the initialization of the J array in `lang_DFDP`.

When $\varepsilon = 0$, the dynamical system

$$\frac{dy}{dt} = f(y, p) \tag{9.5}$$

is unchanged under the coordinate transformation

$$y \mapsto e^{\alpha G} \cdot y \tag{9.6}$$

for

$$G := \begin{pmatrix} 0 & -1 & 0 \\ 1 & 0 & 0 \\ 0 & 0 & 0 \end{pmatrix} \tag{9.7}$$

and arbitrary $\alpha \in \mathbb{R}$. We express this observation by the statement that the vector field in Eq. (9.4) for $\varepsilon = 0$ is *equivariant* under a continuous symmetry generated by the matrix G. Now consider the coordinate transformation

$$y = e^{\omega t G} \cdot \begin{pmatrix} r \cos\theta \\ r \sin\theta \\ z \end{pmatrix} \tag{9.8}$$

obtained from the introduction of cylindrical coordinates in a corotating frame. It follows that $d\theta/dt = 0$,

$$\frac{dr}{dt} = r(z - 0.7), \tag{9.9}$$

and

$$\frac{dz}{dt} = 0.6 + z - \frac{1}{3}z^3 - r^2(1 + \rho z). \tag{9.10}$$

It is straightforward to show that the reduced dynamical system given by Eqs. (9.9)–(9.10) possesses an equilibrium for $z = z^* = 0.7$ and some positive $r = r^*$. In the original set of coordinates, this equilibrium corresponds to a periodic trajectory of the equivariant Langford system with period $T_{\text{ret}} = 2\pi/\omega$.

Consider, instead, the existence of a periodic orbit of the reduced dynamical system with period T_{po} and define the *rotation number* $\varrho = T_{\text{ret}}/T_{\text{po}}$. It follows that there exists an invariant torus of the equivariant Langford system with parallel flow, consisting of either

- torus-covering quasi-periodic trajectories, in the case that ϱ is irrational, or

- a continuous family of periodic orbits, in the case that ϱ is rational.

As it happens, the equilibrium with $z = z^*$ of the reduced dynamical system undergoes a Hopf bifurcation for $\rho = \rho^* \approx 0.615$. A family of asymptotically stable periodic orbits emanates from this bifurcation point for values of $\rho \leq \rho^*$, such that $\varrho \approx 1.54/\omega$ for ρ near ρ^*. In the analysis below, we seek to approximate the corresponding invariant torus of the equivariant Langford system, in the case of an irrational rotation number, and to continue this approximation for variations in the problem parameters, including those that break the symmetry.

9.2.1 Algorithm

Given a dynamical system

$$\frac{dy}{dt} = f(y, p), \, y \in \mathbb{R}^n, \, p \in \mathbb{R}^q, \tag{9.11}$$

an invariant set $\mathbb{T} \in \mathbb{R}^n$ is said to be a 2-*dimensional quasi-periodic invariant torus* if there exist a *torus function* $u : \mathbb{S} \times \mathbb{S} \to \mathbb{T}$ and two frequencies $\omega_1, \omega_2 \in \mathbb{R}$ such that the parallel flow

$$\frac{d\theta_i}{dt} = \omega_i, \, i = 1, 2, \tag{9.12}$$

on $\mathbb{S} \times \mathbb{S}$ is mapped under u onto the flow on $\mathbb{T}$ (cf. Fig. 9.1). In this case, the torus function must satisfy the transport equation

$$\omega_1 \frac{\partial u}{\partial \theta_1}(\theta_1, \theta_2) + \omega_2 \frac{\partial u}{\partial \theta_2}(\theta_1, \theta_2) = f(u(\theta_1, \theta_2), p). \tag{9.13}$$

Now consider the function

$$v(\varphi, \tau) := u(2\pi\varphi + 2\pi\varrho\tau, 2\pi\tau), \tag{9.14}$$

for $\varphi, \tau \in [0, 1]$, where the rotation number ϱ is given by the ratio ω_1/ω_2. It follows that

$$v(\varphi, 0) = u(2\pi\varphi, 0) \tag{9.15}$$

and, by the definition of $\mathbb{S}$,

$$v(\varphi, 1) = u(2\pi\varphi + 2\pi\varrho, 2\pi) = u(2\pi\varphi + 2\pi\varrho, 0) = v(\varphi + \varrho, 0). \tag{9.16}$$

Substitution in Eq. (9.13) finally yields

$$\frac{\partial v}{\partial \tau} = Tf(v, p), \tag{9.17}$$

where $T - 2\pi/\omega_2$.

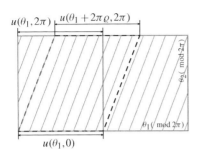
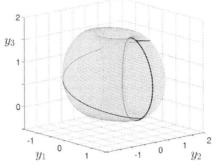

(a) Fundamental domain and characteristic field.   (b) Image of characteristics in phase space.

Figure 9.1. *Two patches of a lift onto $\mathbb{R}^2$ of the fundamental domain of the transport equation in Eq. (9.13) next to each other with a sketch of the characteristic field (a). Applying the method of characteristics with initial condition $u(\theta_1, 0)$ maps the closed curve $u(\cdot, 0)$ onto the closed curve $u(\cdot, 2\pi)$, whereby the parameterization experiences a rotation by $2\pi \varrho$. In state space, this translates to a flow on an invariant torus for which there exists a curve $y(\varphi) = \upsilon(\varphi, 0) := u(2\pi\varphi, 0)$, for $\varphi \in [0, 1]$, that is mapped onto itself under the original flow after time $T = 2\pi/\omega_2$ and under a rotation of the parameterization by ϱ, as illustrated in panel (b).*

Let χ represent a component of $\upsilon(\varphi, 0)$ and consider the truncated Fourier expansion

$$\chi(\varphi) = a_0 + \sum_{k=1}^{N} a_k \cos(2\pi k\varphi) + b_k \sin(2\pi k\varphi) \tag{9.18}$$

for some integer N. It follows that

$$\begin{pmatrix} \chi(0) \\ \chi\left(\frac{1}{2N+1}\right) \\ \vdots \\ \chi\left(\frac{2N}{2N+1}\right) \end{pmatrix} = F^{-1} \cdot \begin{pmatrix} a_0 \\ a_1 \\ b_1 \\ \vdots \\ a_N \\ b_N \end{pmatrix}, \tag{9.19}$$

where the matrix

$$F^{-1} := \begin{pmatrix} 1 & 1 & 0 & \cdots & 1 & 0 \\ 1 & \cos\left(\frac{2\pi}{2N+1}\right) & \sin\left(\frac{2\pi}{2N+1}\right) & \cdots & \cos\left(\frac{2\pi N}{2N+1}\right) & \sin\left(\frac{2\pi N}{2N+1}\right) \\ \vdots & \vdots & \vdots & \ddots & \vdots & \vdots \\ 1 & \cos\left(\frac{4\pi N}{2N+1}\right) & \sin\left(\frac{4\pi N}{2N+1}\right) & \cdots & \cos\left(\frac{4\pi N^2}{2N+1}\right) & \sin\left(\frac{4\pi N^2}{2N+1}\right) \end{pmatrix} \tag{9.20}$$

is the inverse of the *discrete Fourier transform matrix*

$$
F = \frac{1}{2N+1}
\begin{pmatrix}
1 & 1 & \cdots & 1 \\
2 & 2\cos\left(\frac{2\pi}{2N+1}\right) & \cdots & 2\cos\left(\frac{4\pi N}{2N+1}\right) \\
0 & 2\sin\left(\frac{2\pi}{2N+1}\right) & \cdots & 2\sin\left(\frac{4\pi N}{2N+1}\right) \\
\vdots & \vdots & \ddots & \vdots \\
2 & 2\cos\left(\frac{2\pi N}{2N+1}\right) & \cdots & 2\cos\left(\frac{4\pi N^2}{2N+1}\right) \\
0 & 2\sin\left(\frac{2\pi N}{2N+1}\right) & \cdots & 2\sin\left(\frac{4\pi N^2}{2N+1}\right)
\end{pmatrix}.
\tag{9.21}
$$

Consider, similarly, the truncated Fourier expansion of the function χ after a rotation by ϱ:

$$
\chi(\varphi + \varrho) = a_0' + \sum_{k=1}^{N} a_k' \cos(2\pi k\varphi) + b_k' \sin(2\pi k\varphi).
\tag{9.22}
$$

It is straightforward to show that

$$
\begin{pmatrix}
a_0' \\
a_1' \\
b_1' \\
\vdots \\
a_N' \\
b_N'
\end{pmatrix}
= R \cdot
\begin{pmatrix}
a_0 \\
a_1 \\
b_1 \\
\vdots \\
a_N \\
b_N
\end{pmatrix}
\tag{9.23}
$$

in terms of a unique *rotation matrix* R. Indeed, from inspection it follows that

$$
R =
\begin{pmatrix}
1 & & & & & \\
& \cos(2\pi\varrho) & \sin(2\pi\varrho) & & & \\
& -\sin(2\pi\varrho) & \cos(2\pi\varrho) & & & \\
& & & \ddots & & \\
& & & & \cos(2\pi N\varrho) & \sin(2\pi N\varrho) \\
& & & & -\sin(2\pi N\varrho) & \cos(2\pi N\varrho)
\end{pmatrix},
\tag{9.24}
$$

where the omitted entries equal 0.

We proceed to discretize the continuous family $\upsilon(\varphi, \tau)$ by restricting attention to the mesh $\{\varphi_j\}_{j=1}^{2N+1}$, where $\varphi_j := \frac{j-1}{2N+1}$. Following Eqs. (9.15) and (9.16), we require that

$$
(F \otimes I_n) \cdot
\begin{pmatrix}
\upsilon(\varphi_1, 1) \\
\vdots \\
\upsilon(\varphi_{2N+1}, 1)
\end{pmatrix}
= ((R \cdot F) \otimes I_n) \cdot
\begin{pmatrix}
\upsilon(\varphi_1, 0) \\
\vdots \\
\upsilon(\varphi_{2N+1}, 0)
\end{pmatrix}.
\tag{9.25}
$$

Furthermore, from Eq. (9.17), we obtain the differential equations

$$
\frac{d\upsilon}{d\tau}(\varphi_j, \tau) = T_j f\left(\upsilon(\varphi_j, \tau), p\right),
\tag{9.26}
$$

where

$$T_j = T, \, j = 1, \ldots, 2N+1, \tag{9.27}$$

for some continuation variable T.

We finally note that if $(\theta_1, \theta_2) \mapsto u(\theta_1, \theta_2)$ satisfies the transport equation in Eq. (9.13), then so does the function $(\theta_1, \theta_2) \mapsto u(\theta_{1,0} + \theta_1, \theta_{2,0} + \theta_2)$ for arbitrary $\theta_{1,0}, \theta_{2,0} \in \mathbb{S}$. Following the discussion in Chaps. 2 and 8, we may eliminate this lack of uniqueness through the introduction of two suitably formulated phase conditions. The collection of continuation zero problems and discretized boundary conditions then constitute a continuation problem of nominal dimensional deficit $q - 1$. We obtain an initializable restricted continuation problem by introducing at least $q - 1$ monitor functions corresponding to $q - 1$ inactive continuation parameters.

9.2.2   Example

We return to the Langford dynamical system where $n = q = 3$. As shown previously, when $\varepsilon = 0$, an invariant quasi-periodic torus may be obtained by applying the coordinate transformation in Eq. (9.8) to a member of a family of periodic orbits of Eqs. (9.9)–(9.10) emanating from a Hopf bifurcation of an equilibrium with $r = r^* > 0$, $\theta = \theta^* = 0$, and $z = z^* = 0.7$. Let one such periodic orbit be approximated by the functions $r_{\text{per}}(T_{\text{per}}\varphi)$, $\theta_{\text{per}}(T_{\text{per}}\varphi) = 0$, and $z_{\text{per}}(T_{\text{per}}\varphi)$ for some approximate period T_{per}. We obtain a family of approximate curves on the invariant torus by the construction

$$\begin{pmatrix} y_{1,j}(t_k) \\ y_{2,j}(t_k) \\ y_{3,j}(t_k) \end{pmatrix} = \begin{pmatrix} \cos(2\pi t_k) r_{\text{per}}\left(T_{\text{per}}\varphi_j + T_{\text{ret}}t_k\right) \\ \sin(2\pi t_k) r_{\text{per}}\left(T_{\text{per}}\varphi_j + T_{\text{ret}}t_k\right) \\ z_{\text{per}}\left(T_{\text{per}}\varphi_j + T_{\text{ret}}t_k\right) \end{pmatrix} \tag{9.28}$$

for some sequence of sample times $t_k \in [0, 1]$.

By construction, $y_{2,1}(0) = 0$. We generalize this to a phase condition of the form

$$\upsilon_2(\varphi_1, 0) = 0 \tag{9.29}$$

on the discretized family of solution curves discussed in the previous section. Here, the subscript on υ refers to the corresponding component of the state vector. It follows that the initial point on the orbit $\upsilon(\varphi_1, \tau)$ is constrained to the plane $y_2 = 0$. Also, again by construction, we note that $y_{1,1}(0) - y_{1,2}(0) \approx T_{per} r'_{per}(0)/(2N+1)$ provided that N is sufficiently large. Without loss of generality, we choose the phase φ such that $r'_{\text{per}}(\varphi) = 0$ for $\varphi = 0$ and append the corresponding phase condition

$$\upsilon_1(0, \varphi_1) - \upsilon_1(0, \varphi_2) = 0 \tag{9.30}$$

to the continuation problem.

Example 9.1 The `lang_red` function below encodes the reduced dynamical system given by Eqs. (9.9)–(9.10).

```
function y = lang_red(x, p)

x1 = x(1,:);
x2 = x(2,:);
```

```
ro = p(2,:);

y(1,:) = (x2-0.7).*x1;
y(2,:) = 0.6+x2-x2.^3/3-x1.^2.*(1+ro.*x2);

end
```

For $\rho = 0.35$, we approximate the periodic orbit found in this system by a single stage of numerical integration.

```
>> p0      = [3.5; 0.35; 0];
>> T_po    = 5.3;
>> N       = 50;
>> tout    = linspace(0, T_po, 2*N+2);
>> [t x0]  = ode45(@(t,x) lang_red(x, p0), tout, [0.3; 0.4]);
```

We proceed to use the algorithm in Eq. (9.28) to generate initial solution guesses for a family of collocation zero problems on the invariant torus.

```
>> T_ret = 2*pi/p0(1);
>> tt       = linspace(0, 1, 20*(2*N+1))';
>> t1       = T_ret*tt;
>> stt      = sin(tt*2*pi);
>> ctt      = cos(tt*2*pi);
>> coll_args = {};
>> for i=1:2*N+1
     [t xx]   = ode45(@(t,x) lang_red(x, p0), [0 T_ret], x0(i,:));
     xx       = interp1(t, xx, t1);
     x1       = [ctt.*xx(:,1) stt.*xx(:,1) xx(:,2)];
     coll_args = [coll_args {@lang, @lang_DFDX, @lang_DFDP, ...
        t1, x1, [p0; T_ret]}];
   end
```

Finally, we construct the matrices F in Eq. (9.21) and R in Eq. (9.24) and assign $F \otimes I_n$ and $(R \cdot F) \otimes I_n$ to the F and RF fields, respectively, of the user-level function data structure.

```
>> Th   = 2*pi*(0:2*N)/(2*N+1);
>> Th   = kron(1:N, Th');
>> F    = [ones(2*N+1,1) ...
      2*reshape([cos(Th); sin(Th)], [2*N+1 2*N])]'/(2*N+1);
>> Th   = (1:N)*2*pi*T_ret/T_po;
>> SIN  = [zeros(size(Th)); sin(Th)];
>> R    = diag([1, kron(cos(Th), [1, 1])])+ ...
      diag(SIN(:), 1)-diag(SIN(:), -1);
>> data.F  = kron(F, eye(3));
>> data.RF = kron(R*F, eye(3));
```

Specifically, the functions torus_bc and torus_bc_DFDX provide encodings of the boundary conditions and the corresponding Jacobian with respect to the arrays T, x_0, x_1, and p.

```
function fbc = torus_bc(data, T, x0, x1, p)
  fbc = [T-p(4); data.F*x1-data.RF*x0; x0(2); x0(4)-x0(1)];
end


function Jbc = torus_bc_DFDX(data, T, x0, x1, p)

nt = numel(T);
nx = numcl(x0);
```

```
np = numel(p);

J1 = zeros(2,nt+2*nx+np);
J1(1,nt+2) = 1;
J1(2,nt+[1 4]) = [-1 1];

Jbc = [eye(nt), zeros(nt,2*nx+np-1), -ones(nt,1);
    zeros(nx,nt), -data.RF, data.F, zeros(nx,np);
    J1];

end
```

The extract below demonstrates the use of the `msbvp` toolbox and the subsequent application of a 0-dimensional atlas algorithm to locate a single torus approximation.

```
>> prob = msbvp_isol2segs(coco_prob(), '', coll_args{:}, ...
    {'om' 'ro' 'eps' 'T_ret'}, @torus_bc, @torus_bc_DFDX, data);
>> coco(prob, 'run0', [], 0, {'ro' 'T_ret'});
```

	STEP	DAMPING		NORMS			COMPUTATION	TIMES	
IT	SIT	GAMMA	||d||	||f||	||U||	F(x)	DF(x)	SOLVE	
0				7.79e-02	1.00e+02	0.0	0.0	0.0	
1	2	5.00e-01	1.31e+01	7.58e-02	1.00e+02	0.1	1.3	0.9	
2	2	5.00e-01	7.94e+00	5.54e-02	1.00e+02	0.2	1.6	1.1	
3	2	5.00e-01	6.97e+00	4.15e-02	1.00e+02	0.3	1.8	1.3	
4	2	5.00e-01	6.27e+00	3.29e-02	1.00e+02	0.4	2.1	1.5	
5	1	1.00e+00	4.54e+00	2.88e-02	1.00e+02	0.5	2.3	1.7	
6	1	1.00e+00	5.84e-01	7.19e-04	1.00e+02	0.5	2.6	1.9	
7	1	1.00e+00	5.03e-02	4.60e-06	1.00e+02	0.6	2.8	2.1	
8	1	1.00e+00	1.20e-04	2.54e-11	1.00e+02	0.6	3.0	2.3	
9	1	1.00e+00	2.61e-09	3.08e-14	1.00e+02	0.7	3.3	2.4	

STEP	TIME	||U||	LABEL	TYPE	ro	T_ret
0	00:00:07	1.0018e+02	1	EP	3.7375e-01	1.7952e+00

We rely on this torus as the initial guess for continuation with the default 1-dimensional atlas algorithm under additional variations in the problem parameters ε and ρ.

```
>> prob = msbvp_sol2segs(coco_prob(), '', 'run0', 1);
>> coco(prob, 'run_eps', [], 1, {'eps' 'ro' 'T_ret'}, [-0.3 0.3]);
```

	NORMS		COMPUTATION	TIMES	
...	||f||	||U||	F(x)	DF(x)	SOLVE
...	4.81e-14	1.00e+02	0.1	0.0	0.0
...	3.06e-14	1.00e+02	0.0	0.3	0.3

...	LABEL	TYPE	eps	ro	T_ret
...	1	EP	2.6446e-15	3.7375e-01	1.7952e+00
...	2		-4.2855e-03	3.7416e-01	1.7952e+00
...	3		-1.0207e-02	3.7588e-01	1.7952e+00
...	4		-1.8040e-02	3.7943e-01	1.7952e+00
...	5		-2.9839e-02	3.8581e-01	1.7952e+00
...	6	EP	-4.5533e-02	3.9426e-01	1.7952e+00

...	LABEL	TYPE	eps	ro	T_ret
...	7	EP	2.6446e-15	3.7375e-01	1.7952e+00
...	8		4.2071e-03	3.7415e-01	1.7952e+00
...	9		9.7926e-03	3.7573e-01	1.7952e+00
...	10		1.6971e-02	3.7890e-01	1.7952e+00
...	11		2.7977e-02	3.8478e-01	1.7952e+00
...	12	EP	3.9783e-02	3.9126e-01	1.7952e+00

The projection of the solution manifold onto the (ε, ρ)-plane is shown in Fig. 9.2(a). A subset of the invariant tori, corresponding to labeled solutions, together with the associated orbit segments, is shown in Figs. 9.2(b)–9.2(f). ∎

9.3   Multisegment periodic orbits

We proceed to describe a special case of the general multisegment boundary-value problem that applies to the continuation of periodic orbits in hybrid dynamical systems. To this end, consider a sequence $\left\{ y^{(j)} : [0, T_j] \to \mathbb{R}^{n_j} \right\}_{j=1}^{M}$ of M smooth curves and an associated sequence of triplets $\left\{ \left(\mathfrak{m}_j, \mathfrak{e}_j, \mathfrak{r}_j \right) \right\}_{j=1}^{M}$, referred to as the *orbit signature*, such that

$$\frac{dy^{(j)}}{dt} = f\left(y^{(j)}, p; \mathfrak{m}_j \right) \tag{9.31}$$

and

$$h\left(y^{(j)}\left(T_j \right), p; \mathfrak{e}_j \right) = 0 \tag{9.32}$$

for $j = 1, \ldots, M$,

$$g\left(y^{(j)}\left(T_j \right), p; \mathfrak{r}_j \right) = y^{(j+1)}(0) \tag{9.33}$$

for $j = 1, \ldots M - 1$, and

$$g\left(y^{(M)}(T_M), p; \mathfrak{r}_M \right) = y^{(1)}(0) \tag{9.34}$$

for families of smooth functions $f\left(\cdot, \cdot, \mathfrak{m}_j \right) : \mathbb{R}^{n_j} \times \mathbb{R}^q \to \mathbb{R}^{n_j}$ and $h\left(\cdot, \cdot, \mathfrak{e}_j \right) : \mathbb{R}^{n_j} \times \mathbb{R}^q \to \mathbb{R}$ for $j = 1, \ldots, M$, $g\left(\cdot, \cdot, \mathfrak{m}_j \right) : \mathbb{R}^{n_j} \times \mathbb{R}^q \to \mathbb{R}^{n_{j+1}}$ for $j = 1, \ldots, M - 1$, and $g\left(\cdot, \cdot, \mathfrak{m}_M \right) : \mathbb{R}^{n_M} \times \mathbb{R}^q \to \mathbb{R}^{n_1}$ parameterized by

- the sequence of *mode identifiers* $\left\{ \mathfrak{m}_j \right\}_{j=1}^{M}$,

- the sequence of *event identifiers* $\left\{ \mathfrak{e}_j \right\}_{j=1}^{M}$, and

- the sequence of *reset identifiers* $\left\{ \mathfrak{r}_j \right\}_{j=1}^{M}$.

We obtain an extended continuation problem, corresponding to a piecewise-polynomial approximant of a multisegment periodic orbit, by appending to each of the segment-specific collocation zero problems the corresponding *event condition* in Eq. (9.32) and *reset condition* in Eq. (9.33) or (9.34).

As an example, consider a hybrid dynamical system defined by the vector fields

$$f\left(y, p; \mathfrak{left} \right) := \begin{pmatrix} y_1 - y_2 - y_1\sqrt{y_1^2 + y_2^2} \\ y_1 + y_2 - y_2\sqrt{y_1^2 + y_2^2} \end{pmatrix} \tag{9.35}$$

and

$$f\left(y, p; \mathfrak{right} \right) := \begin{pmatrix} \alpha\beta y_1 - (\beta + \gamma) y_2 - (\alpha y_1 - y_2)\sqrt{y_1^2 + y_2^2} \\ \alpha\beta y_2 + (\beta + \gamma) y_1 - (\alpha y_2 + y_1)\sqrt{y_1^2 + y_2^2} \end{pmatrix}, \tag{9.36}$$

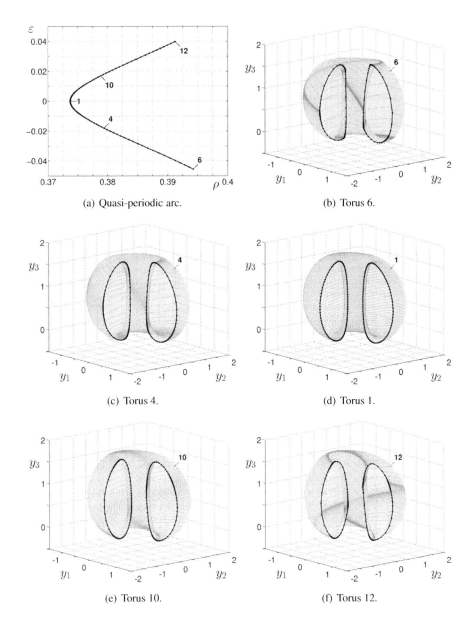

(a) Quasi-periodic arc.   (b) Torus 6.

(c) Torus 4.   (d) Torus 1.

(e) Torus 10.   (f) Torus 12.

Figure 9.2. *A continuation of quasi-periodic invariant tori of the Langford system in Eq.* (9.4) *results in the curve shown in panel* (a), *which is referred to as a* quasi-periodic arc *or a* quasi-periodic hair. *Selected members of this family are shown in panels* (b) *to* (f). *The accumulation of orbits on the torus suggests that this family approaches the vicinity of a* 1:3 *resonance at both ends of the arc. The labels correspond to the session output included in the text.*

the event function

$$h(y, p; \mathfrak{boundary}) := y_1, \qquad (9.37)$$

and the reset function

$$g(y, p; \mathfrak{switch}) := y, \qquad (9.38)$$

where $p = (\alpha, \beta, \gamma)$. An example of a multisegment periodic trajectory is then a two-element sequence of curves

$$\left\{ y^{(j)} : [0, T_j] \to \mathbb{R}^2 \right\}_{j=1}^2 \qquad (9.39)$$

with signature $\{(\mathfrak{left}, \mathfrak{boundary}, \mathfrak{switch}), (\mathfrak{right}, \mathfrak{boundary}, \mathfrak{switch})\}$, where $y_1^{(1)}(t)$ is negative on the interior of its interval of definition and $y_1^{(2)}(t)$ is positive on the interior of its interval of definition. Since the reset function is given by the identity, the two curves join continuously at two points on the y_2 axis. The curve sequence is then equivalent to a two-segment periodic trajectory of the piecewise-smooth dynamical system with vector field $f(y, p; \mathfrak{left})$ for $y_1 < 0$, and $f(y, p; \mathfrak{right})$ for $y_1 > 0$.

The family of event and reset conditions may collectively be described in terms of a family of zero problems of the form

$$f_{bc}\left(v_1^*, v_{N(m+1)}, p; (\mathfrak{e}, \mathfrak{r})\right) = 0, \qquad (9.40)$$

where $v_{N(m+1)}$ denotes the value of the polynomial approximant at the final point on a curve and v_1^* denotes the value of the polynomial approximant at the initial point on the subsequent curve. In the case of the example, it follows that

$$f_{bc}\left(v_1^{(2)}, v_{N_1(m_1+1)}^{(1)}, p; (\mathfrak{boundary}, \mathfrak{switch})\right) := \begin{pmatrix} v_{N_1(m_1+1),1}^{(1)} - v_{1,1}^{(2)} \\ v_{N_1(m_1+1),2}^{(1)} - v_{1,2}^{(2)} \\ v_{N_1(m_1+1),1}^{(1)} \end{pmatrix} \qquad (9.41)$$

and

$$f_{bc}\left(v_1^{(1)}, v_{N_2(m_2+1)}^{(2)}, p; (\mathfrak{boundary}, \mathfrak{switch})\right) := \begin{pmatrix} v_{N_2(m_2+1),1}^{(2)} - v_{1,1}^{(1)} \\ v_{N_2(m_2+1),2}^{(2)} - v_{1,2}^{(1)} \\ v_{N_2(m_2+1),1}^{(2)} \end{pmatrix}, \qquad (9.42)$$

where the second subscript refers to an array component.

9.3.1 Encoding

In this section, we present a partial encoding of an embeddable 'hspo' toolbox that conforms with the paradigm of toolbox construction and task embedding presented in Chaps. 4 and 5. We focus, in particular, on the development of a wrapper to the msbvp_isol2segs constructor that provides an interface suitable for appending the event and reset conditions to the collection of collocation zero problems.

To this end, consider the following 'msbvp'-compatible encoding of the family of event and reset conditions.

```
function y = hspo_bc(data, T, x0, x1, p)

y = [];
for i=1:data.nsegs
  y = [y;
    data.hhan(x1(data.x1_idx{i}), p, data.events{i}); ...
    x0(data.x0_idx{mod(i,data.nsegs)+1})- ...
    data.ghan(x1(data.x1_idx{i}), p, data.resets{i})];
end

end
```

Here, the `data` function data structure is assumed to contain the fields

- `nsegs` equal to M,

- `hhan` equal to a function handle to a vectorized encoding of h,

- `ghan` equal to a function handle to a vectorized encoding of g,

- `events` equal to a cell array whose elements are the sequence $\{\mathfrak{e}_j\}_{j=1}^M$,

- `resets` equal to a cell array whose elements are the sequence $\{\mathfrak{r}_j\}_{j=1}^M$,

- `x0_idx` equal to a cell array whose jth element contains the integer indices of $v_1^{(j)}$ in the `x0` input argument, and

- `x1_idx` equal to a cell array whose jth element contains the integer indices of $v_{N_j(m_j+1)}^{(j)}$ in the `x1` input argument.

The corresponding Jacobian is shown in the encoding of the `hspo_bc_DFDX` function below.

```
function J = hspo_bc_DFDX(data, T, x0, x1, p)

cdim  = data.cdim;
pdim  = data.pdim;
nsegs = data.nsegs;
vals  = [];
for i=1:nsegs
  if ~isempty(data.dhdxhan)
    dhdx = data.dhdxhan(x1(data.x1_idx{i}), p, data.events{i});
  else
    dhdx = coco_ezDFDX('f(x,p)v', data.hhan, x1(data.x1_idx{i}), ...
      p, data.events{i});
  end
  if ~isempty(data.dhdphan)
    dhdp = data.dhdphan(x1(data.x1_idx{i}), p, data.events{i});
  else
    dhdp = coco_ezDFDP('f(x,p)v', data.hhan, x1(data.x1_idx{i}), ...
      p, data.events{i});
  end
  if ~isempty(data.dgdxhan)
    dgdx = data.dgdxhan(x1(data.x1_idx{i}), p, data.resets{i});
  else
    dgdx = coco_ezDFDX('f(x,p)v', data.ghan, x1(data.x1_idx{i}), ...
      p, data.resets{i});
  end
  if ~isempty(data.dgdphan)
    dgdp = data.dgdphan(x1(data.x1_idx{i}), p, data.resets{i});
```

```
    else
      dgdp = coco_ezDFDP('f(x,p)v', data.ghan, x1(data.x1_idx{i}), ...
        p, data.resets{i});
    end
    vals = [vals; dhdx(:); dhdp(:); ...
      ones(data.dim(mod(i,data.nsegs)+1),1); -dgdx(:); -dgdp(:)];
  end
  J = sparse(data.rows, data.cols, vals, nsegs+cdim, nsegs+2*cdim+pdim);

end
```

Here, the `data` function data structure is assumed to contain the additional fields

- `cdim` equal to $\sum_{j=1}^{M} n_j$,

- `dim` equal to the array $\llbracket\ n_1\ \ \cdots\ \ n_M\ \rrbracket$,

- `pdim` equal to q,

- `dhdxhan` equal to a function handle to a vectorized encoding of $\partial_x h$,

- `dhdphan` equal to a function handle to a vectorized encoding of $\partial_p h$,

- `dgdxhan` equal to a function handle to a vectorized encoding of $\partial_x g$,

- `dgdphan` equal to a function handle to a vectorized encoding of $\partial_p g$,

- `rows` equal to the row indices of nontrivial elements of the Jacobian, and

- `cols` equal to the column indices of nontrivial elements of the Jacobian.

The `hspo_isol2segs` constructor, shown with the `coll_func` and `is_empty_or_func` subfunctions below, performs the initial argument parsing required to convert the `varargin` input argument into a format appropriate for the corresponding `'msbvp'` toolbox constructor.

```
function prob = hspo_isol2segs(prob, oid, varargin)

tbid = coco_get_id(oid, 'hspo');
str  = coco_stream(varargin{:});
fhan = str.get;
dfdxhan = cell(1,3);
dfdphan = cell(1,3);
if is_empty_or_func(str.peek)
  dfdxhan = str.get;
  if is_empty_or_func(str.peek)
    dfdphan = str.get;
  end
end
modes  = str.get('cell');
events = str.get('cell');
resets = str.get('cell');
t0     = str.get('cell');
x0     = str.get('cell');
pnames = {};
if iscellstr(str.peek('cell'))
  pnames = str.get('cell');
end
```

```
p0 = str.get;

hspo_arg_check(tbid, fhan, dfdxhan, dfdphan, ...
  modes, events, resets, t0, x0, p0, pnames);
coll = {};
for i=1:numel(modes)
  coll = [coll, {...
    coll_func(fhan{1},    modes{i}), ...
    coll_func(dfdxhan{1}, modes{i}), ...
    coll_func(dfdphan{1}, modes{i}), ...
    t0{i}, x0{i}, p0}];
end
hspo_bc_data = hspo_init_data(fhan, dfdxhan, dfdphan, modes, ...
  events, resets, x0, p0);
prob = msbvp_isol2segs(prob, oid, coll{:}, pnames, ...
  @hspo_bc, @hspo_bc_DFDX, hspo_bc_data);

end

function fhan = coll_func(fhan, mode)

if isa(fhan, 'function_handle')
  fhan = @(x,p) fhan(x,p,mode);
end

end

function flag = is_empty_or_func(x)

flag = all(cellfun('isempty', x) | ...
  cellfun('isclass', x, 'function_handle'));

end
```

Basic error checking is provided by the hspo_arg_check function, shown below.

```
function hspo_arg_check(tbid, fhan, dfdxhan, dfdphan, modes, events, ...
  resets, t0, x0, p0, pnames)

assert(numel(fhan)==3, ...
  '%s: incomplete input for ''fhan''', tbid);
assert(all(cellfun('isclass', fhan, 'function_handle')), ...
  '%s: input for ''fhan'' not function handles', tbid);
assert(numel(dfdxhan)==3, ...
  '%s: incomplete input for ''dfdxhan''', tbid);
assert(numel(dfdphan)==3, ...
  '%s: incomplete input for ''dfdphan''', tbid);
nos = [numel(modes) numel(events) numel(resets) numel(t0) numel(x0)];
assert(~any(diff(nos)), ...
  '%s incompatible segment specification', tbid);
assert(numel(p0)==numel(pnames) || isempty(pnames), ...
  '%s: incompatible number of elements for ''p0'' and ''pnames''', ...
  tbid);

end
```

Finally, the function hspo_init_data, shown below, initializes the function data structure associated with the hspo_bc and hspo_bc_DFDX wrappers.

```
function data = hspo_init_data(fhan, dfdxhan, dfdphan, modes, ...
  events, resets, x0, p0)

data.hhan     = fhan{2};
data.dhdxhan  = dfdxhan{2};
data.dhdphan  = dfdphan{2};
data.ghan     = fhan{3};
data.dgdxhan  = dfdxhan{3};
data.dgdphan  = dfdphan{3};
data.modes    = modes;
data.events   = events;
data.resets   = resets;

nsegs         = numel(events);
data.nsegs    = nsegs;
data.x0_idx   = cell(1,nsegs);
data.x1_idx   = cell(1,nsegs);
cdim          = 0;
dim           = zeros(1,nsegs);
data.dim      = dim;
for i=1:nsegs
  dim(i)          = size(x0{i},2);
  data.dim(i)     = dim(i);
  data.x0_idx{i}  = cdim+(1:dim(i))';
  data.x1_idx{i}  = cdim+(1:dim(i))';
  cdim            = cdim+dim(i);
end
data.cdim = cdim;
rows = [];
cols = [];
off  = 0;
pdim = numel(p0);
data.pdim = pdim;
for i=1:nsegs
  rows = [rows; repmat(off+1, [dim(i)+pdim 1])];
  rows = [rows; repmat(off+1+(1:dim(mod(i,data.nsegs)+1))', ...
    [1+dim(i)+pdim 1])];
  cols = [cols; nsegs+cdim+data.x1_idx{i}; nsegs+2*cdim+(1:pdim)'];
  c2   = repmat(nsegs+cdim+data.x1_idx{i}', ...
    [dim(mod(i,data.nsegs)+1) 1]);
  c3   = repmat(nsegs+2*cdim+(1:pdim), [dim(mod(i,data.nsegs)+1) 1]);
  cols = [cols; nsegs+data.x0_idx{mod(i,data.nsegs)+1}; c2(:); c3(:)];
  off  = off+dim(mod(i,data.nsegs)+1)+1;
end
data.rows = rows;
data.cols = cols;

end
```

9.3.2   Examples

We return to the multisegment boundary-value problem given by the vector fields (9.35) and (9.36), the event function (9.37), and the reset function (9.38). The vector field functions are encoded in the vectorized function shown below.

```
function y = pwlin(x, p, mode)

x1 = x(1,:);
x2 = x(2,:);
```

```
r   = sqrt(x1.^2+x2.^2);

switch mode
  case 'left'
    y(1,:) = (1-r).*x1-x2;
    y(2,:) = x1+(1-r).*x2;
  case 'right'
    al = p(1,:);
    be = p(2,:);
    ga = p(3,:);

    y(1,:) = al.*(be-r).*x1-(ga+be-r).*x2;
    y(2,:) = al.*(be-r).*x2+(ga+be-r).*x1;
end

end
```

The corresponding Jacobians with respect to the state vector and the problem parameters are encoded in the vectorized functions shown below.

```
function J = pwlin_DFDX(x, p, mode)

x1 = x(1,:);
x2 = x(2,:);

r   = sqrt(x1.^2+x2.^2);
rx = x1./r;
ry = x2./r;

J   = zeros(2,2,numel(r));
switch mode
  case 'left'
    J(1,1,:) = 1-r-x1.*rx;
    J(1,2,:) = -1-x1.*ry;
    J(2,1,:) = 1-x2.*rx;
    J(2,2,:) = 1-r- x2.*ry;
  case 'right'
    al   = p(1,:);
    be   = p(2,:);
    ga   = p(3,:);

    al_x = al.*x1-x2;
    al_y = al.*x2+x1;

    J(1,1,:) = al.*be-al.*r-rx.*al_x;
    J(1,2,:) = -be-ga+r-ry.*al_x;
    J(2,1,:) = be+ga-r-rx.*al_y;
    J(2,2,:) = al.*be-al.*r-ry.*al_y;
end

end


function J = pwlin_DFDP(x, p, mode)

x1 = x(1,:);
x2 = x(2,:);

r   = sqrt(x1.^2+x2.^2);

J   = zeros(2,3,numel(r));
```

```
switch mode
  case 'right'
    al = p(1,:);
    be = p(2,:);

    J(1,1,:) = (be-r).*x1;
    J(1,2,:) = al.*x1-x2;
    J(1,3,:) = -x2;
    J(2,1,:) = (be-r).*x2;
    J(2,2,:) = al.*x2+x1;
    J(2,3,:) = x1;
end

end
```

Similarly, the event function h and its Jacobians with respect to x and p are implemented in the functions shown below.

```
function y = pwlin_events(x, p, event)
  y = x(1,:);
end


function J = pwlin_events_DFDX(x, p, event)
  J = [1 0];
end


function J = pwlin_events_DFDP(x, p, event)
  J = zeros(1,3);
end
```

Finally, the functions below implement encodings of the reset function g and its Jacobians with respect to x and p.

```
function y = pwlin_resets(x, p, reset)
  y = x;
end


function J = pwlin_resets_DFDX(x, p, reset)
  J = eye(2);
end


function J = pwlin_resets_DFDP(x, p, reset)
  J = zeros(2,3);
end
```

Example 9.2 Now consider the extract below.

```
>> p0      = [1; 2; 1.5];
>> modes  = {'left' 'right'};
>> events = {'boundary' 'boundary'};
>> resets = {'switch' 'switch'};
>> t0 = linspace(0, pi, 100)';
>> x1 = [-sin(t0) 0.5+cos(t0)];
>> x2 = [ sin(t0) 0.5-cos(t0)];
>> t0 = {t0 0.5*t0};
>> x0 = {x1 x2};
```

```
>> prob = coco_prob();
>> prob = coco_set(prob, 'msbvp.seg1.coll', 'NTST', 10, 'NCOL', 6);
>> prob = coco_set(prob, 'msbvp.seg2.coll', 'NTST', 20, 'NCOL', 4);
>> prob = hspo_isol2segs(prob, '', ...
      {@pwlin, @pwlin_events, @pwlin_resets}, ...
      {@pwlin_DFDX, @pwlin_events_DFDX, @pwlin_resets_DFDX}, ...
      {@pwlin_DFDP, @pwlin_events_DFDP, @pwlin_resets_DFDP}, ...
      modes, events, resets, t0, x0, {'al' 'be' 'ga'}, p0);
>> coco(prob, 'run1', [], 1, 'be', [0 5]);
```

	STEP	DAMPING		NORMS			COMPUTATION	TIMES	
IT	SIT	GAMMA	||d||	||f||	||U||	F(x)	DF(x)	SOLVE	
0				7.95e-01	1.56e+01	0.0	0.0	0.0	
1	2	5.00e-01	1.51e+01	4.36e-01	2.17e+01	0.0	0.0	0.0	
2	1	1.00e+00	3.55e+00	6.87e-02	2.06e+01	0.0	0.0	0.0	
3	1	1.00e+00	4.99e-01	1.55e-03	2.01e+01	0.0	0.0	0.0	
4	1	1.00e+00	2.00e-02	2.69e-06	2.01e+01	0.0	0.0	0.0	
5	1	1.00e+00	2.98e-05	5.74e-12	2.01e+01	0.0	0.1	0.0	
6	1	1.00e+00	3.22e-11	1.18e-14	2.01e+01	0.0	0.1	0.0	

STEP	TIME	||U||	LABEL	TYPE	be
0	00:00:00	2.0117e+01	1	EP	2.0000e+00
10	00:00:00	1.6676e+01	2		1.4606e+00
20	00:00:01	1.2748e+01	3		7.9345e-01
30	00:00:02	9.1859e+00	4		1.1724e-01
32	00:00:02	8.5993e+00	5	EP	0.0000e+00

STEP	TIME	||U||	LABEL	TYPE	be
0	00:00:03	2.0117e+01	6	EP	2.0000e+00
10	00:00:03	2.3665e+01	7		2.5305e+00
20	00:00:04	2.8036e+01	8		3.1630e+00
30	00:00:04	3.2483e+01	9		3.7935e+00
40	00:00:05	3.6987e+01	10		4.4264e+00
49	00:00:06	4.1072e+01	11	EP	5.0000e+00

Here, the initial solution guess u_0 simply equals the unit circle shifted in the y_2 direction by 0.5. Labeled solutions are shown in Fig. 9.3(a).

We may select one of the stored solutions from the initial execution of the coco entry-point function as the starting point for a new continuation run, in which we vary a different parameter. This is demonstrated in the extract below.

```
>> prob = msbvp_sol2segs(coco_prob(), '', 'run1', 9);
>> coco(prob, 'run2', [], 1, 'al', [0 4]);
```

	STEP	DAMPING		NORMS			COMPUTATION	TIMES	
IT	SIT	GAMMA	||d||	||f||	||U||	F(x)	DF(x)	SOLVE	
0				2.28e-14	3.23e+01	0.0	0.0	0.0	
1	1	1.00e+00	2.73e-14	1.37e-14	3.23e+01	0.0	0.0	0.0	

STEP	TIME	||U||	LABEL	TYPE	al
0	00:00:00	3.2277e+01	1	EP	1.0000e+00
10	00:00:00	2.8927e+01	2		8.0331e-01
20	00:00:01	2.4788e+01	3		6.1219e-01
30	00:00:02	2.0661e+01	4		4.1400e-01
40	00:00:02	1.6695e+01	5		1.7071e-01
46	00:00:03	1.4618e+01	6	EP	0.0000e+00

STEP	TIME	||U||	LABEL	TYPE	al
0	00:00:03	3.2277e+01	7	EP	1.0000e+00
10	00:00:04	3.5158e+01	8		1.2954e+00

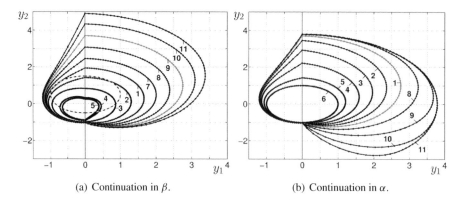

(a) Continuation in β.   (b) Continuation in α.

Figure 9.3. *Families of multisegment periodic orbits from Example 9.2. The initial guess is shown as the dashed closed curve in panel* (a). *The initial correction step converges to orbit 1. Selected members of the family of orbits resulting from a continuation in the parameter β are shown in* (a). *We restart a continuation in α from orbit 9, highlighted in gray. Some members of this family are shown in* (b). *The gray orbit is identical in both panels. The labels correspond to the session output included in the text.*

```
20   00:00:05   3.7453e+01      9            1.8545e+00
30   00:00:05   3.9068e+01     10            2.9120e+00
37   00:00:06   3.9922e+01     11   EP       4.0000e+00
```

Notably, here we use the `msbvp_sol2segs` constructor directly, since all the problem-specific assignments are stored in the solution data array associated with the `msbvp_F` zero function. Representative solutions are shown in Fig. 9.3(b). ∎

As a second example, consider a two-degree-of-freedom, mechanical system consisting of two objects of mass m_1 and m_2, respectively, described by the coupled second-order differential equations

$$m_1 \frac{d^2 z_1}{dt^2} = F_f - F, \quad m_2 \frac{d^2 z_2}{dt^2} = F, \tag{9.43}$$

in terms of their displacements z_1 and z_2 relative to an inertial reference frame. Here, the force

$$F := \frac{\alpha V^2 \sin^2 \omega t}{(d + z_1 - z_2)^2} - k(z_2 - z_1 - l_0) - c\left(\frac{dz_2}{dt} - \frac{dz_1}{dt}\right) \tag{9.44}$$

is obtained from a linear spring of stiffness k and natural length l_0, a linear damper with coefficient c, and an electrostatic attraction between the masses due to an oscillatory voltage excitation described in terms of the permittivity α of free space, the excitation amplitude V, the excitation frequency ω, and the zero-relative-displacement gap distance d. Moreover, F_f denotes a dry friction force acting on the first mass such that

- $F_f = F$ when the mass is in *stick* relative to a substrate and

- F_f achieves a constant magnitude and has a sign opposite to that of $\frac{dz_1}{dt}$ when the mass is in *slip* relative to the substrate.

Finally, we assume the existence of an *equilibrium clearance* δ and a *coefficient of restitution e* such that

$$z_2 - z_1 - l_0 = \delta \text{ and } \frac{dz_2}{dt} - \frac{dz_1}{dt} \geq 0 \qquad (9.45)$$

implies a collisional jump in the velocity of the two masses given by

$$\frac{dz_1}{dt} \mapsto \frac{m_1 - em_2}{m_1 + m_2} \frac{dz_1}{dt} + \frac{m_2(1+e)}{m_1 + m_2} \frac{dz_2}{dt}, \qquad (9.46)$$

$$\frac{dz_2}{dt} \mapsto \frac{m_1(1+e)}{m_1 + m_2} \frac{dz_1}{dt} + \frac{m_2 - em_1}{m_1 + m_2} \frac{dz_2}{dt}. \qquad (9.47)$$

Now let $x_1 = z_1$, $x_2 = \frac{dz_1}{dt}$, $x_3 = z_2 - z_1 - l_0$, $x_4 = \frac{dz_2}{dt} - \frac{dz_1}{dt}$, and $x_5 = \omega t$. It follows that

$$\frac{dx}{dt} = \begin{pmatrix} x_2 \\ \dfrac{F_f - F}{m_1} \\ x_4 \\ \dfrac{F}{m_2} - \dfrac{F_f - F}{m_1} \\ \omega \end{pmatrix}, \qquad (9.48)$$

where

$$F = \frac{\alpha V^2 \sin^2 x_5}{(d - l_0 - x_3)^2} - kx_3 - cx_4 \qquad (9.49)$$

and the impact condition obtained when $x_3 = \delta$ and $x_4 \geq 0$ implies the instantaneous jump

$$x_2 \mapsto x_2 + \frac{m_2(1+e)}{m_1 + m_2} x_4, \qquad (9.50)$$

$$x_4 \mapsto -ex_4. \qquad (9.51)$$

We restrict our attention to the cases of stick, on the one hand, and slip with $x_2 > 0$, on the other. We make the assumptions that stick persists as long as there are no collisions between the masses, and that collisions occur only in stick and with $x_4 \geq 0$. In both stick and slip, it is evident from the equations of motion and the impact conditions that the dynamics in the x_2, x_3, x_4, and x_5 variables may be treated separately from that of x_1. Moreover, in stick, we may similarly omit x_2 from consideration, since this variable remains equal to 0 throughout stick. In slip, however, it is necessary to monitor the value of x_2 in order to determine the cessation of slip and resumption of stick. Finally, we let $m_1 = 5$, $m_2 = 1$, $d - l_0 = 1$, $\delta = 0.5$, and $k = 1$, and we substitute $\alpha V^2 \mapsto V^2$ and $F_f/m_1 \mapsto -\eta$ in slip.

Let $p = (V, c, \eta, \omega, e)$. In stick, the dynamics is thus described by the vector field

$$f(y, p; \text{stick}) := \begin{pmatrix} y_2 \\ F \\ \omega \end{pmatrix}, \qquad (9.52)$$

where

$$F := \frac{V^2 \sin^2 y_3}{(1 - y_1)^2} - cy_2 - y_1, \qquad (9.53)$$

in terms of the state variables $y_1 = x_3$, $y_2 = x_4$, and $y_3 = x_5$. Similarly, in slip with $x_2 > 0$, the dynamics is described by the vector field

$$f(y, p; \text{slip}) := \begin{pmatrix} -\eta - \dfrac{1}{5}F \\ y_3 \\ \eta + \dfrac{6}{5}F \\ \omega \end{pmatrix}, \qquad (9.54)$$

where

$$F := \frac{V^2 \sin^2 y_4}{(1 - y_2)^2} - cy_3 - y_2, \qquad (9.55)$$

in terms of the state variables $y_1 = x_2$, $y_2 = x_3$, $y_3 = x_4$, and $y_4 = x_5$.

In stick, a collision corresponds to a vanishing of the event function

$$h(y, p; \text{collision}) := 0.5 - y_1 \qquad (9.56)$$

and an application of the reset function

$$g(y, p; \text{bounce}) := \begin{pmatrix} \dfrac{(1+e)}{6}y_2 \\ y_1 \\ -ey_2 \\ y_3 \end{pmatrix}, \qquad (9.57)$$

where $y \in \mathbb{R}^3$. In addition, during stick, we may consider a *constant-phase event* given by the vanishing of the event function

$$h(y, p; \text{phase}) := \pi/2 - y_3 \qquad (9.58)$$

and the associated reset function

$$g(y, p; \text{phase}) := \begin{pmatrix} y_1 \\ y_2 \\ y_3 - \pi \end{pmatrix}. \qquad (9.59)$$

Similarly, in stick, we may consider an event obtained at a minimum separation between the objects given by the vanishing of the event function

$$h(y, p; \text{minsep}) := y_2 \qquad (9.60)$$

and the associated reset function

$$g(y, p; \text{turn}) := y, \qquad (9.61)$$

where, again, $y \in \mathbb{R}^3$. Finally, in slip, cessation of slip and resumption of stick occur with the vanishing of the event function

$$h(y, p; \text{rest}) := y_1 \qquad (9.62)$$

and the application of the reset function

$$g\left(y, p; \text{stick}\right) := \begin{pmatrix} y_2 \\ y_3 \\ y_4 \end{pmatrix}, \tag{9.63}$$

where, this time, $y \in \mathbb{R}^4$.

The vector field functions are encoded in the vectorized function below.

```
function y = stickslip(x, p, mode)

switch mode
  case 'stick'
    x1 = x(1,:);
    x2 = x(2,:);
    x3 = x(3,:);
    V  = p(1,:);
    c  = p(2,:);
    w  = p(4,:);

    F  = V.^2.*sin(x3).^2./(1-x1).^2-c.*x2-x1;

    y(1,:) = x2;
    y(2,:) = F;
    y(3,:) = w;
  case 'slip'
    x2 = x(2,:);
    x3 = x(3,:);
    x4 = x(4,:);
    V  = p(1,:);
    c  = p(2,:);
    n  = p(3,:);
    w  = p(4,:);

    F  = V.^2.*sin(x4).^2./(1-x2).^2-c.*x3-x2;

    y(1,:) = -n-F/5;
    y(2,:) = x3;
    y(3,:) = n+6*F/5;
    y(4,:) = w;
  end

end
```

The corresponding Jacobians, with respect to the state vector, are encoded in the vectorized function below.

```
function J = stickslip_DFDX(x, p, mode)

switch mode
  case 'stick'
    x1  = x(1,:);
    x3  = x(3,:);
    V   = p(1,:);
    c   = p(2,:);

    dF1 = 2*V.^2.*sin(x3).^2./(1-x1).^3-1;
    dF2 = -c;
    dF3 = 2*V.^2.*cos(x3).*sin(x3)./(1-x1).^2;
```

```
      J(:,:,:) = zeros(3,3,numel(x1));
      J(1,2,:) = 1;
      J(2,1,:) = dF1;
      J(2,2,:) = dF2;
      J(2,3,:) = dF3;
    case 'slip'
      x1  = x(1,:);
      x2  = x(2,:);
      x4  = x(4,:);
      V   = p(1,:);
      c   = p(2,:);

      dF2 = 2*V.^2.*sin(x4).^2./(1-x2).^3-1;
      dF3 = -c;
      dF4 = 2*V.^2.*cos(x4).*sin(x4)./(1-x2).^2;

      J(:,:,:) = zeros(4,4,numel(x1));
      J(2,3,:) = 1;
      J(1,2,:) = -dF2/5;
      J(1,3,:) = -dF3/5;
      J(1,4,:) = -dF4/5;
      J(3,2,:) = 6*dF2/5;
      J(3,3,:) = 6*dF3/5;
      J(3,4,:) = 6*dF4/5;
  end

end
```

We encode the event and reset functions in the vectorized functions shown below.

```
  function y = stickslip_events(x, p, model)

  switch model
    case 'collision'
      y = 0.5-x(1,:);
    case 'phase'
      y = pi/2-x(3,:);
    case 'minsep'
      y = x(2,:);
    case 'rest'
      y = x(1,:);
  end

end


  function y = stickslip_resets(x, p, mode)

  switch mode
    case 'bounce'
      e = p(5);

      y(1,:) = (1+e)/6*x(2,:);
      y(2,:) = x(1,:);
      y(3,:) = -e*x(2,:);
      y(4,:) = x(3,:);
    case 'phase'
      y(1,:) = x(1,:);
      y(2,:) = x(2,:);
      y(3,:) = x(3,:)-pi;
    case 'turn'
      y = x;
```

```
   case 'stick'
      y = x(2:4,:);
   end

end
```

Example 9.3 Now consider the extract below.

```
>> p0 = [0.59; 0.04; 3.17; 0.92; 0.80];
>> modes  = {'stick' 'stick'};
>> events = {'phase' 'minsep'};
>> resets = {'phase' 'turn'};
>> f  = @(t, x) stickslip(x, p0, modes{1});
>> [t1, x1] = ode45(f, [0  1.5], [0.5; 0; 0]);
>> f  = @(t, x) stickslip(x, p0, modes{2});
>> [t2, x2] = ode45(f, [0  1.5], [0.25; 0; -pi/2]);
>> t0 = {t1 t2};
>> x0 = {x1 x2};
>> prob = hspo_isol2segs(coco_prob(), '', ...
      {@stickslip, @stickslip_events, @stickslip_resets}, ...
      {@stickslip_DFDX, [], []}, modes, events, resets, ...
      t0, x0, {'V' 'c' 'n' 'w' 'e'}, p0);
>> coco(prob, 'run1', [], 1, 'V', [0.5 0.7]);
```

STEP		DAMPING		NORMS			COMPUTATION TIMES		
IT	SIT	GAMMA	||d||	||f||		||U||	F(x)	DF(x)	SOLVE
0				3.62e-01	1.12e+01		0.0	0.0	0.0
1	1	1.00e+00	1.18e+00	7.45e-03	1.13e+01		0.0	0.0	0.0
2	1	1.00e+00	4.20e-02	3.97e-05	1.13e+01		0.0	0.2	0.0
3	1	1.00e+00	2.63e-04	1.69e-09	1.13e+01		0.0	0.2	0.0
4	1	1.00e+00	2.03e-08	4.20e-15	1.13e+01		0.0	0.2	0.0

STEP	TIME	||U||	LABEL	TYPE	V
0	00:00:00	1.1348e+01	1	EP	5.8750e-01
7	00:00:01	1.2355e+01	2	EP	5.0000e-01

STEP	TIME	||U||	LABEL	TYPE	V
0	00:00:01	1.1348e+01	3	EP	5.8750e-01
6	00:00:02	1.0759e+01	4	EP	5.0000e-01

Here, the construction of the initial solution guess u_0 is based on an approximate periodic orbit generated using two instances of forward integration. From the state-space representation of the periodic orbits in Fig. 9.4, we note, in particular, that continuation produces solutions for which x_3 exceeds 0.5, in contradiction to the physical assumption of a collision when x_3 reaches 0.5 followed by a reversal of the relative velocity. It follows that the family of periodic orbits includes an orbit that corresponds to *grazing*, zero-relative-velocity contact between the two objects. We locate this periodic orbit and extract the corresponding time histories using the following commands.

```
>> [data uidx] = coco_get_func_data(prob, 'msbvp', 'data', 'uidx');
>> prob = coco_add_pars(prob, 'graze', uidx(data.x0_idx(1)), 'pos');
>> prob = coco_set_parival(prob, 'pos', 0.5);
>> coco(prob, 'graze_run', [], 0, {'V' 'pos'});
```

STEP		DAMPING		NORMS			COMPUTATION TIMES		
IT	SIT	GAMMA	||d||	||f||		||U||	F(x)	DF(x)	SOLVE
0				3.62e-01	1.12e+01		0.0	0.0	0.0
1	1	1.00e+00	1.18e+00	7.62e-03	1.14e+01		0.0	0.0	0.0

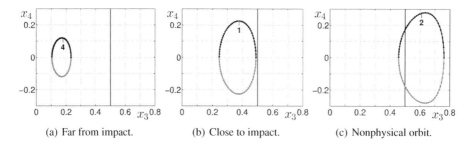

| (a) Far from impact. | (b) Close to impact. | (c) Nonphysical orbit. |

Figure 9.4. *Selected members of a family of nonimpacting two-segment periodic orbits of the stick-slip oscillator considered in Example 9.3 in the coordinates defined on page 237. Orbit 1 is close to impact and orbit 2 is nonphysical, because it penetrates a surface of impact. Consequently, between 1 and 2 there must exist a grazing orbit; see Fig. 9.5. The labels correspond to the session output included in the text.*

```
     2  1  1.00e+00  4.67e-02  6.79e-05  1.14e+01     0.0      0.0      0.0
     3  1  1.00e+00  5.86e-04  4.82e-09  1.14e+01     0.0      0.0      0.0
     4  1  1.00e+00  5.90e-08  2.99e-15  1.14e+01     0.0      0.0      0.0

  STEP      TIME         ||U||   LABEL  TYPE              V          pos
     0  00:00:00   1.1377e+01       1  EP         5.8752e-01   5.0000e-01
```

Here, the `coco_set_parival` utility is used to assign the initial value of 0.5 to the inactive continuation parameter labeled `'pos'`. The corresponding *grazing periodic orbit* is shown in Fig. 9.5(b). ∎

The grazing periodic orbit found in Example 9.3 supports a *resegmentation* obtained by appending a third segment of zero interval length, coincident with the point of grazing contact, and associated with the slip mode identifier, the rest event identifier, and the stick reset identifier.

To explore the implications of segments of zero interval length, recall the rescaled formulation given in Eq. (6.3),

$$\frac{dy}{d\tau} = Tf(y, p),\tag{9.64}$$

and consider, for example, the initial condition

$$y(0) = Y\tag{9.65}$$

for an arbitrary constant Y. It is clear that $y(\tau) \equiv Y$ is a unique solution to this initial-value problem when $T = 0$. It is similarly clear that

$$y_{bp} = \begin{pmatrix} \vdots \\ Y \\ \vdots \end{pmatrix}\tag{9.66}$$

satisfies the corresponding collocation problem combined with the boundary condition

$$y_1 = Y,\tag{9.67}$$

where y_1 denotes the initial base point. The Jacobian of the combined zero problem is now given by the block form

$$\begin{pmatrix} -W' \\ Q \\ \begin{pmatrix} I_n & 0 \end{pmatrix} \end{pmatrix}. \tag{9.68}$$

Nullvectors v_{bp} of this Jacobian thus correspond to solutions to the collocation problem corresponding to

$$\frac{dy}{d\tau} = 0 \tag{9.69}$$

and the initial condition

$$y(0) = 0; \tag{9.70}$$

i.e., by necessity, $v_{bp} = 0$. It follows that the solution (9.66) is a member of a one-parameter family of solutions, parameterized by T, that satisfy (9.64)–(9.65).

A small modification is necessary in the function `coll_init_sol` in order to support the initial construction of the collocation zero problem with $T = 0$. Specifically, consider the encoding below.

```
function sol = coll_init_sol(data, t0, x0, p0)

t0 = t0(:);
T0 = t0(end)-t0(1);
if abs(T0)>eps
  t0 = (t0-t0(1))/T0;
  x0 = interp1(t0, x0, data.tbp)';
else
  x0 = repmat(x0(1,:), size(data.tbp))';
end

sol.u = [x0(:); T0; p0];

end
```

With this modification, it becomes possible to repeat the computations in Sect. 7.3 without the need to first initialize a short trajectory segment.

Example 9.4 We continue the analysis from Example 9.3 and construct a three-segment initial solution guess from the grazing periodic orbit found therein.

```
>> sol = msbvp_read_solution('', 'graze_run', 1);
>> modes  = {'stick' 'stick' 'slip'};
>> events = {'phase' 'collision' 'rest'};
>> resets = {'phase' 'bounce' 'stick'};
>> t0 = {sol{1}.t, sol{2}.t, 0};
>> x0 = {sol{1}.x, sol{2}.x, [0 sol{1}.x(1,:)]};
>> prob = hspo_isol2segs(coco_prob(), '', ...
     {@stickslip, @stickslip_events, @stickslip_resets}, ...
     {@stickslip_DFDX, [], []}, modes, events, resets, ...
     t0, x0, {'V' 'c' 'n' 'w' 'e'}, p0);
>> coco(prob, 'run2', [], 1, 'V', [0.5 0.7]);
```

	STEP	DAMPING		NORMS		COMPUTATION	TIMES	
IT	SIT	GAMMA	$\|\|d\|\|$	$\|\|f\|\|$	$\|\|U\|\|$	F(x)	DF(x)	SOLVE
0				2.77e-03	1.24e+01	0.0	0.0	0.0
1	1	1.00e+00	4.98e-02	2.25e-05	1.24e+01	0.0	0.0	0.0
2	1	1.00e+00	3.76e-04	1.38e-09	1.24e+01	0.0	0.0	0.0

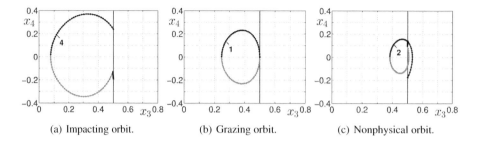

(a) Impacting orbit.   (b) Grazing orbit.   (c) Nonphysical orbit.

Figure 9.5. *Selected members of the family of impacting three-segment periodic orbits of the stick-slip oscillator considered in Example 9.4 in the coordinates defined on page 237. We initialize a computation of this family by resegmenting the grazing orbit from Example 9.3 and performing an initial correction under the additional constraint that the new segment has length 0; that is, it corresponds to the point of grazing contact. We again observe nonphysical orbits penetrating the impact surface. The labels correspond to the session output included in the text.*

```
      3    1  1.00e+00  2.38e-08  3.16e-15  1.24e+01    0.0    0.2    0.0

  STEP      TIME        ||U||   LABEL  TYPE              V
     0  00:00:00   1.2399e+01       1  EP        5.9000e-01
     6  00:00:01   1.2627e+01       2  EP        5.0000e-01

  STEP      TIME        ||U||   LABEL  TYPE              V
     0  00:00:01   1.2399e+01       3  EP        5.9000e-01
     7  00:00:02   1.2479e+01       4  EP        7.0000e-01
```

It is clear from the state-space representations in Fig. 9.5 of the periodic orbits obtained during continuation that the analysis again produces solutions that break the physical assumption of impenetrability of the two objects. Indeed, as seen from a bifurcation diagram showing the value of x_4 at the initial point of the first solution segment as a function of V for each of the two families of periodic orbits, physically viable solutions (i.e., with $x_4 < 0$) exist only for V below a critical value. ∎

9.4 Conclusions

The 'msbvp' toolbox developed in this chapter further demonstrates the principle of task embedding that forms the foundation of the proposed continuation paradigm. With the current implementation, multiple orbit segments may be treated simultaneously with a segment-specific discretization strategy. The general encoding of the boundary conditions also supports a variety of problem-specific segment couplings, including all-to-all coupling in the case of quasi-periodic invariant tori and sequential coupling in the case of periodic orbits in hybrid dynamical systems. We note, in passing, that the latter approach extends naturally to piecewise segmentation of solution trajectories of smooth dynamical systems, for example for monitoring local extrema along a trajectory.

In contrast to smooth systems, hybrid dynamical systems support a phenomenology associated with degenerate, nontransversal intersections of orbit segments with zero-level surfaces of event functions corresponding to nontrivial reset functions. Such instances of *grazing contact* are known to occur in the immediate vicinity, in parameter space, of discontinuity-induced bifurcations—changes in the steady-state behavior that would not occur in the absence of the discontinuity. As seen in Examples 9.3 and 9.4, within the context of boundary-value problems associated with transient dynamics, instances of grazing contact suggest changes to the solution signature. We return to the detection of such special events in Chap. 15.

Exercises

9.1. Consider the dynamical system given by the vector field

$$f(y, p) := \begin{pmatrix} (y_3 - 0.7)\, y_1 - \omega y_2 \\ \omega y_1 + (y_3 - 0.7)\, y_2 \\ 0.6 + y_3 - \frac{1}{3} y_3^3 - \left(y_1^2 + y_2^2\right)(1 + \rho y_3) + \varepsilon y_3 y_1^3 \end{pmatrix},$$

where $p = (\omega, \rho, \varepsilon)$. Show that the differential equations obtained for $\varepsilon = 0$ are unchanged under the coordinate transformation

$$y \mapsto e^{\alpha G} \cdot y,$$

where

$$G := \begin{pmatrix} 0 & -1 & 0 \\ 1 & 0 & 0 \\ 0 & 0 & 0 \end{pmatrix}$$

and $\alpha \in \mathbb{R}$ is arbitrary.

9.2. Suppose that

$$e^{-\alpha G} \cdot f\left(e^{\alpha G} \cdot y, p\right) = f(y, p)$$

for a given vector field f, a given matrix G, and an arbitrary scalar α. Show that the coordinate transformation $y = e^{\omega t G} \cdot \tilde{y}$ implies the equivalence

$$\frac{dy}{dt} = f(y, p) \iff \frac{d\tilde{y}}{dt} = f(\tilde{y}, p) - \omega G \cdot \tilde{y}.$$

9.3. Locate all equilibria of the dynamical system

$$\frac{dr}{dt} = r(z - 0.7)$$

and

$$\frac{dz}{dt} = 0.6 + z - \frac{1}{3} z^3 - r^2 (1 + \rho z).$$

Show that one of these undergoes a Hopf bifurcation when $\rho = \rho^* \approx 0.615$. Use the 'po' toolbox to continue the associated family of periodic orbits and confirm that these orbits exist for $\rho \leq \rho^*$. What is the period in the limit as $\rho \to \rho^*$?

9.4. Explain why an equilibrium point in the reduced dynamical system given by Eqs. (9.9)–(9.10) corresponds to a periodic orbit of the equivariant Langford system obtained by letting $\varepsilon = 0$ in the vector field in Eq. (9.4). What is its period?

9.5. Explain why a periodic orbit in the reduced dynamical system given by Eqs. (9.9)–(9.10) corresponds to an invariant torus of the equivariant Langford system obtained by letting $\varepsilon = 0$ in the vector field in Eq. (9.4). Explain why an irrational rotation number implies a torus-covering quasi-periodic trajectory.

9.6. Consider the transport equation

$$\omega_1 \frac{\partial u}{\partial \theta_1}(\theta_1, \theta_2) + \omega_2 \frac{\partial u}{\partial \theta_2}(\theta_1, \theta_2) = f(u(\theta_1, \theta_2), p)$$

in terms of the positive scalars ω_1 and ω_2. Show that there exists a positive scalar T such that

$$\frac{\partial \upsilon}{\partial \tau} = Tf(\upsilon, p),$$

where

$$\upsilon(\varphi, \tau) := u(2\pi\varphi + 2\pi\varrho\tau, 2\pi\tau).$$

9.7. Show that if $(\theta_1, \theta_2) \mapsto u(\theta_1, \theta_2)$ satisfies the transport equation

$$\omega_1 \frac{\partial u}{\partial \theta_1}(\theta_1, \theta_2) + \omega_2 \frac{\partial u}{\partial \theta_2}(\theta_1, \theta_2) = f(u(\theta_1, \theta_2), p),$$

then so does the function $(\theta_1, \theta_2) \mapsto u(\theta_{1,0} + \theta_1, \theta_{2,0} + \theta_2)$ for arbitrary $\theta_{1,0}$ and $\theta_{2,0}$ in $\mathbb{S}$. Interpret this result in terms of the parameterization of the torus.

9.8. Derive the expression for the matrix F^{-1} given in Eq. (9.20). Show that this matrix is invertible with inverse given by the matrix F in Eq. (9.21). Review the documentation for the MATLAB fft function and propose an alternative way to compute F.

9.9. Derive the expression for the matrix R given in Eq. (9.24). Why must we consider the Kronecker product $(R \cdot F) \otimes I_n$ in Eq. (9.25)?

9.10. Explain the meaning of the conditions in Eq. (9.27) in terms of their implications on the time for different initial conditions on the closed curve $u(2\pi\varphi, 0)$ to return to this curve. Would it be possible to assign a constant value to T in Eq. (9.27), and, if so, what would be the implications to the dimensional deficit of the extended continuation problem?

9.11. Give a geometric interpretation of the phase conditions in Eqs. (9.29)–(9.30) in terms of constraints on the initial condition $\upsilon(0, 0)$ and the initial tangent vector $\partial_\varphi \upsilon(0, 0)$. Do these require updating during continuation?

9.12. Let prob denote the continuation problem structure constructed in Example 9.1. Perform continuation using the following command.

```
coco(prob, 'rpo', [], 1, {'ro', 'om', 'T_ret'}, [0, 1]);
```

Explain the reason for the existence of a 1-dimensional solution manifold to the corresponding restricted continuation problem. Why do we need to release three continuation parameters? Why is it not necessary to release ε? Compare your observations and those in Sect. 9.2 to the numerical results in [Langford, W.F., "Numerical studies of torus bifurcations," in *Numerical Methods for Bifurcation Problems*, Küpper, T.,

Mittlemann, H.D., and Weber, H. (Eds.), Birkhäuser Verlag, Basel, pp. 285–295, 1984].

9.13. Consider the dynamical systems in $\mathbb{R}^2$ governed by the vector fields

$$f(y, p; \mathfrak{left}) := \begin{pmatrix} y_1 - y_2 - y_1\sqrt{y_1^2 + y_2^2} \\ y_1 + y_2 - y_2\sqrt{y_1^2 + y_2^2} \end{pmatrix}$$

and

$$f(y, p; \mathfrak{right}) := \begin{pmatrix} \alpha\beta y_1 - (\beta + \gamma)y_2 - (\alpha y_1 - y_2)\sqrt{y_1^2 + y_2^2} \\ \alpha\beta y_2 + (\beta + \gamma)y_1 - (\alpha y_2 + y_1)\sqrt{y_1^2 + y_2^2} \end{pmatrix}.$$

Write the corresponding differential equations in terms of polar coordinates and use these to find explicit solutions for y_1 and y_2 for each of the two vector fields. Use these solutions to formulate an algebraic continuation problem whose solutions parameterize the periodic orbits found in Example 9.2.

9.14. Consider the implementation of `hspo_init_data` on page 232 and validate the computation of the `rows` and `cols` fields of the toolbox data structure.

9.15. Consider the *impact mapping*

$$\frac{dz_1}{dt} \mapsto \frac{m_1 - em_2}{m_1 + m_2}\frac{dz_1}{dt} + \frac{m_2(1+e)}{m_1 + m_2}\frac{dz_2}{dt},$$

$$\frac{dz_2}{dt} \mapsto \frac{m_1(1+e)}{m_1 + m_2}\frac{dz_1}{dt} + \frac{m_2 - em_1}{m_1 + m_2}\frac{dz_2}{dt},$$

describing an instantaneous change in the rates of change of z_1 and z_2. Show that the linear combination

$$m_1\frac{dz_1}{dt} + m_2\frac{dz_2}{dt}$$

is invariant under the impact mapping. Show that

$$\frac{dz_2}{dt} - \frac{dz_1}{dt} \mapsto -e\left(\frac{dz_2}{dt} - \frac{dz_1}{dt}\right).$$

9.16. Show that $y(\tau) \equiv Y$ is a unique solution to the initial-value problem

$$\frac{dy}{d\tau} = Tf(y, p), \, y(0) = Y,$$

when $T = 0$. Similarly, show that

$$y_{bp} = \begin{pmatrix} \vdots \\ Y \\ \vdots \end{pmatrix}$$

satisfies the collocation problem associated with the initial-value problem

$$\frac{dy}{d\tau} = Tf(y, p), \, y(0) = Y,$$

when $T = 0$.

9.17. Show that the Jacobian matrix

$$\begin{pmatrix} -W' \\ Q \\ \begin{pmatrix} I_n & 0 \end{pmatrix} \end{pmatrix}$$

is nonsingular.

9.18. Using the modified version of the `coll_init_sol` function on page 243, repeat the analysis in Sect. 7.3 by performing the initial continuation from a single point, rather than a short trajectory segment.

9.19. Replace the vector field

$$f(y, p; \mathfrak{stick}) := \begin{pmatrix} y_2 \\ F \\ \omega \end{pmatrix}$$

in Eq. (9.52) with the vector field

$$f(y, p; \mathfrak{stick}) := \begin{pmatrix} 0 \\ y_3 \\ F \\ \omega \end{pmatrix},$$

where $y_1 = x_2$, $y_2 = x_3$, $y_3 = x_4$, and $y_4 = x_5$. Modify the event and reset functions accordingly. Can you repeat the analysis in Example 9.3? Explain your observations.

9.20. Replace the vector field

$$f(y, p; \mathfrak{stick}) := \begin{pmatrix} y_2 \\ F \\ \omega \end{pmatrix}$$

in Eq. (9.52) with the vector field

$$f(y, p; \mathfrak{stick}) := \begin{pmatrix} -y_1 \\ y_3 \\ F \\ \omega \end{pmatrix},$$

where $y_1 = x_2$, $y_2 = x_3$, $y_3 = x_4$, and $y_4 = x_5$. Modify the event and reset functions accordingly. Can you repeat the analysis in Example 9.3? Explain your observations.

9.21. Explain the screen output in Example 9.3 in terms of the shape of the corresponding solution manifold. Review the theory of electromechanical *pull-in instabilities* in deformable capacitive drives and relate these to the observed geometry.

9.22. Review the analysis in [Zhao, X., Reddy, C.K., and Nayfeh, A., "Nonlinear dynamics of an electrically driven impact microactuator," *Nonlinear Dynamics*, 40(3), pp. 227–239, 2005] and use the `'hspo'` toolbox to reproduce some subset of the results therein.

9.23. Verify that the `'hspo'` toolbox supports continuation of single-segment periodic trajectories.

9.24. Explain the physical meaning of the hybrid dynamical system given by the vector field

$$f(y, p; \mathfrak{flight}) = \begin{pmatrix} y_2 \\ -ky_1 - cy_2 + A\cos y_3 \\ \omega \end{pmatrix},$$

event functions

$$h(y, p; \mathfrak{collision}) = 1 - y_1,$$
$$h(y, p; \mathfrak{phase}) = \pi - y_3,$$
$$h(y, p; \mathfrak{minsep}) = y_2,$$

and reset functions

$$g(y, p; \mathfrak{bounce}) = \begin{pmatrix} y_1 \\ -ey_2 \\ y_3 \end{pmatrix},$$

$$g(y, p; \mathfrak{phase}) = \begin{pmatrix} y_1 \\ y_2 \\ y_3 - 2\pi \end{pmatrix},$$

$$g(y, p; \mathfrak{turn}) = y,$$

where $p = (c, k, A, \omega, e)$. Show that, for $A = A(c, k, \omega, e)$, there exist two multisegment periodic orbits with signatures

$$\{(\mathfrak{flight}, \mathfrak{phase}, \mathfrak{phase}), (\mathfrak{flight}, \mathfrak{minsep}, \mathfrak{turn})\}$$

and

$$\{(\mathfrak{flight}, \mathfrak{collision}, \mathfrak{bounce}), (\mathfrak{flight}, \mathfrak{minsep}, \mathfrak{turn})\},$$

respectively, that coincide in state space. Prove that each such trajectory is a regular point on a unique solution manifold and comment on the physical validity of nearby points on this manifold. Hint: the parameter e corresponds to a coefficient of restitution associated with normal impacts between an object and a stationary wall.

9.25. Generalize the result of the previous exercise to the grazing contact of a trajectory segment on an arbitrary multisegment periodic orbit with the zero-level surface of the corresponding event function. In particular, show that under suitable nondegeneracy conditions, such an orbit is a regular point on a unique solution manifold associated with the given signature.

9.26. How would you modify the 'hspo' toolbox to support continuation of two-point boundary-value problems in hybrid dynamical systems?

9.27. Review the continuation paradigm for a problem of optimal control considered in [Cerf, M., Haberkorn, T., and Trélat, E., "Continuation from a flat to a round Earth model in the coplanar orbit transfer problem," *Optimal Control Applications and Methods*, 33(6), pp. 654–670, 2011]. Deploy some version of the 'hspo' toolbox to reproduce the numerical results reported therein.

9.28. The 'hspo' toolbox is naturally restricted to trajectories with finite signatures. It is quite common, however, to encounter periodic trajectories in hybrid dynamical systems with countable, but not finite, signatures and finite total duration. Review the analysis of such orbits in mechanical systems with impacts in [Nordmark, A. and Piiroinen, P., "Simulation and stability analysis of impacting systems with complete chattering," *Nonlinear Dynamics*, 58(1–2), pp. 85–106, 2009] and implement a toolbox capable of performing continuation of such orbits following the scheme proposed therein.

9.29. Write a wrapper to the `'msbvp'` toolbox to handle three-point boundary-value problems of the form considered in [Yao, Q., "Successive iteration and positive solution for nonlinear second-order three-point boundary value problems," *Computers and Mathematics with Applications*, 50(3–4), pp. 433–444, 2005] and apply your toolbox to the examples treated therein.

Chapter 10

The Variational Collocation Problem

We conclude our study of toolbox templates by providing toolboxes for the computation of the Jacobian of the flow of an autonomous, smooth dynamical system along a flow trajectory. The final result constitutes a *sensitivity matrix* whose eigenvalues and eigenvectors provide a comprehensive description of the linearized finite-time behavior of nearby flow trajectories. In the case of a periodic orbit, the eigenvalues (known as *Floquet multipliers*) uniquely determine the *Lyapunov stability* properties of the orbit, at least as long as they all lie away from the unit circle in the complex plane. As shown below, similar observations apply to the case of periodic orbits of hybrid dynamical systems.

We provide two distinct encodings of the `'var_coll'` toolbox. In the first case, the sensitivity matrix is obtained a posteriori, subsequent to finding a solution to the collocation continuation problem. This implementation illustrates the use of nonembedded monitor functions, and of function data to pass auxiliary solution information across toolbox boundaries. In the second implementation, we embed a *variational zero problem* together with the collocation continuation problem, enabling the imposition of additional constraints on the eigenstructure of the sensitivity matrix. We illustrate the application of this second implementation to the continuation of a connecting orbit between an equilibrium point and a periodic orbit. In both cases, each call to the `'var_coll'` toolbox is designed to append a variational collocation problem to one previously instantiated segment object, rather than embedding a call to a `'coll'` constructor. It follows that the `'var_coll'` toolbox may be employed in parallel with any toolbox that embeds one or several calls to the `'coll'` toolbox.

10.1 The first variational equation

Consider a known solution $y(t) \in \mathbb{R}^n$ on the interval $[0, T]$, for some known scalar T, to the ordinary differential equation

$$\frac{dy}{dt} = f(y, p), \tag{10.1}$$

where the vector field $f : \mathbb{R}^n \times \mathbb{R}^q \to \mathbb{R}^n$ is again parameterized by a vector of problem parameters $p \in \mathbb{R}^q$. Suppose that f is continuously differentiable in its first argument and let $Y(t) \in \mathbb{R}^{n \times n}$ denote a solution on the interval $[0, T]$ to the *first variational equation*

$$\frac{dY}{dt} = \partial_y f(y, p) \cdot Y. \tag{10.2}$$

It is then straightforward to show that $Y(t)$ is invertible for all $t \in [0, T]$ provided that it is invertible for some $t \in [0, T]$. In the special case that $Y(0) = I_n$, $Y(t)$ is called the *fundamental solution* of Eq. (10.2). For a general $n \times k$ matrix C, $Y(t) \cdot C$ is the unique solution to Eq. (10.2) that evaluates to $Y(0) \cdot C$ at $t = 0$. In particular, for arbitrary, invertible $Y(0)$, the matrix-valued function $Y(t) \cdot \left(Y(0)\right)^{-1}$ is the fundamental solution of Eq. (10.2).

10.1.1   Perturbations

Now consider a curve $\varepsilon \mapsto \tilde{y}_0(\varepsilon)$, for $\varepsilon \in (-\varepsilon_0, \varepsilon_0)$ and $0 < \varepsilon_0 \ll 1$, such that $\tilde{y}_0(0) = y(0)$. By the fundamental theory of differential equations, there exists a unique function $\tilde{y}(\varepsilon, t)$, for $t \in [0, T]$ and $\varepsilon \in (-\varepsilon_0, \varepsilon_0)$, such that $\tilde{y}(\varepsilon, 0) = \tilde{y}_0(\varepsilon)$ and

$$\frac{\partial \tilde{y}}{\partial t} = f(\tilde{y}, p) \tag{10.3}$$

holds everywhere. For each fixed ε, $\tilde{y}(\varepsilon, t)$ then corresponds to the solution to Eq. (10.1) with initial condition $\tilde{y}_0(\varepsilon)$ at $t = 0$. It follows, for example, that $\tilde{y}|_{\varepsilon=0}(t) = y(t)$. We refer to $\tilde{y}_0(\varepsilon)$ as a 1-dimensional family of *perturbations* to the initial condition $y(0)$ at $t = 0$. For fixed t, $\tilde{y}(\varepsilon, t)$ is the corresponding 1-dimensional family of perturbations to $y(t)$ after elapsed time t.

For fixed t, the partial derivative

$$\left. \frac{\partial \tilde{y}}{\partial \varepsilon} \right|_{\varepsilon=0} \tag{10.4}$$

denotes the *tangent vector* to the curve $\varepsilon \mapsto \tilde{y}(\varepsilon, t)$ at $\tilde{y}(0, t) = y(t)$. In particular,

$$\left. \frac{\partial \tilde{y}}{\partial \varepsilon} \right|_{\varepsilon=t=0} = \left. \frac{\partial \tilde{y}_0}{\partial \varepsilon} \right|_{\varepsilon=0}. \tag{10.5}$$

Since

$$\left. \frac{\partial^2 \tilde{y}}{\partial t \partial \varepsilon} \right|_{\varepsilon=0} = \left. \frac{\partial^2 \tilde{y}}{\partial \varepsilon \partial t} \right|_{\varepsilon=0} = \partial_y f\left(\tilde{y}|_{\varepsilon=0}, p\right) \cdot \left. \frac{\partial \tilde{y}}{\partial \varepsilon} \right|_{\varepsilon=0} = \partial_y f(y, p) \cdot \left. \frac{\partial \tilde{y}}{\partial \varepsilon} \right|_{\varepsilon=0}, \tag{10.6}$$

it follows from the previous section that

$$\left. \frac{\partial \tilde{y}}{\partial \varepsilon} \right|_{\varepsilon=0} (t) = Y(t) \cdot \left(Y(0)\right)^{-1} \cdot \left. \frac{\partial \tilde{y}_0}{\partial \varepsilon} \right|_{\varepsilon=0} \tag{10.7}$$

for an arbitrary nonsingular solution $Y(t)$ to Eq. (10.2). The *sensitivity matrix* $Y(T) \cdot \left(Y(0)\right)^{-1}$ thus maps the initial tangent vector at $t = 0$ to the final tangent vector at $t = T$.

We refer to the eigenvalues of the sensitivity matrix as the *sensitivity gains* of the linearization of the vector field f about the known solution $y(t)$. For example, if

$$\left. \frac{\partial \tilde{y}_0}{\partial \varepsilon} \right|_{\varepsilon=0} \tag{10.8}$$

is a real eigenvector of $Y(T) \cdot (Y(0))^{-1}$ with real eigenvalue λ, then it follows that the corresponding tangent vector at $t = T$ equals

$$\lambda \left. \frac{\partial \tilde{y}_0}{\partial \varepsilon} \right|_{\varepsilon=0}. \tag{10.9}$$

In this case, the sensitivity gain λ determines the response to perturbations in the initial condition in the direction of the corresponding eigenvector.

Example 10.1 Differentiation of Eq. (10.1) with respect to t yields

$$\frac{d^2 y}{dt^2} = \partial_y f(y, p) \cdot \frac{dy}{dt}. \tag{10.10}$$

Moreover,

$$\frac{dy}{dt}(0) = f(y(0), p). \tag{10.11}$$

It follows that

$$f(y(t), p) = \frac{dy}{dt}(t) = Y(t) \cdot (Y(0))^{-1} \cdot f(y(0), p) \tag{10.12}$$

for an arbitrary nonsingular solution $Y(t)$ to Eq. (10.2). In the special case that $y(T) = y(0)$, the column matrix $f(y(0), p)$ is an eigenvector of the corresponding sensitivity matrix with sensitivity gain equal to 1. ∎

10.1.2 Discretization

We proceed to derive a discretization of the first variational equation in Eq. (10.2) that is compatible with the collocation discretization of a solution segment introduced in Chap. 6. To this end, consider again the linear transformation

$$t = t(\tau) := T\tau, \tau \in [0, 1], \tag{10.13}$$

such that $v(\tau) := y(t(\tau))$ and $\Delta(\tau) := Y(t(\tau))$ are solutions on the interval $[0, 1]$ to

$$\frac{dv}{d\tau} = Tf(v, p) \tag{10.14}$$

and

$$\frac{d\Delta}{d\tau} = T\partial_y f(v, p) \cdot \Delta, \tag{10.15}$$

respectively. Given the positive integer N, we again consider the partition

$$0 = \tau_1 < \cdots < \tau_j = \frac{j-1}{N} < \cdots < \tau_{N+1} = 1 \tag{10.16}$$

and the linear transformation

$$\tau = \tau^{(j)}(\sigma) := \tau_j + \frac{(1+\sigma)}{2}(\tau_{j+1} - \tau_j), \sigma \in [-1, 1], \tag{10.17}$$

on the interval $[\tau_j, \tau_{j+1}]$ for $1 \leq j < N$. It follows that $v^{(j)}(\sigma) := v\left(\tau^{(j)}(\sigma)\right)$ and $\Delta^{(j)}(\sigma) := \Delta\left(\tau^{(j)}(\sigma)\right)$ satisfy the equations

$$\frac{dv^{(j)}}{d\sigma} = \frac{T}{2N} f\left(v^{(j)}, p\right) \tag{10.18}$$

and

$$\frac{d\Delta^{(j)}}{d\sigma} = \frac{T}{2N} \partial_y f\left(v^{(j)}, p\right) \cdot \Delta^{(j)}, \tag{10.19}$$

respectively, on the interval $[-1, 1]$. As in Chap. 6,

$$\tau^{(j)}(1) = \tau_{j+1} = \tau^{(j+1)}(-1) \tag{10.20}$$

for every $1 \leq j < N$. Continuity of the original solutions on the interval $[0, T]$ then implies that

$$v^{(j)}(1) = v^{(j+1)}(-1) \tag{10.21}$$

and

$$\Delta^{(j)}(1) = \Delta^{(j+1)}(-1) \tag{10.22}$$

for $1 \leq j < N$.

We proceed to approximate $\Delta^{(j)}$ for each $j \in \{1, \ldots, N\}$ by the matrix-valued polynomial of degree m

$$g_{\Delta,j}(\sigma) = \sum_{i=1}^{m+1} \mathcal{L}_i(\sigma) \Delta_{(m+1)(j-1)+i} \tag{10.23}$$

expressed in terms of the Lagrange polynomials

$$\mathcal{L}_i(\sigma) = \prod_{k=1, k \neq i}^{m+1} \frac{\sigma - \sigma_k}{\sigma_i - \sigma_k}, i = 1, \ldots, m+1, \tag{10.24}$$

for the partition

$$-1 = \sigma_1 < \cdots < \sigma_i < \cdots < \sigma_{m+1} = 1 \tag{10.25}$$

introduced in Chap. 6. It follows that $g_{\Delta,j}$ interpolates the values $\Delta_{(m+1)(j-1)+i} \in \mathbb{R}^{n \times n}$ attained at $\sigma = \sigma_i$ for $i \in \{1, \ldots, m+1\}$.

Recall the notation $v_{(m+1)(j-1)+i+1} \in \mathbb{R}^n$ for the approximate value of the solution y to the differential equation in Eq. (10.1) at the base point

$$t_{(m+1)(j-1)+i} := T\tau^{(j)}(\sigma_i) \tag{10.26}$$

for $j = 1, \ldots, N$ and $i = 1, \ldots, m+1$. It follows that the polynomial interpolant

$$g_{v,j}(\sigma) = \sum_{i=1}^{m+1} \mathcal{L}_i(\sigma) v_{(m+1)(j-1)+i} \tag{10.27}$$

approximates the function $v^{(j)}(\sigma)$ for $\sigma \in [-1, 1]$. The *variational collocation problem* is now given by

- the imposition of *variational collocation conditions* corresponding to the equality (10.19) on the polynomial interpolants $g_{\Delta,j}$ at the collocation nodes z_l, i.e.,

$$0 = \frac{dg_{\Delta,j}}{d\sigma}(z_l) - \frac{T}{2N}\partial_y f\left(g_{\upsilon,j}(z_l), p\right) \cdot g_{\Delta,j}(z_l) \tag{10.28}$$

for $l = 1, \ldots, m$ and $j = 1, \ldots, N$, and

- the imposition of *variational continuity conditions* on the concatenation of the polynomial interpolants $g_{\Delta,j}$ at the interior end points, i.e.,

$$0 = g_{\Delta,j+1}(\sigma_1) - g_{\Delta,j}(\sigma_{m+1}) = \Delta_{(m+1)j+1} - \Delta_{(m+1)(j-1)+m+1} \tag{10.29}$$

for $j = 1, \ldots, N-1$.

Let υ_{bp} and Δ_{bp} denote the $N(m+1)n \times 1$ and $N(m+1)n \times n$ matrices

$$\begin{pmatrix} \vdots \\ \upsilon_{(m+1)(j-1)+1} \\ \vdots \\ \upsilon_{(m+1)j} \\ \vdots \end{pmatrix} \tag{10.30}$$

and

$$\begin{pmatrix} \vdots \\ \Delta_{(m+1)(j-1)+1} \\ \vdots \\ \Delta_{(m+1)j} \\ \vdots \end{pmatrix}, \tag{10.31}$$

respectively. Further, recall the $Nmn \times N(m+1)n$, $Nmn \times N(m+1)n$, and $(N-1)n \times N(m+1)n$ matrices W, W', and Q defined in Sect. 6.3. It follows that the matrix products $W \cdot \Delta_{bp}$ and $W' \cdot \Delta_{bp}$ contain the values of the polynomial interpolants $g_{\Delta,j}$ and their derivatives, respectively, at the corresponding collocation nodes. The variational collocation problem is now given by the linear equations

$$\left(\frac{T}{2N}\mathfrak{diag}\left(\partial_y f\left(\mathfrak{vec}_n\left(W \cdot \upsilon_{bp}\right), 1_{1,Nm} \otimes p\right)\right) \cdot W - W'\right) \cdot \Delta_{bp} = 0, \tag{10.32}$$

corresponding to the variational collocation conditions, and

$$Q \cdot \Delta_{bp} = 0, \tag{10.33}$$

corresponding to the variational continuity conditions.

10.1.3 Encoding

We provide below an encoding of the variational collocation problem in the form of the `'var_coll'` toolbox. In contrast with the toolboxes developed in the previous two chapters, the `'var_coll'` toolbox does not embed an instance of the `'coll'` toolbox. Instead, it assumes that the toolbox data structure associated with such an instance is already available in the continuation problem structure. The variational collocation problem may thus be added at will and at run-time, independently of the construction of a segment object.

We obtain a unique solution Δ_{bp} to the variational collocation problem in Eqs. (10.32)–(10.33) provided that we append an additional n^2 scalar conditions on the entries of Δ_{bp}. In this initial encoding of the `'var_coll'` toolbox, we simply require that $\Delta_1 = I_n$, corresponding to the initial condition $Y(0) = I_n$. The matrix Δ_{bp} then contains a discretized approximation of the fundamental solution to the corresponding first variational equation.

Consider, for example, the `var_coll_init_data` function, shown below.

```
function data = var_coll_init_data(prob, segoid)

data.tbid    = coco_get_id(segoid, 'coll');
fdata        = coco_get_func_data(prob, data.tbid, 'data');

dim          = fdata.dim;
data.dim     = dim;
data.M1_idx  = fdata.xbp_idx(end-dim+(1:dim));
data.row     = [eye(dim), zeros(dim, fdata.xbp_idx(end)-dim)];

end
```

Here, the segment object instance identifier `segoid` is used to identify a unique `'coll'` toolbox instance identifier and to extract the corresponding toolbox data structure.

The following COCO-compatible encoding of the `var_coll_seg` function solves the variational collocation problem and stores the result in the `M` field of the `'var_coll'` toolbox data structure.

```
function [data y] = var_coll_seg(prob, data, u)

fdata = coco_get_func_data(prob, data.tbid, 'data');

x = u(fdata.xbp_idx);
T = u(fdata.T_idx);
p = u(fdata.p_idx);

xx = reshape(fdata.W*x, fdata.x_shp);
pp = repmat(p, fdata.p_rep);

if isempty(fdata.dfdxhan)
  dxode = coco_ezDFDX('f(x,p)v', fdata.fhan, xx, pp);
else
  dxode = fdata.dfdxhan(xx, pp);
end
dxode = sparse(fdata.dxrows, fdata.dxcols, dxode(:));
dxode = (0.5*T/fdata.coll.NTST)*dxode*fdata.W-fdata.Wp;

data.M = [data.row; dxode; fdata.Q]\data.row';

y = [];

end
```

Here, the u input argument contains the subset of the vector of continuation variables containing the discretization υ_{bp}, interval length T, and problem parameters p of the corresponding segment object.

We append the var_coll_seg function to the continuation problem structure by declaring a *nonembedded monitor function* with function type 'regular', as shown in the encoding of the var_coll_add function below.

```
function prob = var_coll_add(prob, segoid)

data = var_coll_init_data(prob, segoid);
uidx = coco_get_func_data(prob, data.tbid, 'uidx');
tbid = coco_get_id(segoid, 'var');
prob = coco_add_func(prob, tbid, @var_coll_seg, data, ...
   'regular', {}, 'uidx', uidx);

end
```

This choice of function type excludes the monitor function from the extended continuation problem and eliminates the need to provide exact or approximate first derivatives of the components of the monitor function with respect to the continuation variables.

10.1.4 Floquet multipliers

In this section, we consider the application of the variational collocation problem to the computation of the *Floquet multipliers* of a single-segment periodic trajectory in a smooth dynamical system. Since in this case $f(y(T), p) = f(y(0), p)$, it follows from Example 10.1 that one of the eigenvalues of the sensitivity matrix must equal 1. The Floquet multipliers are the remaining sensitivity gains. When these all lie within the unit circle, the periodic trajectory can be shown to be asymptotically stable to perturbations in initial conditions, whereas it is unstable if any one Floquet multiplier has a magnitude greater than 1.

From the encoding of the 'var_coll' toolbox, it follows that the solution Δ_{bp} of the variational collocation problem, combined with the appropriate boundary condition, is stored in the M field of the toolbox data structure. The following encoding of the po_mult_eigs_bddat slot function uses this information to compute the sensitivity gains.

```
function [data res] = po_mult_eigs_bddat(prob, data, command, varargin)

switch command
  case 'init'
    res   = {data.mnames};
  case 'data'
    fdata = coco_get_func_data(prob, data.tbid, 'data');
    M1    = fdata.M(fdata.M1_idx,:);
    res   = {eig(full(M1))};
end

end
```

The function makes use of the M1_idx field of the 'var_coll' toolbox data structure, assigned in var_coll_init_data to contain the array

$$\llbracket\ nN(m+1)-n+1 \quad \cdots \quad nN(m+1)\ \rrbracket.\tag{10.34}$$

The contents of the `mnames` and `tbid` fields are initialized in the `po_mult_add` function, shown below.

```
function prob = po_mult_add(prob, segoid)

data.tbid   = coco_get_id(segoid, 'var');
data.mnames = coco_get_id(segoid, 'multipliers');
prob = coco_add_slot(prob, data.tbid, @po_mult_eigs_bddat, data, ...
   'bddat');

end
```

Since the `po_mult_eigs_bddat` function is associated with the `'bddat'` signal, the content of the `res` output argument is appended to the output from the `coco` entry-point function.

Example 10.2 Consider the linear vector field

$$f(y,p) = \begin{pmatrix} y_2 \\ -y_2 - p_1 y_1 + \cos y_3 \\ 1 \end{pmatrix}\tag{10.35}$$

for $y \in \mathbb{R}^2 \times \mathbb{S}$, corresponding to a harmonically excited, linear, single-degree-of-freedom oscillator with excitation period 2π. Here, periodic trajectories correspond to the steady-state response of the oscillator to the harmonic excitation. We give `'coll'`-compatible encodings of f and the Jacobian matrices $\partial_y f$ and $\partial_p f$ in the `linode`, `linode_DFDX`, and `linode_DFDP` functions, shown below.

```
function y = linode(x, p)

x1 = x(1,:);
x2 = x(2,:);
x3 = x(3,:);
p1 = p(1,:);

y(1,:) = x2;
y(2,:) = -x2-p1.*x1+cos(x3);
y(3,:) =  1;

end
```

```
function J = linode_DFDX(x, p)

x1 = x(1,:);
x3 = x(3,:);
p1 = p(1,:);

J = zeros(3,3,numel(x1));
J(1,2,:) = 1;
J(2,1,:) = -p1;
J(2,2,:) = -1;
J(2,3,:) = -sin(x3);

end
```

```
function J = linode_DFDP(x, p)

x1 = x(1,:);

J = zeros(3,1,numel(x1));
J(2,1,:) = -x1;

end
```

From the general solution expressed in terms of the initial condition $y(0)$, it is straightforward to show that the Floquet multipliers equal

$$e^{(-1\pm\sqrt{1-4p_1})\pi}. \tag{10.36}$$

For example, when $p_1 = 1$, there exists a periodic trajectory given by

$$y(t) = \begin{pmatrix} \sin t \\ \cos t \\ t \end{pmatrix} \tag{10.37}$$

whose Floquet multipliers equal

$$e^{\left(-1\pm i\sqrt{3}\right)\pi} \approx 0.0288 \pm 0.0322i. \tag{10.38}$$

In the extract below, we rely on the 'bvp' toolbox to construct a two-point boundary-value problem whose solutions approximate periodic orbits of the dynamical system given by the vector field f.

```
>> [t0 x0]   = ode45(@(t,x) linode(x, 1), [0 2*pi], [0; 1; 0]);
>> coll_args = {@linode, @linode_DFDX, @linode_DFDP, t0, x0, 'p', 1};
>> prob = bvp_isol2seg(coco_prob(), '', coll_args{:}, @lin_bc);
```

The boundary conditions are encoded in the 'bvp'-compatible function `lin_bc`, shown below.

```
function fbc = lin_bc(T, x0, x1, p)
  fbc = [x1(1:2)-x0(1:2); x1(3)-x0(3)-2*pi; x0(1)];
end
```

We proceed to append an instance of the 'var_coll' toolbox to the continuation problem structure. The `po_mult_add` function is then invoked to include an instance of the `po_mult_eigs_bddat` slot function with the continuation problem structure.

```
>> prob = var_coll_add(prob, 'bvp.seg');
>> prob = po_mult_add(prob, 'bvp.seg');
>> coco(prob, 'run', [], 1, 'p', [0.2 2]);
```

STEP		DAMPING		NORMS			COMPUTATION TIMES		
IT	SIT	GAMMA	||d||	||f||		||U||	F(x)	DF(x)	SOLVE
0				1.86e-04		2.74e+01	0.0	0.0	0.0
1	1	1.00e+00	8.30e-05	8.56e-16		2.74e+01	0.0	0.0	0.1
2	1	1.00e+00	2.44e-15	2.88e-16		2.74e+01	0.0	0.0	0.1

STEP	TIME	||U||	LABEL	TYPE	p
0	00:00:00	2.7391e+01	1	EP	1.0000e+00

```
      10  00:00:01   3.0375e+01       2              3.8778e-01
      13  00:00:01   3.1042e+01       3  EP          2.0000e-01

    STEP      TIME        ||U||   LABEL  TYPE                 p
       0  00:00:01   2.7391e+01       4  EP          1.0000e+00
      10  00:00:02   2.4132e+01       5              1.6122e+00
      15  00:00:02   2.2702e+01       6  EP          2.0000e+00
```

In this case, the Floquet multipliers are stored in the `'bvp.seg.multipliers'` column of the output from the `coco` entry-point function. ∎

10.1.5 Multisegment periodic orbits

In the case of multisegment periodic orbits, the algorithm developed in this chapter may be applied to each segment in order to compute a discretized solution to the corresponding variational collocation problem. In this case, however, the stability properties of the periodic orbit are determined by a suitable composition of each individual flow, while accounting for changes in the interval duration along each segment for nearby initial conditions, as well as the nontrivial action of the reset functions that determine the connectivity between segments.

We recall the definition of a multisegment periodic orbit as a sequence $\{y^{(j)} : [0, T_j] \to \mathbb{R}^{n_j}\}_{j=1}^{M}$ of M smooth curves and an associated orbit signature $\{(\mathfrak{m}_j, \mathfrak{e}_j, \mathfrak{r}_j)\}_{j=1}^{M}$ such that

$$\frac{dy^{(j)}}{dt} = f\left(y^{(j)}, p; \mathfrak{m}_j\right) \tag{10.39}$$

and

$$h\left(y^{(j)}(T_j), p; \mathfrak{e}_j\right) = 0 \tag{10.40}$$

for $j = 1, \ldots, M$,

$$g\left(y^{(j)}(T_j), p; \mathfrak{r}_j\right) = y^{(j+1)}(0) \tag{10.41}$$

for $j = 1, \ldots, M - 1$, and

$$g\left(y^{(M)}(T_M), p; \mathfrak{r}_M\right) = y^{(1)}(0) \tag{10.42}$$

for families of smooth functions functions $f\left(\cdot, \cdot, \mathfrak{m}_j\right) : \mathbb{R}^{n_j} \times \mathbb{R}^q \to \mathbb{R}^{n_j}$ and $h\left(\cdot, \cdot, \mathfrak{e}_j\right) : \mathbb{R}^{n_j} \times \mathbb{R}^q \to \mathbb{R}$ for $j = 1, \ldots, M$, $g\left(\cdot, \cdot, \mathfrak{m}_j\right) : \mathbb{R}^{n_j} \times \mathbb{R}^q \to \mathbb{R}^{n_{j+1}}$ for $j = 1, \ldots, M - 1$, and $g\left(\cdot, \cdot, \mathfrak{m}_M\right) : \mathbb{R}^{n_M} \times \mathbb{R}^q \to \mathbb{R}^{n_1}$ parameterized by

- the sequence of mode identifiers $\left\{\mathfrak{m}_j\right\}_{j=1}^{M}$,

- the sequence of event identifiers $\left\{\mathfrak{e}_j\right\}_{j=1}^{M}$, and

- the sequence of reset identifiers $\left\{\mathfrak{r}_j\right\}_{j=1}^{M}$.

For each $j \in \{1, \ldots, M\}$, let $\tilde{y}_0^{(j)}$ again denote a curve parameterized by $\varepsilon \in (-\varepsilon_0, \varepsilon_0)$, for $0 < \varepsilon \ll 1$, such that $\tilde{y}_0^{(j)}(0) = y^{(j)}(0)$. From the theory of differential equations, it follows that there exists a function $\tilde{y}^{(j)}(\varepsilon, t)$ for t in some neighborhood of $[0, T_j]$ such that $\tilde{y}^{(j)}(\varepsilon, 0) = \tilde{y}_0^{(j)}(\varepsilon)$ and

$$\frac{\partial \tilde{y}^{(j)}}{\partial t} = f(\tilde{y}^{(j)}, p; \mathfrak{m}_j). \tag{10.43}$$

For each fixed t, $\varepsilon \mapsto \tilde{y}^{(j)}(\varepsilon, t)$ is a curve through the point $\tilde{y}^{(j)}(0, t) = y^{(j)}(t)$. By the analysis in Sect. 10.1.1, the corresponding family of tangent vectors are related by the expression

$$\frac{\partial \tilde{y}^{(j)}}{\partial \varepsilon}\bigg|_{\varepsilon=0} (t) = Y^{(j)}(t) \cdot \left(Y^{(j)}(0)\right)^{-1} \cdot \frac{\partial \tilde{y}_0^{(j)}}{\partial \varepsilon}\bigg|_{\varepsilon=0}, \tag{10.44}$$

where $Y^{(j)}$ is a solution to the corresponding first variational equation

$$\frac{dY^{(j)}}{dt} = \partial_y f\left(y^{(j)}, p; \mathfrak{m}_j\right) \cdot Y^{(j)}. \tag{10.45}$$

Now let

$$E^{(j)}(\varepsilon, t) := h\left(\tilde{y}^{(j)}(\varepsilon, t), p; \mathfrak{e}_j\right). \tag{10.46}$$

It follows from Eq. (10.40) that $E^{(j)}\left(0, T_j\right) = 0$. Suppose that

$$\partial_y h\left(y^{(j)}\left(T_j\right), p; \mathfrak{e}_j\right) \cdot f\left(y^{(j)}\left(T_j\right), p; \mathfrak{m}_j\right) \neq 0. \tag{10.47}$$

Since

$$\partial_t E^{(j)}(\varepsilon, t) = \partial_y h\left(\tilde{y}^{(j)}(\varepsilon, t), p; \mathfrak{e}_j\right) \cdot f\left(\tilde{y}^{(j)}(\varepsilon, t), p; \mathfrak{m}_j\right), \tag{10.48}$$

the implicit function theorem then implies the existence of an open interval $I_\varepsilon^{(j)}$ containing 0 and an open interval $I_t^{(j)}$ containing T_j such that

$$E^{(j)}(\varepsilon, t) = 0 \text{ for } (\varepsilon, t) \in I_\varepsilon^{(j)} \times I_t^{(j)} \tag{10.49}$$

if and only if $t = t^{(j)}(\varepsilon)$ for a unique scalar-valued smooth function $t^{(j)} : \mathbb{R} \to \mathbb{R}$. We conclude that

$$\varepsilon \mapsto \tilde{y}^{(j)}\left(\varepsilon, t^{(j)}(\varepsilon)\right) \tag{10.50}$$

is a curve parameterized by $\varepsilon \in I_\varepsilon^{(j)}$ on the zero-level surface of the event function $h\left(\cdot, p, \mathfrak{e}_j\right)$.

From the equality

$$h\left(\tilde{y}^{(j)}(\varepsilon, t^{(j)}(\varepsilon)), p; \mathfrak{e}_j\right) = 0, \tag{10.51}$$

it follows by implicit differentiation that

$$\frac{dt^{(j)}}{d\varepsilon}\bigg|_{\varepsilon=0} = -\frac{\partial_y h\left(y^{(j)}\left(T_j\right), p; \mathfrak{e}_j\right)}{\partial_y h\left(y^{(j)}\left(T_j\right), p; \mathfrak{e}_j\right) \cdot f\left(y^{(j)}\left(T_j\right), p; \mathfrak{m}_j\right)} \cdot \frac{\partial \tilde{y}^{(j)}}{\partial \varepsilon}\bigg|_{\varepsilon=0}\left(T_j\right). \tag{10.52}$$

The tangent vector to the curve $\varepsilon \mapsto \tilde{y}^{(j)}\left(\varepsilon, t^{(j)}(\varepsilon)\right)$ at $\varepsilon = 0$ is then given by

$$\frac{d}{d\varepsilon} \tilde{y}^{(j)}\left(\varepsilon, t^{(j)}(\varepsilon)\right)\bigg|_{\varepsilon=0}$$
$$= \left(I_n - \frac{f\left(y^{(j)}\left(T_j\right), p; \mathfrak{m}_j\right) \cdot \partial_y h\left(y^{(j)}\left(T_j\right), p; \mathfrak{e}_j\right)}{\partial_y h\left(y^{(j)}\left(T_j\right), p; \mathfrak{e}_j\right) \cdot f\left(y^{(j)}\left(T_j\right), p; \mathfrak{m}_j\right)}\right) \cdot \frac{\partial \tilde{y}^{(j)}}{\partial \varepsilon}\bigg|_{\varepsilon=0}\left(T_j\right). \tag{10.53}$$

The further application of the reset function $g\left(\cdot,p,\mathfrak{r}_j\right)$ finally results in the curve

$$\varepsilon \mapsto g\left(\tilde{y}^{(j)}\left(\varepsilon,t^{(j)}(\varepsilon)\right),p;\mathfrak{r}_j\right). \qquad (10.54)$$

The corresponding tangent vector at $\varepsilon = 0$ is then given by the product

$$\frac{d}{d\varepsilon}g\left(\tilde{y}^{(j)}\left(\varepsilon,t^{(j)}(\varepsilon)\right),p;\mathfrak{r}_j\right)\bigg|_{\varepsilon=0} = \partial_y g\left(y^{(j)}(T_j),p;\mathfrak{r}_j\right)\cdot\frac{d}{d\varepsilon}\,\tilde{y}^{(j)}\left(\varepsilon,t^{(j)}(\varepsilon)\right)\bigg|_{\varepsilon=0}.$$
$$(10.55)$$

Now consider the imposition of the equality

$$\tilde{y}_0^{(j+1)}(\varepsilon) = g\left(\tilde{y}^{(j)}\left(\varepsilon,t^{(j)}(\varepsilon)\right),p;\mathfrak{r}_j\right) \qquad (10.56)$$

for $j = 1,\ldots,M-1$. This is compatible with the interpretation of the curve $\varepsilon \mapsto \tilde{y}^{(j)}(\varepsilon,t)$ for fixed t as the result of forward transport of a family of perturbations $\varepsilon \mapsto \tilde{y}_0^{(1)}(\varepsilon)$ to the initial condition along the first curve segment. The *composite sensitivity matrix* is then given by the matrix product

$$\prod_{j=M}^{1}\partial_y g\left(y^{(j)}(T_j),p;\mathfrak{r}_j\right)\cdot\left(I_n - \frac{f\left(y^{(j)}(T_j),p;\mathfrak{m}_j\right)\cdot\partial_y h\left(y^{(j)}(T_j),p;\mathfrak{e}_j\right)}{\partial_y h\left(y^{(j)}(T_j),p;\mathfrak{e}_j\right)\cdot f\left(y^{(j)}(T_j),p;\mathfrak{m}_j\right)}\right)$$
$$\cdot Y^{(j)}(T_j)\cdot\left(Y^{(j)}(0)\right)^{-1}. \qquad (10.57)$$

Since $g\left(y^{(M)}(T_M),p;\mathfrak{r}_M\right) = y^{(1)}(0)$, it follows that the composite sensitivity matrix has an eigenvalue equal to 0 corresponding to the eigenvector $f\left(y^{(1)}(0),p;\mathfrak{m}_1\right)$. As in the case of a single-segment periodic orbit, the remaining eigenvalues again provide information about the stability properties of the multisegment periodic orbit under forward integration with the given signature. The following encoding of the `hspo_mult_eigs_bddat` slot function and the auxiliary function `hspo_P` uses information stored with segment-specific instances of the `'var_coll'` toolbox to compute the eigenvalues of the composite sensitivity matrix.

```
function [data res] = hspo_mult_eigs_bddat(prob, data, command, ...
  varargin)

switch command
  case 'init'
    res = {data.mnames};
  case 'data'
    [fdata uidx] = coco_get_func_data(prob, data.msid, 'data', 'uidx');
    chart = varargin{1};
    u     = chart.x(uidx);
    P     = hspo_P(prob, data, u([fdata.x1_idx; fdata.p_idx]));
    M = P{1};
    for i=2:data.nsegs
      M = P{i}*M;
    end
    res = {eig(M)};
end

end
```

```
function P = hspo_P(prob, data, u)

P = cell(1,data.nsegs);
p = u(data.p_idx);
for i=1:data.nsegs
  fdata = coco_get_func_data(prob, data.vids{i}, 'data');
  M1    = fdata.M(fdata.M1_idx,:);
  dim   = fdata.dim;

  fdata = coco_get_func_data(prob, data.cids{i}, 'data');
  x     = u(data.x1_idx{i});
  fs    = fdata.fhan(x, p);
  if ~isempty(data.dhdxhan)
    hx = data.dhdxhan(x, p, data.events{i});
  else
    hx = coco_ezDFDX('f(x,p)v', data.hhan, x, p, data.events{i});
  end
  if ~isempty(data.dgdxhan)
    gx = data.dgdxhan(x, p, data.resets{i});
  else
    gx = coco_ezDFDX('f(x,p)v', data.ghan, x, p, data.resets{i});
  end
  P{i}  = gx*(eye(dim)-(fs*hx)/(hx*fs))*M1;
end

end
```

The required contents of the fields of the `data` function data structure are initialized in the `hspo_mult_add` function, shown below.

```
function prob = hspo_mult_add(prob, segsoid)

cids  = {};
vids  = {};
msid  = coco_get_id(segsoid, 'msbvp');
fdata = coco_get_func_data(prob, msid, 'data');
data  = fdata.bc_data;
for i=1:data.nsegs
  soid = coco_get_id(msid, sprintf('seg%d', i));
  cids = [cids {coco_get_id(soid, 'coll')}];
  vids = [vids {coco_get_id(soid, 'var')}];
end
data.msid   = msid;
data.cids   = cids;
data.vids   = vids;
data.p_idx  = numel(fdata.x1_idx)+(1:numel(fdata.p_idx));
data.mnames = coco_get_id(msid, 'multipliers');
prob = coco_add_slot(prob, data.mnames, @hspo_mult_eigs_bddat, ...
  data, 'bddat');

end
```

Since the `hspo_mult_eigs_bddat` function is associated with the `'bddat'` signal, the contents of its `res` output argument are appended to the output from the `coco` entry-point function.

Example 10.3 We consider again the piecewise-smooth dynamical system from Example 9.2 and the continuation of two-segment periodic orbits with signature

$$\{(\mathfrak{left}, \mathfrak{boundary}, \mathfrak{switch}), (\mathfrak{right}, \mathfrak{boundary}, \mathfrak{switch})\}. \tag{10.58}$$

The extract below shows the creation of one instance of the `'var_coll'` toolbox per segment and the subsequent use of the `hspo_mult_add` function in order to append the `hspo_mult_eigs_bddat` slot function to the continuation problem structure.

```
>> p0     = [1; 2; 1.5];
>> modes  = {'left' 'right'};
>> events = {'boundary' 'boundary'};
>> resets = {'switch' 'switch'};
>> t0 = linspace(0, pi, 100)';
>> x1 = [-sin(t0) 0.5+cos(t0)];
>> x2 = [ sin(t0) 0.5-cos(t0)];
>> t0 = {t0 0.5*t0};
>> x0 = {x1 x2};
>> prob = coco_prob();
>> prob = coco_set(prob, 'msbvp.seg1.coll', 'NTST', 10, 'NCOL', 6);
>> prob = coco_set(prob, 'msbvp.seg2.coll', 'NTST', 20, 'NCOL', 4);
>> prob = hspo_isol2segs(prob, '', ...
     {@pwlin, @pwlin_events, @pwlin_resets}, ...
     {@pwlin_DFDX, @pwlin_events_DFDX, @pwlin_resets_DFDX}, ...
     {@pwlin_DFDP, @pwlin_events_DFDP, @pwlin_resets_DFDP} ...
     modes, events, resets, t0, x0, {'al' 'be' 'ga'}, p0);
>> prob = var_coll_add(prob, 'msbvp.seg1');
>> prob = var_coll_add(prob, 'msbvp.seg2');
>> prob = hspo_mult_add(prob, '');
>> coco(prob, 'run', [], 1, 'be', [1 2]);
```

STEP		DAMPING		NORMS			COMPUTATION	TIMES	
IT	SIT	GAMMA	||d||	||f||	||U||	F(x)	DF(x)	SOLVE	
0				4.79e-02	1.56e+01	0.0	0.0	0.0	
1	1	1.10e-01	9.09e-01	4.24e-02	1.68e+01	0.0	0.0	0.0	
2	1	1.80e-01	5.56e-01	3.22e-02	1.81e+01	0.0	0.0	0.0	
3	1	3.25e-01	3.08e-01	2.01e-02	1.94e+01	0.0	0.0	0.0	
4	1	6.84e-01	1.46e-01	5.90e-03	2.03e+01	0.0	0.0	0.0	
5	1	1.00e+00	4.07e-02	2.34e-04	2.02e+01	0.0	0.0	0.0	
6	1	1.00e+00	2.17e-03	6.96e-07	2.01e+01	0.0	0.1	0.0	
7	1	1.00e+00	7.72e-06	8.15e-12	2.01e+01	0.0	0.1	0.0	
8	1	1.00e+00	6.98e-11	5.76e-16	2.01e+01	0.0	0.1	0.0	

STEP	TIME	||U||	LABEL	TYPE	be
0	00:00:00	2.0117e+01	1	EP	2.0000e+00
10	00:00:01	1.6676e+01	2		1.4606e+00
17	00:00:01	1.3920e+01	3	EP	1.0000e+00

In this case, the eigenvalues are stored in the `'msbvp.multipliers'` field of the output of the `coco` entry-point function.   ∎

10.2   A variational zero problem

The implementation of the `'var_coll'` toolbox in the previous section enables the computation of the discretized solution Δ_{bp} to the first variational equation only after the successful location of a solution to the corresponding collocation zero problem. As a result, it is not possible to include additional zero functions expressed in terms of Δ_{bp} in the extended continuation problem. Such additional conditions could be desirable, for example, in order to continue embedded submanifolds defined by the eigenstructure of the sensitivity matrix. To support such computations, we provide below an alternative implementation of the `'var_coll'` toolbox, in which the variational collocation problem and the initial condition

$\Delta_1 = I_n$ are included in the zero problem of the extended continuation problem, stored in the continuation problem structure.

10.2.1 Encoding

The combined *variational zero problem* is now given by the vanishing of the zero function

$$\Phi : u \mapsto \begin{pmatrix} \mathfrak{vec}\left(V \cdot \Delta_{bp}\right) \\ \mathfrak{vec}\left(Q \cdot \Delta_{bp}\right) \\ \mathfrak{vec}(R \cdot \Delta_{bp} - I_n) \end{pmatrix}, \tag{10.59}$$

where

$$V := \frac{T}{2N} \mathfrak{diag}\left(\partial_y f\left(\mathfrak{vec}_n\left(W \cdot v_{bp}\right), 1_{1,Nm} \otimes p\right)\right) \cdot W - W', \tag{10.60}$$

$$R = \begin{pmatrix} I_n & 0 \end{pmatrix}, \tag{10.61}$$

and $u = \left(v_{bp}, T, p, \Delta_{bp}\right)$. The family of zero functions Φ is then encoded in the function var_coll_F, shown below.

```
function [data y] = var_coll_F(prob, data, u)

fdata = coco_get_func_data(prob, data.coll_id, 'data');

ubp = u(data.ubp_idx);
T   = u(fdata.T_idx);
x   = u(fdata.xbp_idx);
p   = u(fdata.p_idx);

Mbp = reshape(ubp, data.u_shp);
xx  = reshape(fdata.W*x, fdata.x_shp);
pp  = repmat(p, fdata.p_rep);

ode = fdata.dfdxhan(xx, pp);
ode = sparse(fdata.dxrows, fdata.dxcols, ode(:));
ode = (0.5*T/fdata.coll.NTST)*ode*fdata.W-fdata.Wp;
ode = ode*Mbp;
cnt = fdata.Q*Mbp;
bcd = data.R*Mbp-data.Id;

y = [ode(:); cnt(:); bcd(:)];

end
```

Here, the u input argument contains the subset of the vector of continuation variables corresponding to v_{bp}, T, p, and $\mathfrak{vec}\left(\Delta_{bp}\right)$, in that order, associated with a single segment object. The 'var_coll' toolbox data structure is here assumed to contain copies of the xbp_idx, T_idx, p_idx, x_shp, p_rep, dxrows, dxcols, W, Wp, and Q fields of the corresponding 'coll' toolbox data structure. In addition, the data input argument contains the fields

- R equal to the $n \times nN(m+1)$ matrix R and

- Id equal to I_n.

In contrast to the implementation of the 'coll' toolbox and the encoding of the 'var_coll' toolbox in the first part of this chapter, the encoding of var_coll_F requires that the dfdxhan field contain a function handle to a vectorized encoding of the Jacobian $\partial_y f$.

We proceed to consider the Jacobian of the variational zero problem with respect to the elements of v_{bp}, T, p, and $\mathfrak{vec}\left(\Delta_{bp}\right)$. By the linearity with respect to Δ_{bp}, the Jacobian with respect to the components of $\mathfrak{vec}\left(\Delta_{bp}\right)$ is simply given by the matrix

$$\begin{pmatrix} I_n \otimes V \\ I_n \otimes Q \\ I_n \otimes R \end{pmatrix}. \tag{10.62}$$

Since the variational continuity conditions $Q \cdot \Delta_{bp} = 0$ and the initial condition $\Delta_1 = I_n$ are independent of v_{bp}, T, and p, the corresponding components of the Jacobian of Φ equal 0. It remains to determine the Jacobians of $\mathfrak{vec}\left(V \cdot \Delta_{bp}\right)$ with respect to v_{bp}, T, and p. As an example, the Jacobian of $\mathfrak{vec}\left(V \cdot \Delta_{bp}\right)$, with respect to T, is clearly given by the expression

$$\mathfrak{vec}\left(\frac{1}{2N}\mathfrak{diag}\big(\partial_y f\left(\mathfrak{vec}_n\left(W \cdot v_{bp}\right), 1_{1,Nm} \otimes p\right)\big) \cdot W \cdot \Delta_{bp}\right). \tag{10.63}$$

Now let $\partial_{py} f$ denote the function

$$(y,p) \mapsto \left[\!\left[\; \partial_{p_1 y} f(y,p) \quad \cdots \quad \partial_{p_q y} f(y,p) \;\right]\!\right], \tag{10.64}$$

where

$$\partial_{p_i y} f(y,p) := \begin{pmatrix} \frac{\partial^2 f_1}{\partial p_i \partial y_1}(y,p) & \cdots & \frac{\partial^2 f_1}{\partial p_i \partial y_n}(y,p) \\ \vdots & \ddots & \vdots \\ \frac{\partial^2 f_n}{\partial p_i \partial y_1}(y,p) & \cdots & \frac{\partial^2 f_n}{\partial p_i \partial y_n}(y,p) \end{pmatrix}. \tag{10.65}$$

The Jacobian of $\mathfrak{vec}\left(V \cdot \Delta_{bp}\right)$ with respect to p is then given by the matrix

$$\mathfrak{vec}_{n^2 Nm}(B), \tag{10.66}$$

where

$$B = \frac{T}{2N}\left(\; B_1 \cdot W \cdot \Delta_{bp} \quad \cdots \quad B_q \cdot W \cdot \Delta_{bp} \;\right) \tag{10.67}$$

and B_i denotes the $nNm \times nN(m+1)$ matrix

$$\mathfrak{diag}\big(\partial_{p_i y} f\left(\mathfrak{vec}\left(W \cdot v_{bp}\right), 1_{1,Nm} \otimes p\right)\big), \tag{10.68}$$

defined in terms of the vectorized extension of $\partial_{p_i y} f$. We obtain a vectorized evaluation of the matrix in Eq. (10.67) by first computing the matrix product

$$\frac{T}{2N}\begin{pmatrix} B_1 \\ \vdots \\ B_q \end{pmatrix} \cdot W \cdot \Delta_{bp} \tag{10.69}$$

and then applying the array reshaping operation

$$\begin{pmatrix} A_1 \\ \vdots \end{pmatrix} \longmapsto \left(A_1 \quad \cdots \quad \right). \tag{10.70}$$

For small integers q, the matrix in Eq. (10.67) may also be obtained, efficiently and without the need for array manipulations, directly from the matrix product

$$\frac{T}{2N} \left(\begin{array}{ccc} B_1 & \cdots & B_q \end{array} \right) \cdot \left(I_q \otimes \left(W \cdot \Delta_{bp} \right) \right). \tag{10.71}$$

Example 10.4 Let $\{A_i\}_{i=1}^k$ denote a sequence of $nNm \times n$ matrices. The array reshaping operation in Eq. (10.70) followed by the application of $\mathfrak{vec}_{nNm}$ may then be defined a priori in terms of n, N, m, and k. In particular, consider the 1-dimensional arrays

$$r = 1_{k,1} \otimes \mathfrak{vec}_{nNm} \left(\left[\!\left[\begin{array}{ccc} 1 & \cdots & n^2Nm \end{array} \right]\!\right] \right) \tag{10.72}$$

and

$$c = 1_{nNm,n} \otimes \mathfrak{vec}_1 \left(\left[\!\left[\begin{array}{ccc} 1 & \cdots & k \end{array} \right]\!\right] \right). \tag{10.73}$$

It follows that the (r_i, c_i) entry of the matrix $\mathfrak{vec}_{nNm}(A)$, where

$$A = \left(\begin{array}{ccc} A_1 & \cdots & A_k \end{array} \right), \tag{10.74}$$

equals the ith element of the array

$$\left(\begin{array}{c} A_1 \\ \vdots \\ A_k \end{array} \right). \quad \blacksquare \tag{10.75}$$

The Jacobian of $\mathfrak{vec}\left(V \cdot \Delta_{bp} \right)$ with respect to υ_{bp} is given, similarly, by the matrix

$$\mathfrak{vec}_{n^2Nm} \left(\begin{array}{ccc} A_1 & \cdots & A_{nN(m+1)} \end{array} \right), \tag{10.76}$$

where A_i denotes the $nNm \times n$ matrix

$$\frac{T}{2N} B_i \cdot W \cdot \Delta_{bp} \tag{10.77}$$

and

$$B_{n(i-1)+j} := \frac{\partial}{\partial \upsilon_{i,j}} \mathfrak{diag} \left(\partial_y f \left(\mathfrak{vec} \left(W \cdot \upsilon_{bp} \right), 1_{1,Nm} \otimes p \right) \right) \tag{10.78}$$

for $i = 1, \ldots, N(m+1)$ and $j = 1, \ldots, n$. As before, we obtain a vectorized evaluation of the matrix in Eq. (10.76) by first computing the matrix product

$$\frac{T}{2N} \left(\begin{array}{c} B_1 \\ \vdots \\ B_{nN(m+1)} \end{array} \right) \cdot W \cdot \Delta_{bp} \tag{10.79}$$

and then applying the array reshaping operation in Eq. (10.70).

Denote by $\partial_{yy} f$ the vectorized extension of the function

$$(y, p) \mapsto \left[\!\left[\begin{array}{ccc} \partial_{y_1 y} f(y, p) & \cdots & \partial_{y_n y} f(y, p) \end{array} \right]\!\right], \tag{10.80}$$

where

$$\partial_{y_i y} f(y,p) := \begin{pmatrix} \frac{\partial^2 f_1}{\partial y_i \partial y_1}(y,p) & \cdots & \frac{\partial^2 f_1}{\partial y_i \partial y_n}(y,p) \\ \vdots & \ddots & \vdots \\ \frac{\partial^2 f_n}{\partial y_i \partial y_1}(y,p) & \cdots & \frac{\partial^2 f_n}{\partial y_i \partial y_n}(y,p) \end{pmatrix}. \tag{10.81}$$

We then obtain the matrix

$$\begin{pmatrix} B_1 \\ \vdots \\ B_{nN(m+1)} \end{pmatrix} \tag{10.82}$$

by reshaping the content of the matrix

$$\mathfrak{diag}\Big(\big(\mathfrak{vec}_{n^2} \circ \partial_{yy} f\big)\big(\mathfrak{vec}\big(W \cdot \upsilon_{bp}\big), 1_{1,Nm} \otimes p\big) \Big) \cdot W, \tag{10.83}$$

where $\mathfrak{vec}_{n^2} \circ \partial_{yy} f$ denotes the vectorized extension of the composition

$$(y,p) \mapsto \mathfrak{vec}_{n^2}\big(\partial_{yy} f(y,p)\big). \tag{10.84}$$

Example 10.5 The array reshaping operation that obtains the matrix in Eq. (10.82) from that in Eq. (10.83) may again be defined a priori in terms of n, N, and m. Specifically, consider the 1-dimensional arrays

$$r = 1_{n,1} \otimes \mathfrak{vec}_n\big(\llbracket\, 1 \quad \cdots \quad n^2 N^2 m(m+1) \,\rrbracket \big) \tag{10.85}$$

and

$$c = 1_{n,nN(m+1)} \otimes \mathfrak{vec}_1\big(\llbracket\, 1 \quad \cdots \quad nNm \,\rrbracket \big). \tag{10.86}$$

Then the (r_i, c_i) entry of the matrix in Eq. (10.82) equals the ith element of the array in Eq. (10.83). ∎

The Jacobian of Φ with respect to the components of the u input argument is now encoded in the function var_coll_DFDU, shown below.

```
function [data J] = var_coll_DFDU(prob, data, u)

fdata = coco_get_func_data(prob, data.coll_id, 'data');

NTST = fdata.coll.NTST;

ubp = u(data.ubp_idx);
T   = u(fdata.T_idx);
x   = u(fdata.xbp_idx);
p   = u(fdata.p_idx);

Mbp = reshape(ubp, data.u_shp);
xx  = reshape(fdata.W*x, fdata.x_shp);
pp  = repmat(p, fdata.p_rep);

dfdx = fdata.dfdxhan(xx, pp);
dfdx = sparse(fdata.dxrows, fdata.dxcols, dfdx(:));

dfdxdx = data.dfdxdxhan(xx, pp);
dfdxdx = sparse(data.dxdxrows1, data.dxdxcols1, dfdxdx(:));
```

```
dxode   = dfdxdx*fdata.W;
dxode   = sparse(data.dxdxrows2, data.dxdxcols2, dxode(:))*fdata.W*Mbp;
dxode   = (0.5*T/NTST)*sparse(data.dxdxrows3, data.dxdxcols3, dxode(:));

dTode   = (0.5/NTST)*dfdx*fdata.W*Mbp;

dfdxdp  = data.dfdxdphan(xx, pp);
dpode   = sparse(data.dxdprows, data.dxdpcols, dfdxdp(:));
dpode   = dpode*kron(speye(fdata.pdim), fdata.W*Mbp);
dpode   = (0.5*T/NTST)*reshape(dpode, data.dxdp_shp);

duode   = kron(data.Id, (0.5*T/NTST)*dfdx*fdata.W-fdata.Wp);

J = [dxode, dTode(:), dpode, duode; data.jac];

end
```

Here, the `dfdxdxhan` and `dfdxdphan` fields of the toolbox data structure are assumed to contain function handles to vectorized encodings of the functions $\partial_{yy}f : \mathbb{R}^n \times \mathbb{R}^q \to \mathbb{R}^{n \times n \times n}$ and $\partial_{py}f : \mathbb{R}^n \times \mathbb{R}^q \to \mathbb{R}^{n \times n \times q}$, respectively. We precompute row and column index arrays for the array reshaping operations employed in `var_coll_DFDU` in the encoding of the `var_coll_init_data` function, shown below.

```
function data = var_coll_init_data(prob, data)

fdata = coco_get_func_data(prob, data.coll_id, 'data');

NTST    = fdata.coll.NTST;
NCOL    = fdata.coll.NCOL;
dim     = fdata.dim;
pdim    = fdata.pdim;
xbpnum  = (NCOL+1)*NTST;
xbpdim  = dim*(NCOL+1)*NTST;
ubpdim  = dim^2*(NCOL+1)*NTST;
xcnnum  = NCOL*NTST;
xcndim  = dim*NCOL*NTST;
ucndim  = dim^2*NCOL*NTST;

data.dim     = dim;
data.M1_idx  = xbpdim-dim+(1:dim)';
data.ubp_idx = xbpdim+1+pdim+(1:ubpdim)';
data.u_shp   = [xbpdim dim];
data.R       = [speye(dim) sparse(dim, xbpdim-dim)];
data.Id      = eye(dim);
data.jac     = [sparse(dim^2*NTST, xbpdim+1+pdim) ...
                [kron(eye(dim), fdata.Q); kron(eye(dim), data.R)]];

rows = reshape(1:ucndim, [dim^2 xcnnum]);
data.dxdxrows1 = repmat(rows, [dim 1]);
data.dxdxcols1 = repmat(1:xcndim, [dim^2 1]);

rows = reshape(1:xbpdim*xcndim, [dim xbpnum*xcndim]);
data.dxdxrows2 = repmat(rows, [dim 1]);
data.dxdxcols2 = repmat(1:xcndim, [dim xbpdim]);

rows = reshape(1:ucndim, [xcndim dim]);
data.dxdxrows3 = repmat(rows, [xbpdim 1]);
data.dxdxcols3 = repmat(1:xbpdim, [xcndim dim]);

rows = reshape(1:xcndim, [dim xcnnum]);
data.dxdprows  = repmat(rows, [dim*pdim 1]);
```

```
cols = permute(reshape(1:xcndim*pdim, [dim xcnnum pdim]), [1 3 2]);
data.dxdpcols = repmat(cols(:)', [dim 1]);
data.dxdp_shp = [ucndim pdim];

end
```

The variational collocation zero problem is finally appended to the extended continuation problem stored in the `prob` continuation problem structure in the modified `var_coll_add` function, shown below.

```
function prob = var_coll_add(prob, segoid, dfdxdx, dfdxdp)

tbid = coco_get_id(segoid, 'var');
data.coll_id  = coco_get_id(segoid, 'coll');
data.dfdxdxhan = dfdxdx;
data.dfdxdphan = dfdxdp;

data = var_coll_init_data(prob, data);
M0   = var_coll_init_sol(prob, data);
uidx = coco_get_func_data(prob, data.coll_id, 'uidx');
prob = coco_add_func(prob, tbid, @var_coll_F, @var_coll_DFDU, ...
  data, 'zero', 'uidx', uidx, 'u0', M0);
prob = coco_add_slot(prob, tbid, @coco_save_data, data, 'save_full');

end
```

Here, the call to the `var_coll_init_sol` function, shown below, provides an initial solution guess for Δ_{bp} using the algorithm in Sect. 10.1.3.

```
function M0 = var_coll_init_sol(prob, data)

[fdata u0] = coco_get_func_data(prob, data.coll_id, 'data', 'u0');

x = u0(fdata.xbp_idx);
T = u0(fdata.T_idx);
p = u0(fdata.p_idx);

xx = reshape(fdata.W*x, fdata.x_shp);
pp = repmat(p, fdata.p_rep);

if isempty(fdata.dfdxhan)
  dxode = coco_ezDFDX('f(x,p)v', fdata.fhan, xx, pp);
else
  dxode = fdata.dfdxhan(xx, pp);
end
dxode = sparse(fdata.dxrows, fdata.dxcols, dxode(:));
dxode = (0.5*T/fdata.coll.NTST)*dxode*fdata.W-fdata.Wp;

M0 = [data.R; dxode; fdata.Q]\data.R';

end
```

10.2.2   A heteroclinic connection

In the example in Sect. 7.3.1, a shooting method was developed in the context of continuation in order to find a solution to a two-point boundary-value problem. In the first step of this algorithm, the length of the time interval on which the trajectory was defined was introduced as a continuation parameter. Continuation in this parameter enabled the initial construction of a solution trajectory to an initial-value problem across the entire domain of definition $[0, 1]$. In the end, we were concerned not with the intermediate solution trajectories generated during continuation, but only with the final trajectory. This final trajectory served, in turn, as the starting point of a continuation in the value of the first component of the solution at $t = 1$. Here, too, we were interested only in the end result of the computation, rather than the many candidate trajectories for which the value of the first component of the solution at $t = 1$ deviated from the desired value set by the boundary condition.

We refer to an algorithm that performs continuation solely in order to generate an initial solution to some zero problem of interest as a *homotopy algorithm* and to the family of intermediate solutions as a *homotopy*. Additional examples of homotopy analyses were given in Sects. 7.3.2 and 7.3.3. In this section, we employ several stages of a homotopy algorithm coupled with the application of the modified 'var_coll' toolbox in order to construct an initial solution guess for continuation along a family of connecting orbits between an equilibrium point and a saddle-type periodic orbit in the *Lorentz dynamical system* given by the vector field

$$f(y, p) := \begin{pmatrix} s(y_2 - y_1) \\ (r - y_3)y_1 - y_2 \\ y_1 y_2 - b y_3 \end{pmatrix}, \qquad (10.87)$$

where $p := (s, r, b)$. Vectorized encodings of the vector field f and the Jacobians $\partial_y f$, $\partial_p f$, $\partial_{py} f$, and $\partial_{yy} f$ are provided in the functions shown below.

```
function y = lorentz(x, p)

x1 = x(1,:);
x2 = x(2,:);
x3 = x(3,:);
s  = p(1,:);
r  = p(2,:);
b  = p(3,:);

y(1,:) = -s.*x1+s.*x2;
y(2,:) = -x1.*x3+r.*x1-x2;
y(3,:) = x1.*x2-b.*x3;

end
```

```
function J = lorentz_DFDX(x, p)

x1 = x(1,:);
x2 = x(2,:);
x3 = x(3,:);
s  = p(1,:);
r  = p(2,:);
b  = p(3,:);

J = zeros(3,3,numel(s));
J(1,1,:) = -s;
J(1,2,:) = s;
J(2,1,:) = r-x3;
J(2,2,:) = -1;
```

```
J(2,3,:) = -x1;
J(3,1,:) = x2;
J(3,2,:) = x1;
J(3,3,:) = -b;

end


function J = lorentz_DFDP(x, p)

x1 = x(1,:);
x2 = x(2,:);
x3 = x(3,:);

J = zeros(3,3,numel(x1));
J(1,1,:) = -x1+x2;
J(2,2,:) = x1;
J(3,3,:) = -x3;

end


function J = lorentz_DFDXDX(x, p)

s = p(1,:);

J = zeros(3,3,3,numel(s));
J(2,3,1,:) = -1;
J(3,2,1,:) = 1;
J(3,1,2,:) = 1;
J(2,1,3,:) = -1;

end


function J = lorentz_DFDXDP(x, p)

s = p(1,:);

J = zeros(3,3,3,numel(s));
J(1,1,1,:) = -1;
J(1,2,1,:) = 1;
J(2,1,2,:) = 1;
J(3,3,3,:) = -1;

end
```

For $r > 1$ and $s > 0$, the *trivial equilibrium* of the Lorentz dynamical system at $y = 0$ is a saddle with a 2-dimensional stable manifold $\mathcal{W}_0^s$ and a 1-dimensional unstable manifold $\mathcal{W}_0^u$. The latter is tangential at $y = 0$ to the unstable eigenspace $\mathcal{E}_0^u$, spanned by the eigenvector

$$v_0^u = \begin{pmatrix} 1 - s + \sqrt{(1-s)^2 + 4rs} \\ -2r \\ 0 \end{pmatrix} \tag{10.88}$$

corresponding to the eigenvalue

$$\lambda_0^u = \frac{1}{2}(\sqrt{(1-s)^2 + 4rs} - 1 - s). \tag{10.89}$$

In addition, as long as $b > 0$ and $r > 1$, a pair of nontrivial equilibria may be found at

$$y^* = \left(\pm\sqrt{b(r-1)}, \pm\sqrt{b(r-1)}, r-1 \right). \tag{10.90}$$

For $s = 10$, $b = 8/3$, and $r = r^* = 470/19$, these undergo subcritical Hopf bifurcations and families of periodic orbits emanate from y^* as r is reduced below r^*. As an example, at $r = r^*$, the equilibrium obtained with the minus sign is associated with a pair of purely imaginary eigenvalues $\pm i\omega$, where $\omega = 4\sqrt{110/19}$, with corresponding eigenvectors $v :=$ $v_R \pm iv_I$, where

$$v_R = \left(-\frac{20}{9}\sqrt{\frac{38}{1353}}, \frac{2}{9}\sqrt{\frac{38}{1353}}, 1 \right)^T \tag{10.91}$$

and

$$v_I = \left(-\frac{19}{9}\sqrt{\frac{5}{123}}, -\frac{35}{9}\sqrt{\frac{5}{123}}, 0 \right)^T. \tag{10.92}$$

For $r \approx r^*$, the periodic orbits are unstable with a single Floquet multiplier λ_{per}^s inside the unit circle and corresponding to the eigenvector v_{per}^s. Each periodic orbit is thus associated with a 2-dimensional *stable manifold* $\mathcal{W}_{\mathrm{per}}^s$ and a 2-dimensional *unstable manifold* $\mathcal{W}_{\mathrm{per}}^u$ that intersect along the orbit and consist of nearby orbits that approach the periodic orbit at an exponential rate in forward or backward time, respectively. At the initial point $y_{\mathrm{per}}(0)$ along the periodic orbit, the tangent space to the stable manifold $\mathcal{W}_{\mathrm{per}}^s$ is spanned by the vector field $f\left(y_{\mathrm{per}}(0), p \right)$ and v_{per}^s.

Examples of the forward integration of a trajectory emanating from an initial condition in $\mathcal{E}_0^u$ at close proximity to the trivial equilibrium are shown in Figs. 10.1 and 10.2 for different values of r. As indicated in the captions, these are suggestive of the existence of a connecting orbit between the trivial equilibrium and a member of the family of unstable periodic orbits for some $r \approx 24$. In particular, we note a characteristic change in the spiraling behavior of the trajectory in the immediate vicinity of the nontrivial equilibrium and along the corresponding unstable manifold.

The following describes several stages of homotopy analysis used to initialize the subsequent continuation of a family of such heteroclinic orbits under simultaneous variations in at least two problem parameters.

Stage I: Continue periodic orbits from Hopf bifurcation

Consider first the following encoding of the `povar_isol2orb` constructor.

```
function prob = povar_isol2orb(prob, oid, varargin)

str = coco_stream(varargin{:});
prob = po_isol2orb(prob, oid, str);
oid  = coco_get_id(oid, 'po.seg');
dfdxdxhan = str.get;
dfdxdphan = str.get;
prob = var_coll_add(prob, oid, dfdxdxhan, dfdxdphan);

end
```

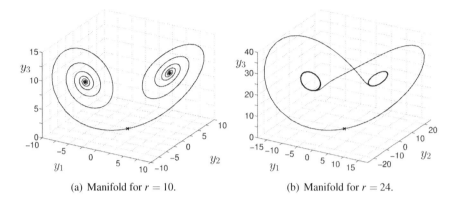

(a) Manifold for $r = 10$.   (b) Manifold for $r = 24$.

Figure 10.1. *Orbits of the Lorentz system given by the vector field in Eq. (10.87) starting in the unstable eigenspace of the equilibrium at 0, tracing the unstable manifold. The orbits in (a) approach equilibria, while the orbits in (b) seem to approach periodic orbits. This approach is shown in more detail in Fig. 10.2.*

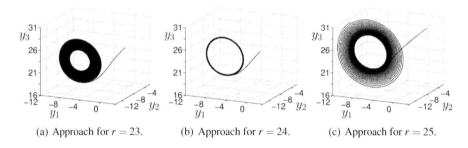

(a) Approach for $r = 23$.   (b) Approach for $r = 24$.   (c) Approach for $r = 25$.

Figure 10.2. *A close-up of the potential approach and exit of $\mathcal{W}_0^u$ to and from a periodic orbit, as observed in Fig. 10.1. In panel (a), the manifold exits spiraling inward, while the exit is outward in (c). This corresponds to a switch of approach from an orbit inside a stable manifold of a periodic orbit to an approach from an orbit outside. This suggests that, in between these parameter values, there exists an orbit approaching on the stable manifold—a heteroclinic connection between the equilibrium at 0 and a periodic orbit of saddle type.*

This function collects the construction of a periodic orbit object, together with the corresponding variational collocation zero problem, in a single function call. The extract below shows the application of this function to continuation along the family of periodic orbits emanating from the Hopf bifurcation at $r = r^*$.

```
>> s  = 10;
>> b  = 8/3;
>> r  = 470/19;
>> p0 - [s; r; b];
>> eq = [-sqrt(b*(r-1)) -sqrt(b*(r-1)) r-1];
>> om = 4*sqrt(110/19);
>> re = [-20/9*sqrt(38/1353) 2/9*sqrt(38/1353) 1];
>> im = [-19/9*sqrt(5/123) -35/9*sqrt(5/123) 0];
>> t0 = (0:2*pi/100:2*pi)'/om;
```

```
>> x0 = repmat(eq, size(t0))+0.01*(cos(om*t0)*re-sin(om*t0)*im);
>> coll_args = {@lorentz, @lorentz_DFDX, @lorentz_DFDP, t0, x0, ...
    {'s' 'r' 'b'}, p0};
>> var_args = {@lorentz_DFDXDX, @lorentz_DFDXDP};
>> prob = povar_isol2orb(coco_prob(), '', coll_args{:}, var_args{:});
>> coco(prob, 'runHopf', [], 1, 'r', [24 25]);
```

STEP		DAMPING			NORMS		COMPUTATION		TIMES
IT	SIT	GAMMA	||d||	||f||		||U||	F(x)	DF(x)	SOLVE
0				1.14e-04		1.90e+02	0.0	0.0	0.0
1	1	1.00e+00	5.63e-05	1.48e-12		1.90e+02	0.0	0.0	0.0
2	1	1.00e+00	1.13e-09	6.76e-14		1.90e+02	0.0	0.1	0.0

STEP	TIME	||U||	LABEL	TYPE	r
0	00:00:00	1.8959e+02	1	EP	2.4737e+01
10	00:00:01	1.8945e+02	2		2.4722e+01
20	00:00:03	1.8894e+02	3		2.4667e+01
30	00:00:04	1.8807e+02	4		2.4572e+01
40	00:00:06	1.8687e+02	5		2.4441e+01
50	00:00:08	1.8536e+02	6	EP	2.4275e+01

STEP	TIME	||U||	LABEL	TYPE	r
0	00:00:08	1.8959e+02	7	EP	2.4737e+01
10	00:00:09	1.8947e+02	8		2.4723e+01
20	00:00:11	1.8898e+02	9		2.4669e+01
30	00:00:12	1.8813e+02	10		2.4575e+01
40	00:00:14	1.8693e+02	11		2.4442e+01
50	00:00:16	1.8542e+02	12	EP	2.4275e+01

Here, each solution point is associated with a discretization υ_{bp} and period T of the corresponding periodic orbit, as well as a discretization Δ_{bp} of the fundamental solution to the corresponding variational equation. For example, the following sequence of commands assigns the Floquet multiplier λ_{per}^s and the corresponding eigenvector v_{per}^s, associated with the 6th labeled solution point, to the lam0 and vec0 variables, respectively.

```
[data sol] = coco_read_solution('', 'runHopf', 6);
M       = reshape(sol.x(data.ubp_idx), data.u_shp);
M1      = M(data.M1_idx,:);
[v, d]  = eig(M1);
ind     = find(abs(diag(d))<1);
vec0    = -v(:,ind);
lam0    = d(ind,ind);
```

As seen in Fig. 10.3(a), the trivial equilibrium and the corresponding periodic orbit are found on opposite sides of the plane Σ through the point $y_\Sigma := (20, 20, 30)$ and perpendicular to the unit vector $n_\Sigma := (0, -1, 1)^T / \sqrt{2}$. It follows that a heteroclinic connection between these must intersect Σ in at least one point.

Stage II: Grow orbit in $\mathcal{W}_0^u$

Given our preliminary observations of the relationship between $\mathcal{W}_0^u$ and $\mathcal{W}_{per}^s$, we proceed to select one of the periodic orbits found in Stage I as the basis for further analysis. In particular, we seek to construct a composite continuation problem that includes

- a periodic orbit object,

- a segment object associated with $\mathcal{W}_0^u$,

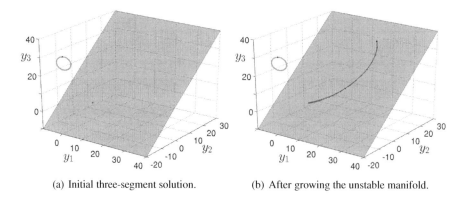

(a) Initial three-segment solution.   (b) After growing the unstable manifold.

Figure 10.3. *Construction of an initial approximation of an orbit connecting an equilibrium and a periodic orbit in the Lorentz system given by the vector field in Eq. (10.87) following the homotopy approach described in Sect. 10.2.2. A state-space representation of the three-segment solution, consisting of a periodic orbit (gray) and two zero-length segments (black dots), that is used to initialize Stage II of the homotopy is shown in panel* (a), *together with the hyperplane* Σ *that separates the periodic orbit from the equilibrium. In Stage* II, *we grow an orbit in* $\mathcal{W}_0^u$ *until it terminates on* Σ, *as shown in* (b). *In the subsequent Stage* III *we grow an orbit in* $\mathcal{W}_{\mathrm{per}}^s$ *in a similar way; see Fig.* 10.4.

- a segment object associated with $\mathcal{W}_{\mathrm{per}}^s$, and

- boundary conditions that constrain the segment end points accordingly.

In this and the subsequent stages, we generate homotopies of solutions to this composite continuation problem.

   We may construct a new periodic orbit object, together with the corresponding variational collocation zero problem, from a point on the solution manifold of an extended continuation problem that is compatible with the `povar_isol2orb` constructor, using the `povar_sol2orb` constructor, shown below.

```
function [prob data sol] = povar_sol2orb(prob, oid, varargin)

str   = coco_stream(varargin{:});
run   = str.get;
if ischar(str.peek)
  soid = str.get;
else
  soid = oid;
end
lab = str.get;

stbid = coco_get_id(soid, 'po.seg.var');
[data sol] = coco_read_solution(stbid, run, lab);
prob = po_sol2orb(prob, oid, run, soid, lab);
toid = coco_get_id(oid, 'po.seg');
prob = var_coll_add(prob, toid, data.dfdxdxhan, data.dfdxdphan);

end
```

As shown in the following encoding of the `riess_start_1` function, the `data` and `sol` output arguments can now be used to add two additional segment objects with zero interval length and based at points displaced relative to the trivial equilibrium $y = 0$ and relative to $y_{per}(0)$ by a scalar multiple of v_0^u and v_{per}^s, respectively.

```
function prob = riess_start_1(prob, run, lab)

[prob data sol] = povar_sol2orb(prob, '', run, lab);

fdata = coco_get_func_data(prob, data.coll_id, 'data');
eps0  = [0.1; 0.1];
p0    = sol.x(fdata.p_idx);

s  = p0(1);
r  = p0(2);
t0 = 0;
x0 = eps0(1)*[(1-s+sqrt((1-s)^2+4*r*s))/2/r; 1; 0]';
coll_args = {fdata.fhan, fdata.dfdxhan, fdata.dfdphan, t0, x0, p0};
prob = coll_isol2seg(prob, 'col1', coll_args{:});

M    = reshape(sol.x(data.ubp_idx), data.u_shp);
M1   = M(data.M1_idx,:);
[v, d] = eig(M1);
ind  = find(abs(diag(d))<1);
vec0 = -v(:,ind);
lam0 = d(ind,ind);

t0 = 0;
x0 = sol.x(fdata.xbp_idx(end-data.dim+1:end))'+eps0(2)*vec0';
coll_args = {fdata.fhan, fdata.dfdxhan, fdata.dfdphan, t0, x0, p0};
prob = coll_isol2seg(prob, 'col2', coll_args{:});

data.nrm = [0, -1, 1]/sqrt(2);
data.pt0 = [20; 20; 30];

prob = riess_close_het_1(prob, data, vec0, lam0, eps0);

end
```

We include in the function body the assignment of the coordinates of the point y_Σ and the unit vector n_Σ to the `pt0` and `norm` fields of the `data` structure. The corresponding initial solution guess is shown in Fig. 10.3(a).

The `riess_close_het_1` function, shown below, begins by providing gluing conditions for the problem parameters associated with each segment object.

```
function prob = riess_close_het_1(prob, data, vec, lam, eps)

[data1 uidx1] = coco_get_func_data(prob, 'col1.coll', 'data', 'uidx');
[data2 uidx2] = coco_get_func_data(prob, 'col2.coll', 'data', 'uidx');
[data3 uidx3] = coco_get_func_data(prob, 'po.seg.var', 'data', 'uidx');
[data4 uidx4] = coco_get_func_data(prob, data3.coll_id, 'data', 'uidx');

prob = coco_add_glue(prob, 'shared', ...
  [uidx1(data1.p_idx); uidx1(data1.p_idx)], ...
  [uidx2(data2.p_idx); uidx4(data4.p_idx)]);

evsdata = struct('tbid', 'po.seg.var');
prob = coco_add_func(prob, 'evs', @var_evs, evsdata, 'zero', ...
```

```
      'uidx', uidx3(data3.ubp_idx), 'u0', [vec; lam]);
uidx = coco_get_func_data(prob, 'evs', 'uidx');
data.vec_idx = uidx(end-3:end-1);
data.lam_idx = uidx(end);

prob = coco_add_func(prob, 'bcs1', @eig_bcs, [], 'zero', 'uidx', ...
      [uidx1(data1.x0_idx); uidx2(data2.x1_idx); ...
      uidx4(data4.xbp_idx(end-data.dim+1:end)); ...
      uidx4(data4.p_idx(1:2)); data.vec_idx], 'x0', eps);
uidx = coco_get_func_data(prob, 'bcs1', 'uidx');
data.eps_idx = uidx(end-1:end);

prob = coco_add_func(prob, 'bcs2', @proj_bcs, data, 'inactive', ...
      {'sg1' 'sg2'}, 'uidx', [uidx1(data1.x1_idx); uidx2(data2.x0_idx)]);

prob = coco_add_pars(prob, 'pars', ...
      [data.eps_idx; uidx1(data1.T_idx); uidx2(data2.T_idx)], ...
      {'eps1' 'eps2' 'T1' 'T2'});

prob = coco_add_slot(prob, 'riess_save_1', @coco_save_data, data, ...
      'save_full');

end
```

The zero functions encoded in the var_evs function, shown below, impose the conditions that vec and lam correspond to a unit eigenvector and the corresponding eigenvalue of the sensitivity matrix associated with the periodic orbit.

```
function [data y] = var_evs(prob, data, u)

fdata = coco_get_func_data(prob, data.tbid, 'data');

M  = reshape(u(1:end-4), fdata.u_shp);
M1 = M(fdata.M1_idx,:);

vec = u(end-3:end-1);
lam = u(end);

y = [M1*vec-lam*vec; vec'*vec-1];

end
```

The zero functions encoded in eig_bcs, shown below, impose boundary conditions on the two additional segment objects to ensure that they approximate orbits in $\mathcal{W}_0^u$ and $\mathcal{W}_{\text{per}}^s$, respectively.

```
function [data y] = eig_bcs(prob, data, u)

x10 = u(1:3);
x20 = u(4:6);
x30 = u(7:9);
s   = u(10);
r   = u(11);
vec = u(12:14);
eps = u(15:16);

evec = [(1-s+sqrt((1-s)^2+4*r*s))/2/r; 1; 0];
```

```
y = [x10-eps(1).*evec; x20-(x30+eps(2)*vec)];

end
```

Finally, the components of the `proj_bcs` function, shown below, monitor the distances from the end point on the orbit segment in W_0^u and the starting point on the orbit segment in W_{per}^s, respectively, to Σ.

```
function [data y] = proj_bcs(prob, data, u)

x1 = u(1:3);
x2 = u(4:6);

y = [data.nrm*(x1-data.pt0); data.nrm*(x2-data.pt0)];

end
```

It follows that the nominal dimensional deficit of the restricted continuation problem obtained from a call to `riess_start_1` is -2, implying that at least three continuation parameters must be released in order to enable continuation with a 1-dimensional atlas algorithm.

In the extract below, we obtain a problem with nominal dimensional deficit of 1 by releasing the continuation parameters `'T1'` and `'sg1'` associated with the interval length of the orbit segment in W_0^u and the distance from the starting point of this orbit segment to Σ, respectively, and the continuation parameter `'T2'` associated with the interval length of the orbit segment in W_{per}^s. As the distance from the end points of the latter to Σ must remain fixed, it follows that the continuation parameter `'T2'` remains constant during continuation within numerical accuracy.

```
>> prob = riess_start_1(coco_prob(), 'runHopf', 6);
>> prob = coco_set(prob, 'cont', 'ItMX', 500);
>> prob = coco_set(prob, 'cont', 'NPR', 50);
>> cont_args = {{'sg1' 'T1'  'T2'}, {[-30 0] [0 1]}};
>> coco(prob, 'run1', [], 1, cont_args{:});
```

```
...    NORMS              COMPUTATION TIMES
...    ||f||      ||U||     F(x)   DF(x)  SOLVE
... 4.09e-12  2.78e+02     0.0     0.0    0.0
... 9.88e-14  2.78e+02     0.0     0.3    0.0

... LABEL  TYPE        sg1            T1            T2
...     1  EP    -7.1418e+00   1.6897e-16  -2.8069e-14
...     2        -1.2097e+01   4.1889e-01  -2.9122e-13
...     3        -1.4309e+01   4.9522e-01  -6.3957e-13
...     4        -1.1711e+01   5.4086e-01  -9.2310e-13
...     5        -5.0642e+00   5.7457e-01  -1.2016e-12
...     6  EP     0.0000e+00   5.9189e-01  -1.4207e-12
```

Here, the values assigned to the `'ItMX'` and `'NPR'` settings of the `'cont'` core toolbox allow for at most 500 successful continuation steps along the solution manifold with data printed to screen and saved to disk every 50 steps. It follows, by construction, that the solution associated with the solution label 6 approximates a collection of three state-space trajectories such that the end point of the trajectory in W_0^u lies in Σ. The corresponding state-space representation is shown in Fig. 10.3(b).

Stage III: Grow orbit in $\mathcal{W}_{\mathrm{per}}^s$

Now consider the following encoding of the `riess_restart_1` function.

```
function prob = riess_restart_1(prob, run, lab)

prob = povar_sol2orb(prob, '', run, lab);
prob = coll_sol2seg(prob, 'col1', run, lab);
prob = coll_sol2seg(prob, 'col2', run, lab);

[data sol] = coco_read_solution('riess_save_1', run, lab);
eps = sol.x(data.eps_idx);
vec = sol.x(data.vec_idx);
lam = sol.x(data.lam_idx);

prob = riess_close_het_1(prob, data, vec, lam, eps);

end
```

This reconstructs the continuation problem structure from a previously obtained solution so as to allow for additional homotopy analysis under variations in other continuation parameters. As before, the nominal dimensional deficit remains equal to -2. In the extract below, we perform continuation along a 1-dimensional solution manifold obtained by releasing the continuation parameters `'sg2'`, `'T1'`, and `'T2'`.

```
>> prob = riess_restart_1(coco_prob(), 'run1', 6);
>> prob = coco_set(prob, 'cont', 'ItMX', 500);
>> prob = coco_set(prob, 'cont', 'NPR', 50);
>> cont_args = {{'sg2' 'T2' 'T1'}, {[0 30] [0 1]}};
>> coco(prob, 'run2', [], 1, cont_args{:});
```

	NORMS		COMPUTATION TIMES		
...	$\|\|f\|\|$	$\|\|U\|\|$	F(x)	DF(x)	SOLVE
...	3.89e-12	2.96e+02	0.0	0.0	0.0
...	1.11e-13	2.96e+02	0.0	0.3	0.0

	LABEL	TYPE	sg2	T2	T1
...	1	EP	1.6905e+01	-1.4253e-12	5.9189e-01
...	2		1.6807e+01	1.7993e-01	5.9189e-01
...	3		1.7545e+01	3.1632e-01	5.9189e-01
...	4		2.2253e+01	3.7652e-01	5.9189e-01
...	5		2.4331e+01	4.1257e-01	5.9189e-01
...	6		2.1280e+01	4.3919e-01	5.9189e-01
...	7		1.3829e+01	4.6043e-01	5.9189e-01
...	8		4.5417e+00	4.7855e-01	5.9189e-01
...	9	EP	-2.4425e-15	4.8690e-01	5.9189e-01

Here `'sg2'` corresponds to the distance from the starting point of the trajectory in $\mathcal{W}_{\mathrm{per}}^s$ to Σ. It follows that the solution associated with the solution label 9 approximates a collection of three state-space trajectories such that the end point of the orbit segment in $\mathcal{W}_0^u$ and the starting point of the orbit segment in $\mathcal{W}_{\mathrm{per}}^s$ both lie in Σ, as shown in Fig. 10.4(a).

Stage IV: Sweep family of orbits in $\mathcal{W}_{\mathrm{per}}^s$

A connecting orbit from the trivial equilibrium to the periodic orbit must lie in $\mathcal{W}_{\mathrm{per}}^s$ and must approach arbitrarily closely to every point on the periodic orbit. If $\mathcal{W}_{\mathrm{per},0}^s$ denotes a curve on $\mathcal{W}_{\mathrm{per}}^s$ through $y_{\mathrm{per}}(0)$ and tangential at $y_{\mathrm{per}}(0)$ to v_{per}^s, then the connecting orbit

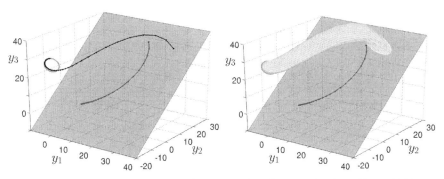

(a) Growing one orbit in the stable manifold.   (b) After sweeping the stable manifold.

Figure 10.4. *Panel* (a) *shows the three-segment solution after completing Stage* III, *i.e., growing an orbit in* $\mathcal{W}^s_{\text{per}}$ *starting at the solution shown in Fig.* 10.3(b). *Here, the end point of the orbit segment in* $\mathcal{W}^u_0$ *and the starting point of the orbit segment in* $\mathcal{W}^s_{\text{per}}$ *both terminate on* Σ. *Although* $\mathcal{W}^s_{\text{per}}$ *is 2-dimensional, the connecting orbit is unique. To obtain an initial approximation of the connecting orbit, we first sweep* $\mathcal{W}^s_{\text{per}}$ *in Stage* IV *and compute a set of orbit segments that cover the manifold sufficiently densely* (b). *From this family of orbits we select the one that terminates closest to the end point of the segment in* $\mathcal{W}^u_0$; *see Fig.* 10.5.

must intersect $\mathcal{W}^s_{\text{per},0}$ at countably many points accumulating on $y_{\text{per}}(0)$. To locate such an intersection, we may reconstruct the continuation problem from the final element of the homotopy obtained in Stage III and release the continuation parameters 'eps2', 'T1', and 'T2', while allowing the former to vary on the interval $[10^{-6}, 0.1]$, as shown in the extract below.

```
>> prob = riess_restart_1(coco_prob(), 'run2', 9);
>> prob = coco_set(prob, 'cont', 'ItMX', 300);
>> cont_args = {{'eps2' 'T1' 'T2'}, [1e-6 1e-1]};
>> coco(prob, 'run3', [], 1, cont_args{:});
```

```
...    NORMS                 COMPUTATION TIMES
...    ||f||      ||U||      F(x)   DF(x)  SOLVE
... 3.86e-12  3.13e+02       0.0    0.0    0.0
... 1.07e-13  3.13e+02       0.0    0.3    0.0

... LABEL  TYPE          eps2            T1            T2
...     1  EP      1.0000e-01    5.9189e-01    4.8690e-01
...     2          7.6379e-02    5.9189e-01    5.0488e-01
...     3          5.3320e-02    5.9189e-01    5.2942e-01
...     4          3.5855e-02    5.9189e-01    5.5738e-01
...     5          2.3489e-02    5.9189e-01    5.8833e-01
...     6          1.5354e-02    5.9189e-01    6.2091e-01
...     7          1.0268e-02    5.9189e-01    6.5342e-01
...     8          7.1172e-03    5.9189e-01    6.8480e-01
...     9          5.1192e-03    5.9189e-01    7.1477e-01
...    10          3.8011e-03    5.9189e-01    7.4347e-01
...    11          2.8902e-03    5.9189e-01    7.7124e-01
...    12          2.2266e-03    5.9189e-01    7.9862e-01
...    13          1.7115e-03    5.9189e-01    8.2641e-01
...    14          1.2827e-03    5.9189e-01    8.5543e-01
```

...	15		9.5289e-04	5.9189e-01	8.8160e-01
...	16		7.4684e-04	5.9189e-01	8.9970e-01
...	17		6.0634e-04	5.9189e-01	9.1336e-01
...	18		4.9870e-04	5.9189e-01	9.2523e-01
...	19		4.1099e-04	5.9189e-01	9.3648e-01
...	20		3.3728e-04	5.9189e-01	9.4770e-01
...	21		2.7445e-04	5.9189e-01	9.5930e-01
...	22		2.2062e-04	5.9189e-01	9.7157e-01
...	23		1.7460e-04	5.9189e-01	9.8478e-01
...	24		1.3552e-04	5.9189e-01	9.9922e-01
...	25		1.0272e-04	5.9189e-01	1.0152e+00
...	26		7.5655e-05	5.9189e-01	1.0331e+00
...	27		5.3818e-05	5.9189e-01	1.0534e+00
...	28		3.6725e-05	5.9189e-01	1.0767e+00
...	29		2.3885e-05	5.9189e-01	1.1037e+00
...	30		1.4780e-05	5.9189e-01	1.1351e+00
...	31	EP	8.8314e-06	5.9189e-01	1.1711e+00

We provide two views of $\mathcal{W}_{\mathrm{per}}^s$ in Figs. 10.4(b) and 10.5(a).

Stage V: Reduce Lin gap to 0

We proceed to restart continuation from the $k_{\min}$ element in the homotopy in Stage IV for which the trajectory segment in $\mathcal{W}_{\mathrm{per}}^s$ intersects Σ at the smallest distance from the intersection of the trajectory segment in $\mathcal{W}_0^u$ with Σ. Our objective is to generate a final homotopy under variations in r whose last element corresponds to a zero separation between the two points of intersection. It is clear that this homotopy must allow for variations in the interval lengths associated with the two trajectory segments. Now let the unit vector v_{gap} be parallel to the initial separation between the two points of intersection. The homotopy must

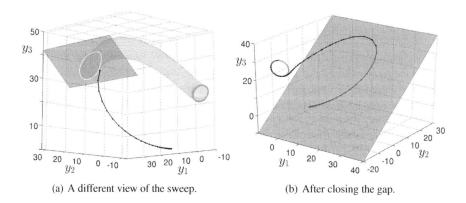

(a) A different view of the sweep.   (b) After closing the gap.

Figure 10.5. *Panel* (a) *shows a different view of Fig.* 10.4(b), *the result of a sweep of* $\mathcal{W}_{\mathrm{per}}^s$. *The intersection with* Σ *is highlighted. We compute the point of the intersecting curve that is closest to the end point of the segment in* $\mathcal{W}_0^u$ *and initialize Stage* V *of the homotopy, i.e., the closing of the Lin gap. The resulting connecting orbit after closing the gap is shown in panel* (b).

allow for variations in the projection of the actual separation on v_{gap}, while maintaining a zero projection on a transversal direction. Here, we let such a transversal direction be represented by the unit vector v_{phase}, parallel to the separation between the points of intersection of W^s_{per} with Σ for the $(k_{min} - 1)$th and $(k_{min} + 1)$th elements of the homotopy in Stage IV.

The determination of the integer k_{min} and the construction of an associated composite continuation problem is implemented in the initial sequence of assignments in the `riess_start_2` function, shown below.

```
function prob = riess_start_2(prob, run)

bd    = coco_bd_read(run);
labs = coco_bd_labs(bd, 'ALL');
endpoints = [];
labels = [];
for lab=labs
  sol       = coll_read_solution('col2', run, lab);
  endpoints = [endpoints; sol.x(1,:)];
  labels    = [labels; lab];
end
sol       = coll_read_solution('col1', run, 1);
pt        = repmat(sol.x(end,:), [size(endpoints, 1) 1]);
[m1 i1] = min(sqrt(sum((endpoints-pt).*(endpoints-pt), 2)));

prob = riess_restart_1(prob, run, labels(i1));

vgap        = endpoints(i1,:)-pt(i1,:);
data.gapvec = vgap/norm(vgap);
vphase      = endpoints(i1+1,:)-endpoints(i1-1,:);
data.vphase = vphase/norm(vphase);

prob = riess_close_het_2(prob, data);

end
```

The function body includes the assignment of the *Lin gap vector* v_{gap} and the phase vector v_{phase} to the `gapvec` and `vphase` fields of the `data` structure. The `riess_close_het_2` function, shown below, provides for the imposition of the constraints on the projections of the separation between the points of intersection with Σ onto v_{gap} and v_{phase}.

```
function prob = riess_close_het_2(prob, data)

[data1 uidx1] = coco_get_func_data(prob, 'col1.coll', 'data', 'uidx');
[data2 uidx2] = coco_get_func_data(prob, 'col2.coll', 'data', 'uidx');

prob = coco_add_func(prob, 'gap', @lingap, data, 'inactive', ...
  'lingap', 'uidx', [uidx1(data1.x1_idx); uidx2(data2.x0_idx)]);
prob = coco_add_func(prob, 'phase', @linphase, data, 'zero', ...
  'uidx', [uidx1(data1.x1_idx); uidx2(data2.x0_idx)]);
prob = coco_add_slot(prob, 'riess_save_2', @coco_save_data, data, ...
  'save_full');

end
```

The encoding associates the `lingap` monitor function, shown below, with the initially inactive continuation parameter labeled `'lingap'`.

```
function [data y] = lingap(prob, data, xp)
  y = data.gapvec*(xp(1:3)-xp(4:6));
end
```

In contrast, the vanishing projection onto v_{phase} is encoded in the zero function `linphase`, shown below.

```
function [data y] = linphase(prob, data, xp)
  y = data.vphase*(xp(1:3)-xp(4:6));
end
```

The extract below shows the result of a homotopy analysis under variations in 'lingap', 'T1', 'T2', 'eps2', and 'r'.

```
>> prob = riess_start_2(coco_prob(), 'run3');
>> prob = coco_set(prob, 'cont', 'ItMX', 500);
>> cont_args = {{'lingap' 'r' 'eps2' 'T1' 'T2'}, [-1 0]};
>> coco(prob, 'run4', [], 1, cont_args{:});
```

```
...    NORMS                COMPUTATION TIMES
...     ||f||     ||U||    F(x)  DF(x)  SOLVE
... 6.49e-01  3.01e+02    0.0   0.0    0.0
... 8.13e-02  3.05e+02    0.0   0.3    0.0
... 1.57e-03  3.05e+02    0.0   0.5    0.0
... 2.51e-07  3.05e+02    0.0   0.7    0.0
... 1.70e-13  3.05e+02    0.0   0.9    0.0
... 1.05e-13  3.05e+02    0.0   1.1    0.0
```

```
... LABEL   TYPE        lingap           r        eps2           T1 ...
...     1   EP      -8.7354e-01   2.4466e+01   3.9372e-03   5.8915e-01 ...
...     2   EP      -1.0000e+00   2.4505e+01   3.9911e-03   5.8859e-01 ...
```

```
... LABEL   TYPE        lingap           r        eps2           T1 ...
...     3   EP      -8.7354e-01   2.4466e+01   3.9372e-03   5.8915e-01 ...
...     4           -6.3696e-01   2.4379e+01   3.7680e-03   5.9038e-01 ...
...     5           -3.6893e-01   2.4262e+01   3.4937e-03   5.9208e-01 ...
...     6           -1.1918e-01   2.4130e+01   3.1810e-03   5.9400e-01 ...
...     7   EP       0.0000e+00   2.4058e+01   3.0173e-03   5.9504e-01 ...
```

It follows that the element of the homotopy with solution label 7 corresponds to an approximate connecting orbit between the equilibrium point and the periodic orbit, as shown in Fig. 10.5(b).

Stage VI: Continuation in problem parameters

We conclude the analysis by again constraining the 'lingap' continuation parameter and releasing the constraint on a second problem parameter, as shown in the extract below.

```
>> prob = riess_restart_2(coco_prob(), 'run4', 7);
>> prob = coco_set(prob, 'cont', 'ItMX', 500);
>> prob = coco_set(prob, 'cont', 'NPR', 50);
>> cont_args = {{'r' 'b' 'eps1' 'eps2' 'T2'}, [20 30]};
>> coco(prob, 'run5', [], 1, cont_args{:});
```

```
...    NORMS                COMPUTATION TIMES
...     ||f||     ||U||    F(x)  DF(x)  SOLVE
```

```
... 3.61e-12  2.97e+02    0.0    0.0    0.0
... 1.22e-13  2.97e+02    0.0    0.4    0.0

... LABEL  TYPE           r           b        eps1          eps2 ...
...     1  EP      2.4058e+01  2.6667e+00  1.0000e-01  3.0173e-03 ...
...     2          2.2380e+01  2.3137e+00  1.2829e-01  9.6797e-04 ...
...     3          2.0665e+01  1.9674e+00  1.6684e-01  1.9265e-04 ...
...     4  EP      2.0000e+01  1.8413e+00  1.8522e-01  9.5323e-05 ...

... LABEL  TYPE           r           b        eps1          eps2 ...
...     5  EP      2.4058e+01  2.6667e+00  1.0000e-01  3.0173e-03 ...
...     6          2.5755e+01  3.0171e+00  7.8217e-02  6.1635e-03 ...
...     7          2.7560e+01  3.3673e+00  6.0578e-02  9.9836e-03 ...
...     8          2.9427e+01  3.6989e+00  4.6769e-02  1.4158e-02 ...
...     9  EP      3.0000e+01  3.7945e+00  4.3241e-02  1.5486e-02 ...
```

This relies on a reconstruction of the composite continuation problem, from the final element of the homotopy in Stage V, using the `riess_restart_2` function, shown below.

```
function prob = riess_restart_2(prob, run, lab)

prob = riess_restart_1(prob, run, lab);
data = coco_read_solution('riess_save_2', run, lab);
prob = riess_close_het_2(prob, data);

end
```

10.3   Candidate boundary conditions

Recall that we obtain a unique solution to the first variational equation in Eq. (10.2) by appending a set of n^2 additional independent conditions on the entries of $Y(t)$ to the differential equation. In the previous sections, the initial condition $Y(0) = I_n$ was employed with apparent success, providing satisfactory estimates for the sensitivity gains and the local eigenstructure. As shown below, however, significant care is necessary in the selection of the additional conditions in order to guarantee robust performance.

For any solution to the first variational equation, by the properties of the determinant operator $|\cdot|$, Eq. (10.2) implies that

$$\frac{d}{dt} \ln\left(|Y|\right) = \operatorname{Tr}\left(\partial_y f(y, p)\right), \tag{10.93}$$

where Tr denotes the matrix trace operator. It follows that

$$\left|Y(t)\right| = e^{\int_0^t \operatorname{Tr}(\partial_y f(y(s), p)) ds} \left|Y(0)\right|. \tag{10.94}$$

By definition, the determinant of the sensitivity matrix is thus given by

$$\left|Y(T) \cdot \left(Y(0)\right)^{-1}\right| = e^{\int_0^T \operatorname{Tr}(\partial_y f(y(s), p)) ds}. \tag{10.95}$$

Now let v denote an eigenvector of the sensitivity matrix with corresponding sensitivity gain λ. It follows that $w = \left(Y(0)\right)^{-1} \cdot v$ satisfies the *generalized eigenvalue problem*

$$\left(Y(T) - \lambda Y(0)\right) \cdot w = 0. \tag{10.96}$$

Ideally, we seek to impose conditions on the entries of $Y(t)$ that guarantee, at least, that $Y(T)$ and $Y(0)$ have bounded entries across the relevant portion of the solution manifold. By estimating these bounds, we may control any loss of numerical accuracy in the analysis of the generalized eigenvalue problem in Eq. (10.96).

Example 10.6 The unique solution to the first variational equation in Eq. (10.2) that results from the additional boundary condition

$$Y(0) = I_n \qquad (10.97)$$

equals the corresponding fundamental solution. In this case, the sensitivity matrix is simply given by $Y(T)$ and its determinant is given by

$$\left| Y(T) \right| = e^{\int_0^T \mathrm{Tr}(\partial_y f(y(s),p)) ds}. \qquad (10.98)$$

The entries of $Y(T)$ clearly grow rapidly with the integral in the exponent.

Alternatively, consider the unique solution that results from the additional boundary condition

$$Y(T) = I_n. \qquad (10.99)$$

In this case, $\left| Y(T) \right| = 1$ and, consequently,

$$\left| Y(0) \right| = e^{-\int_0^T \mathrm{Tr}(\partial_y f(y(s),p)) ds}. \qquad (10.100)$$

With this choice, it is the entries of $Y(0)$ that must grow rapidly with the negative of the integral in the exponent. ∎

The candidate boundary conditions given in Example 10.6 guarantee that $Y(t)$ is nonsingular for all $t \in [0, T]$. However, the potential large growth in the entries of $Y(T)$ and $Y(0)$, respectively, is a possible source of loss of numerical accuracy. Consider, as an alternative, the boundary condition

$$Y(0) + Y(T) = 2I_n. \qquad (10.101)$$

In this case, bounded entries of $Y(0)$ must imply bounded entries of $Y(T)$. There still remains the possibility, however, that the entries of both matrices are very large or that there does not exist a solution to Eq. (10.2) that satisfies Eq. (10.101).

Example 10.7 Suppose that, for some critical choice of problem parameters p and solution $y(t)$, there exists an eigenvector v of the sensitivity matrix corresponding to the eigenvalue -1, i.e., such that

$$Y(T) \cdot \left(Y(0)\right)^{-1} \cdot v = -v \qquad (10.102)$$

or, equivalently,

$$\left(Y(0) + Y(T)\right) \cdot w = 0, \qquad (10.103)$$

where $w = \left(Y(0)\right)^{-1} \cdot v$. The function $Y(t) \cdot w$ is the unique solution to the first variational equation that evaluates to $Y(0) \cdot w$ at $t = 0$. It follows that $Y(t) \cdot w$ must lie in the kernel to the combined linear problem obtained from the first variational equation and the boundary condition in Eq. (10.101). Such a nonempty kernel is equivalent to a singular point on the corresponding solution manifold. For parameter values p and solutions $y(t)$ close to the critical choice, the combined linear problem must be very ill conditioned, resulting in a highly error-prone evaluation of $Y(0)$ and $Y(T)$. ∎

More generally, consider a linear discrete condition of the form

$$\sum_{i=1}^{s} a_i Y(t_i) = I_n, \tag{10.104}$$

where $\{a_i\}_{i=1}^{s}$ denotes a sequence in $\mathbb{R}$, such that

$$\sum_{i=1}^{s} a_i = 1, \tag{10.105}$$

and $\{t_i\}_{i=1}^{s}$ denotes a sequence of nodal points on the interval $[0, T]$ such that $t_1 = 0$ and $t_s = T$. It is possible to show that no choice of a_i's or t_i's can ensure that the linear condition is immune to the type of singularity observed in Example 10.7. Indeed, for each such choice, there exist countably many pairs $(a, b) \in \mathbb{R}^2$ such that every solution to the linear differential equation

$$\frac{dY}{dt} = \begin{pmatrix} a & -b \\ b & a \end{pmatrix} \cdot Y \tag{10.106}$$

lies in the kernel to the linear operator

$$Y(t) \mapsto \sum_{i=1}^{s} a_i Y(t_i). \tag{10.107}$$

Similar observations apply to linear integral conditions of the form

$$\int_0^T a(t) Y(t) \, dt = I_n \tag{10.108}$$

for some function $a(t)$ such that

$$\int_0^T a(t) \, dt = 1. \tag{10.109}$$

These observations imply that it is not possible to put bounds on the condition number of the coefficient matrix used to solve for the discretization Δ_{bp}, when relying on linear conditions on the entries of Δ_{bp}. We return to alternative formulations that might support such bounds in the epilogue to this book.

10.4 Conclusions

We argued briefly in the introduction to this chapter for the value of the sensitivity matrix in describing the local warping and stretching of families of perturbations to a given initial condition under the flow of a dynamical system. In particular, the eigenstructure of the sensitivity matrix was shown to be related to properties of *stability* to such perturbations, providing necessary as well as sufficient conditions for asymptotic stability or instability of periodic orbits, in smooth and hybrid dynamical systems. An alternative use of sensitivity matrices arises through the introduction of problem parameters among the state variables and the artificial extension of the vector field, as suggested by the augmented dynamical system

$$\frac{dz}{dt} = g(z), \qquad (10.110)$$

obtained by the substitution $z := (\ y \quad p\ )^T$ and $g(z) := (\ f(y,p) \quad 0\ )^T$. It follows that the effects of perturbations to the problem parameters on the flow may be explored by means of applying the theory of this chapter to Eq. (10.110) in lieu of the original differential equation in Eq. (10.1). From this follows the use of the 'var_coll' toolbox to design studies, where parameter values must be chosen to optimize certain functions of the solution $y(t)$.

The embedding of the variational zero problem into the extended continuation problem encoded in the second version of the 'var_coll' toolbox enables the imposition of arbitrary constraints on the components of the sensitivity matrix (or, for that matter, of the full discretization Δ_{bp}). In the application to the continuation of a connecting orbit in Sect. 10.2.2, we relied on this functionality to track the eigenvalue λ_{per}^s and the associated eigenvector v_{per}^s, and to use the latter in the imposition of a boundary condition on an orbit segment. Other potential uses include the continuation of local bifurcations of a periodic orbit associated with Floquet multipliers located on the unit circle in the complex plane. In these cases, however, so-called *minimally extended continuation problems* may sometimes be found that eliminate the need to solve for the entire discretization Δ_{bp}. As an example, in the case of a period-doubling bifurcation, it suffices to append the discretization of the differential equation

$$\frac{dv}{dt} = \partial_y f(y,p) \cdot v, t \in [0, T], \qquad (10.111)$$

and the algebraic conditions

$$v(T) + v(0) = 0, \qquad (10.112)$$

$$v(0)^T \cdot v(0) - 1 = 0 \qquad (10.113)$$

to the collocation continuation problem.

Exercises

10.1. Consider the first variational equation

$$\frac{dY}{dt} = \partial_y f(y,p) \cdot Y$$

associated with a solution $y(t)$ to the differential equation

$$\frac{dy}{dt} = f(y,p)$$

on the interval $t \in [0, T]$. Show that the solution $Y(t)$ is invertible for all $t \in [0, T]$ provided that it is invertible for some $t \in [0, T]$.

10.2. Suppose that $Y(t) \in \mathbb{R}^{n \times n}$ is an invertible solution to the first variational equation in the previous exercise for $t \in [0, T]$. Let C denote an arbitrary $n \times k$ matrix and show that $Y(t) \cdot C$ is the unique solution to the first variational equation that evaluates to $Y(0) \cdot C$ at $t = 0$. Show, in particular, that $Y(t) \cdot (Y(0))^{-1}$ is the fundamental solution to the first variational equation.

10.3. Consider the function $\tilde{y}(\varepsilon, t)$ introduced in Sect. 10.1.1. Show that

$$\left. \frac{\partial \tilde{y}}{\partial \varepsilon} \right|_{\varepsilon = t = 0} = \left. \frac{\partial \tilde{y}_0}{\partial \varepsilon} \right|_{\varepsilon = 0}$$

and

$$\left. \frac{\partial^2 \tilde{y}}{\partial t \partial \varepsilon} \right|_{\varepsilon = 0} = \left. \frac{\partial^2 \tilde{y}}{\partial \varepsilon \partial t} \right|_{\varepsilon = 0} = \partial_y f \left(\tilde{y}|_{\varepsilon = 0}, p \right) \cdot \left. \frac{\partial \tilde{y}}{\partial \varepsilon} \right|_{\varepsilon = 0} = \partial_y f(y, p) \cdot \left. \frac{\partial \tilde{y}}{\partial \varepsilon} \right|_{\varepsilon = 0}.$$

10.4. Suppose that $Y(t)$ is an invertible solution to the first variational equation. Use the result of the previous exercise to prove that

$$\left. \frac{\partial \tilde{y}}{\partial \varepsilon} \right|_{\varepsilon = 0} (t) = Y(t) \cdot (Y(0))^{-1} \cdot \left. \frac{\partial \tilde{y}_0}{\partial \varepsilon} \right|_{\varepsilon = 0}.$$

10.5. Suppose that $y(t)$ is a solution to the differential equation

$$\frac{dy}{dt} = f(y, p)$$

on the interval $t \in [0, T]$. Show that

$$\frac{dy}{dt}(0) = f(y(0), p)$$

and

$$\frac{d^2 y}{dt^2} = \partial_y f(y, p) \cdot \frac{dy}{dt}.$$

10.6. Suppose that $Y(t)$ is an invertible solution to the first variational equation. Use the result of the previous exercise to show that

$$f(y(t), p) = Y(t) \cdot (Y(0))^{-1} \cdot f(y(0), p).$$

10.7. Verify the vectorized form of the variational collocation problem in Eqs. (10.32)–(10.33).

10.8. Find the general solution to the differential equation

$$\frac{dy}{dt} = f(y, p),$$

where

$$f(y, p) = \begin{pmatrix} y_2 \\ -y_2 - p_1 y_1 + \cos y_3 \\ 1 \end{pmatrix}$$

for $y \in \mathbb{R}^2 \times \mathbb{S}$ in terms of the initial conditions $y_{1,0}$, $y_{2,0}$, and $y_{3,0}$. Derive the corresponding first variational equation and show that this is satisfied by the Jacobian of the general solution with respect to the initial conditions.

10.9. Find the special choice of initial conditions that correspond to a periodic orbit for the dynamical system in the previous exercise. Show that the corresponding Floquet multipliers equal

$$e^{(-1\pm\sqrt{1-4p_1})\pi}.$$

10.10. Explain the meaning of the boundary conditions in the function `lin_bc` in Example 10.2.

10.11. Consider the function

$$E(\varepsilon,t) := h(\tilde{y}(\varepsilon,t),p)$$

for some smooth function h and a solution $\tilde{y}(\varepsilon,t)$ to the initial-value problem

$$\frac{dy}{dt} = f(y,p), \ y(0) = y_0(\varepsilon)$$

for $\varepsilon \in (-\varepsilon_0, \varepsilon_0)$ and $0 < \varepsilon_0 \ll 1$. Suppose that $h(\tilde{y}(0,T),p) = 0$ and

$$\partial_y h(\tilde{y}(0,T),p) \cdot f(\tilde{y}(0,T),p) \neq 0$$

for some T. Show that there exist an open interval $I_\varepsilon \ni 0$ and an open interval $I_t \ni T$ such that

$$E(\varepsilon,t) = 0 \text{ for } (\varepsilon,t) \in I_\varepsilon \times I_t$$

if and only if $t = t(\varepsilon)$ for a unique scalar-valued function $t : \mathbb{R} \to \mathbb{R}$.

10.12. Derive the derivatives with respect to ε given in Eqs. (10.52)–(10.53).

10.13. Explain the significance of the composite sensitivity matrix given in Eq. (10.57).

10.14. Consider the piecewise-smooth vector field analyzed in Example 10.3. Show that, when $\alpha = \beta = 1$, the two-segment periodic orbit with the desired signature is given by the sequence $\{y^{(j)} : [0,T_j] \to \mathbb{R}^2\}_{j=1}^2$, where

$$y^{(1)}(t) = \begin{pmatrix} -\sin t \\ \cos t \end{pmatrix},$$

$$y^{(2)}(t) = \begin{pmatrix} \sin\gamma t \\ -\cos\gamma t \end{pmatrix},$$

$T_1 = \pi$, and $T_2 = \pi/\gamma$. Derive the corresponding first variational equations and verify that their fundamental solutions are given by

$$Y^{(1)}(t) = \begin{pmatrix} \cos t & -e^{-t}\sin t \\ \sin t & e^{-t}\cos t \end{pmatrix}$$

and

$$Y^{(2)}(t) = \begin{pmatrix} \cos\gamma t & e^{-t}\left(-\sin\gamma t + \left(e^t - 1\right)\cos\gamma t\right) \\ \sin\gamma t & e^{-t}\left(\cos\gamma t + \left(e^t - 1\right)\sin\gamma t\right) \end{pmatrix}.$$

Show that -1 is an eigenvalue of each corresponding sensitivity matrix. Find the eigenvalues of the composite sensitivity matrix and compare your prediction to the output from `hspo_mult_eigs_bddat`.

10.15. Repeat the analysis of the previous exercise for arbitrary parameter values. Hint: let $y_1 = r\cos\theta$ and $y_2 = r\sin\theta$ and derive differential equations for r and θ for each curve segment.

10.16. Suppose that the matrix B consists of n columns. Show that

$$\text{vec}(C \cdot B) = (I_n \otimes C) \cdot \text{vec}(B)$$

for any matrix C with as many columns as the number of rows of B. Use this result to show that the Jacobian of the variational zero functions with respect to $\text{vec}\left(\Delta_{bp}\right)$ equals

$$I_n \otimes \begin{pmatrix} \frac{T}{2N}\text{diag}\left(\partial_y f\left(\text{vec}_n\left(W \cdot v_{bp}\right), 1_{1,Nm} \otimes p\right)\right) \cdot W - W' \\ Q \\ \left(\begin{array}{cc} I_n & 0 \end{array}\right) \end{pmatrix}.$$

10.17. Show that the Jacobian of the zero functions corresponding to the variational collocation conditions with respect to p may be given alternatively by the matrix

$$\frac{T}{2N}\left(\begin{array}{ccc} B_1 & \cdots & B_q \end{array}\right) \cdot \left(I_q \otimes \left(W \cdot \Delta_{bp}\right)\right)$$

or the matrix obtained by a suitable array operation on the matrix

$$\frac{T}{2N}\begin{pmatrix} B_1 \\ \vdots \\ B_q \end{pmatrix} \cdot W \cdot \Delta_{bp}.$$

Which formula would you choose? Does it depend on circumstances? Which formula is implemented in `var_coll_DFDU`? Modify this encoding (and the initialization of the `'var_coll'` toolbox data structure in `var_coll_init_data`) to use the other formula.

10.18. Explain the encoding of the Jacobian of the variational zero functions with respect to v_{bp} in `var_coll_DFDU`. Can you simplify this implementation further?

10.19. Verify the statements regarding the existence and stability of equilibria in the Lorentz dynamical system in Sect. 10.2.2.

10.20. Determine the dimensional deficits of the restricted continuation problems constructed in the various stages of homotopy analysis in Sect. 10.2.2. Explain the choice of active continuation parameters in each stage.

10.21. Suppose that $Y(t)$ is a solution to the first variational equation associated with a vector field f and a solution $y(t)$ to the corresponding dynamical system. Show that

$$\frac{d}{dt}\ln\left(|Y|\right) = \text{Tr}\left(\partial_y f\left(y, p\right)\right),$$

where Tr denotes the matrix trace operator, and conclude that

$$\left|Y(t)\right| = e^{\int_0^t \text{Tr}\left(\partial_y f(y(s),p)\right)ds}\left|Y(0)\right|.$$

10.22. Consider the differential equation

$$\frac{dy}{dt} = f(y,\epsilon)$$

in terms of the vector field

$$f(y,\epsilon) := \begin{pmatrix} y_2 \\ \left(\frac{1}{2} - y_2^2\right)\epsilon y_2 - y_1 \end{pmatrix}.$$

Use the `'po'` toolbox and the first version of the `'var_coll'` toolbox to continue a family of periodic orbits under variations in ϵ on the interval $(-10, 10)$. What do you observe about the Floquet multipliers?

10.23. Modify the first version of the `'var_coll'` toolbox to use the $\Delta_{N(m+1)} = I_n$ boundary condition. Repeat the analysis in the previous exercise. What do you observe this time?

10.24. Modify the first version of the `'var_coll'` toolbox to use the $\Delta_1 + \Delta_{N(m+1)} = 2I_n$ boundary condition. Repeat the analysis in the previous exercise. What do you observe this time?

10.25. Apply the modified version of the `'var_coll'` toolbox from the previous exercise to the hybrid periodic orbits in Example 10.3. What happens when $\beta \to 1$? Verify that, in this limit, the coefficient matrix denoted by J and used in `var_coll_seg` to solve for the discretization Δ_{bp} for the first curve segment is singular with nullvector obtained by evaluating the tangent vector

$$\begin{pmatrix} \cos t & \sin t \end{pmatrix}^T$$

at each of the nodal points. Can you explain this observation? Can you make a similar observation about the second curve segment?

10.26. Show that the eigenvalues of the matrix $Y(T) \cdot \left(Y(0)\right)^{-1}$ equal those of the matrix $\left(Y(0)\right)^{-1} \cdot Y(T)$ as well as the generalized eigenvalues associated with the equation

$$Y(T) \cdot v = \lambda Y(0) \cdot v.$$

10.27. Review [Lust, K., "Improved numerical Floquet multipliers," *International Journal of Bifurcation and Chaos in Applied Sciences and Engineering*, 11(9), pp. 2389–2410, 2001] and apply the `'var_coll'` toolbox to any of the examples treated therein.

10.28. Review alternative methods for computing the eigenvalues of the composite sensitivity matrix in [Gusev, S., Johansson, S., Kågström, B., Shiriaev, A., and Varga, A., "A numerical evaluation of solvers for the periodic Riccati equation," *BIT Numerical Mathematics*, 50, pp. 301–329, 2010] and modify `hspo_mult_eigs_bddat` accordingly.

10.29. The implementation of Lin's method for continuation of connecting orbits between equilibria and periodic orbits in the Lorentz dynamical system, described in Sect. 10.2.2, reproduces the corresponding methodology and results in [Krauskopf, B. and Riess, T., "A Lin's method approach to finding and continuing heteroclinic connections involving periodic orbits," *Nonlinearity*, 21, pp. 1655–1690, 2008]. Review this paper and use COCO to investigate the other examples discussed therein.

10.30. Consider the linear condition

$$\sum_{i=1}^{s} a_i Y(t_i) = I_n,$$

where $\{a_i\}_{i=1}^{s}$ denotes a sequence in $\mathbb{R}$, such that

$$\sum_{i=1}^{s} a_i = 1,$$

and $\{t_i\}_{i=1}^{s}$ denotes a sequence of nodal points on the interval $[0, T]$ such that $t_1 = 0$, $t_s = T$, and $t_i = k_i \sigma$ for $k_i \in \mathbb{Z}$ and some scalar σ. Show that the *exponential sum*

$$\sum_{i=1}^{s} a_i e^{u k_i \sigma}, u \in \mathbb{C},$$

has countably many complex roots of the form

$$u = \frac{1}{\sigma}(\ln z_i + 2m\pi i), m \in \mathbb{Z},$$

for some collection z_i, $i = 1, \ldots, k_s$, of complex numbers.

10.31. Suppose that $u = u_R + i u_I$ is a root of the exponential sum in the previous exercise. Show that no solution of the linear differential equation

$$\frac{dY}{dt} = \begin{pmatrix} u_R & -u_I \\ u_I & u_R \end{pmatrix} \cdot Y$$

satisfies the linear condition

$$\sum_{i=1}^{s} a_i Y(t_i) = I_n.$$

10.32. Review the discussion about exponential sums and exponential integrals in [Langer, R.E., "On the zeros of exponential sums and integrals," *Bulletin of the American Mathematical Society*, 37(4), pp. 213–239, 1931] and comment on the use of linear conditions in the entries of Δ_{bp} to supplement the variational equations.

Part III

Atlas Algorithms

Chapter 11

Covering Manifolds

The example calls to the `coco` entry-point function shown in previous chapters all rely on default implementations of atlas algorithms included with COCO. The core offers great flexibility in modifying these or substituting entirely different atlas algorithms that are better suited for a particular problem class. In this and the next chapters, we illustrate the theory and minimal implementation of basic atlas algorithms that provide a manifold cover with a minimum of bells and whistles. We indicate along the way possible improvements or alternatives that bring a degree of sophistication to these minimal implementations.

11.1 Theory and terminology

11.1.1 The tangent space

Recall, from Sect. 2.2, the general case of a continuously differentiable (extended/restricted/reduced) continuation problem $\Phi(u) = 0$, in terms of a family of functions $\Phi : \mathbb{R}^n \to \mathbb{R}^m$ for $n > m > 0$. Let u^* denote a regular solution point, i.e., such that $\Phi(u^*) = 0$ and the Jacobian $\partial_u \Phi(u^*)$ has maximal rank m. By the implicit function theorem, there exists a locally unique $(n - m)$-dimensional manifold through u^* of solutions to the continuation problem that can be parameterized by some choice of $n - m$ components of u.

Consider the collection of all smooth *curve segments* $\gamma : [-\varepsilon, \varepsilon] \mapsto \mathbb{R}^n$, for some sufficiently small ε, such that $\gamma(0) = u^*$ and

$$\Phi(\gamma(t)) \equiv 0, t \in [-\varepsilon, \varepsilon]. \tag{11.1}$$

For each such curve segment, we refer to $\gamma'(0)$ as the *tangent vector* corresponding to this curve segment at the point u^*. The space of all such tangent vectors is the *tangent space* $\mathcal{T}_{u^*}$ to the solution manifold at the point u^*.

We claim that the tangent space $\mathcal{T}_{u^*}$ coincides with the *nullspace* $\mathcal{N}\left[\partial_u \Phi(u^*)\right]$ of the Jacobian $\partial_u \Phi(u^*)$. Indeed, by differentiation of Eq. (11.1) with respect to t and evaluation at $t = 0$, we obtain

$$\partial_u \Phi\left(u^*\right) \cdot \gamma'(0) = 0, \tag{11.2}$$

i.e., that $\mathcal{T}_{u^*} \subseteq \mathcal{N}\left[\partial_u \Phi(u^*)\right]$.

To establish the opposite inclusion, suppose that $v \in \mathcal{N}\left[\partial_u \Phi(u^*)\right]$ and let $T^\perp$ denote an $n \times m$ matrix that spans the orthogonal complement to the nullspace. It follows that the $m \times m$ matrix

$$\partial_u \Phi\left(u^*\right) \cdot T^\perp \tag{11.3}$$

is invertible. Consider now the zero problem

$$\Phi\left(u^* + tv + T^\perp \cdot \lambda\right) = 0 \tag{11.4}$$

in the continuation variables $(t, \lambda) \in \mathbb{R} \times \mathbb{R}^m$. This has a regular solution point at $(0,0)$, since the corresponding Jacobian

$$\partial_u \Phi\left(u^*\right) \cdot \left(\begin{array}{cc} v & T^\perp \end{array}\right) = \left(\begin{array}{cc} 0 & \partial_u \Phi(u^*) \cdot T^\perp \end{array}\right) \tag{11.5}$$

has full rank m. By the implicit function theorem, there exists a locally unique 1-dimensional manifold through $(0,0)$ of solutions to the zero problem Eq. (11.4). Let $s \mapsto (t(s), \lambda(s))$, for s on some sufficiently small neighborhood of 0, denote a curve segment on this solution manifold such that $t(0) = 0$ and $\lambda(0) = 0$. Since the corresponding tangent vector

$$\left(\begin{array}{c} t'(0) \\ \lambda'(0) \end{array}\right) \tag{11.6}$$

must lie in the nullspace to the Jacobian in Eq. (11.5), it follows that $\lambda'(0) = 0$. Without loss of generality, choose the parameterization s such that $t'(0) = 1$. By Eq. (11.4), the function

$$\gamma : s \mapsto u^* + t(s)v + T^\perp \cdot \lambda(s) \tag{11.7}$$

is a smooth curve segment through $\gamma(0) = u^*$ on the solution manifold to the original continuation problem. Moreover, its tangent vector at u^* is given by $\gamma'(0) = v$. We conclude that $\mathcal{N}\left[\partial_u \Phi(u^*)\right] \subseteq \mathcal{T}_{u^*}$, and the claim follows.

Since $\partial_u \Phi(u^*)$ is assumed to be of maximal rank m, it follows that the tangent space is an $(n-m)$-dimensional vector space.

Example 11.1 We illustrate this theory with the zero function

$$\Phi(u) := u_1^2 + (u_2 - 1)^2 - 1, u \in \mathbb{R}^3, \tag{11.8}$$

from Eq. (2.24). In this case, the Jacobian

$$\partial_u \Phi(u) = \left(\begin{array}{ccc} 2u_1 & 2(u_2 - 1) & 0 \end{array}\right) \tag{11.9}$$

has full rank at all points for which u_1 and $u_2 - 1$ do not both equal zero. Since no point of the form $(0, 1, u_3)$ is a solution point of the zero problem $\Phi(u) = 0$, it follows that all solution points are regular. Near any such point, there thus exists a local 2-dimensional solution manifold that can be parameterized by some choice of two components of the vector of continuation variables u.

Now let u^* denote any solution point of the zero problem $\Phi(u) = 0$ corresponding to Eq. (11.8). The 2-dimensional tangent space $\mathcal{T}_{u^*}$ is then spanned by the columns of the matrix

$$T = \left(\begin{array}{cc} 0 & 1 - u_2^* \\ 0 & u_1^* \\ 1 & 0 \end{array}\right), \tag{11.10}$$

since these provide a basis for the nullspace $\mathcal{N}\left[\partial_u\Phi(u^*)\right]$. In this case, we may choose

$$T^\perp = \begin{pmatrix} u_1^* \\ u_2^* - 1 \\ 0 \end{pmatrix} \tag{11.11}$$

such that

$$\partial_u\Phi(u)\cdot T^\perp = 2u_1^{*2} + 2\left(u_2^* - 1\right)^2 = 2 \neq 0. \tag{11.12}$$

Now let

$$v = \begin{pmatrix} a\left(1 - u_2^*\right) \\ au_1^* \\ b \end{pmatrix}. \tag{11.13}$$

Substitution and straightforward calculation then shows that

$$\Phi\left(u^* + tv + \lambda(t)T^\perp\right) = 0 \tag{11.14}$$

and $\lambda(0) = 0$ if and only if $\lambda = \sqrt{1 - a^2t^2} - 1$ for $at \in [-1,1]$. ∎

11.1.2 A local manifold cover

Let $V \in \mathbb{R}^{n\times(n-m)}$ be an arbitrary matrix for which the square matrix

$$\begin{pmatrix} \partial_u\Phi(u^*) \\ V^T \end{pmatrix} \tag{11.15}$$

is invertible and for which $V^T \cdot V = I_{n-m}$. It follows that the columns of V constitute an orthonormal basis for $\text{span}\{V\}$ in $\mathbb{R}^n$. Now suppose that the columns of the matrix $V^\perp \in \mathbb{R}^{n\times m}$ constitute a basis for the orthogonal complement of $\text{span}\{V\}$ in $\mathbb{R}^n$ such that $V^T \cdot V^\perp = 0$. The invertibility of (11.15) follows if and only if $\mathcal{T}_{u^*} \cap \text{span}\left\{V^\perp\right\} = \{0\}$ or, equivalently, if and only if the square matrix $\partial_u\Phi(u^*)\cdot V^\perp$ is nonsingular. In this case, $\mathcal{T}_{u^*}$ is spanned by the columns of the matrix

$$\begin{pmatrix} \partial_u\Phi(u^*) \\ V^T \end{pmatrix}^{-1} \cdot \begin{pmatrix} 0 \\ I_{n-m} \end{pmatrix}, \tag{11.16}$$

where I_{n-m} denotes the $(n-m)\times(n-m)$ identity matrix. In the special case that the columns of V constitute an orthonormal basis of $\mathcal{N}\left[\partial_u\Phi(u^*)\right]$, the product in Eq. (11.16) simply evaluates to V.

Now consider the function $E : \mathbb{R}^n \times \mathbb{R}^m \to \mathbb{R}^m$, where

$$E(\tilde{u},\lambda) := \Phi\left(\tilde{u} + V^\perp \cdot \lambda\right). \tag{11.17}$$

It then holds that

$$E\left(u^*,0\right) = 0 \tag{11.18}$$

and

$$\partial_\lambda E\left(u^*,0\right) = \partial_u \Phi\left(u^*\right) \cdot V^\perp. \tag{11.19}$$

The implicit function theorem implies that there exists an open ball U in $\mathbb{R}^n$, with center at u^*, and an open ball Λ in $\mathbb{R}^m$, with center at the origin, such that

$$E\left(\tilde{u},\lambda\right) = 0 \text{ for } (\tilde{u},\lambda) \in U \times \Lambda \tag{11.20}$$

if and only if $\lambda = \lambda(\tilde{u})$ for a unique vector-valued smooth function $\lambda : \mathbb{R}^n \to \mathbb{R}^m$. In particular, $\lambda(u^*) = 0$. From Eq. (11.15) it follows that we may choose U such that

$$\Phi\left(\tilde{u} + V^\perp \cdot \lambda(\tilde{u})\right) = 0 \tag{11.21}$$

and such that the square matrix

$$\begin{pmatrix} \partial_u \Phi\left(\tilde{u} + V^\perp \cdot \lambda(\tilde{u})\right) \\ V^T \end{pmatrix} \tag{11.22}$$

is again invertible for all $\tilde{u} \in U$. In particular, for a given $\tilde{u}$, the $(n-m)$-dimensional tangent space at the point $\tilde{u} + V^\perp \cdot \lambda(\tilde{u})$ on the solution manifold is spanned by the columns of the unique solution to the linear equation

$$\begin{pmatrix} \partial_u \Phi\left(\tilde{u} + V^\perp \cdot \lambda(\tilde{u})\right) \\ V^T \end{pmatrix} \cdot T = \begin{pmatrix} 0 \\ I_{n-m} \end{pmatrix}. \tag{11.23}$$

For given $\tilde{u}$ and V, the point $\tilde{u} + V^\perp \cdot \lambda(\tilde{u})$ is the locally unique solution to the closed continuation problem

$$\begin{pmatrix} \Phi(u) \\ V^T \cdot (u - \tilde{u}) \end{pmatrix} = 0, \tag{11.24}$$

where we have appended the $(n-m)$-dimensional *projection condition*

$$V^T \cdot (u - \tilde{u}) = 0 \tag{11.25}$$

to the original continuation problem. Since the projection condition is invariant under the translation $\tilde{u} \mapsto \tilde{u} + V^\perp \cdot \mu$ for arbitrary μ, it follows that the solution manifold through the point u^* may be described locally by the image of the function

$$\rho \mapsto \tilde{u} + V \cdot \rho + V^\perp \cdot \lambda(\tilde{u} + V \cdot \rho) \tag{11.26}$$

for some fixed $\tilde{u} \in U$ and ρ on some neighborhood in $\mathbb{R}^{n-m}$ containing 0. In particular, for every $\tilde{u} \in U$, there exists a unique $\rho = \rho^*$ such that

$$\tilde{u} + V \cdot \rho^* + V^\perp \cdot \lambda\left(\tilde{u} + V \cdot \rho^*\right) = u^*, \tag{11.27}$$

namely

$$\rho^* = V^T \cdot \left(u^* - \tilde{u}\right). \tag{11.28}$$

As a special case, $\rho^* = 0$ when

$$\tilde{u} + V^\perp \cdot \lambda(\tilde{u}) = u^*. \tag{11.29}$$

This is always the case when $\tilde{u} = u^*$.

Example 11.2 Consider the solution manifold to the zero problem in Example 11.1 near $u^* = (1, 1, 0)$ and let

$$V = \begin{pmatrix} 0 & \sin\theta \\ 0 & \cos\theta \\ 1 & 0 \end{pmatrix} \tag{11.30}$$

such that $V^T \cdot V = I_2$. The single column of the matrix

$$V^\perp = \begin{pmatrix} -\cos\theta \\ \sin\theta \\ 0 \end{pmatrix} \tag{11.31}$$

then spans the orthogonal complement to span$\{V\}$. The matrix

$$\begin{pmatrix} \partial_u \Phi(u^*) \\ V^T \end{pmatrix} \tag{11.32}$$

is invertible provided that $\theta \neq \pm\pi/2 + 2n\pi$. Indeed, in this case,

$$\partial_u \Phi(u^*) \cdot V^\perp = -2\cos\theta \neq 0, \tag{11.33}$$

and the tangent space $\mathcal{T}_{u^*}$ is spanned by the columns of the matrix

$$\begin{pmatrix} \partial_u \Phi(u^*) \\ V^T \end{pmatrix}^{-1} \cdot \begin{pmatrix} 0 & 0 \\ 1 & 0 \\ 0 & 1 \end{pmatrix} = \begin{pmatrix} 0 & 0 \\ 0 & \sec\theta \\ 1 & 0 \end{pmatrix}. \tag{11.34}$$

Now suppose that $\theta = \pi/4$ and let

$$\tilde{u} = \begin{pmatrix} 1 \\ 1 + 2\varepsilon \\ \varepsilon \end{pmatrix} \tag{11.35}$$

for some $\varepsilon \ll 1$. The solution to the closed continuation problem

$$\begin{pmatrix} \Phi(u) \\ V^T \cdot (u - \tilde{u} - \varepsilon V \cdot \rho) \end{pmatrix} = 0 \tag{11.36}$$

is then given by the following expansion in ε:

$$u = \tilde{u} + \varepsilon V \cdot \rho + \varepsilon \left(\lambda_0 + \varepsilon\lambda_1 + \varepsilon^2\lambda_2 + \mathcal{O}\left(\varepsilon^3\right) \right) V^\perp, \tag{11.37}$$

where

$$\lambda_0 = \rho_2, \tag{11.38}$$

$$\lambda_1 = \sqrt{2}\left(\sqrt{2} + \rho_2\right)^2, \tag{11.39}$$

$$\lambda_2 = 2\left(\sqrt{2} + \rho_2\right)^3. \tag{11.40}$$

In this case, $u = u^*$ for $\rho_1 = \rho_1^* = -1$ and $\rho_2 = \rho_2^* = -\sqrt{2}$. ■

11.1.3   A curve segment

Now let s be a unit vector in $\mathbb{R}^{n-m}$, chosen such that $\|\rho^*\|\,s$ is parallel to ρ^*. By definition, $V \cdot s$ is a unit vector in span$\{V\}$ and, e.g.,

$$\tilde{u} + \|\rho^*\| \, V \cdot s + V^\perp \cdot \lambda \left(\tilde{u} + \|\rho^*\| \, V \cdot s \right) = u^* \tag{11.41}$$

if $\|\rho^*\|\,s = \rho^*$. It follows that the function $\Gamma : h \mapsto \Gamma_h$, where

$$\Gamma_h := \tilde{u} + h V \cdot s + V^\perp \cdot \lambda (\tilde{u} + h V \cdot s), \tag{11.42}$$

for some fixed $\tilde{u} \in U$ and h on some interval containing 0, is a 1-dimensional curve segment through u^* on the solution manifold. In particular, the image of Γ coincides with the intersection of the solution manifold with the $(m+1)$-dimensional affine subspace $\tilde{u} + \mathrm{span}\{V \cdot s\} \oplus \mathrm{span}\{V^\perp\}$. In the special case that $m = n-1$, the image of Γ naturally agrees with a portion of the actual solution manifold.

   Suppose that the columns of the matrix T_h constitute a basis for the tangent space at Γ_h. Since, by assumption, $V^T \cdot V = I_{n-m}$, it follows by expansion that

$$V^T \cdot \left(T_h \cdot \left(V^T \cdot T_h \right)^{-1} \cdot s - V \cdot s \right) = 0, \tag{11.43}$$

from which we obtain

$$T_h \cdot \left(V^T \cdot T_h \right)^{-1} \cdot s - V \cdot s \in \mathrm{span}\left\{ V^\perp \right\}. \tag{11.44}$$

Since $T_h \cdot \left(V^T \cdot T_h \right)^{-1} \cdot s$ belongs to the tangent space of the solution manifold at Γ_h, it follows that the vector $T_h \cdot \left(V^T \cdot T_h \right)^{-1} \cdot s$ is tangent to Γ at the point Γ_h. We say that the vector-valued function $h \mapsto T_h \cdot \left(V^T \cdot T_h \right)^{-1} \cdot s$ is a curve of tangent vectors that are *positively oriented* relative to the vector $V \cdot s$ (since their projection along $V \cdot s$ equals $V \cdot s$). In the special case that T_h denotes the unique solution to the equation

$$\begin{pmatrix} \partial_u \Phi(\Gamma_h) \\ V^T \end{pmatrix} \cdot T = \begin{pmatrix} 0 \\ I_{n-m} \end{pmatrix}, \tag{11.45}$$

it follows that $T_h \cdot \left(V^T \cdot T_h \right)^{-1} \cdot s = T_h \cdot s$.

   We refer to $\tilde{u}$, V, s, and h as the *base point*, *tangent matrix*, *direction vector*, and *step size*, respectively, of the segment-specific projection condition

$$V^T \cdot (u - \tilde{u}) - h s = 0, \tag{11.46}$$

obtained by substituting $\tilde{u} \mapsto \tilde{u} + h V \cdot s$ in Eq. (11.25). The geometry implied by the projection condition is illustrated in Fig. 11.1. For given $\tilde{u}$, V, s, and h, we refer to an initial solution guess to the corresponding closed continuation problem in Eq. (11.24), and to the algorithm that produces such a guess, as a *predictor*. An example is the *linear predictor* $\tilde{u} + h V \cdot s$ which satisfies the projection condition by construction. Given a predictor, a *corrector* (typically a nonlinear iterative equation solver) must be applied to compute an actual point on the solution manifold to the original continuation problem.

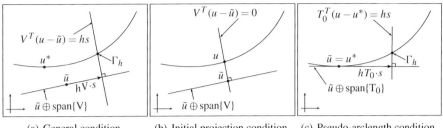

(a) General condition.   (b) Initial projection condition.   (c) Pseudo-arclength condition.

Figure 11.1. *Geometric interpretation of the projection condition in Eq. (11.46). The point Γ_h on the manifold projects orthogonally, with respect to the affine subspace $\tilde{u} \oplus \mathrm{span}\{V\}$, onto the point $\tilde{u} + hV \cdot s$, as illustrated in (a). The same condition for $hV \cdot s = 0$ reduces to Eq. (11.25) and allows for computing a first solution point on the manifold from an initial guess $\tilde{u}$ if a suitable V is available (b). For $\tilde{u} = u^*$ and $V = T_0$, we obtain the well-known pseudo-arclength condition (c).*

Example 11.3 We continue the analysis in Example 11.2. Let

$$s = \frac{\rho^*}{\|\rho^*\|} = \begin{pmatrix} -\frac{1}{\sqrt{3}} \\ -\sqrt{\frac{2}{3}} \end{pmatrix}. \tag{11.47}$$

The corresponding curve segment on the solution manifold is then given by

$$\Gamma_h = \begin{pmatrix} 1 - \frac{2h^2}{3} + \frac{4h^3}{3\sqrt{3}} \\ 1 - \frac{2h}{\sqrt{3}} + \frac{2h^2}{3} - \frac{4h^3}{3\sqrt{3}} \\ -\frac{h}{\sqrt{3}} \end{pmatrix} + \mathcal{O}\left(h^4\right), \tag{11.48}$$

where we have shifted the parameter h relative to Eq. (11.42) such that $\Gamma_0 = u^*$. The tangent space to the solution manifold at Γ_h is then spanned by the columns of the matrix

$$T_h = \begin{pmatrix} \partial_u \Phi(\Gamma_h) \\ V^T \end{pmatrix}^{-1} \begin{pmatrix} 0 & 0 \\ 1 & 0 \\ 0 & 1 \end{pmatrix}$$

$$= \begin{pmatrix} 0 & 2\sqrt{\frac{2}{3}}h - 2\sqrt{2}h^2 + 8\sqrt{\frac{2}{3}}h^3 \\ 0 & \sqrt{2} - 2\sqrt{\frac{2}{3}}h + 2\sqrt{2}h^2 - 8\sqrt{\frac{2}{3}}h^3 \\ 1 & 0 \end{pmatrix} + \mathcal{O}\left(h^4\right). \tag{11.49}$$

It is now straightforward to show that the tangent vector to the curve segment at Γ_h may be obtained from the product $T_h \cdot s$. ∎

11.1.4 Initialization

Suppose that the point u^* is unknown. We seek to locate an initial point on the solution manifold by considering the closed continuation problem in Eq. (11.24), with $\tilde{u}$ equal to the initial solution guess. Since $\mathcal{T}_{u^*}$ is unknown, there is clearly no a priori way to choose V such that $\mathcal{T}_{u^*} \cap \text{span}\{V^\perp\} = 0$, other than to observe that this is generically true and to require the invertibility of the matrix

$$
\begin{pmatrix} \partial_u \Phi(\tilde{u}) \\ V^T \end{pmatrix}.
$$

(11.50)

Instead, the choice of tangent space V is often guided by consideration of the implications of the projection condition in Eq. (11.25) on the components of u. Suppose, for example, that Φ represents the restriction associated with an index set $\mathbb{I}$ of an extended continuation problem formulated in terms of a set of continuation variables and continuation parameters. In this case, one may choose V to span a coordinate hyperplane corresponding to a subset of the active continuation parameters $\mu_{\mathbb{J}}$.

Provided that the initial guess $\tilde{u}$ is sufficiently close to the solution manifold, we may obtain a candidate tangent matrix V by making use of an LU-factorization of the Jacobian $\partial_u \Phi(\tilde{u})$. Specifically, there exist unique matrices $U_1 \in \mathbb{R}^{m \times m}$, $U_2 \in \mathbb{R}^{m \times (n-m)}$, $L \in \mathbb{R}^{m \times m}$, and $P \in \mathbb{R}^{n \times n}$ such that

$$
\partial_u \Phi(\tilde{u}) \cdot P = L \cdot \begin{pmatrix} U_1 & U_2 \end{pmatrix},
$$

(11.51)

where L is a lower triangular matrix, U_1 is an upper triangular matrix with a unit diagonal, and P is a permutation matrix. Since U_1 is invertible, there exists a matrix $Y \in \mathbb{R}^{m \times (n-m)}$ such that $U_2 = U_1 \cdot Y$. It follows that

$$
\partial_u \Phi(\tilde{u}) \cdot P \cdot \begin{pmatrix} Y \\ -I_{n-m} \end{pmatrix} = L \cdot U_1 \cdot \begin{pmatrix} I_m & Y \end{pmatrix} \cdot \begin{pmatrix} Y \\ -I_{n-m} \end{pmatrix} = 0,
$$

(11.52)

i.e., that $\mathcal{N}[\partial_u \Phi(\tilde{u})]$ is given by

$$
\text{span}\left\{ P \cdot \begin{pmatrix} Y \\ -I_{n-m} \end{pmatrix} \right\}.
$$

(11.53)

Example 11.4 In the case of the zero function in Eq. (11.8) and with

$$
\tilde{u} = \begin{pmatrix} 1 \\ 1 + 2\varepsilon \\ \varepsilon \end{pmatrix},
$$

(11.54)

it follows that the desired decomposition of $\partial_u \Phi(\tilde{u})$ is obtained with $P = I_3$, $L = (2)$, $U_1 = (1)$, and $U_2 = \begin{pmatrix} 2\varepsilon & 0 \end{pmatrix}$. It now follows that $Y = \begin{pmatrix} 2\varepsilon & 0 \end{pmatrix}$, and thus that $\mathcal{N}[\partial_u \Phi(\tilde{u})]$ is spanned by the columns of the matrix

$$
V = \begin{pmatrix} 2\varepsilon & 0 \\ -1 & 0 \\ 0 & -1 \end{pmatrix}.
$$

(11.55)

For sufficiently small ε, this choice of tangent matrix ensures the invertibility of the matrix in Eq. (11.50). ∎

11.1.5 Continuation

Now suppose that the point u^* is known and that the columns of T^* constitute a basis for $\mathcal{T}_{u^*}$. Denote by $t \in \mathcal{T}_{u^*}$ a given unit vector such that $t = T^* \cdot \sigma$ for some vector σ. The vector t is (i) tangent at u^* to the curve segment Γ, generated by the base point $\tilde{u}$, tangent matrix V, and direction vector s, and (ii) is positively oriented relative to $V \cdot s$ provided that

$$s = \frac{V^T \cdot t}{\left\| V^T \cdot t \right\|} \tag{11.56}$$

and, if $\tilde{u}$ is chosen such that $V^T \cdot u^* \neq V^T \cdot \tilde{u}$, provided that

$$s \parallel V^T \cdot \left(u^* - \tilde{u} \right). \tag{11.57}$$

Indeed, in this case,

$$T^* \cdot \left(V^T \cdot T^* \right)^{-1} \cdot s = T^* \cdot \left(V^T \cdot T^* \right)^{-1} \cdot \frac{V^T \cdot T^* \cdot \sigma}{\left\| V^T \cdot t \right\|} = \frac{t}{\left\| V^T \cdot t \right\|}. \tag{11.58}$$

Under these conditions, the curve segment Γ is said to be a *continuation* of the point u^* in the direction given by t along the solution manifold. In the special case that $\mathrm{span}\{V\} = \mathcal{T}_{u^*}$, it holds that $\left\| V^T \cdot t \right\| = 1$ and $V \cdot s = t$. It then follows that the vectors $T \cdot \left(V^T \cdot T \right)^{-1} \cdot s$ are positively oriented relative to the vector t along the entire curve segment Γ.

In the event that $V^T \cdot u^* \neq V^T \cdot \tilde{u}$, Eqs. (11.57)–(11.58) imply that t is parallel to the vector

$$T^* \cdot \left(V^T \cdot T^* \right)^{-1} \cdot V^T \cdot \left(u^* - \tilde{u} \right). \tag{11.59}$$

Since $V^T \cdot T^*$ is invertible,

$$u^* - \tilde{u} - T^* \cdot \sigma' \in \mathrm{span}\left(V^\perp \right) \tag{11.60}$$

for

$$\sigma' = (V^T \cdot T^*)^{-1} \cdot V^T \cdot (u^* - \tilde{u}). \tag{11.61}$$

We conclude that

$$t \parallel T^* \cdot \sigma' \tag{11.62}$$

and, consequently, that

$$\sigma \parallel \sigma'. \tag{11.63}$$

In the special case that $u^* - \tilde{u} = T^* \cdot \sigma'$, it follows that

$$t \parallel u^* - \tilde{u}. \tag{11.64}$$

Example 11.5 Consider again the regular solution point

$$u^* = \begin{pmatrix} 1 \\ 1 \\ 0 \end{pmatrix} \tag{11.65}$$

to the zero problem in Eq. (11.8). As shown in Example 11.1, the tangent space $\mathcal{T}_{u^*}$ is spanned by the columns of the matrix

$$T^* = \begin{pmatrix} 0 & 0 \\ 0 & 1 \\ 1 & 0 \end{pmatrix}. \tag{11.66}$$

Now let $\sigma = \begin{pmatrix} \frac{1}{\sqrt{2}} & \frac{1}{\sqrt{2}} \end{pmatrix}^T$ such that $t = \begin{pmatrix} 0 & \frac{1}{\sqrt{2}} & \frac{1}{\sqrt{2}} \end{pmatrix}^T$. Moreover, suppose now that

$$\tilde{u} = \begin{pmatrix} 1+\varepsilon \\ 1+2\varepsilon \\ \varepsilon \end{pmatrix} \tag{11.67}$$

for some $\varepsilon \neq 0$. It follows that

$$u^* - \tilde{u} - \varepsilon k t \in \mathrm{span}\left(V^\perp\right) \tag{11.68}$$

for $k \neq 0$ provided that

$$V^\perp = v \begin{pmatrix} 2 \\ \sqrt{2}k+4 \\ \sqrt{2}k+2 \end{pmatrix} \tag{11.69}$$

for any $v \neq 0$. In this case, $\partial_u \Phi(u^*) \cdot V^\perp = 4v \neq 0$, by construction. Now let

$$w_1 := \begin{pmatrix} 0 \\ \sqrt{2}k+2 \\ -\sqrt{2}k-4 \end{pmatrix} \tag{11.70}$$

and

$$w_2 := \begin{pmatrix} -2k^2 - 6\sqrt{2}k - 10 \\ \sqrt{2}k+4 \\ \sqrt{2}k+2 \end{pmatrix} \tag{11.71}$$

and consider the matrix

$$V = \begin{pmatrix} \frac{w_1}{\|w_1\|} & \frac{w_2}{\|w_2\|} \end{pmatrix}. \tag{11.72}$$

It is straightforward to show that $V^T \cdot V^\perp = 0$ and $V^T \cdot V = I_{n-m}$. Moreover,

$$\sigma' := \left(V^T \cdot T^*\right)^{-1} \cdot V^T \cdot \left(u^* - \tilde{u}\right) = \varepsilon k \sigma \tag{11.73}$$

and

$$V^T \cdot t \parallel V^T \cdot \left(u^* - \tilde{u}\right), \tag{11.74}$$

consistent with Eqs. (11.56)–(11.57). ∎

In the special case that the base point $\tilde{u}$ of the projection condition in Eq. (11.46) is chosen to coincide with u^* and that $\mathrm{span}\{V\} = \mathcal{T}_{u^*}$, the *pseudo-arclength predictor* $u^* + hV \cdot s$ lies at a distance h from the original solution point u^*. Moreover, at this predictor, the norm of the residual of the original continuation problem is $\mathcal{O}\left(h^2\right)$. For sufficiently small h, we consequently expect the distance from u^* to the new solution point $u^* + hV \cdot s + V^\perp \cdot \lambda (u^* + hV \cdot s)$ to be close to h.

11.1.6 Charts and atlases

Following the above discussion, we associate with each point u on the solution manifold

- a matrix T whose columns constitute an orthonormal basis of the corresponding tangent space $\mathcal{T}_u$;

- a set Σ of coordinate vectors, in the basis given by the columns of T, of candidate unit tangent vectors at u along curve segments on the solution manifold; and

- a scalar R corresponding to a radius of validity of the local cover of the solution manifold.

We refer to the collection $\{u, T, \Sigma, R\}$ as a *chart* based at the *base point u*. A family of charts is then an *atlas*. We obtain a *cover* of some portion of the solution manifold by computing an atlas whose base points are sufficiently dense on this portion of the manifold.

An *atlas algorithm* generates an atlas from a given initial solution guess $\tilde{u}$ and a given matrix V. At each stage of construction, such an algorithm proceeds from a *base chart* $\{u, T, \Sigma, R\}$ in the atlas through a phase of *expansion* followed by a phase of *consolidation*. In the expansion phase, the algorithm *constructs* a sequence of charts along a curve segment Γ, which is a continuation of the base point u along the solution manifold in the direction given by $T \cdot \varrho(\sigma)$ for some $\sigma \in \Sigma$ and some endomorphism ϱ on the unit sphere in $\mathbb{R}^{n-m}$. In the consolidation phase, the algorithm *merges* these charts into the atlas by modifying each of the sets Σ.

We consider below the abstract notion of a *curve segment* as a container for all relevant information pertaining to an actual curve segment Γ on the solution manifold, including the corresponding projection condition and a sequence of charts along the curve segment originating with some base chart. Curve segments are initialized, modified, and reinitialized by the atlas algorithm at various stages during expansion and consolidation.

11.2 A finite-state machine

We restrict our attention to atlas algorithms realized in terms of *event-driven finite-state machines*. In particular, we associate one or several possible state transitions with each current state. The choice of successor state then depends on conditions computed at the triggering of such a transition. With each state $\mathfrak{s}$, we implement a *transition function* $\mathfrak{F}_\mathfrak{s}$ that operates on the existing atlas and the current curve segment in order to determine the appropriate successor state. In the general case, transitions may be triggered by user-initiated actions, say, key strokes or mouse clicks. Here, and in the default implementation in COCO, state transitions are instead triggered in an endless loop until a terminal state is reached.

11.2.1 States and transitions

In their simplest form, we conceptualize atlas algorithms in terms of five internal states and the associated state transitions, as illustrated in Fig. 11.2(a). Here, the loop, consisting

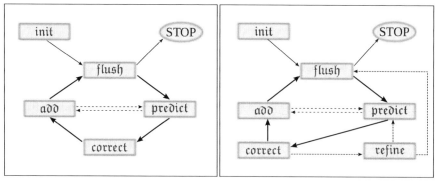

(a) FSM with basic loop.   (b) FSM with additional refine state.

Figure 11.2. *In its simplest form an* atlas algorithm *may be implemented as an event-driven finite-state machine (FSM) consisting of five states* (a), *as explained in Sect. 11.2.1. The loop* flush, predict, correct, *and* add *is entered from* init *via* flush. *A terminal exit of the loop may be triggered in the* flush *state upon meeting some stop condition. The variant described in Sect. 11.2.3 and shown in* (b) *contains the additional state* refine *in order to allow for handling of a failure of convergence in* correct.

of the flush, predict, correct, and add states, is initialized with content from the init state and executed repeatedly until a suitable interrupt condition is triggered. A more detailed description, in terms of the notation and terminology introduced in the previous section, may be given as follows.

The init state:

The atlas algorithm is initialized in the init state with a solution guess $\tilde{u}$. The transition function $\mathfrak{F}_{\text{init}}$ initializes a curve segment by

- constructing a tangent matrix V corresponding to the projection condition

$$V^T \cdot (u - \tilde{u}) = 0; \tag{11.75}$$

- locating a point u^* on the solution manifold that satisfies the projection condition; and

- assigning the corresponding chart $\{u^*, T, \Sigma, R\}$ to the curve segment.

The predict state:

The atlas algorithm enters the predict state with a partial atlas constructed in the immediately preceding flush state. The transition function $\mathfrak{F}_{\text{predict}}$ initializes a curve segment by

- assigning some chart $\{u^*, T, \Sigma, R\}$ from the atlas as a new base chart;

- assigning the base point $\tilde{u}$, tangent matrix V, direction vector s, and step size h, corresponding to the projection condition

$$V^T \cdot (u - \tilde{u}) - hs = 0; \tag{11.76}$$

and

- initializing the corrector algorithm with a suitable predictor.

For continuation along a vector $t = T \cdot \varrho(\sigma)$, for some coordinate vector $\sigma \in \Sigma$ and some function $\varrho : \mathbb{R}^{n-m} \times \mathbb{R}^{n-m}$, the function $\mathfrak{F}_{\mathrm{predict}}$ requires that $\tilde{u}$, V, and s satisfy the *compatibility conditions*

$$s = \frac{V^T \cdot t}{\| V^T \cdot t \|} \tag{11.77}$$

and, if $V^T \cdot u^* \neq V^T \cdot \tilde{u}$,

$$s \parallel V^T \cdot \left(u^* - \tilde{u} \right). \tag{11.78}$$

The correct state:

The atlas algorithm enters the correct state with a projection condition and a predictor constructed in the immediately preceding predict state. The transition function $\mathfrak{F}_{\mathrm{correct}}$ applies a corrector algorithm to locate a new point u^* on the solution manifold that also satisfies the projection condition.

The add state:

The atlas algorithm enters the add state with a partial curve segment, constructed in the preceding predict state, and with a new solution point u^* along this curve segment, computed in the preceding correct state. The transition function $\mathfrak{F}_{\mathrm{add}}$ appends a new chart $\{u^*, T, \Sigma, R\}$ to the curve segment.

The flush state:

The atlas algorithm enters the flush state with a partial atlas and, possibly, a curve segment, finalized in the preceding init or add state. The transition function $\mathfrak{F}_{\mathrm{flush}}$ merges the charts in the curve segment, if available, with the atlas by modifying the corresponding sets Σ.

11.2.2 Choices

Atlas algorithms constructed according to the structure outlined in the previous section differ in the detailed implementations of the transition functions $\mathfrak{F}_{\mathrm{init}}$, $\mathfrak{F}_{\mathrm{predict}}$, $\mathfrak{F}_{\mathrm{correct}}$, $\mathfrak{F}_{\mathrm{add}}$, and $\mathfrak{F}_{\mathrm{flush}}$.

Algorithm-specific choices include the initial construction of the tangent matrix V and the initial chart on the solution manifold by $\mathfrak{F}_{\mathrm{init}}$. As suggested in Sect. 11.1.4, V may be chosen in order to constrain the values of a subset of the elements of u (say, a subset of the active continuation parameters $\mu_{\mathbb{J}}$ in a restricted continuation problem). Alternatively, the columns of V may be chosen to constitute a basis for the nullspace of the Jacobian $\partial_u \Phi(\tilde{u})$, as obtained from the algorithm shown in Sect. 11.1.4. Finally, the coordinate set Σ assigned to the initial chart may be chosen to restrict continuation to a particular direction along the solution manifold or to allow for continuation in multiple directions.

Algorithm-specific choices are also made by the $\mathfrak{F}_{\mathrm{predict}}$ transition function. These include the choice of base point $\tilde{u}$, tangent matrix V, direction vector s, step size h, and predictor. The base point $\tilde{u}$ of the projection condition may be chosen to coincide with a previously located point on the solution manifold or with an extrapolated point along an approximation to a curve segment on the solution manifold. Similarly, the tangent matrix V may be chosen using a basis of the tangent space to the solution manifold at a previously located point or using an estimated basis of the tangent space at some extrapolated point along an approximation to a curve segment on the solution manifold. The step size h may be chosen such that the distance from the base point u^* of the base chart to the point on the solution manifold located in the subsequent corrector step is approximately R, thereby reducing redundant covering of the solution manifold to a minimum. Finally, the predictor may be given by the linear predictor, e.g., the pseudo-arclength predictor, or by some point not in $\tilde{u} \oplus \text{span}\{V\}$.

Different atlas algorithms may naturally employ different corrector algorithms, for example, depending on the degree of smoothness of the continuation problem. Similarly, algorithms may differ in the construction of the matrix T, scalar R, and coordinate set Σ by $\mathfrak{F}_{\mathrm{add}}$. For example, whereas the default value for R may be given by a user-defined option provided to the algorithm, subsequent values for R may be chosen adaptively during continuation in order to accommodate constraints on the properties of the overall atlas. As in the case of $\mathfrak{F}_{\mathrm{init}}$, the choice of the coordinate set Σ constrains the directions available for further continuation in subsequent stages of execution.

Algorithms may also differ in the details of the merge operation. A challenge of particular concern in the case of a multidimensional manifold is the need to avoid unnecessarily retracing those parts of the manifold that have already been appropriately covered by previously computed charts. To this end, algorithms may differ in the modification of the set Σ by $\mathfrak{F}_{\mathrm{flush}}$, as well as in the selection of a value for σ and the choice of function ϱ by $\mathfrak{F}_{\mathrm{predict}}$. In the former case, it may be necessary to first determine the extent to which a neighborhood of the base point of a newly computed chart is already covered by previously computed charts, and then to omit from Σ those directions that would result in a solution point within a previously covered region of the manifold.

11.2.3 Variations

In addition to the algorithm-specific choices outlined above, atlas algorithms may also differ in the state flow implemented in the various transition functions. This includes allowing for alternative flow paths through the finite-state machine, providing for the insertion of additional states, and refining the existing states further. For example, algorithms may differ in the conditions that trigger a transition from flush to the terminal state.

Consider the two additional state transitions indicated by the dashed arrows between the add and predict states shown in Fig. 11.2(b). Here, the transition from predict to add bypasses the transition to the correct state, and the associated application of the corrector algorithm, in the event that the predictor already satisfies the restricted continuation problem to a desired accuracy. The alternative transition from add to predict, on the other hand, allows for the design of atlas algorithms that support appending arbitrary numbers of charts to the current curve segment, prior to merging this curve segment with the existing atlas. In this case, the $\mathfrak{F}_{\mathrm{predict}}$ transition function should initialize a curve segment only if one does not already exist.

We may account for the possible failure to converge in the corrector algorithm by enabling a state transition from the correct state directly to the flush state. A more refined treatment follows by allowing for a modification to the predictor and a repeated application

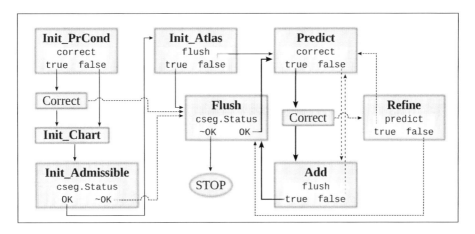

Figure 11.3. *The atlas algorithm implemented in* COCO *has a five-step initialization sequence described in Sect. 11.2.3 and is otherwise identical to the finite-state machine shown in Fig. 11.2(b). The names of states that interface with class methods of the* AtlasBase *class, introduced in Sect. 11.3, are typeset in bold face. This algorithm allows an* AtlasBase *subclass to alter the execution path via a boolean flag included in the class methods' return argument lists or by the content of* cseg.Status *in the case of the* init_admissible *and* flush *class methods, as indicated in the boxes.*

of the corrector algorithm. To this end, consider the insertion of the refine state in the modified state transition diagram in Fig. 11.2(b). Here, in the event that the corrector algorithm fails to converge, the transition function $\mathfrak{F}_{correct}$ results in a transition to the refine state. The transition function $\mathfrak{F}_{refine}$ then seeks to modify the algorithm-specific choices available to the construction of a predictor in the $\mathfrak{F}_{predict}$ transition function. If this is successful, the finite-state machine returns to the predict state, whereas the failure to make any such modifications results in a transition to the flush state.

A more sophisticated implementation of the init state is illustrated in Fig. 11.3. Here, we allow for premature transitions to the flush state of the original transition diagram if

- the corrector algorithm, when invoked by the $\mathfrak{F}_{correct}$ transition function, fails to converge to an initial point on the solution manifold or

- the newly located chart is deemed inadmissible by the $\mathfrak{F}_{init\_admissible}$ transition function.

The state transition diagram further supports a transition from the init_atlas state to the predict state of the original transition diagram, thus bypassing the flush state. Finally, we note that the $\mathfrak{F}_{init\_prcond}$ transition function allows for bypassing the correct state. Such a transition would be warranted, for example, if the initial solution guess lies on or near a singular point of a continuation problem.

11.3 An object-oriented implementation

The COCO core contains a skeletal implementation of an atlas algorithm for smooth manifolds, in terms of a finite-state machine with discrete states and associated transition functions. Default functionality is encoded in the `AtlasBase` and `CurveSegment` core classes. As described below, a functional algorithm requires a subclass implementation of the abstract class methods declared in `AtlasBase`. Combined with the possibility to (partially) override the default class methods of `AtlasBase`, such a subclass implementation provides for customization based on different principles of manifold covering.

Individual charts are represented within the COCO finite-state machine by `chart` structures. Fields reserved for use by the COCO core algorithms include `x`, `t`, `TS`, `R`, `pt`, `pt_type`, `ep_flag`, `p`, `e`, `ignore_at`, `ignore_evs`, and `private`. Any number of additional fields may be added to a `chart` structure by a particular implementation of an atlas algorithm in order to store data computed during continuation in the corresponding chart.

11.3.1 The AtlasBase class

When encoded in a single file, every `CurveSegment`-compatible subclass of the `AtlasBase` class must be at least of the minimal form shown below.

```
classdef atlas_subclass < AtlasBase

  properties (Access=private)
    ...
  end

  methods (Access=private)

    function atlas = atlas_subclass(prob, cont, dim)
      atlas = atlas@AtlasBase(prob);
      ...
    end

  end

  methods (Static)

    function [prob cont atlas] = create(prob, cont, dim)
      atlas = atlas_subclass(prob, cont, dim);
      prob  = CurveSegment.add_prcond(prob, dim);
      ...
    end

  end

  methods (Access=public)

    function [prob atlas cseg correct] = init_prcond(atlas, prob, chart)
      ...
      [prob cseg] = CurveSegment.create_initial(prob, chart);
      correct     = cseg.correct;
    end

    function [prob atlas cseg flush] = init_atlas(atlas, prob, cseg)
      ...
      flush = ...;
    end

    function [prob atlas cseg correct] = predict(atlas, prob, cseg)
      ...
      correct = ...;
    end
```

```
    function [prob atlas cseg flush] = add_chart(atlas, prob, cseg)
        ...
        flush = ...;
    end

  end

end
```

In addition, the `AtlasBase` class contains predefined class methods that may be used as is by a derived subclass or shadowed appropriately. These include the `flush` class method with function declaration

```
function [prob atlas cseg] = flush(atlas, prob, cseg)
```

as well as the `refine_step`, `init_chart`, and `init_admissible` class methods with function declarations

```
function [prob atlas cseg predict] = refine_step(atlas, prob, cseg)

function [prob atlas cseg] = init_chart(atlas, prob, cseg)

function [prob atlas cseg] = init_admissible(atlas, prob, cseg, S)
```

These methods are described further in Chaps. 12 and 14, respectively.

The class constructor `atlas_subclass` instantiates an instance `atlas` of the subclass. This constructor is called only indirectly through a call to the `create` static construction method. The latter allows for additional output arguments, e.g., the inclusion of the modified continuation problem structure `prob`. The input argument `dim` of the `create` static construction method contains the intended dimensional deficit of the restricted continuation problem, passed as an argument to the COCO entry-point function. Finally, the `cont` argument contains a structure whose fields correspond to properties assigned by the `coco_set` utility in calls of the form

```
prob = coco_set(prob, 'cont', 'property', value);
```

As shown in the next chapter, this mechanism may be used to assign toolbox- or user-specific values for optional settings that parameterize the execution of the atlas algorithm.

In the COCO implementation, the `init_prcond`, `init_chart`, `init_admissible`, `init_atlas`, `flush`, `predict`, `add_chart`, and `refine_step` class methods are invoked by the corresponding $\mathfrak{F}_{init\_prcond}$, $\mathfrak{F}_{init\_chart}$, $\mathfrak{F}_{init\_admissible}$, $\mathfrak{F}_{init\_atlas}$, $\mathfrak{F}_{flush}$, $\mathfrak{F}_{predict}$, $\mathfrak{F}_{add}$, and $\mathfrak{F}_{refine}$ transition functions, respectively. In each case, `prob` and `atlas` refer to the continuation problem structure and to the class instance, respectively, both of which may be updated within the corresponding function body. The final output argument for each of the `init_prcond`, `init_atlas`, `predict`, `add_chart`, and `refine_step` class methods determines the flow through the atlas algorithm, as suggested in Fig. 11.3.

11.3.2   The CurveSegment class

The `CurveSegment` core class provides a default implementation of an abstract curve segment, in terms of key properties and behaviors. As an example, the `add_prcond` class method must be invoked in the `create` static construction method of the `AtlasBase` subclass in order to append a skeletal representation of the projection condition to the extended continuation problem encoded in `prob`. During continuation, the corresponding family of zero functions evaluates to the actual projection condition, once content has been defined for the corresponding base point $\tilde{u}$, tangent matrix V, direction vector s, and step size h.

Each instance of the `CurveSegment` class includes

- a `ptlist` property equal to a 1-dimensional cell array of charts along the curve segment;

- a `src_chart` property equal to a parent chart along the curve segment that may be modified by the finite-state machine during execution;

- a `curr_chart` property equal to a working copy of the current chart that may be modified in any state of the finite-state machine during execution;

- a `prcond` property equal to a structure containing the base point $\tilde{u}$, tangent matrix V, direction vector s, and step size h in the `x`, `TS`, `s`, and `h` fields;

- a `Status` property equal to an integer flag representing the admissibility of the curve segment; and

- a `correct` property equal to a boolean flag (with default value of `true`) indicating whether or not the corrector algorithm should be applied to the closed continuation problem.

The `cseg` input and output arguments of the `init_atlas`, `add_chart`, `predict`, and `flush` class methods represent instances of `CurveSegment`.

We first instantiate a `CurveSegment` class instance in the `init_prcond` class method of an `AtlasBase` subclass. As shown in the previous section, this is accomplished by the following call to the static construction method `create_initial`.

```
[prob cseg] = CurveSegment.create_initial(prob, chart);
```

Here, the `chart` input argument of the `init_prcond` class method contains problem-specific data used to initialize the atlas algorithm. Specifically, its `x` field contains the initial solution guess u_0, whereas its `t` field is either a zero vector or a vector whose nonzero components have been assigned in one or several calls to `coco_add_func`. The `create_initial` class method instantiates a class instance by assigning

- an empty array to the `ptlist` property;

- the content of `chart.x` to the `x` field of the `prcond` property;

- a suitably constructed tangent matrix V to the `TS` field of the `prcond` property;

- a default vector in $\mathbb{R}^{n-m}$ to the `s` field of the `prcond` property;

- the value 0 to the `h` field of the `prcond` property;

- a copy of `chart` to the `src_chart` property; and

- a partial copy of `chart` to the `curr_chart` property that inherits the `x`, `R`, and `pt` fields, the image of a user-defined mapping of the content of the `private` field, as well as the content of any nonreserved fields of the `chart` structure.

We may reinstantiate a `CurveSegment` class instance in the `predict` class method of the `AtlasBase` class. This is accomplished by a call to the `CurveSegment.create` static construction method, as shown below.

```
[prob cseg]  = CurveSegment.create(prob, chart, prcond, x);
```

This differs from the call to the `create_initial` static construction method in that

- the `chart` argument is assigned as a single element of the array stored in the `ptlist` property;

- the `prcond` argument is assigned to the `prcond` property; and

- the `x` argument is assigned to the `x` field of the `curr_chart` property.

In both instances, the content of the `x` field of the `curr_chart` property provides an initial solution guess for the subsequent application of the corrector algorithm. Upon convergence, this algorithm updates the `x` field of the `curr_chart` property to contain the successfully located point u^* on the solution manifold.

11.4 Conclusions

We discuss in the next several chapters minimal subclasses of the `AtlasBase` class, without some of the wizardry that would accompany a sophisticated atlas algorithm (such as the default implementations in COCO). In all cases, we assume that the corrector algorithm is the default nonlinear iterative Newton solver implemented in COCO.

Although we often illustrate the atlas algorithms in the context of simple algebraic problems, they apply equally well to any of the problems formulated in Part II of this text. We encourage the reader to experiment accordingly. In this context, it is important to call attention to the emission of the `'update'` signal by `CurveSegment.create_initial` and `CurveSegment.create`, following the construction of a `CurveSegment` class instance. This signal triggers the execution of all corresponding slot functions included in the `prob` continuation problem structure. As an example, it follows that toolbox data structures may change following a call to `CurveSegment.create` in the `predict` class method. To avoid unintended changes to the parameterization of the zero problem, we restrict the use of slot functions associated with the `'update'` signal to cases where the `cseg.ptlist{1}` chart is again a local cover of the solution manifold after the update.

Exercises

11.1. Show that the matrix

$$\begin{pmatrix} \partial_u \Phi(u^*) \\ V^T \end{pmatrix}$$

in Eq. (11.15) is invertible if and only if $\mathcal{T}_{u^*} \cap \mathrm{span}\{V^\perp\} = 0$ or, equivalently, if and only if the square matrix $\partial_u \Phi(u^*) \cdot V^\perp$ is nonsingular.

11.2. Suppose that the matrix V has been chosen so that the matrix in the previous exercise is invertible. Show that $\mathcal{T}_{u*}$ is spanned by the columns of the matrix

$$\begin{pmatrix} \partial_u \Phi(u^*) \\ V^T \end{pmatrix}^{-1} \cdot \begin{pmatrix} 0 \\ I_{n-m} \end{pmatrix}.$$

11.3. Consider the function $E : \mathbb{R}^n \times \mathbb{R}^m \to \mathbb{R}^m$, where

$$E(\tilde{u}, \lambda) := \Phi\left(\tilde{u} + V^\perp \cdot \lambda\right)$$

for $n = 3$ and each of the cases when $m = 1$ and $m = 2$. Draw a schematic diagram of the geometry associated with the zero problem and the affine space through $\tilde{u}$ spanned by the columns of $V^\perp$. Use this to make plausible the conclusions obtained from applying the implicit function theorem to E at the point $\tilde{u} = u^*$ and $\lambda = 0$.

11.4. Verify that the point $\tilde{u} + V^\perp \cdot \lambda(\tilde{u})$ is the locally unique solution to the closed continuation problem

$$\begin{pmatrix} \Phi(u) \\ V^T \cdot (u - \tilde{u}) \end{pmatrix} = 0$$

under the conditions described in Sect. 11.1.2.

11.5. Show that the projection condition

$$V^T \cdot (u - \tilde{u}) = 0$$

is invariant under the translation $\tilde{u} \mapsto \tilde{u} + V^\perp \cdot \mu$ for arbitrary μ. Explain why this implies that the solution manifold through the point u^* may be described locally by the image of the function

$$\rho \mapsto \tilde{u} + V \cdot \rho + V^\perp \cdot \lambda(\tilde{u} + V \cdot \rho)$$

for some fixed $\tilde{u} \in U$ and ρ on some neighborhood in $\mathbb{R}^{n-m}$ containing 0.

11.6. Show that

$$\tilde{u} + V \cdot \rho^* + V^\perp \cdot \lambda\left(\tilde{u} + V \cdot \rho^*\right) = u^*$$

implies that

$$\rho^* = V^T \cdot \left(u^* - \tilde{u}\right).$$

11.7. Show that the curve segment Γ parameterized by

$$h \mapsto \tilde{u} + hV \cdot s + V^\perp \cdot \lambda(\tilde{u} + hV \cdot s)$$

for some fixed $\tilde{u} \in U$ coincides with the intersection of the solution manifold with the $(m+1)$-dimensional affine subspace $\tilde{u} + \text{span}\{V \cdot s\} \oplus \text{span}\{V^\perp\}$. Sketch this curve segment in the cases when $n = 3$ and $m = 1$ or $m = 2$.

11.8. Verify Eq. (11.43) and explain why it implies that

$$T_h \cdot \left(V^T \cdot T_h\right)^{-1} \cdot s - V \cdot s \in \text{span}\left\{V^\perp\right\}.$$

Show that $T_h \cdot \left(V^T \cdot T_h\right)^{-1} \cdot s$ is tangent to Γ at Γ_h.

11.9. What are the implications to the values of the active continuation parameters $\mu_{\mathbb{J}}$, during initialization, of choosing V to span a coordinate hyperplane corresponding to a subset of these parameters?

11.10. Review the LU-decomposition in an introductory text on linear algebra and confirm that Eq. (11.51) is correct. How does this formulation differ from the implementation of the LU-decomposition in MATLAB?

11.11. Verify the following implementation of the LU-decomposition-based initialization algorithm in the text.

```
function X = nullspace(A)
  [L U P]  = lu(A');
  [m n] = size(A);
  Y   = L(1:m, 1:m)'\L(m+1:end, 1:m)';
  X   = P'*[Y; -speye(n-m)];
  X   = orth(full(X));
end
```

11.12. Explain why the nullspace $\mathcal{N}\left[\partial_u \Phi\left(\tilde{u}\right)\right]$ is spanned by the columns of the matrix

$$V = P \cdot \begin{pmatrix} Y \\ -I_{n-m} \end{pmatrix}$$

in Eq. (11.53). Show that this choice ensures the invertibility of the matrix in Eq. (11.50).

11.13. Derive the conditions given in Eqs. (11.56)–(11.57) from the discussion in the previous sections.

11.14. Why it is important to orient tangent vectors such that the conditions in Eqs. (11.56)–(11.57) are satisfied? What are the implications of this choice to the $\mathcal{C}^1$ interpolation of curve segments?

11.15. Explain why the vectors $T \cdot \left(V^T \cdot T\right)^{-1} \cdot s$ given in Sect. 11.1.5 are positively oriented relative to the vector t along the entire curve segment Γ when $\text{span}\{V\} = \mathcal{T}_{u^*}$.

11.16. Let the columns of the $m \times (n-m)$ matrix T^* constitute a basis of the tangent space to the solution manifold at a point u^*. Pick an arbitrary $m \times (n-m)$ matrix V such that $V^T \cdot T^*$ is invertible and let

$$t = \frac{T^* \cdot (V^T \cdot T^*)^{-1} \cdot s}{\|T^* \cdot (V^T \cdot T^*)^{-1} \cdot s\|}$$

for some unit vector $s \in \mathbb{R}^{n-m}$. Show that

$$s = \frac{V^T \cdot t}{\|V^T \cdot t\|}.$$

11.17. Suggest other choices for the matrix V in Example 11.5 that satisfy the compatibility conditions in Eqs. (11.56)–(11.57). What is the relationship between any two such choices?

11.18. What can you say about the selection of the tangent matrix V in Example 11.5 in the case that $\varepsilon = 0$?

11.19. Repeat the sequence of computations in Examples 11.1 through 11.5 with a zero function $\Phi : \mathbb{R}^3 \to \mathbb{R}^2$. Comment on the differences between this case and that considered in Examples 11.1 through 11.5.

11.20. Repeat the sequence of computations in Examples 11.1 through 11.5 with a zero function $\Phi : \mathbb{R}^4 \to \mathbb{R}^2$. Comment on the differences between this case and that considered in the previous exercise.

11.21. Why is the norm of the residual of the original zero problem $\mathcal{O}\left(h^2\right)$ at the pseudo-arclength predictor?

11.22. Redraw the schematic representation of a generic atlas algorithm and include in your graphic the `AtlasBase` class methods, together with elements representing their input and output arguments.

11.23. Enumerate several distinct examples of program flow that might result from different boolean values for the final output arguments of the `init_prcond`, `init_atlas`, `predict`, `add_chart`, and `refine_step` class methods of the `AtlasBase` class.

Chapter 12

Single-Dimensional Atlas Algorithms

We restrict our attention to the special case in which the solution manifold and all associated tangent spaces are 1-dimensional. Specifically, we consider a continuously differentiable continuation problem $\Phi(u) = 0$, in terms of a family of functions $\Phi : \mathbb{R}^n \to \mathbb{R}^{n-1}$ for $n \geq 2$. If, as before, u^* denotes a regular solution point, i.e., such that $\Phi(u^*) = 0$ and the Jacobian $\partial_u \Phi(u^*)$ has maximal rank $n - 1$, then the implicit function theorem establishes the existence of a locally unique 1-dimensional manifold through u^* of solutions to the continuation problem.

Now let $v \in \mathbb{R}^n$ be an arbitrary unit vector for which the square matrix

$$\begin{pmatrix} \partial_u \Phi(u^*) \\ v^T \end{pmatrix} \tag{12.1}$$

is invertible. In this case, the 1-dimensional tangent space $\mathcal{T}_{u^*}$ to the solution manifold at u^* is spanned by the vector

$$t^* := \begin{pmatrix} \partial_u \Phi(u^*) \\ v^T \end{pmatrix}^{-1} \cdot \begin{pmatrix} 0 \\ 1 \end{pmatrix}. \tag{12.2}$$

It follows that $v^T \cdot t^* = 1$; that is, t^* is positively oriented relative to v. If, in particular, $v \in \mathcal{N}\left[\partial_u \Phi(u^*)\right]$, then $\mathcal{T}_{u^*}$ is spanned by v.

We arrive at the 1-dimensional projection condition

$$v^T \cdot (u - \tilde{u}) - hs = 0 \tag{12.3}$$

by a suitable choice of base point $\tilde{u}$, direction vector s, and step size h. From Eqs. (11.56)–(11.57), it follows that the corresponding curve segment Γ is a continuation of the point u^* in the direction $t = t^*/\|t^*\|$ provided that

$$s = \frac{v^T \cdot t}{\|v^T \cdot t\|} = 1. \tag{12.4}$$

Similarly, Γ is a continuation of the point u^* in the direction $t = -t^*/\|t^*\|$ provided that

$$s = \frac{v^T \cdot t}{\|v^T \cdot t\|} = -1. \tag{12.5}$$

In the special case of locating an initial point on the solution manifold, the reduced projection condition

$$v^T \cdot (u - \tilde{u}) = 0 \tag{12.6}$$

implies that the separation $u - \tilde{u}$ is orthogonal to v. In particular, if $v_i = 0$ for $i \neq k$ and $v_k = 1$, for some integer k, then $u_k = \tilde{u}_k$ must hold for the initial point on the solution manifold.

12.1 An advancing local cover

Let u^* denote a point on the solution manifold, with local tangent space spanned by the unit vector t^*. In this section, we implement a *pseudo-arclength-based* atlas algorithm for which

$$\tilde{u} = u^*, \tag{12.7}$$

$$v = t^*, \tag{12.8}$$

and the predictor is given by $u^* + Rst^*$ for s equal to 1 or -1 and some scalar R. For any chart based at a point Γ_h along the corresponding curve segment, the tangent space at Γ_h is then spanned by the vector

$$t_h := \left(\begin{array}{c} \partial_u \Phi(\Gamma_h) \\ t^{*T} \end{array} \right)^{-1} \cdot \left(\begin{array}{c} 0 \\ 1 \end{array} \right). \tag{12.9}$$

Since $t^{*T} \cdot t_h = 1$, we ensure that continuation proceeds consistently in one direction along the solution manifold by fixing the value of s at the outset of the analysis. Moreover, since $t = st^*$, it follows that $\Sigma = \{s\}$ for every chart on the solution manifold.

We initialize the atlas algorithm by assigning $\tilde{u} = u_0$ for some initial solution guess u_0 and by selecting a suitable initial tangent matrix v. Following the location of each successive point u^* on the solution manifold and the construction of an associated chart $\{u^*, t^*, s, R\}$, continuation proceeds in the direction of st^*, as shown above. As this implementation tracks only the most recently located point on the solution manifold, we refer to it as providing an *advancing local cover*.

12.1.1 Encoding

We demonstrate the implementation of the pseudo-arclength-based advancing-local-cover algorithm as a derived subclass `atlas_1d_min` of the `AtlasBase` core class, as shown in the following class declaration.

```
classdef atlas_1d_min < AtlasBase

  properties (Access=private)
    base_chart = struct();
    cont       = struct();
  end

  methods (Access=private)
    function atlas = atlas_1d_min(prob, cont, dim)
      assert(dim==1, '%s: wrong manifold dimension', mfilename);
      atlas      = atlas@AtlasBase(prob);
      atlas.cont = atlas.get_settings(cont);
```

```
      end
   end

   methods (Static)
      function [prob cont atlas] = create(prob, cont, dim)
         atlas = atlas_1d_min(prob, cont, dim);
         prob  = CurveSegment.add_prcond(prob, dim);
      end
   end

   methods (Static, Access=private)
      cont = get_settings(cont)
   end

   methods (Access=public)
      [prob atlas cseg correct] = init_prcond(atlas, prob, chart)
      [prob atlas cseg flush]   = init_atlas (atlas, prob, cseg)
      [prob atlas cseg]         = flush       (atlas, prob, cseg)
      [prob atlas cseg correct] = predict     (atlas, prob, cseg)
      [prob atlas cseg flush]   = add_chart   (atlas, prob, cseg)
   end

end
```

For passing information through the finite-state machine, we rely on the `base_chart` property of an instance of `atlas_1d_min` and the `curr_chart` and `ptlist` properties of an instance of `CurveSegment`. The `atlas_1d_min` class further includes a `cont` instance property to store settings that parameterize the execution of the finite-state machine. For example, in the present implementation, it is assumed that the scalar radius R is constant across all charts and equal to the value stored in `cont.h`. Moreover, the algorithm is designed to trigger a transition to the terminal state when

- the number of successfully executed continuation steps equals the value stored in `cont.PtMX` or

- the corrector algorithm sets the `Status` property of the current `cseg` instance to the `cseg.CorrectionFailed` integer constant, indicating lack of convergence of the corrector algorithm.

We provide reflection and detail on each of the class methods in the discussion below.

As described in the previous chapter, the class constructor `atlas_1d_min` is called only indirectly by the finite-state machine through a call to the `create` class method. The construction of the `atlas` instance of the `atlas_1d_min` subclass further includes a call to the `get_settings` class method, shown below.

```
function cont = get_settings(cont)

defaults.h    = 0.1;
defaults.PtMX = 50 ;
cont          = coco_merge(defaults, cont);

end
```

Here, the `coco_merge` utility is used to provide for default values for the scalar radius R and the maximum number of successfully executed continuation steps, in case these have not been set previously by the `coco_set` utility.

The following encoding of the `predict` class method is consistent with the pseudo-arclength algorithm described above.

```
function [prob atlas cseg correct] = predict(atlas, prob, cseg)

chart       = atlas.base_chart;
prcond      = struct('x', chart.x, 'TS', chart.TS, ...
                     's', chart.s, 'h', chart.R);
xp          = chart.x+chart.R*(chart.TS*chart.s);
[prob cseg] = CurveSegment.create(prob, chart, prcond, xp);
correct     = true;

end
```

Here, the `CurveSegment.create` static construction method is used to instantiate a class instance `cseg` and to assign content to its `prcond`, `ptlist`, and `curr_chart` properties. In particular, suppose that the base point u^*, tangent-space basis t^*, coordinate set $\Sigma = \{\sigma\}$, and scalar radius R have been stored in the `x`, `TS`, `s`, and `R` fields, respectively, of the `base_chart` property of the `atlas` instance. The call to `CurveSegment.create` then assigns the base point $\tilde{u} = u^*$, the tangent matrix $v = t^*$, the direction vector $s = \sigma$, and the step size $h = R$ to the `x`, `TS`, `s`, and `h` fields, respectively, of the `cseg.prcond` property. Similarly, the method assigns the pseudo-arclength predictor $u^* + R\sigma t^*$, the coordinate set $\{\sigma\}$, the scalar radius R, and the content of the `pt` field of the `chart` input argument to the `x`, `s`, `R`, and `pt` fields, respectively, of the `curr_chart` property. Finally, the `cseg.ptlist` property is instantiated to a cell array whose single element is given by the `chart` input argument. The `correct` output argument of the `predict` class method is finally assigned the boolean value `true` in order to indicate a transition to the correct state of the finite-state machine.

Next consider the following encoding of the `add_chart` class method.

```
function [prob atlas cseg flush] = add_chart(atlas, prob, cseg)

chart     = cseg.curr_chart;
chart.pt  = chart.pt+1;
if chart.pt>=atlas.cont.PtMX
  chart.pt_type = 'EP';
  chart.ep_flag = 1;
end
[prob cseg] = cseg.add_chart(prob, chart);
flush       = true;

end
```

In the event of convergence of the corrector algorithm, this extracts the most recently located point u^* on the solution manifold from the `curr_chart` property of the `cseg` class instance. After some bookkeeping, meant to track the number of successfully executed continuation steps and to label special solution points, the class method invokes the `CurveSegment.add_chart` class method. This updates the `TS` field of the `chart` input argument and appends the resultant structure to the `cseg.ptlist` cell array. Finally, the `flush` output argument is assigned the boolean value `true` in order to indicate a transition to the flush state of the finite-state machine.

The encoding below of the `flush` class method illustrates the use of the `cseg.Status` field to trigger the transition of the finite-state machine to the terminal state.

```
function [prob atlas cseg] = flush(atlas, prob, cseg)

[prob atlas cseg] = atlas.flush@AtlasBase(prob, cseg);
if cseg.Status==cseg.CurveSegmentOK
  atlas.base_chart = cseg.ptlist{end};
  if atlas.base_chart.pt>=atlas.cont.PtMX
    cseg.Status = cseg.BoundaryPoint;
  end
end

end
```

The implementation first invokes the `flush` method of the `AtlasBase` parent class. This flushes the content of the `cseg.ptlist` property, as appropriate, to the data structure that becomes the output of the call to the `coco` entry-point function and, possibly, to storage and screen. In the event that the most recent application of the corrector algorithm failed to converge, the `cseg.Status` property equals the `cseg.CorrectionFailed` integer constant. Suppose, instead, that the most recent application of the corrector algorithm converged. In this case, the `cseg.Status` property equals the `cseg.CurveSegmentOK` integer constant provided that either

- the `pt` field of the last element of the `cseg.ptlist` property is nonpositive or

- the `ep_flag` field of the last element of the `cseg.ptlist` property equals zero.

If neither condition holds, the `cseg.Status` property equals the `cseg.BoundaryPoint` integer constant. The finite-state machine returns to the predict state if the `cseg.Status` equals the `cseg.CurveSegmentOK` integer constant upon exiting the `flush` class method. It follows that continuation proceeds as long as the corrector algorithm converges and the number of successfully located points on the solution manifold does not equal the integer assigned to the `cont.PtMX` property of the `atlas` instance.

Following some initial bookkeeping, the `init_prcond` class method, shown below, invokes the `CurveSegment.create_initial` class method. This call instantiates an initial `CurveSegment` class object and assigns content to its `prcond`, `ptlist`, and `curr_chart` properties.

```
function [prob atlas cseg correct] = init_prcond(atlas, prob, chart)

chart.R       = 0;
chart.pt      = -1;
chart.pt_type = 'IP';
chart.ep_flag = 1;
[prob cseg]   = CurveSegment.create_initial(prob, chart);
correct       = cseg.correct;

end
```

In particular, suppose that the initial solution guess u_0 and tangent direction t_0 are stored in the x and t fields, respectively, of the `chart` input argument to the `init_prcond` class method. The `CurveSegment.create_initial` method then assigns

- an empty array to the `ptlist` property;

- the base point $\tilde{u} = u_0$ to the x field of the `prcond` property;

- a suitably constructed tangent matrix v to the TS field of the `prcond` property;

- the step size $h = 0$ to the h field of the prcond property;

- a copy of chart to the src_chart property; and

- a partial copy of chart to the curr_chart property that inherits the x, R, and pt fields.

The correct field of the cseg class instance then equals true, unless $t_0 \neq 0$. In the former case, the corrector algorithm is applied to the closed continuation problem in order to update the x field of the curr_chart property to a successfully located point on the solution manifold.

The following encoding of the init_atlas class method should now be largely self-explanatory.

```
function [prob atlas cseg flush] = init_atlas(atlas, prob, cseg)

chart            = cseg.curr_chart;
chart.pt         = 0;
chart.R          = atlas.cont.h;
chart.s          = sign(atlas.cont.PtMX);
atlas.cont.PtMX  = abs(atlas.cont.PtMX);
chart.pt_type    = 'EP';
chart.ep_flag    = 1;
[prob cseg]      = cseg.add_chart(prob, chart);
flush            = true;

end
```

We note, in particular, the initialization of the pt field of the chart structure in support of the subsequent incremental changes in the add_chart class method of atlas_1d_min. Finally, the sign of the integer originally stored in cont.PtMX is here used to assign content to the set Σ associated with each chart.

12.1.2 Functionality and flow test

We now illustrate the functionality implemented in the basic atlas algorithm atlas_1d_min described above. To this end, we first edit the local project settings by modifying the coco_project_opts function as shown below.

```
function prob = coco_project_opts(prob)

prob = coco_set(prob, 'cont', 'linsolve', 'recipes');
prob = coco_set(prob, 'cont', 'corrector', 'recipes');
prob = coco_set(prob, 'cont', 'atlas', @atlas_1d_min.create);

end
```

Here, the coco_set utility is used to assign the atlas_1d_min.create class method to the atlas property of the finite-state machine. We use this modified encoding throughout this chapter and make the appropriate modifications in subsequent chapters.

Consider now the COCO-compatible function shown below.

```
function [data y] = circle(prob, data, u)
```

```
if u(1)<1 && isfield(data, 'MX')
  y = u(1)^2;
else
  y = (u(1)-1)^2+u(2)^2-1;
end

end
```

As long as $u_1 \geq 1$ or the MX field of the data input argument is not defined, the function implements the zero function

$$\Phi(u) = (u_1 - 1)^2 + u_2^2 - 1. \qquad (12.10)$$

The extract below shows the result of applying the basic atlas algorithm to the continuation of solutions to this zero problem, given the initial guess $u_0 = (1.5, 1)$.

```
>> prob = coco_add_func(coco_prob(), 'circle', @circle, [], ...
     'zero', 'u0', [1.5; 1]);
>> prob = coco_set(prob, 'cont', 'PtMX', 30);
>> coco(prob, 'run', [], 1);
```

STEP	DAMPING		NORMS			COMPUTATION TIMES		
IT SIT	GAMMA	||d||	||f||	||U||	F(x)	DF(x)	SOLVE	
0			2.50e-01	1.80e+00	0.0	0.0	0.0	
1 1	1.00e+00	1.12e-01	1.25e-02	1.71e+00	0.0	0.0	0.0	
2 1	1.00e+00	6.21e-03	3.86e-05	1.70e+00	0.0	0.0	0.0	
3 1	1.00e+00	1.93e-05	3.72e-10	1.70e+00	0.0	0.0	0.0	
4 1	1.00e+00	1.86e-10	1.39e-16	1.70e+00	0.0	0.0	0.0	

STEP	TIME	||U||	LABEL	TYPE
0	00:00:00	1.7013e+00	1	EP
10	00:00:00	1.9972e+00	2	
20	00:00:01	1.8025e+00	3	
30	00:00:02	1.1651e+00	4	EP

By inspection of the cseg class instance, constructed during execution in the init_prcond class method, one finds that the initial tangent matrix is an element of $\mathcal{N}[\partial_u \Phi(u_0)]$.

The following continued extract shows the changes to the screen output that result from associating the continuation parameters 'x' and 'y' with the first and second components, respectively, of the vector u.

```
>> prob = coco_add_pars(prob, '', [1 2], {'x' 'y'});
>> coco(prob, 'run', [], 1, {'x' 'y'});
```

STEP	DAMPING		NORMS			COMPUTATION TIMES		
IT SIT	GAMMA	||d||	||f||	||U||	F(x)	DF(x)	SOLVE	
0			2.50e-01	2.55e+00	0.0	0.0	0.0	
1 1	1.00e+00	1.77e-01	1.56e-02	2.46e+00	0.0	0.0	0.0	
2 1	1.00e+00	1.26e-02	7.97e-05	2.45e+00	0.0	0.0	0.0	
3 1	1.00e+00	6.51e-05	2.12e-09	2.45e+00	0.0	0.0	0.0	
4 1	1.00e+00	1.73e-09	0.00e+00	2.45e+00	0.0	0.0	0.0	

STEP	TIME	||U||	LABEL	TYPE	x	y
0	00:00:00	2.4495e+00	1	EP	1.5000e+00	8.6603e-01
10	00:00:00	2.7878e+00	2		1.9429e+00	3.3302e-01
20	00:00:00	2.7806e+00	3		1.9330e+00	-3.5993e-01
30	00:00:01	2.4289e+00	4	EP	1.4749e+00	-8.8002e-01

In this case, the initial tangent matrix is found to equal $v = \begin{pmatrix} 0 & 0 & 1 & 0 \end{pmatrix}^T$. As a consequence, the μ_1 continuation parameter (and, consequently, the u_1 coordinate) is held fixed at 1.5 during the initial application of the corrector.

Next consider the following extract.

```
>> data.MX = true;
>> prob = coco_add_func(coco_prob(), 'func', @circle, data, ...
   'zero', 'u0', [1.5; 1]);
>> prob = coco_add_pars(prob, '', [1 2], {'x' 'y'});
>> coco(prob, 'run', [], 1, {'x' 'y'});
```

STEP	DAMPING		NORMS			COMPUTATION TIMES		
IT SIT	GAMMA	||d||	||f||	||U||	F(x)	DF(x)	SOLVE
0			2.50e-01	2.55e+00	0.0	0.0	0.0
1 1	1.00e+00	1.77e-01	1.56e-02	2.46e+00	0.0	0.0	0.0
2 1	1.00e+00	1.26e-02	7.97e-05	2.45e+00	0.0	0.0	0.0
3 1	1.00e+00	6.51e-05	2.12e-09	2.45e+00	0.0	0.0	0.0
4 1	1.00e+00	1.73e-09	0.00e+00	2.45e+00	0.0	0.0	0.0

STEP	TIME	||U||	LABEL	TYPE	x	y
0	00:00:00	2.4495e+00	1	EP	1.5000e+00	8.6603e-01
10	00:00:00	2.7878e+00	2		1.9429e+00	3.3302e-01
20	00:00:00	2.7806e+00	3		1.9330e+00	-3.5993e-01
30	00:00:01	2.4289e+00	4		1.4749e+00	-8.8002e-01
36	00:00:01	2.0690e+00	5	MX	1.0702e+00	-9.9753e-01

As this assigns the value `true` to the `MX` field of the `data` input argument, the corresponding zero function is given by

$$\Phi(u) = \begin{cases} u_1^2 + u_2^2 - 1 & \text{when } u_1 > 1, \\ u_1^2 & \text{otherwise.} \end{cases} \tag{12.11}$$

It follows that the corrector algorithm must fail to converge as soon as $u_1 < 1$. This lack of convergence is indicated by the `MX` point type assigned to the last labeled solution obtained on the solution manifold.

Consider next the zero function

$$\Phi(u) = u_1^2 + u_2^2 + 1, \tag{12.12}$$

encoded below.

```
function [data y] = empty(prob, data, u)
  y = u(1)^2+u(2)^2+1;
end
```

In this case, the corrector algorithm fails to converge already in the initial application of the corrector, independently of the initial solution guess, as seen in the extract below.

```
>> prob = coco_add_func(coco_prob(), 'func', @empty, [], ...
   'zero', 'u0', [1; 1]);
>> coco(prob, 'run', [], 1);
```

STEP	DAMPING		NORMS			COMPUTATION TIMES		
IT SIT	GAMMA	||d||	||f||	||U||	F(x)	DF(x)	SOLVE
0			3.00e+00	1.41e+00	0.0	0.0	0.0
1 1	1.00e+00	1.06e+00	1.13e+00	3.54e-01	0.0	0.0	0.0
2 3	2.50e-01	1.59e+00	1.00e+00	4.42e-02	0.0	0.0	0.0

```
  3   8  3.91e-03  1.13e+01  1.00e+00  4.44e-02  0.0   0.0   0.0
  4   8  7.81e-03  1.13e+01  1.00e+00  4.39e-02  0.0   0.0   0.0
  5   8  3.91e-03  1.14e+01  1.00e+00  4.54e-02  0.0   0.0   0.0
  6   8  7.81e-03  1.10e+01  1.00e+00  4.08e-02  0.1   0.0   0.0
  7   8  3.91e-03  1.23e+01  1.00e+00  5.50e-02  0.1   0.0   0.0
  8   8  7.81e-03  9.11e+00  1.00e+00  1.61e-02  0.1   0.0   0.0
  9   8  3.91e-03  3.10e+01  1.05e+00  2.26e-01  0.1   0.0   0.0
 10   4  1.25e-01  2.32e+00  1.00e+00  6.43e-02  0.1   0.0   0.0

 STEP     TIME       ||U||   LABEL  TYPE
  -1   00:00:00   1.4142e+00     1   MX
```

As an additional test of the flow through the minimal atlas algorithm, consider the COCO-compatible function shown below.

```
function [data y] = singular(prob, data, u)
  y = 1;
end
```

Not only are there no solutions to the corresponding zero problem, but the Jacobian is everywhere singular. The following extract shows that the singular nature of the Jacobian and the lack of convergence trigger a transition of the finite-state machine to the terminal state.

```
>> prob = coco_add_func(coco_prob(), 'func', @singular, [], ...
     'zero', 'u0', [1; 1]);
>> coco(prob, 'run', [], 1);
```

STEP	DAMPING			NORMS			COMPUTATION TIMES	
IT SIT	GAMMA	$\|\|d\|\|$		$\|\|f\|\|$	$\|\|U\|\|$	F(x)	DF(x)	SOLVE
0				1.00e+00	1.41e+00	0.0	0.0	0.0

```
Warning: Matrix is singular to working precision.
  1   8  3.91e-03     Inf      NaN      Inf   0.0   0.0   0.0
Warning: Matrix is singular to working precision.
  2   8  3.91e-03     NaN      NaN      NaN   0.0   0.0   0.0
Warning: Matrix is singular to working precision.
  3   8  3.91e-03     NaN      NaN      NaN   0.0   0.0   0.0
  4   8  3.91e-03     NaN      NaN      NaN   0.0   0.0   0.0
  5   8  3.91e-03     NaN      NaN      NaN   0.1   0.0   0.0
  6   8  3.91e-03     NaN      NaN      NaN   0.1   0.0   0.0
  7   8  3.91e-03     NaN      NaN      NaN   0.1   0.0   0.0
  8   8  3.91e-03     NaN      NaN      NaN   0.1   0.0   0.0
  9   8  3.91e-03     NaN      NaN      NaN   0.1   0.0   0.0
 10   8  3.91e-03     NaN      NaN      NaN   0.1   0.0   0.0

 STEP     TIME       ||U||   LABEL  TYPE
  -1   00:00:00   1.4142e+00     1   MX
```

Finally, consider the encoding shown below of the function

$$\Phi(u) := \left(u_1^2 + u_2^2\right)^2 + u_2^2 - u_1^2. \tag{12.13}$$

```
function [data y] = lemniscate(opts, data, u)
  y = (u(1)^2+u(2)^2)^2+u(2)^2-u(1)^2;
end
```

The corresponding solution manifold has a branch point located at $u = (0,0)$, where two 1-dimensional branches intersect transversally. As seen in the extract below, the default construction of the initial tangent matrix implemented in CurveSegment fails to support continuation from this branch point, since the default assignment of true to the correct property of the cseg class instance results in a singular Jacobian

$$\begin{pmatrix} \partial_u \Phi(0) \\ v^T \end{pmatrix}, \tag{12.14}$$

independently of v.

```
>> prob = coco_add_func(coco_prob(), 'lemniscate', @lemniscate, ...
     [], 'zero', 'u0', [0; 0]);
>> prob = coco_add_pars(prob, '', [1 2], {'x' 'y'});
>> coco(prob, 'run', [], 1, {'x' 'y'});
```

STEP		DAMPING		NORMS		COMPUTATION TIMES		
IT	SIT	GAMMA	||d||	||f||	||U||	F(x)	DF(x)	SOLVE
0				0.00e+00	0.00e+00	0.0	0.0	0.0

Warning: Matrix is singular to working precision.

IT	SIT	GAMMA	||d||	||f||	||U||	F(x)	DF(x)	SOLVE
1	8	3.91e-03	NaN	NaN	NaN	0.0	0.0	0.0
2	8	3.91e-03	NaN	NaN	NaN	0.0	0.0	0.0
3	8	3.91e-03	NaN	NaN	NaN	0.0	0.0	0.0
4	8	3.91e-03	NaN	NaN	NaN	0.1	0.0	0.0
5	8	3.91e-03	NaN	NaN	NaN	0.1	0.0	0.0
6	8	3.91e-03	NaN	NaN	NaN	0.1	0.0	0.0
7	8	3.91e-03	NaN	NaN	NaN	0.1	0.0	0.0
8	8	3.91e-03	NaN	NaN	NaN	0.1	0.0	0.0
9	8	3.91e-03	NaN	NaN	NaN	0.1	0.0	0.0
10	8	3.91e-03	NaN	NaN	NaN	0.1	0.0	0.0

STEP	TIME	||U||	LABEL	TYPE	x	y
-1	00:00:00	0.0000e+00	1	MX	---	---

We may avoid the resultant singularity by providing an explicit expression for the initial tangent direction, as shown in the following continued extract.

```
>> prob = coco_add_func(coco_prob(), 'lemniscate', @lemniscate, ...
     [], 'zero', 'u0', [0; 0], 't0', [1; 1]);
>> prob = coco_add_pars(prob, '', [1 2], {'x' 'y'});
>> coco(prob, 'run', [], 1, {'x' 'y'});
```

STEP	TIME	||U||	LABEL	TYPE	x	y
0	00:00:00	0.0000e+00	1	EP	0.0000e+00	0.0000e+00
10	00:00:00	9.7655e-01	2		5.9338e-01	3.5317e-01
20	00:00:00	1.3977e+00	3		9.8254e-01	-1.0658e-01
30	00:00:01	6.9171e-01	4		3.8501e-01	-3.0166e-01
40	00:00:01	3.0464e-01	5		-1.5581e-01	1.4874e-01
50	00:00:02	1.2173e+00	6	EP	-8.0305e-01	3.0982e-01

12.2   Adaptation and accelerated convergence

In this section, we consider modifications to the basic atlas algorithm in support of using a higher-order predictor and a more refined handling of the failure of the corrector algorithm to converge.

12.2.1   A two-step algorithm

Let u^* again denote a point on the solution manifold such that the unit vector t^* spans the corresponding tangent space. Then, for $\theta \in [0,1]$, we let the tangent matrix v of the projection condition be given by the unit vector aligned with the vector

$$\begin{pmatrix} \partial_u \Phi\left(u^* + \theta\, R s t^*\right) \\ t^{*T} \end{pmatrix}^{-1} \cdot \begin{pmatrix} 0 \\ 1 \end{pmatrix}. \tag{12.15}$$

Moreover, the corresponding step size h is given by the scalar $R\, v^T \cdot t^*$ and the new predictor is given by the linear predictor $u^* + h s v$. We clearly regain the pseudo-arclength atlas algorithm when $\theta = 0$. We encode this construction in the modified `predict` class method, shown below.

```
function [prob atlas cseg correct] = predict(atlas, prob, cseg)

chart   = atlas.base_chart;
prcond  = struct('x', chart.x, 'TS', chart.TS, ...
                 's', chart.s, 'h', chart.R);
th      = atlas.cont.theta;
if th>=0.5 && th<=1
  xp          = chart.x+(th*chart.R)*(chart.TS*chart.s);
  [prob cseg] = CurveSegment.create(prob, chart, prcond, xp);
  [prob ch2]  = cseg.update_TS(prob, cseg.curr_chart);
  h           = chart.R*chart.TS'*ch2.TS;
  xp          = chart.x+h*(ch2.TS*chart.s);
  prcond      = struct('x', chart.x, 'TS', ch2.TS, ...
                       's', chart.s, 'h', h);
else
  xp          = chart.x+chart.R*(chart.TS*chart.s);
end
[prob cseg] = CurveSegment.create(prob, chart, prcond, xp);
correct     = true;

end
```

Here, in the case that $\theta < 1/2$, the implementation reduces to the simple pseudo-arclength-based atlas algorithm. For $\theta \in [1/2, 1]$, the implementation relies on a temporary instance of the `CurveSegment` class, as well as the `update_TS` class method, in order to compute the tangent matrix v.

Example 12.1 Consider, as a special case, the zero function

$$\Phi(u) := (u_1 - 1)^2 + u_2^2 - 1 \tag{12.16}$$

and let $u^* = \begin{pmatrix} 1 + \cos\varphi & \sin\varphi \end{pmatrix}^T$ for some angle $\varphi \in [0, 2\pi)$. Since

$$\partial_u \Phi(u) = \begin{pmatrix} 2(u_1 - 1) & 2u_2 \end{pmatrix}, \tag{12.17}$$

it follows that $\mathcal{N}\left[\partial_u \Phi(u^*)\right]$ is spanned by the unit vector

$$t^* := \begin{pmatrix} \sin\varphi \\ -\cos\varphi \end{pmatrix}. \tag{12.18}$$

Now let v be a unit vector aligned with the vector

$$\left(\begin{array}{c} \partial_u \Phi(u^* + \theta Rst^*) \\ t^{*T} \end{array} \right)^{-1} \cdot \left(\begin{array}{c} 0 \\ 1 \end{array} \right) = \left(\begin{array}{c} \sin\varphi - \theta Rs \cos\varphi \\ -\cos\varphi - \theta Rs \sin\varphi \end{array} \right). \tag{12.19}$$

It follows that

$$\Phi\left(u^* + R\left(t^{*T} \cdot v \right) sv \right) = \frac{R^2 \left(1 + \theta \left(R^2 \left(\theta - 2\theta^2 \right) - 2 \right) \right)}{\left(1 + R^2 \theta^2 \right)^2}. \tag{12.20}$$

Notably, this quantity equals R^2 for $\theta = 0$, 0 for $\theta = 1/2$ and $-R^2/(1 + R^2)$ for $\theta = 1$. In the case that $\theta = 1/2$, it follows that the linear predictor $u^* + R\left(t^{*T} \cdot v \right) sv$ lies on the solution manifold. By the local nature of the argument, it follows that $\theta = 1/2$ gives $\mathcal{O}\left(h^3 \right)$ prediction errors. ∎

12.2.2   Step-size adaptation

We proceed to modify the atlas algorithm to vary adaptively the scalar radius R in response to successful convergence, or lack thereof, of the corrector algorithm. To accommodate a failure to converge, subsequent to locating successfully at least one point on the solution manifold, we consider the finite-state machine represented by the state transition diagram in Fig. 11.2(b). This includes the additional state refine, from which the atlas algorithm subsequently re-enters the predict state, in order to reconstruct a predictor and reinitialize the corrector algorithm. The refine_step class method, shown below, is invoked by the $\mathfrak{F}_{\mathfrak{refine}}$ transition function.

```
function [prob atlas cseg predict] = refine_step(atlas, prob, cseg)

predict = false;
chart   = cseg.ptlist{1};
R       = chart.R;
if R > atlas.cont.hmin
  chart.R = max(atlas.cont.hfred*R, atlas.cont.hmin);
  atlas.base_chart = chart;
  predict = true;
end

end
```

Here, the hmin and hfred fields of the atlas.cont structure contain the minimum values of the scalar radius R and a multiplicative factor used to decrease R, respectively. The predict output argument is set to true, indicating a transition to predict, only in the case that a reduction in the value of R occurs within the function body.

The modified add_chart class method, shown below, provides for additional adaptive changes to the scalar radius R.

```
function [prob atlas cseg flush] = add_chart(atlas, prob, cseg)

chart    = cseg.curr_chart;
chart.pt = chart.pt+1;
if chart.pt>=atlas.cont.PtMX
```

```
    chart.pt_type = 'EP';
    chart.ep_flag = 1;
  end
  [prob cseg] = cseg.add_chart(prob, chart);
  flush       = true;

  al = subspace(cseg.ptlist{1}.TS, cseg.ptlist{end}.TS);
  R  = atlas.base_chart.R;
  if al>atlas.cont.almax
    if R>atlas.cont.hmin
      atlas.base_chart.R = max(atlas.cont.hfred*R, atlas.cont.hmin);
      flush              = false;
    end
  elseif al<=atlas.cont.almax/2
    cseg.ptlist{end}.R = min(atlas.cont.hfinc*R, atlas.cont.hmax);
  end

end
```

Here, R is increased by a multiplicative factor stored in the `hfinc` field of the `atlas.cont` structure, up to a maximum stored in the `hmax` field, provided that the new tangent vector deviates from the previous tangent vector by an angle less than half the value stored in the `almax` field of the `atlas.cont` structure. In contrast, if the angle exceeds the value stored in the `almax` field, then either

- the radius R is greater than the minimum scalar radius, in which case R is decreased and the continuation step is repeated with the same base chart, or

- the radius R is less than or equal to the minimum scalar radius, in which case the chart is accepted.

In neither case does continuation terminate. We provide default values for the various parameters introduced here in the modified `get_settings` class method, shown below.

```
function cont = get_settings(cont)

defaults.h     = 0.1;
defaults.PtMX  = 50;
defaults.theta = 0.5;
defaults.hmax  = 0.1;
defaults.hmin  = 0.01;
defaults.hfinc = 1.1;
defaults.hfred = 0.5;
defaults.almax = 10;
cont           = coco_merge(defaults, cont);
cont.almax     = cont.almax*pi/180;

end
```

The modified `atlas_1d_min` class declaration includes the function declaration

```
[prob atlas cseg predict] = refine_step(atlas, prob, cseg)
```

among the nonstatic, public class methods.

Example 12.2 Consider the zero function

$$\Phi(u) := 4(u_1 - 1)^2 + u_2^2 - 1 \qquad (12.21)$$

and the corresponding encoding in the `ellipse` function, shown below.

```
function [data y] = ellipse(prob, data, u)
  y = (2*(u(1)-1))^2+u(2)^2-1;
end
```

The variations in the step size that are realized by the modified `atlas_1d_min` atlas algorithm are evident in the `'StepSize'` column of the `bd` output from the call to the `coco` entry-point function, as extracted below using the `coco_bd_col` utility.

```
>> prob = coco_add_func(coco_prob(), 'ellipse', @ellipse, [], ...
     'zero', 'u0', [0.7; 0.8]);
>> prob = coco_set(prob, 'cont', 'PtMX', 20, 'hmin', 0.065);
>> bd = coco(prob, 'run', [], 1);

     STEP   DAMPING                    NORMS                  COMPUTATION TIMES
    IT SIT    GAMMA    ||d||       ||f||       ||U||     F(x)   DF(x)  SOLVE
     0                          2.22e-16  1.06e+00      0.0    0.0    0.0
     1    1  1.00e+00  7.70e-17  6.16e-17  1.06e+00      0.0    0.0    0.0

    STEP      TIME        ||U||    LABEL  TYPE
       0  00:00:00   1.0630e+00       1  EP
      10  00:00:00   1.5222e+00       2
      20  00:00:01   1.5001e+00       3  EP
>> coco_bd_col(bd, 'StepSize')

ans =

  Columns 1 through 7

    0.1000    0.1000    0.0650    0.0650    0.0650    0.0650    0.0650

  Columns 8 through 14

    0.0650    0.0650    0.0650    0.0650    0.0650    0.0715    0.0787

  Columns 15 through 21

    0.0865    0.0952    0.1000    0.1000    0.1000    0.1000    0.1000
```

Here, the minimum step size is reached after two successful continuation steps. ∎

12.3   An expanding-boundary algorithm

In the introduction to atlas algorithms in Chap. 11, an atlas was simply described as a collection of charts covering a portion of the solution manifold. The generation of an atlas was described in terms of a phase of expansion, corresponding to the construction of a curve segment emanating from some base chart on the solution manifold, followed by a phase of consolidation, in which the charts in the curve segment are merged into the atlas.

At any point during the execution of the 1-dimensional atlas algorithms in the previous sections of this chapter, the atlas was represented by a single base chart at the head of an advancing local cover along the solution manifold. In the expansion phase, a single additional chart was generated in the "forward" direction of propagation. In the consolidation

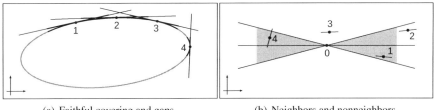

(a) Faithful covering and gaps.   (b) Neighbors and nonneighbors.

Figure 12.1. *Every chart based at $\tilde{u}$ and with tangent matrix V defines an affine subspace $A := \tilde{u} \oplus \mathrm{span}\{V\}$ that approximates the manifold M locally. We define the local cover $C := \{v \in M \mid \mathrm{err}(v, A) < \mathrm{TOL}\}$, where err is some measure of the approximation error and TOL is a required accuracy. An important requirement of an atlas algorithm is that the intersection of local covers of neighboring charts be nonempty, i.e., that the respective local covers overlap. In panel (a) the local covers of charts 1–3 form a sequence of overlapping covers, while the local cover of chart 4 is disconnected. We say that the covering defined by charts 1–4 has a gap. A simple heuristic to prevent the occurrence of gaps is a test that ensures that the approximation error between neighboring charts is sufficiently small on a local cover with radius greater than $R/2$ such that overlapping local covers result. Panel (b) illustrates the conditions in Eqs. (12.22)–(12.26) that require that the base point of a neighboring chart lie within the shaded area and that the angle between the tangent spaces be bounded. According to this condition, chart 1 is a neighbor of chart 0, while charts 2–4 are not.*

phase, this additional chart was substituted for the base chart and the process then repeated until termination.

In this section, we consider an alternative implementation of a 1-dimensional atlas algorithm, in which the atlas is represented by charts on its *boundary* enclosing an expanding 1-dimensional volume on the solution manifold. In this case, expansion proceeds from such a boundary chart, along an "outward" direction, away from the interior of this volume. Consolidation is, in turn, achieved by reconstructing the atlas boundary, while accounting for any newly generated charts on the solution manifold.

In the first version of the advancing-local-cover algorithm, coverage along the solution manifold was ensured only as far as was supported by the convergence of the corrector algorithm. Other than the restriction imposed by the step size, no condition was applied to the succession of points located on the solution manifold in order to guarantee a suitable coverage of the solution manifold by the corresponding charts. Along segments of high local curvature, sparser coverage would thus result from the absence of step-size control. The possibility of gaps in a representation of the solution manifold by straight-line segments of length R, centered on each solution point and tangent to the solution manifold, could not be excluded (cf. Fig. 12.1(a)).

In the second version of the advancing-local-cover algorithm, coverage along the solution manifold was regulated by varying the projection condition step size so as to ensure a finer coverage in regions of high local curvature. By suitable selection of the parameterization of the atlas algorithm in the form of the relative angle α_{max} (stored in the `almax` field of the `atlas.cont` structure), consecutive charts were made to overlap to the extent that gaps would be unlikely in the segmental representation of the atlas.

In the *expanding-boundary algorithm*, implemented in this section, we further restrict the criteria for neighboring charts to require that

- each corresponding base point lie inside a cone centered on the base point of the other chart, aligned along the tangent vector of the other chart, and with opening angle $2\alpha_{max}$, and

- that the projection of each base point onto the tangent vector of the other chart lie a distance less than R from the other base point.

These criteria are illustrated in Fig. 12.1(b). Specifically, suppose that u_1 and u_2 denote the base points of two charts with tangent spaces spanned by the unit vectors t_1 and t_2, respectively. Then, if $du = u_2 - u_1$, the two charts are said to be *neighbors* provided that

$$|t_1^T \cdot du| < R, \tag{12.22}$$

$$|t_2^T \cdot du| < R, \tag{12.23}$$

$$\left\| du - t_1 \left(t_1^T \cdot du \right) \right\| < |t_1^T \cdot du| \tan\alpha_{max}, \tag{12.24}$$

$$\left\| du - t_2 \left(t_2^T \cdot du \right) \right\| < |t_2^T \cdot du| \tan\alpha_{max}, \tag{12.25}$$

and, as before,

$$t_1^T \cdot t_2 > \cos\alpha_{max}. \tag{12.26}$$

Notably, this set of criteria may apply not only to consecutive charts but to any two charts obtained during continuation. Consider, in particular, the case of a closed solution manifold. Then, provided that the step size h is chosen appropriately, full coverage of the manifold is obtained once the two boundary charts at either end of the expanding volume become neighbors.

12.3.1 Encoding

We encode abstract support for the expanding-boundary algorithm described above in the following declaration of the `atlas_1d_min` subclass of the `AtlasBase` class.

```
classdef atlas_1d_min < AtlasBase

  properties (Access=private)
    boundary = {};
    cont     = struct();
  end

  methods (Access=private)
    function atlas = atlas_1d_min(prob, cont, dim)
      assert(dim==1, '%s: wrong manifold dimension', mfilename);
      atlas      = atlas@AtlasBase(prob);
      atlas.cont = atlas.get_settings(cont);
    end
  end
```

```
methods (Static)
  function [prob cont atlas] = create(prob, cont, dim)
    atlas = atlas_1d_min(prob, cont, dim);
    prob  = CurveSegment.add_prcond(prob, dim);
  end
end

methods (Static, Access=private)
  cont = get_settings(cont)
end

methods (Access=public)
  [prob atlas cseg correct] = init_prcond(atlas, prob, chart)
  [prob atlas cseg flush]   = init_atlas (atlas, prob, cseg)
  [prob atlas cseg]         = flush       (atlas, prob, cseg)
  [prob atlas cseg correct] = predict     (atlas, prob, cseg)
  [prob atlas cseg flush]   = add_chart   (atlas, prob, cseg)
end

methods (Access=private)
  flag = isneighbor(atlas, chart1, chart2)
  [atlas cseg] = merge(atlas, cseg)
end

end
```

Here, the `isneighbor` and `merge` class methods are introduced in order to collect functionality particular to the determination of proximity between charts and to the atlas boundary reconstruction during consolidation, respectively. The encoding of `isneighbor`, shown below, implements the criteria introduced in Eqs. (12.22)–(12.26).

```
function flag = isneighbor(atlas, chart1, chart2)

al  = atlas.cont.almax;
ta  = tan(al);
R   = atlas.cont.h;
x1  = chart1.x;
x2  = chart2.x;
dx  = x2-x1;
x1s = chart1.TS*(chart1.TS'*dx);
x2s = chart2.TS*(chart2.TS'*dx);
dst = [norm(x1s), norm(x2s), norm(dx-x1s), norm(dx-x2s), ...
       subspace(chart1.TS, chart2.TS)];
dstmx = [R, R, ta*norm(x1s), ta*norm(x2s), al];
flag = all(dst<dstmx);

end
```

The `init_prcond` class method is identical to those shown in the previous sections and repeated here for completeness.

```
function [prob atlas cseg correct] = init_prcond(atlas, prob, chart)

chart.R       = 0;
chart.pt      = -1;
chart.pt_type = 'IP';
chart.ep_flag = 1;
[prob cseg]   = CurveSegment.create_initial(prob, chart);
correct       = cseg.correct;

end
```

We account for the two possible directions of propagation along the solution manifold by letting $\Sigma = \{-1, 1\}$ for each chart constructed along the solution manifold, as shown in the following encoding of the `init_atlas` class method.

```
function [prob atlas cseg flush] = init_atlas(atlas, prob, cseg)

    chart           = cseg.curr_chart;
    chart.pt        = 0;
    chart.R         = atlas.cont.h;
    chart.s         = [1, -1]*sign(atlas.cont.PtMX);
    atlas.cont.PtMX = abs(atlas.cont.PtMX);
    chart.pt_type   = 'EP';
    chart.ep_flag   = 1;
    [prob cseg]     = cseg.add_chart(prob, chart);
    flush           = true;

end
```

Finally, the encoding of the `add_chart` class method, shown below, illustrates the use of the `isneighbor` class method in order to trigger the subsequent termination of the atlas algorithm in case a gap is found between two consecutive charts.

```
function [prob atlas cseg flush] = add_chart(atlas, prob, cseg)

    chart     = cseg.curr_chart;
    chart.pt = chart.pt+1;
    if chart.pt>=atlas.cont.PtMX
      chart.pt_type = 'EP';
      chart.ep_flag = 1;
    end
    [prob cseg] = cseg.add_chart(prob, chart);
    flush        = true;

    if ~atlas.isneighbor(cseg.ptlist{1}, cseg.ptlist{end})
      cseg.ptlist{end}.pt_type = 'GAP';
      cseg.ptlist{end}.ep_flag = 2;
      cseg.Status              = cseg.CurveSegmentCorrupted;
    end

end
```

To accommodate the reconstruction of the atlas boundary associated with the consolidation phase, the encoding of the `atlas_1d_min` subclass replaces the `base_chart` property of the previous implementations (which held only a single chart) with the `boundary` property, which is initialized to an empty array. As seen in the following encoding of the `predict` class method, each row of the `boundary` array contains a boundary chart `chart`, an "outward" predictor `xp`, an "outward" direction vector s, and a step size h.

```
function [prob atlas cseg correct] = predict(atlas, prob, cseg)

    [chart xp s h] = atlas.boundary{1,:};
    prcond          = struct('x', chart.x, 'TS', chart.TS, 's', s, 'h', h);
    th              = atlas.cont.theta;
    if th>=0.5 && th<=1
      xp            = chart.x+(th*h)*(chart.TS*s);
      [prob cseg]   = CurveSegment.create(prob, chart, prcond, xp);
      [prob ch2]    = cseg.update_TS(prob, cseg.curr_chart);
```

```
  h            = h*(ch2.TS'*chart.TS);
  xp           = chart.x+h*(ch2.TS*s);
  prcond       = struct('x', chart.x, 'TS', ch2.TS, 's', s, 'h', h);
end
[prob cseg] = CurveSegment.create(prob, chart, prcond, xp);
correct      = true;

end
```

The `flush` class method, shown below, first invokes the `merge` class method to reconstruct the atlas boundary and then flags the atlas algorithm for termination in the case that the boundary is empty (as would eventually be the case for a closed solution manifold) or the maximum number of successfully executed continuation steps has been reached.

```
function [prob atlas cseg] = flush(atlas, prob, cseg)

if cseg.Status==cseg.CurveSegmentOK
  [atlas cseg] = atlas.merge(cseg);
end
[prob atlas cseg] = atlas.flush@AtlasBase(prob, cseg);
if cseg.Status==cseg.CurveSegmentOK
  if isempty(atlas.boundary) || ...
       (atlas.boundary{1,1}.pt>=atlas.cont.PtMX)
    cseg.Status = cseg.BoundaryPoint;
  end
end

end
```

Notably, in the event that the `Status` property of the `cseg` class instance is set to the `cseg.CurveSegmentCorrupted` integer constant by the `add_chart` class method, this property is unchanged during the call to `AtlasBase.flush`.

The detailed atlas boundary reconstruction is implemented in the following encoding of the `merge` class method.

```
function [atlas cseg] = merge(atlas, cseg)

chart = cseg.ptlist{end};
R      = atlas.cont.h;
h      = atlas.cont.Rmarg*R;
nb     = cell(2,4);
for k=1:2
  sk         = chart.s(k);
  xk         = chart.x+h*(chart.TS*sk);
  nb(k,:) = {chart, xk, sk, h};
end
for i=size(atlas.boundary,1):-1:1
  chart2 = atlas.boundary{i,1};
  if atlas.isneighbor(chart, chart2)
    x2 = atlas.boundary{i,2};
    if norm(chart.TS'*(x2-chart.x))<R
      atlas.boundary(i,:) = [];
    end
    for k=size(nb,1):-1:1
      x1 = nb{k,2};
      if norm(chart2.TS'*(x1-chart2.x))<R
        nb(k,:) = [];
      end
    end
  end
```

```
      end
   end
   atlas.boundary = [nb; atlas.boundary];
   if isempty(atlas.boundary)
      chart.pt_type   = 'EP';
      chart.ep_flag   = 1;
      cseg.ptlist{end} = chart;
   end

   end
```

Here, each new chart is associated with two outward pseudo-arclength predictors corresponding to $s = 1$ and $s = -1$, respectively, and represented by the two rows of the nb cell array. The algorithm then uses the isneighbor class method to identify and remove those rows of the existing boundary cell array for which the corresponding predictors are considered to be covered by the new chart, and vice versa. We note, in particular, the use of the Rmarg field of the atlas.cont structure to make likely the outcome that a chart constructed from each predictor will be considered a neighbor of the original chart.

The encoding of the get_settings class method, shown below, now completes the definition of the atlas_1d_min subclass.

```
   function cont = get_settings(cont)

   defaults.h       = 0.1;
   defaults.PtMX    = 50;
   defaults.theta   = 0.5;
   defaults.almax   = 10;
   defaults.Rmarg   = 0.95;
   cont             = coco_merge(defaults, cont);
   cont.almax       = cont.almax*pi/180;

   end
```

12.3.2   Functionality and flow test

We illustrate the functionality encoded in the expanding-boundary algorithm by considering continuation along a closed solution manifold, as well as along a high-curvature manifold with an incompatible choice of the algorithm parameters. Consider, for example, the zero function

$$\Phi(u) := (u_1 - 1)^2 + u_2^2 - 1 \qquad (12.27)$$

and the encoding in the circle function, shown previously. The extract below shows the termination of the continuation algorithm when, after 93 successfully completed execution steps, the atlas boundary becomes empty, thus avoiding redundant coverage of any portion of the solution manifold.

```
   >> prob = coco_add_func(coco_prob(), 'circle', @circle, [], ...
         'zero', 'u0', [1.5; 1]);
   >> prob = coco_add_pars(prob, '', [1 2], {'x' 'y'});
   >> prob = coco_set(prob, 'cont', 'PtMX', 100);
   >> coco(prob, 'run', [], 1, {'x' 'y'});
```

	STEP	DAMPING		NORMS		COMPUTATION TIMES		
IT SIT		GAMMA	||d||	||f||	||U||	F(x)	DF(x)	SOLVE
0				2.50e-01	2.55e+00	0.0	0.0	0.0

```
    1   1   1.00e+00   1.77e-01   1.56e-02   2.46e+00   0.0   0.0   0.0
    2   1   1.00e+00   1.26e-02   7.97e-05   2.45e+00   0.0   0.0   0.0
    3   1   1.00e+00   6.51e-05   2.12e-09   2.45e+00   0.0   0.0   0.0
    4   1   1.00e+00   1.73e-09   0.00e+00   2.45e+00   0.0   0.0   0.0
```

```
 STEP      TIME       ||U||   LABEL  TYPE           x            y
    0   00:00:00   2.4495e+00      1  EP     1.5000e+00   8.6603e-01
   10   00:00:01   2.7787e+00      2         1.9303e+00   3.6692e-01
   20   00:00:01   2.7975e+00      3         1.9566e+00  -2.9151e-01
   30   00:00:01   2.5040e+00      4         1.5675e+00  -8.2336e-01
   40   00:00:02   1.9309e+00      5         9.3205e-01  -9.9769e-01
   50   00:00:02   1.1421e+00      6         3.2608e-01  -7.3880e-01
   60   00:00:03   2.2575e-01      7         1.2740e-02  -1.5912e-01
   70   00:00:03   7.1579e-01      8         1.2809e-01   4.8966e-01
   80   00:00:04   1.5774e+00      9         6.2204e-01   9.2582e-01
   90   00:00:04   2.2628e+00     10         1.2801e+00   9.5997e-01
   93   00:00:05   2.4220e+00     11  EP     1.4665e+00   8.8451e-01
```

Alternatively, consider again the zero function

$$\Phi(u) := 4(u_1 - 1)^2 + u_2^2 - 1 \tag{12.28}$$

and the encoding in the `ellipse` function, shown previously. As seen in the extract below, the default choice of the step size h, the angle α_{max}, and the safety factor, stored in the `Rmarg` field of the `atlas.cont` structure, results in an excessive change to the tangent direction and the subsequent termination of the finite-state machine.

```
>> prob = coco_add_func(coco_prob(), 'ellipse', @ellipse, [], ...
     'zero', 'u0', [0.6; 0.5]);
>> prob = coco_add_pars(prob, '', [1 2], {'x' 'y'});
>> coco(prob, 'run', [], 1, {'x' 'y'});
```

```
  STEP   DAMPING            NORMS                COMPUTATION TIMES
 IT SIT   GAMMA    ||d||    ||f||      ||U||    F(x)  DF(x)  SOLVE
    0                      1.10e-01   1.10e+00  0.0   0.0    0.0
    1   1   1.00e+00   1.56e-01   1.21e-02   1.21e+00   0.0   0.0   0.0
    2   1   1.00e+00   1.40e-02   9.84e-05   1.20e+00   0.0   0.0   0.0
    3   1   1.00e+00   1.16e-04   6.72e-09   1.20e+00   0.0   0.0   0.0
    4   1   1.00e+00   7.92e-09   2.22e-16   1.20e+00   0.0   0.0   0.0
```

```
 STEP      TIME       ||U||   LABEL  TYPE           x            y
    0   00:00:00   1.2000e+00      1  EP     6.0000e-01   6.0000e-01
    6   00:00:00   1.7550e+00      2  GAP    8.1907e-01   9.3223e-01
    7   00:00:00   1.8462e+00         GAP    8.7537e-01   9.6844e-01
```

12.4 Conclusions

The minimal 1-dimensional atlas algorithms considered in this chapter illustrate the interface to continuation afforded by `AtlasBase` and `CurveSegment`. They serve further as templates for independent development. There is evidently room for more sophisticated implementations of step-size adaptation, as well as algorithms that make use of multiple points along a curve segment and nonlinear extrapolation, as may be warranted by the nature of the problem class.

The COCO core framework for covering solution manifolds affords the developer great algorithm flexibility through the selection of the base point $\tilde{u}$, the tangent matrix V, and the direction vector s, as well as by making innovative use of the `correct` and `flush` output arguments of the corresponding class methods in order to direct the flow of the finite-state machine.

In the next chapter, we generalize the methodology developed here to the case of 2-dimensional solution manifolds. Although many of the class method implementations shown in the previous sections are reused with only minor changes, the additional dimensionality poses unique challenges. These challenges are principally associated with the choice of atlas representation and the merge algorithm employed during the consolidation phase.

Exercises

12.1. What happens with the condition that the direction vector must be parallel to $V^T \cdot (u^* - \tilde{u})$, when this is nonzero, in the case of 1-dimensional solution manifolds?

12.2. Consider the atlas algorithm in Sect. 12.1. Seek to identify the elements of the encoding that ensure that

   1. the first and last points on the solution branch are denoted by `'EP'` in the output from the `coco` entry-point function.

   2. the scalar radius R is constant across all charts and equals the value stored in `atlas.cont.h`.

   3. the algorithm interrupts when the number of successfully executed continuation steps equals the value stored in `atlas.cont.PtMX`.

   4. the algorithm interrupts when the corrector sets the `Status` property of the `cseg` class instance to the `CurveSegment.CorrectionFailed` integer constant.

12.3. Consider the atlas algorithm in Sect. 12.1. Seek to identify the elements of the encoding that ensure that

   1. the tangent vector t' at a newly located point u' is aligned with the previous tangent vector t.

   2. the direction vector s equals 1 or -1.

   3. the predictor is a pseudo-arclength predictor.

   4. `MX` appears in the `TYPE` column of the screen output in the absence of convergence of the corrector algorithm.

12.4. Consider the atlas algorithm in Sect. 12.1. Seek to identify the elements of the encoding that ensure that

   1. the base point $\tilde{u}$ of the projection condition equals a previously located point u on the solution manifold.

   2. the tangent matrix V of the projection condition equals the tangent vector t at a previously located point u on the solution manifold.

   3. the default step size h for the projection condition equals 0.1.

12.5. Consider the atlas algorithm in Sect. 12.1. What happens if you

    1. change the `correct` output argument of the `init_prcond` class method to `false`?

    2. change the `flush` output argument of the `init_atlas` class method to `false`?

    3. change the `flush` output argument of the `add_chart` class method to `false`?

    4. remove the `atlas.base_chart.pt>=atlas.cont.PtMX` test from the `flush` class method?

    5. remove the conditional block, shown below, from the `flush` class method?

```
if cseg.Status==cseg.CurveSegmentOK
  .
  .
  .
end
```

12.6. Use the content of the `sol1.mat` files produced by the tests of the algorithm flow in Sect. 12.1 to verify the statements made regarding the initial tangent matrix and the resultant effect on the components of u in the initial application of the corrector.

12.7. Use the `coco_view_log` utility to verify the observations in Example 12.1 by applying the atlas algorithm in Sect. 12.2 to the zero problem

$$(u_1 - 1)^2 + u_2^2 - 1 = 0.$$

12.8. Consider the atlas algorithm in Sect. 12.2:

    1. Is it possible to have `chart.R` less than `atlas.cont.hmin` or larger than `atlas.cont.hmax`?

    2. What happens if the `R` field of the `atlas.base_chart` property is less than or equal to `atlas.cont.hmin` in the `refine_step` class method?

    3. Is the `refine_step` class method called if the corrector algorithm fails to converge on an initial point on the solution manifold?

    4. What is the consequence of the assignment `flush = false` during step-size adaptation in the `add_chart` class method?

12.9. Provide a complete suite of flow tests for the atlas algorithm in Sect. 12.3. Make sure that your examples test all relevant branches of execution.

12.10. What may be a consequence of setting the `Rmarg` field of the `atlas.cont` structure equal to 1?

12.11. Comment on the ability of the atlas algorithm in Sect. 12.3 to cover the solution manifold of the zero problem

$$u_1 - \sin \frac{1}{u_2} = 0.$$

12.12. Given two points u_0 and u_1 along a curve segment on the solution manifold with corresponding tangent spaces spanned by t_0 and t_1, respectively, show that

$$h(s) = u_0 + st_0 + s^2 (3u_1 - 3u_0 - 2t_0 - t_1) + s^3 (t_1 + t_0 - 2u_1 + 2u_0)$$

is the unique cubic polynomial for which

$$h(0) = u_0, h(1) = u_1$$

and

$$h'(0) = t_0, h'(1) = t_1.$$

12.13. Consider the *Hermite polynomial* h introduced in the previous exercise. Let $t_0 = k_0 \tau_0$ and $t_1 = k_1 \tau_1$ for two given vectors τ_0 and τ_1. Show that k_0 and k_1 minimize the function

$$F(k_1, k_2) := |||H(0)|||^2 + |||H(1)|||^2,$$

where $H : s \mapsto h(s) - u_0 - s t_0$ and

$$|||H(s)|||^2 := \|H(s)\|_2^2 + \|H'(s)\|_2^2 + \|H''(s)\|_2^2 + \|H'''(s)\|_2^2$$

provided that

$$\begin{pmatrix} 26\tau_0^T \cdot \tau_0 & 23\tau_0^T \cdot \tau_1 \\ 23\tau_0^T \cdot \tau_1 & 28\tau_1^T \cdot \tau_1 \end{pmatrix} \begin{pmatrix} k_0 \\ k_1 \end{pmatrix} = \begin{pmatrix} 49\tau_0^T \cdot (u_1 - u_0) \\ 51\tau_1^T \cdot (u_1 - u_0) \end{pmatrix}.$$

12.14. Show that the coefficient matrix in the linear equation in the previous exercise is invertible provided that $\|\tau_0\| = \|\tau_1\| = 1$.

12.15. Consider the final encoding of the `add_chart` class method in Sect. 12.3. Let A, B, C, and D denote the coefficients of the Hermite polynomial, considered in the previous exercises, that interpolates between `cseg.plist{end}` and `chart`. Insert a sequence of commands immediately before the command

```
[prob cseg] = cseg.add_chart(prob, chart);
```

to compute these coefficients and to store them in the `A`, `B`, `C`, and `D` fields of `chart`.

12.16. Consider the modifications to `add_chart` discussed in the previous exercise. Write a slot function that responds to the `'bddat'` signal for appending the `A`, `B`, `C`, and `D` fields of the `chart` argument to the output of the `coco` entry-point function. Add a call to `coco_add_slot` to define this slot function in the static construction method `create` of the `atlas_1d_min` class.

Chapter 13

Multidimensional Manifolds

Although the abstract framework for covering manifolds developed in Chap. 11 is independent of manifold dimension, the analysis of multidimensional manifolds demands careful attention to matters of geometry, abstract data structures, and algorithm design. The treatment in this chapter is meant as an introduction to the topic from the perspective of basic algorithm development, with only minimal attention paid to computational efficiency or handling of exceptional cases, however intriguing.

A challenge of particular concern in the case of a multidimensional manifold is the need to avoid unnecessarily retracing those parts of the manifold that have already been appropriately covered by previously computed charts. To this end, care must be taken in the assignment of the coordinate set Σ during the consolidation phase, subsequent to locating a new point on the solution manifold. With higher dimensionality also comes the possibility of voids in the manifold cover that may be accounted for in the analysis only by a faithful abstract representation of the atlas and its boundary. The solutions proposed in this chapter for the case of 2-dimensional manifolds provide a basis for a generalized treatment of manifolds of arbitrary dimension. Even without such a generalization (the dimensional curse quickly catches up with the best of intentions), the 2-dimensional framework serves as a useful and significant advance from continuation along 1-dimensional manifolds.

As shown below, the 1-dimensional atlas algorithms developed in Chap. 12 may be adapted to the case of a multidimensional manifold by making the appropriate changes to the class methods discussed there. As advertised, we demonstrate the realization of 2-dimensional atlas algorithms in terms of a succession of modifications to the 1-dimensional expanding-boundary algorithm, first introduced in Sect. 12.3. We leave to the reader the option of implementing 2-dimensional atlas algorithms that generalize the advancing-local-cover algorithm or that implement alternative solutions available in the literature.

13.1 A point-cloud algorithm

A few minimal changes to the expanding-boundary algorithm introduced in Sect. 12.3 enable its application to multidimensional solution manifolds. These changes reflect the need to provide a chart-independent, global counter of the number of successfully executed continuation steps, as well as to account for the multiplicity of possible continuation directions.

13.1.1 Encoding

A candidate 2-dimensional atlas algorithm is encoded in the following declaration of the `atlas_2d_min` subclass of `AtlasBase` and the definition of its class methods.

```
classdef atlas_2d_min < AtlasBase

  properties (Access=private)
    boundary = {};
    next_pt  = 0;
    cont     = struct();
  end

  methods (Access=private)
    function atlas = atlas_2d_min(prob, cont, dim)
      assert(dim==2, '%s: wrong manifold dimension', mfilename);
      atlas      = atlas@AtlasBase(prob);
      atlas.cont = atlas.get_settings(cont);
    end
  end

  methods (Static)
    function [prob cont atlas] = create(prob, cont, dim)
      atlas = atlas_2d_min(prob, cont, dim);
      prob  = CurveSegment.add_prcond(prob, dim);
    end
  end

  methods (Static, Access=private)
    cont = get_settings(cont)
  end

  methods (Access=public)
    [prob atlas cseg correct] = init_prcond(atlas, prob, chart)
    [prob atlas cseg flush]   = init_atlas (atlas, prob, cseg)
    [prob atlas cseg]         = flush       (atlas, prob, cseg)
    [prob atlas cseg correct] = predict     (atlas, prob, cseg)
    [prob atlas cseg flush]   = add_chart   (atlas, prob, cseg)
  end

  methods (Access=private)
    flag = isneighbor(atlas, chart1, chart2)
    [atlas cseg] = merge(atlas, cseg)
  end

end
```

Here, the `init_prcond` and `isneighbor` class methods are identical to the encodings in Sect. 12.3. The following modified encodings of the `init_atlas`, `add_chart`, and `flush` class methods account for the introduction of the `next_pt` property of the `atlas` class instance to track the number of successfully executed continuation steps, independently of the ordinal number of the current base chart.

```
function [prob atlas cseg flush] = init_atlas(atlas, prob, cseg)

chart         = cseg.curr_chart;
chart.pt      = atlas.next_pt;
atlas.next_pt = atlas.next_pt+1;
chart.R       = atlas.cont.h;
chart.s       = atlas.cont.s0;
```

```
chart.pt_type  = 'EP';
chart.ep_flag  = 1;
[prob cseg]    = cseg.add_chart(prob, chart);
flush          = true;

end


function [prob atlas cseg flush] = add_chart(atlas, prob, cseg)

chart        = cseg.curr_chart;
chart.pt     = atlas.next_pt;
atlas.next_pt = atlas.next_pt+1;
if chart.pt>=atlas.cont.PtMX
  chart.pt_type = 'EP';
  chart.ep_flag = 1;
end
[prob cseg] = cseg.add_chart(prob, chart);
flush       = true;

if ~atlas.isneighbor(cseg.ptlist{1}, cseg.ptlist{end})
  cseg.ptlist{end}.pt_type = 'GAP';
  cseg.ptlist{end}.ep_flag = 2;
  cseg.Status              = cseg.CurveSegmentCorrupted;
end

end


function [prob atlas cseg] = flush(atlas, prob, cseg)

if cseg.Status==cseg.CurveSegmentOK
  [atlas cseg] = atlas.merge(cseg);
end
[prob atlas cseg] = atlas.flush@AtlasBase(prob, cseg);
if cseg.Status==cseg.CurveSegmentOK
  if isempty(atlas.boundary) || (atlas.next_pt>atlas.cont.PtMX)
    cseg.Status = cseg.BoundaryPoint;
  end
end

end
```

As before, the R and s fields of the `cseg.curr_chart` structure are inherited by construction from the corresponding base chart. The s0 field of the `atlas.cont` structure is initialized in the `get_settings` class method, shown below.

```
function cont = get_settings(cont)

defaults.h     = 0.1;
defaults.PtMX  = 50;
defaults.theta = 0.5;
defaults.almax = 10;
defaults.Rmarg = 0.95;
defaults.Ndirs = 6;
cont           = coco_merge(defaults, cont);
cont.almax     = cont.almax*pi/180;
al             = (0:cont.Ndirs-1)*(2*pi/cont.Ndirs);
cont.s0        = [cos(al); sin(al)];

end
```

This provides for a uniformly distributed collection of directions in the tangent space associated with each solution point that reduces to the 1-dimensional implementation in Sect. 12.3 when the `cont.Ndirs` field equals 2.

As seen in the following encoding, the generalization of the `merge` class method to the case of a 2-dimensional solution manifold is now immediate.

```
function [atlas cseg] = merge(atlas, cseg)

chart = cseg.ptlist{end};
R     = atlas.cont.h;
h     = atlas.cont.Rmarg*R;
nb    = cell(atlas.cont.Ndirs,4);
for k=1:atlas.cont.Ndirs
  sk       = chart.s(:,k);
  xk       = chart.x+h*(chart.TS*sk);
  nb(k,:) = {chart, xk, sk, h};
end
for i=size(atlas.boundary,1):-1:1
  chart2 = atlas.boundary{i,1};
  if atlas.isneighbor(chart, chart2)
    x2 = atlas.boundary{i,2};
    if norm(chart.TS'*(x2-chart.x))<R
      atlas.boundary(i,:) = [];
    end
    for k=size(nb,1):-1:1
      x1 = nb{k,2};
      if norm(chart2.TS'*(x1-chart2.x))<R
        nb(k,:) = [];
      end
    end
  end
end
atlas.boundary = [nb; atlas.boundary];
if isempty(atlas.boundary)
  chart.pt_type    = 'EP';
  chart.ep_flag    = 1;
  cseg.ptlist{end} = chart;
end

end
```

Here, the `atlas.boundary` property again contains a cell array that is modified during the consolidation phase of the atlas algorithm.

Finally, we generalize the two-step algorithm from Sect. 12.2 for the construction of a projection condition and a suitable predictor for continuation from a base point u^* in the direction of a unit vector $T^* \cdot \sigma$. Specifically, let $\tilde{u} = u^*$ and suppose that h represents a nominal step size. Then, the tangent matrix V is obtained by applying an orthonormalization algorithm to the columns of the matrix

$$\begin{pmatrix} \partial_u \Phi(u^* + \theta h T^* \cdot \sigma) \\ T^{*T} \end{pmatrix}^{-1} \cdot \begin{pmatrix} 0 \\ I_{n-m} \end{pmatrix} \tag{13.1}$$

for some $\theta \in [0,1]$. Given V, the direction vector is now given by

$$s = \frac{V^T \cdot T^* \cdot \sigma}{\| V^T \cdot T^* \cdot \sigma \|}, \tag{13.2}$$

and the step size equals $h\|V^T \cdot T^* \cdot \sigma\|$. Finally, the corresponding linear predictor is given by

$$u^* + h\|V^T \cdot T^* \cdot \sigma\|V \cdot s. \tag{13.3}$$

These definitions are implemented in the `predict` class method, shown below.

```
function [prob atlas cseg correct] = predict(atlas, prob, cseg)

[chart xp s h] = atlas.boundary{1,:};
prcond          = struct('x', chart.x, 'TS', chart.TS, 's', s, 'h', h);
th              = atlas.cont.theta;
if th>=0.5 && th<=1
  xp            = chart.x+(th*h)*(chart.TS*s);
  [prob cseg]   = CurveSegment.create(prob, chart, prcond, xp);
  [prob ch2]    = cseg.update_TS(prob, cseg.curr_chart);
  s             = h*(ch2.TS'*chart.TS*s);
  h             = norm(s);
  s             = s/h;
  xp            = chart.x+h*(ch2.TS*s);
  prcond        = struct('x', chart.x, 'TS', ch2.TS, 's', s, 'h', h);
end
[prob cseg] = CurveSegment.create(prob, chart, prcond, xp);
correct     = true;

end
```

13.1.2 Interpretation

There is little to raise concern in the implementation of the 2-dimensional atlas algorithm, outside of the naïve generalization of the 1-dimensional consolidation algorithm to two dimensions. The implications of the encoding of the `merge` class method to the ability of the atlas algorithm to generate an efficient cover of a portion of the solution manifold, however, are less immediate.

A key deficiency of the encoding is the failure of the `atlas.boundary` property to represent faithfully the 1-dimensional boundary of a 2-dimensional "volume" on the solution manifold. The data structure lacks information about the relative arrangement of charts in the atlas, as well as about the role played by the predictors, stored in the rows of the `atlas.boundary` property, in spanning the atlas boundary. A related shortcoming is the restricted application of the `isneighbor` class method, during consolidation, to elements of `atlas.boundary`, with the tacit, and flawed, assumption that these enumerate exhaustively the actual atlas boundary.

Exploration of the functionality and flow of the `atlas_2d_min` atlas algorithm establishes the implications of these observations to the character of the implementation. Consider, for example, the possibility of the algorithm terminating prematurely, i.e., before reaching the maximally allowed number of successful continuation steps, without having achieved a complete cover of the solution manifold. In the 1-dimensional expanding-boundary atlas algorithm, this would occur only in the event of a failure of the corrector algorithm to converge, or in case two consecutive charts failed to qualify as neighbors. Indeed, since the `atlas.boundary` property in this implementation always remains nonempty for an open solution manifold, the algorithm would repeatedly return to the `predict` state to initialize yet another curve segment.

In the 2-dimensional algorithm described above, a failure to converge, or of two consecutive charts to qualify as neighbors, again constitutes possible reasons for premature termination of the atlas algorithm. In this case, however, experiments reveal the possibility of premature termination due to an empty `atlas.boundary` cell array, even for a closed solution manifold.

A further concern is the redundant coverage of regions of the solution manifold by multiple charts. In the 1-dimensional atlas algorithm in Sect. 12.3, this possibility is excluded by design for an open solution manifold. For a closed solution manifold, the use of a constant step size guarantees termination of the atlas algorithm when the two boundary charts become neighbors prior to the occurrence of redundant coverage.

In the 2-dimensional algorithm implemented in the previous section, redundant coverage is found to be the norm, rather than the exception. Notably, each chart is removed from the `atlas.boundary` property once all its associated linear predictors have been found to approximate points on the solution manifold within regions covered by other charts in `atlas.boundary`. When this happens, the existence of a chart is forgotten by the algorithm. Regions on the solution manifold may thus be encountered repeatedly during execution, resulting in a cloud of base points across the solution manifold, rather than a sparse, but efficient coverage with a minimal set of charts.

We proceed to remedy these shortcomings successively in the following several sections. The modifications enable a systematic and exhaustive coverage of a 2-dimensional solution manifold that avoids gaps. It further provides for a well-defined graph representation of the atlas in terms a continuous 1-dimensional boundary and a relative ordering of its boundary and interior charts.

13.2   A chart network

To eliminate the possibility of redundant coverage of portions of the solution manifold, we provide below a modified encoding of the `atlas_2d_min` subclass that relies on a network representation of an atlas. The class declaration is here modified by replacing the property declaration by the following lines.

```
properties (Access=private)
    boundary = [];
    charts   = {};
    next_pt  = 0;
    cont     = struct();
end
```

Throughout execution, we retain all previously computed charts in the `charts` cell array property of the class instance and enumerate these in order of construction by the ordinal number assigned to the `id` field of each `chart` structure. The network representation follows by assigning an integer array to the `nb` field of each `chart` structure, containing each of the ordinal numbers associated with its neighboring charts.

The `init_prcond` and `isneighbor` class methods are again identical to the encodings in Sect. 12.3. Moreover, the `flush` class method is identical to the encoding in the previous section. Finally, only minor modifications apply to the encodings of the `init_atlas` and `add_chart` class methods, shown below.

```
function [prob atlas cseg flush] = init_atlas(atlas, prob, cseg)

chart         = cseg.curr_chart;
chart.pt      = atlas.next_pt;
atlas.next_pt = atlas.next_pt+1;
chart.R       = atlas.cont.h;
chart.s       = atlas.cont.s0;
chart.id      = chart.pt+1;
chart.bv      = atlas.cont.bv0;
chart.nb      = atlas.cont.nb0;
chart.pt_type = 'EP';
chart.ep_flag = 1;
[prob cseg]   = cseg.add_chart(prob, chart);
flush         = true;

end


function [prob atlas cseg flush] = add_chart(atlas, prob, cseg)

chart         = cseg.curr_chart;
chart.pt      = atlas.next_pt;
atlas.next_pt = atlas.next_pt+1;
chart.id      = chart.pt+1;
chart.bv      = atlas.cont.bv0;
chart.nb      = atlas.cont.nb0;
if chart.pt>=atlas.cont.PtMX
  chart.pt_type = 'EP';
  chart.ep_flag = 1;
end
[prob cseg] = cseg.add_chart(prob, chart);
flush       = true;

if ~atlas.isneighbor(cseg.ptlist{1}, cseg.ptlist{end})
  cseg.ptlist{end}.pt_type = 'GAP';
  cseg.ptlist{end}.ep_flag = 2;
  cseg.Status              = cseg.CurveSegmentCorrupted;
end

end
```

Here, for each `chart` structure, the `bv` field contains an integer array referencing the sub-set of the `s` field that constitutes the corresponding coordinate set Σ. By construction, the `boundary` property of the `atlas` class instance contains the ordinal numbers associated with all charts with nonempty `bv` fields, since the latter provide directions of atlas expansion.

As shown in the following modified encoding of the `get_settings` class method, the `bv` field is initialized to a sequence of integers enumerating all the entries of the `s` field.

```
function cont = get_settings(cont)

defaults.h     = 0.1;
defaults.PtMX  = 50;
defaults.theta = 0.5;
defaults.almax = 10;
defaults.Rmarg = 0.95;
defaults.Ndirs = 6;
cont           = coco_merge(defaults, cont);
```

```
cont.almax     = cont.almax*pi/180;
al             = (0:cont.Ndirs-1)*(2*pi/cont.Ndirs);
cont.s0        = [cos(al); sin(al)];
cont.bv0       = 1:cont.Ndirs;
cont.nb0       = [];

end
```

That the nb field is initially empty is consistent with its interpretation as an array of ordinal numbers associated with all neighboring charts.

The modified encoding of the predict class method, shown below, now extracts a base chart from the atlas.charts cell array by relying on the integer references in the atlas.boundary array.

```
function [prob atlas cseg correct] = predict(atlas, prob, cseg)

chart  = atlas.charts{atlas.boundary(1)};
s      = chart.s(:,chart.bv(1));
h      = atlas.cont.Rmarg*chart.R;
prcond = struct('x', chart.x, 'TS', chart.TS, 's', s, 'h', h);
th     = atlas.cont.theta;
if th>=0.5 && th<=1
    xp         = chart.x+(th*h)*(chart.TS*s);
    [prob cseg] = CurveSegment.create(prob, chart, prcond, xp);
    [prob ch2]  = cseg.update_TS(prob, cseg.curr_chart);
    s           = h*(ch2.TS'*chart.TS*s);
    h           = norm(s);
    s           = s/h;
    xp          = chart.x+h*(ch2.TS*s);
    prcond      = struct('x', chart.x, 'TS', ch2.TS, 's', s, 'h', h);
else
    xp          = chart.x+h*(chart.TS*s);
end
[prob cseg] = CurveSegment.create(prob, chart, prcond, xp);
correct     = true;

end
```

Notably, from the encoding of the flush class method, it follows that the predict class method is called only in case the atlas.boundary array is nonempty.

A dramatic departure from the atlas algorithm implemented in the previous section is evident in the modified encoding of the merge class method, shown below.

```
function [atlas cseg] = merge(atlas, cseg)

chart        = cseg.ptlist{end};
R            = atlas.cont.h;
h            = atlas.cont.Rmarg*R;
v            = @(i,ch) ch.x+h*(ch.TS*ch.s(:,i));
nbfunc       = @(x) atlas.isneighbor(chart, x);
close_charts = find(cellfun(nbfunc, atlas.charts));
for k=close_charts
    chartk       = atlas.charts{k};
    vx = arrayfun(@(i) v(i,chart), chart.bv, 'UniformOutput', false);
    idx = cellfun(@(x) (norm(chartk.TS'*(x-chartk.x))<R), vx);
    chart.bv(idx) = [];
    chart.nb      = [chart.nb, chartk.id];
    vx = arrayfun(@(i) v(i,chartk), chartk.bv, 'UniformOutput', false);
```

```
   idx = cellfun(@(x) (norm(chart.TS'*(x-chart.x))<R), vx);
   chartk.bv(idx)  = [];
   chartk.nb       = [chartk.nb, chart.id];
   atlas.charts{k} = chartk;
 end
 atlas.charts   = [atlas.charts, {chart}];
 atlas.boundary = [chart.id, atlas.boundary];
 bd_charts      = atlas.charts(atlas.boundary);
 idx            = cellfun(@(x) ~isempty(x.bv), bd_charts);
 atlas.boundary = atlas.boundary(idx);
 if isempty(atlas.boundary)
   chart.pt_type     = 'EP';
   chart.ep_flag     = 1;
   cseg.ptlist{end}  = chart;
 end

 end
```

Here, the `isneighbor` class method is applied to detect neighbors to the chart stored in the `chart` structure among all charts stored in the `atlas.charts` cell array, rather than only among those referenced by elements of `atlas.boundary`. The commands

```
 atlas.charts   = [atlas.charts, {chart}];
 atlas.boundary = [chart.id, atlas.boundary];
 bd_charts      = atlas.charts(atlas.boundary);
 idx            = cellfun(@(x) ~isempty(x.bv), bd_charts);
 atlas.boundary = atlas.boundary(idx);
```

append the newly computed chart to `atlas.charts` and prepend its ordinal number to `atlas.boundary`. The application of the `@(x) ~isempty(x.bv)` anonymous function to the cell array of charts referenced by `atlas.boundary` finally allows for the elimination of all elements of `atlas.boundary` that reference `chart` structures whose bv field is empty.

The anonymous function `@(i,ch) ch.x+h*(ch.TS*ch.s(:,i))` returns the ith pseudo-arclength predictor associated with the `ch` chart structure. The commands

```
 vx  = arrayfun(@(i) v(i,chart), chart.bv, 'UniformOutput', false);
 idx = cellfun(@(x) (norm(chartk.TS'*(x-chartk.x))<R), vx);
 chart.bv(idx)   = [];
```

thus eliminate those elements of the bv fields of the `chart` structure that reference predictors whose projections onto the tangent space of the `chartk` structure lie a distance less than R from the base point of the `chartk` structure. Similarly, the commands

```
 vx  = arrayfun(@(i) v(i,chartk), chartk.bv, 'UniformOutput', false);
 idx = cellfun(@(x) (norm(chart.TS'*(x-chart.x))<R), vx);
 chartk.bv(idx)  = [];
```

eliminate those elements of the bv fields of the `chartk` structure that reference predictors whose projections onto the tangent space of the `chart` structure lie a distance less than R from the base point of the `chart` structure.

Example 13.1 We illustrate the application of the modified `atlas_2d_min` atlas algorithm to the covering of the closed solution manifold corresponding to the zero problem

$$(u_1 - 1)^2 + u_2^2 + u_3^2 - 1 = 0, u \in \mathbb{R}^3, \tag{13.4}$$

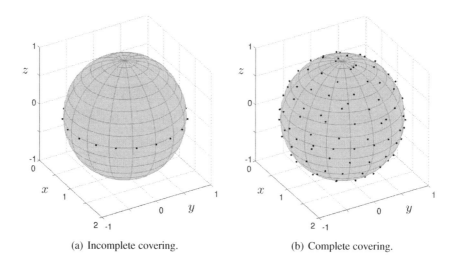

(a) Incomplete covering.   (b) Complete covering.

Figure 13.1. *Coverings of a unit sphere obtained in Example 13.1 with the modified* atlas_2d_min *atlas algorithm from Sect. 13.2. The first run results in an incomplete covering shown in panel* (a)*: the algorithm terminates prematurely after covering a strip around the equator. The second run, with a different choice for the number of continuation directions for each chart, results in a complete cover* (b)*. The dots mark the position of the base points of the charts in the final atlas.*

encoded in a COCO-compatible function with function handle @sphere. As in the previous chapter, we first edit the local project settings in the encoding of the coco_project_opts function in order to identify the atlas algorithm to the COCO finite-state machine.

As seen in the extract below, and from the sequence of base points in Fig. 13.1(a), the algorithm may still terminate prematurely.

```
>> prob = coco_add_func(coco_prob(), 'sphere', @sphere, [], ...
   'zero', 'u0', [2;0;0]);
>> prob = coco_add_pars(prob, '', [1 2 3], {'x' 'y' 'z'});
>> prob = coco_set(prob, 'cont', 'h', 0.5, 'almax', 35);
>> coco(prob, 'sphere1', [], 2, {'x' 'y' 'z'});
```

```
...    NORMS                COMPUTATION TIMES
...    ||f||      ||U||    F(x)   DF(x)   SOLVE
... 0.00e+00   2.83e+00    0.0    0.0     0.0
... 0.00e+00   2.83e+00    0.0    0.0     0.0

... LABEL   TYPE           x              y              z
...     1   EP      2.0000e+00     0.0000e+00     0.0000e+00
...     2           1.7269e-02     1.8504e-01     0.0000e+00
...     3   EP      1.9573e+00     2.8913e-01     0.0000e+00
```

In contrast, Fig. 13.1(b) demonstrates that an efficient and complete covering of the spherical surface is obtained for a different parameterization of the atlas algorithm.

```
>> prob = coco_set(prob, 'cont', 'Ndirs', 4, 'PtMX', 200);
>> coco(prob, 'sphere2', [], 2, {'x' 'y' 'z'});
```

```
...    NORMS                COMPUTATION TIMES
...    ||f||       ||U||     F(x)   DF(x)   SOLVE
... 0.00e+00   2.83e+00     0.0     0.0     0.0
... 0.00e+00   2.83e+00     0.0     0.0     0.0

... LABEL   TYPE           x              y              z
...     1   EP     2.0000e+00    0.0000e+00    0.0000e+00
...     2          1.7269e-02    1.8504e-01    0.0000e+00
...     3          1.9496e+00   -5.4435e-02   -3.0874e-01
...     4          2.6535e-01   -6.2622e-01   -2.6103e-01
...     5          1.5875e+00    2.2930e-01   -7.7603e-01
...     6          2.8602e-01   -5.5133e-01   -4.3158e-01
...     7          9.9167e-01    8.7937e-01   -4.7606e-01
...     8          1.5161e+00    7.7880e-01   -3.5657e-01
...     9          1.9369e-01    5.8864e-01   -5.7964e-02
...    10          2.9168e-01   -5.5517e-01    4.3597e-01
...    11          1.8241e+00   -5.3581e-01    1.8371e-01
...    12          1.1467e+00   -6.5235e-01    7.4358e-01
...    13          1.6866e+00   -4.0382e-01    6.0454e-01
...    14          1.4418e+00    8.4067e-01    3.1322e-01
...    15          4.5678e-01    8.0229e-01    2.4747e-01
...    16   EP     7.7592e-03    2.2833e-02    1.2222e-01
```

∎

The linear traversal of the `close_charts` array encoded in the **for**-loop in the `merge` class method reflects the lack of information about the actual atlas boundary in the properties of the `atlas` class instance. Given such information, it would suffice to first detect neighbors to the chart stored in the `chart` structure among the boundary charts and to then repeat this process recursively among the neighbors to the boundary charts. The following modified encoding of the `merge` class method implements such a recursive paradigm.

```
function [atlas cseg] = merge(atlas, cseg)

chart        = cseg.ptlist{end};
nbfunc       = @(x) atlas.isneighbor(chart, x);
close_charts = find(cellfun(nbfunc, atlas.charts));
checked      = chart.id;
while ~isempty(close_charts)
  [atlas chart checked] = ...
    atlas.merge_recursive(chart, close_charts(1), checked);
  close_charts = setdiff(close_charts, checked);
end
atlas.charts   = [atlas.charts, {chart}];
atlas.boundary = [chart.id, atlas.boundary];
bd_charts      = atlas.charts(atlas.boundary);
idx            = cellfun(@(x) ~isempty(x.bv), bd_charts);
atlas.boundary = atlas.boundary(idx);
if isempty(atlas.boundary)
  chart.pt_type    = 'EP';
  chart.ep_flag    = 1;
  cseg.ptlist{end} = chart;
end

end
```

Here, the variable `checked` tracks the ordinal numbers of those charts in `atlas.charts` that have already been processed during the consolidation phase. The workhorse of the `merge` class method is now the repeated call to the recursive `merge_recursive` class method, shown below, until all elements stored initially in `close_charts` have been processed.

```
function [atlas chart1 checked] = ...
  merge_recursive(atlas, chart1, k, checked)

checked(end+1) = k;
chartk = atlas.charts{k};
if atlas.isneighbor(chart1, chartk)
  R             = atlas.cont.h;
  h             = atlas.cont.Rmarg*R;
  v             = @(i,ch) ch.x+h*(ch.TS*ch.s(:,i));
  v1  = arrayfun(@(i) v(i,chart1), chart1.bv, 'UniformOutput', false);
  idx = cellfun(@(x) (norm(chartk.TS'*(x-chartk.x))<R), v1);
  chart1.bv(idx)   = [];
  chart1.nb        = [chart1.nb, chartk.id];
  vk  = arrayfun(@(i) v(i,chartk), chartk.bv, 'UniformOutput', false);
  idx = cellfun(@(x) (norm(chart1.TS'*(x-chart1.x))<R), vk);
  chartk.bv(idx)   = [];
  chartk.nb        = [chartk.nb, chart1.id];
  atlas.charts{k} = chartk;
  check = setdiff(chartk.nb, checked);
  while ~isempty(check)
    [atlas chart1 checked] = ...
      atlas.merge_recursive(chart1, check(1), checked);
    check = setdiff(chartk.nb, checked);
  end
end

end
```

We modify the class declaration to support this recursive implementation by including the following function declaration among the nonstatic private class methods.

```
[atlas chart1 checked] = merge_recursive(atlas, chart1, k, checked)
```

13.3   Henderson's algorithm

We seek next to eliminate the possibility of premature termination of the atlas algorithm and the concomitant failure to cover all desired portions of the solution manifold. Specifically, we modify the `atlas_2d_min` atlas algorithm and a subset of its class methods to include a piecewise-polygonal path approximation of the 1-dimensional boundary curve among the properties of a class instance. Our approach follows a methodology originally developed in the context of manifold continuation by Michael Henderson.

13.3.1   Interior and boundary charts

In the implementation below, we associate with each chart a convex polygon P with vertices ϖ_i, $i = 1, \ldots, k$, in the corresponding tangent space, ordered according to some default orientation of the polygonal boundary. In order to reflect the changes to the atlas boundary that result from consolidation, we allow for dynamic changes to the number of vertices, as well as to the corresponding direction vectors

$$s_i := \frac{\varpi_i}{\|\varpi_i\|} \tag{13.5}$$

and normalized distances

$$v_i := \|\varpi_i\|, \tag{13.6}$$

during execution. We aim to have the polygon P represent the domain of the solution manifold covered by this chart (and, except for possible minimal overlap at the boundary, only by this chart).

For a given scalar $\tilde{R} < R$, the initialization of each new chart $\{u, T, \Sigma, R\}$ on the solution manifold during the expansion phase is accompanied by the construction of a polygon P exterior at all points to the circle of radius $\tilde{R}$. In the implementation below, we simply initialize P to a regular polygon with $n_{\mathrm{vx}} \geq 3$ vertices whose inscribed circle is of radius R and centered at the origin. It follows that

$$s_i = \left(\begin{array}{c} \cos\left(2\pi(i-1)/n_{\mathrm{vx}}\right) \\ \sin\left(2\pi(i-1)/n_{\mathrm{vx}}\right) \end{array} \right) \tag{13.7}$$

and

$$v_i = v := \frac{R}{\cos\left(\pi/n_{\mathrm{vx}}\right)} \tag{13.8}$$

for $i = 1, \ldots, n_{\mathrm{vx}}$.

During consolidation, we repeatedly decimate the extent of each polygon P by retaining only that portion that lies on one side of each of a succession of straight lines in the tangent space to the chart, as shown in Fig. 13.2. Specifically, given a suitably chosen point $u' \neq u$ on the portion of the solution manifold covered by the chart $\{u, T, \Sigma, R\}$, consider the orthogonal projection

$$\varphi\left(u'\right) := T \cdot T^T \cdot \left(u' - u\right) \tag{13.9}$$

onto the tangent space $\mathcal{T}$ of the solution manifold at the point u. Then, the corresponding polygon P is replaced by the convex polygon

$$P \setminus \left\{ r \in \mathcal{T}, \|r\| > \left\| r - \varphi\left(u'\right) \right\| \right\} \tag{13.10}$$

obtained by subtracting the half space of points in the tangent space that lie closer to $\varphi\left(u'\right)$ than to 0. This operation is then repeated for all u' from some collection of points distinct from u.

The set operation implied by Eq. (13.10) is equivalent to eliminating all vertices that violate the inequality

$$r^T \cdot \varphi\left(u'\right) \leq \frac{1}{2} \left\| \varphi\left(u'\right) \right\|^2 \tag{13.11}$$

and by introducing new vertices along any intersections of the edges of P with the straight line given by

$$r^T \cdot \varphi\left(u'\right) = \frac{1}{2} \left\| \varphi\left(u'\right) \right\|^2. \tag{13.12}$$

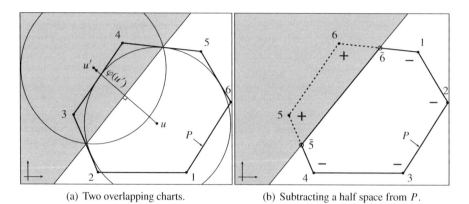

(a) Two overlapping charts.   (b) Subtracting a half space from P.

Figure 13.2. *Updating a polygon P associated with a chart $\{u, T, \Sigma, R\}$ during consolidation. The overlapping chart $\{u', T', \Sigma', R\}$ is projected onto the affine subspace $u \oplus \mathrm{span}\{T\}$, which is divided into two half spaces along a straight line normal to the vector $\varphi(u')$ and separating u and u', as shown in panel (a). In a subsequent step, the polygon is updated to the intersection of P with the half space containing u according to Eqs. (13.12)–(13.15), as illustrated in panel (b) after renumbering. Here, the sign of the expression $r^T \cdot \varphi\left(u'\right) - \frac{1}{2}\left\|\varphi\left(u'\right)\right\|^2$ is indicated next to the vertices.*

Suppose, in particular, that at least one of the vertices of P violates (13.11). Without loss of generality, consider a renumbering of the vertices such that

$$\varpi_1^T \cdot \varphi\left(u'\right) - \frac{1}{2}\left\|\varphi\left(u'\right)\right\|^2 \leq 0 < \varpi_k^T \cdot \varphi\left(u'\right) - \frac{1}{2}\left\|\varphi\left(u'\right)\right\|^2 \tag{13.13}$$

and such that j denotes the smallest integer, for which

$$\varpi_j^T \cdot \varphi\left(u'\right) - \frac{1}{2}\left\|\varphi\left(u'\right)\right\|^2 \leq 0 < \varpi_{j+1}^T \cdot \varphi\left(u'\right) - \frac{1}{2}\left\|\varphi\left(u'\right)\right\|^2. \tag{13.14}$$

The modified polygon given by Eq. (13.10) is then described by a new sequence of vertices $\tilde{\varpi}_i$, $i = 1, \ldots, j+2$, where

$$\tilde{\varpi}_i := \begin{cases} \varpi_i, & i = 1, \ldots, j, \\[2mm] \varpi_j - \dfrac{\varpi_j^T \cdot \varphi(u') - \frac{1}{2}\|\varphi(u')\|^2}{(\varpi_{j+1}-\varpi_j)^T \cdot \varphi(u')}\left(\varpi_{j+1}-\varpi_j\right), & i = j+1, \\[3mm] \varpi_k - \dfrac{\varpi_k^T \cdot \varphi(u') - \frac{1}{2}\|\varphi(u')\|^2}{(\varpi_1-\varpi_k)^T \cdot \varphi(u')}\left(\varpi_1-\varpi_k\right), & i = j+2. \end{cases} \tag{13.15}$$

By construction, it follows that $0 \in P$ throughout continuation.

The selection of the collection of points u' is designed to establish the association between the properties of a polygon P and the membership of the corresponding chart in the interior or boundary of the atlas. Specifically, if at least one of the polygonal vertices of a polygon P lies outside of the circle of radius $\tilde{R}$ following consolidation, then the corresponding chart is said to be a *boundary chart*. On the other hand, if all the polygonal vertices of the polygon P lie on or inside this circle, then the chart is said to be an *interior chart*. The choice of points u' during consolidation is thus guided by our desire to have

- a subset of the edges of each of the polygons associated with boundary charts approximate the atlas boundary and

- all interior charts cover regions of the solution manifold that are interior to the atlas boundary.

We accommodate these design objectives by selecting the points u' from the collection of base points of all charts that are considered sufficiently close to the chart represented by the polygon P. Each new polygonal edge generated through the subtraction of the corresponding half spaces may then be associated with a particular point u' and its corresponding chart.

13.3.2 Encoding

We proceed to explore the design choices implied by the previous discussion in a modified encoding of the `atlas_2d_min` subclass and its class methods. We begin by modifying the group of nonstatic private class methods in the class declaration, as follows.

```
methods (Access=private)
    flag = isneighbor(atlas, chart1, chart2)
    flag = isclose(atlas, chart1, chart2)
    [atlas cseg] = merge(atlas, cseg)
    [atlas chart1 checked] = merge_recursive(atlas, chart1, k, checked)
    chart = subtract_half_space(atlas, chart, test, phi, flag, NB)
end
```

No changes are made to the earlier encodings of the `init_prcond`, `flush`, `predict`, and `isneighbor` class methods. The following modifications to the `init_atlas` and `add_chart` class methods include the initialization of the v field of each `chart` structure, corresponding to the array of normalized distances v_i, $i = 1, \ldots, k$.

```
function [prob atlas cseg flush] = init_atlas(atlas, prob, cseg)

chart          = cseg.curr_chart;
chart.pt       = atlas.next_pt;
atlas.next_pt  = atlas.next_pt+1;
chart.R        = atlas.cont.h;
chart.s        = atlas.cont.s0;
chart.id       = chart.pt+1;
chart.bv       = atlas.cont.bv0;
chart.nb       = atlas.cont.nb0;
chart.v        = atlas.cont.v0;
chart.pt_type  = 'EP';
chart.ep_flag  = 1;
[prob cseg]    = cseg.add_chart(prob, chart);
flush          = true;

end
```

```
function [prob atlas cseg flush] = add_chart(atlas, prob, cseg)

chart          = cseg.curr_chart;
chart.pt       = atlas.next_pt;
atlas.next_pt  = atlas.next_pt+1;
chart.id       = chart.pt+1;
```

```
chart.s         = atlas.cont.s0;
chart.bv        = atlas.cont.bv0;
chart.nb        = atlas.cont.nb0;
chart.v         = atlas.cont.v0;
if chart.pt>=atlas.cont.PtMX
  chart.pt_type = 'EP';
  chart.ep_flag = 1;
end
[prob cseg] = cseg.add_chart(prob, chart);
flush       = true;

if ~atlas.isneighbor(cseg.ptlist{1}, cseg.ptlist{end})
  cseg.ptlist{end}.pt_type = 'GAP';
  cseg.ptlist{end}.ep_flag = 2;
  cseg.Status              = cseg.CurveSegmentCorrupted;
end

end
```

In this implementation, the `nb` field of each `chart` structure associates an integer with each polygonal edge, corresponding to the ordinal number of the nearby chart used in the construction of this edge. In particular, boundary edges are represented by a 0 in this field. The default assignments are encoded in the `get_settings` class method, shown below.

```
function cont = get_settings(cont)

defaults.h     = 0.1;
defaults.PtMX  = 50;
defaults.theta = 0.5;
defaults.almax = 10;
defaults.Rmarg = 0.95;
defaults.Ndirs = 6;
cont           = coco_merge(defaults, cont);
cont.almax     = cont.almax*pi/180;
cont.Ndirs     = max(3, ceil(cont.Ndirs));
al             = (0:cont.Ndirs-1)*(2*pi/cont.Ndirs);
cont.s0        = [cos(al); sin(al)];
cont.bv0       = 1:cont.Ndirs;
cont.nb0       = zeros(1,cont.Ndirs);
r1             = cont.h/cos(pi/cont.Ndirs);
cont.v0        = r1*ones(cont.Ndirs,1);

end
```

Here, the command

```
cont.Ndirs     = max(3, ceil(cont.Ndirs));
```

guarantees the initial construction of a nondegenerate regular polygon. Furthermore, the radius $\tilde{R}$ is here represented by the product of the content of the `cont.h` and `cont.Rmarg` fields, consistent with the construction of a pseudo-arclength predictor in the `predict` class method.

For a given chart represented by the polygon P, the proximity of charts that warrant the inclusion of the corresponding base points in the modification of P is now determined by the encoding of the `isclose` class method, shown below.

```
function flag = isclose(atlas, chart1, chart2)

al   = atlas.cont.almax;
```

```
R     = atlas.cont.h;
ta    = tan(al);
t2a   = tan(2*al);
x1    = chart1.x;
x2    = chart2.x;
dx    = x2-x1;
phi1 = chart1.TS'*dx;
phi2 = chart2.TS'*dx;
x1s   = chart1.TS*(phi1);
x2s   = chart2.TS*(phi2);
dst   = [norm(x1s), norm(x2s), norm(dx-x1s), norm(dx-x2s), ...
         subspace(chart1.TS, chart2.TS)];
n1mx  = ta*min(R,norm(x1s))+t2a*max(0,norm(x1s)-R);
n2mx  = ta*min(R,norm(x2s))+t2a*max(0,norm(x2s)-R);
dstmx = [2*R, 2*R, n1mx, n2mx, 2*al];
flag  = false;
if all(dst<dstmx);
  test1 = chart1.v.*(chart1.s'*phi1)-norm(phi1)^2/2;
  test2 = chart2.v.*(chart2.s'*phi2)+norm(phi2)^2/2;
  flag  = any(test1>0) && any(test2<0);
end

end
```

Here, we generalize the criteria encoded in the `isneighbor` class method to include charts that could have been obtained by two consecutive steps of continuation, each of which would satisfy the bounds on the various norms and angles given in Eqs. (12.22)–(12.26). The commands

```
test1 = chart1.v.*(chart1.s'*phi1)-norm(phi1)^2/2;
test2 = chart2.v.*(chart2.s'*phi2)+norm(phi2)^2/2;
flag  = any(test1>0) && any(test2<0);
```

guarantee that the condition encoded in the `isclose` class method is symmetric, i.e., that modifications to the polygon associated with one chart are accompanied by modifications to the polygon associated with the other chart.

The modified encodings of the `merge` and `merge_recursive` class methods, shown below, generalize the recursive traversal introduced in the previous section to the Henderson consolidation algorithm.

```
function [atlas cseg] = merge(atlas, cseg)

chart        = cseg.ptlist{end};
nbfunc       = @(x) atlas.isclose(chart, x);
bd_charts    = atlas.charts(atlas.boundary);
idx          = cellfun(nbfunc, bd_charts);
close_charts = atlas.boundary(idx);
checked      = [0, chart.id];
while ~isempty(close_charts)
  [atlas chart checked] = ...
    atlas.merge_recursive(chart, close_charts(1), checked);
  close_charts = setdiff(close_charts, checked);
end
atlas.charts    = [atlas.charts, {chart}];
atlas.boundary  = [atlas.boundary, chart.id];
bd_charts       = atlas.charts(atlas.boundary);
idx             = cellfun(@(x) ~isempty(x.bv), bd_charts);
atlas.boundary  = atlas.boundary(idx);
```

```
  if isempty(atlas.boundary)
    chart.pt_type  = 'EP';
    chart.ep_flag  = 1;
    cseg.ptlist{end} = chart;
  end

end


function [atlas chart1 checked] = ...
  merge_recursive(atlas, chart1, k, checked)

checked(end+1) = k;
chartk = atlas.charts{k};
if atlas.isclose(chart1, chartk)
  dx     = chartk.x-chart1.x;
  phi1   = chart1.TS'*dx;
  phik   = chartk.TS'*(-dx);
  test1  = chart1.v.*(chart1.s'*phi1)-norm(phi1)^2/2;
  testk  = chartk.v.*(chartk.s'*phik)-norm(phik)^2/2;
  flag1  = (test1>0);
  flagk  = (testk>0);
  chart1 = ...
    atlas.subtract_half_space(chart1, test1, phi1, flag1, chartk.id);
  chartk = ...
    atlas.subtract_half_space(chartk, testk, phik, flagk, chart1.id);
  atlas.charts{k} = chartk;
  check = setdiff(chartk.nb, checked);
  while ~isempty(check)
    [atlas chart1 checked] = ...
      atlas.merge_recursive(chart1, check(1), checked);
    check = setdiff(chartk.nb, checked);
  end
end

end
```

Here, the `subtract_half_space` class method, shown below, implements the construction in Eq. (13.15).

```
function chart = subtract_half_space(atlas, chart, test, phi, flag, NB)

k       = find(flag & ~circshift(flag, -1), 1);
flag    = circshift(flag, -k(1));
test    = circshift(test, -k(1));
chart.s = circshift(chart.s, [0, -k(1)]);
chart.v = circshift(chart.v, -k(1));
chart.nb = circshift(chart.nb, [0, -k(1)]);
j       = find(~flag & circshift(flag, -1), 1);
vx1     = chart.v(j)*chart.s(:,j);
vx2     = chart.v(j+1)*chart.s(:,j+1);
nvx1    = vx1-test(j)/((vx2-vx1)'*phi)*(vx2-vx1);
vx1     = chart.v(end)*chart.s(:,end);
vx2     = chart.v(1)*chart.s(:,1);
nvx2    = vx1-test(end)/((vx2-vx1)'*phi)*(vx2-vx1);
chart.s = [chart.s(:,1:j), nvx1/norm(nvx1), nvx2/norm(nvx2)];
chart.v = [chart.v(1:j); norm(nvx1); norm(nvx2)];
chart.nb = [chart.nb(1:j+1), NB];
chart.bv = find(chart.v>atlas.cont.Rmarg*chart.R);

end
```

The designation of a chart as a boundary or interior chart is ensured in this encoding by the assignment

```
chart.bv = find(chart.v>atlas.cont.Rmarg*chart.R);
```

Similarly, the assignment

```
chart.nb = [chart.nb(1:j+1), NB];
```

encodes a distinction between interior and exterior edges of each polygon that support the construction of a polygonal path approximation to the atlas boundary.

13.3.3 Functionality and flow test

We enable exploration of the atlas generated during execution by saving the content of `atlas.charts` and `atlas.boundary` to disk every time a chart is stored to disk. The final modified encoding of the `atlas_2d_min` class declaration, shown below, achieves this by appending a slot function corresponding to the `'save_bd'` signal to the continuation problem structure.

```
classdef atlas_2d_min < AtlasBase

  properties (Access=private)
    boundary = [];
    charts   = {};
    next_pt  = 0;
    cont     = struct();
  end

  methods (Access=private)
    function atlas = atlas_2d_min(prob, cont, dim)
      assert(dim==2, '%s: wrong manifold dimension', mfilename);
      atlas      = atlas@AtlasBase(prob);
      atlas.cont = atlas.get_settings(cont);
    end
  end

  methods (Static)
    function [prob cont atlas] = create(prob, cont, dim)
      atlas = atlas_2d_min(prob, cont, dim);
      prob  = CurveSegment.add_prcond(prob, dim);
      prob  = coco_add_slot(prob, 'atlas', @atlas.save_atlas, ...
        [], 'save_bd');
    end

    function [data res] = save_atlas(prob, data, varargin)
      res.charts   = prob.atlas.charts;
      res.boundary = prob.atlas.boundary;
    end
  end

  methods (Static, Access=private)
    cont = get_settings(cont)
  end

  methods (Access=public)
    [prob atlas cseg correct] = init_prcond(atlas, prob, chart)
    [prob atlas cseg flush]   = init_atlas (atlas, prob, cseg)
```

```
        [prob atlas cseg]          = flush      (atlas, prob, cseg)
        [prob atlas cseg correct] = predict    (atlas, prob, cseg)
        [prob atlas cseg flush]   = add_chart  (atlas, prob, cseg)
    end

    methods (Access=private)
        flag = isneighbor(atlas, chart1, chart2)
        flag = isclose(atlas, chart1, chart2)
        [atlas cseg] = merge(atlas, cseg)
        [atlas chart1 checked] = merge_recursive(atlas, chart1, k, checked)
        chart = subtract_half_space(atlas, chart, test, phi, flag, NB)
    end

end
```

The data saved by the `save_atlas` slot function can be accessed, after execution of the `coco` entry-point function, using the `coco_bd_read` utility. In this case, the command

```
atlas = coco_bd_read(runid, 'atlas');
```

extracts a structure with the fields `boundary` and `charts`, as generated during a continuation run with run identifier `runid`.

Example 13.2 We illustrate the functionality of the 2-dimensional atlas algorithm by generating a partial cover of the solution manifold for the zero function $\Phi : \mathbb{R}^3 \to \mathbb{R}$, given by

$$\Phi : u \mapsto u_1^4 + u_2^4 + u_3^4 - u_1^2 - u_2^2 - u_3^2 \tag{13.16}$$

and encoded in a COCO-compatible function with function handle `@pillow`. The extract below shows continuation along the solution manifold for some choice of parameterization of the atlas algorithm.

```
>> prob = coco_add_func(coco_prob(), 'pillow', @pillow, [], ...
     'zero', 'u0', [1; 0; 0]);
>> prob = coco_add_pars(prob, '', [1 2 3], {'x' 'y' 'z'});
>> prob = coco_set(prob, 'cont', 'PtMX', 1000, 'NPR', 200);
>> prob = coco_set(prob, 'cont', 'h', 0.15, 'almax', 30);
>> coco(prob, 'pillow', [], 2, {'x' 'y' 'z'});

...    NORMS              COMPUTATION TIMES
...     ||f||      ||U||    F(x)   DF(x)   SOLVE
... 0.00e+00  1.41e+00     0.0     0.0     0.0
... 0.00e+00  1.41e+00     0.0     0.0     0.0

... LABEL   TYPE              x               y               z
...      1   EP       1.0000e+00     0.0000e+00      0.0000e+00
...      2            1.1361e+00     7.8287e-01     -4.0676e-01
...      3            1.0061e+00     4.3805e-01     -1.0615e+00
...      4            4.0796e-01    -3.7590e-02     -1.0605e+00
...      5            3.9145e-01    -2.2503e-01      1.0743e+00
...      6   EP       2.7979e-01     5.6750e-01     -1.1114e+00
```

A graphical representation of the base points of the first 4 and 21 successfully located charts, respectively, as well as of the polygonal representations of each boundary, is shown in Fig. 13.3. The partial cover of the 2-dimensional surface computed during continuation is shown in Fig. 13.4, including a close-up of a portion of the surface to highlight the local geometry of the atlas. ∎

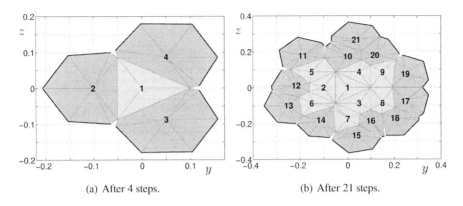

(a) After 4 steps.   (b) After 21 steps.

Figure 13.3. *Initial steps of covering the manifold of Example* 13.2 *with the 2-dimensional atlas algorithm from Sect.* 13.3. *After four steps we obtain an atlas with one interior and three boundary charts* (a), *where each chart has all other charts as neighbors. The piecewise-linear approximation of the atlas boundary is the union of the bold line segments. The algorithm continues to grow the atlas and expands the boundary outwards* (b).

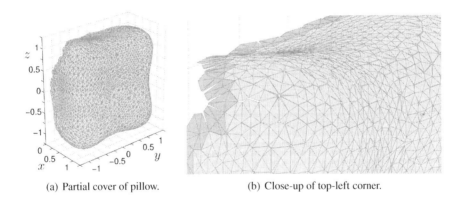

(a) Partial cover of pillow.   (b) Close-up of top-left corner.

Figure 13.4. *After executing* 1,000 *steps of the algorithm from Sect.* 13.3, *we obtain the partial cover of the manifold of Example* 13.2 *shown in panel* (a). *The local geometry of this partial cover is visualized in the close-up of the top-left corner, shown in panel* (b).

13.4 Conclusions

As with the 1-dimensional atlas algorithms described in the previous chapter, the development of 2-dimensional atlas algorithms in this chapter scratches the surface of design objectives and solutions. Obvious shortcomings include the absence of step-size adaptation, an unnecessarily expensive chart search during consolidation, and a brute-force approach to the choice of continuation direction during expansion. The sequence of successively more sophisticated encodings of the `atlas_2d_min` subclass, introduced in this chapter, should enable as well as inspire such further developments, including algorithms that take advantage of opportunities to parallelize steps of atlas construction.

The final 2-dimensional atlas algorithm implemented in Sect. 13.3 derives from the work of Michael Henderson and the reliance on a *Voronoi*-like representation of the atlas. This approach generalizes naturally to even higher dimensions, where polygons are replaced by multidimensional polyhedra and the philosophy of subtracting a half space carries over without modification. Atlas algorithms that support such multiparameter continuation offer a computational framework for design and multiobjective optimization.

In the absence of support for bounded computational domains, the examples used to illustrate the 2-dimensional atlas algorithm in this chapter terminated either after reaching the maximum number of successfully executed continuation steps or after achieving a complete cover of the solution manifold. In the next chapter, we introduce the minimal modifications to the 1-dimensional and 2-dimensional atlas algorithms developed thus far that are required to support locating charts on the boundary of the computational domain, as well as initiating continuation from a base chart on such a boundary.

Exercises

13.1. Explain the use of the `next_pt` property of the `atlas_2d_min` class in Sect. 13.1. Show that the encoding correctly terminates the finite-state machine when the number of successfully executed continuation steps equals the integer stored in the `cont.PtMX` field of the `atlas` instance. Should we have already introduced this property in the 1-dimensional atlas algorithm in Sect. 12.3?

13.2. Document and explain the differences in the encodings of the `merge` class methods in Sects. 12.3 and 13.1.

13.3. Consider the encoding of the `isneighbor` class method in Sect. 12.3. Provide a geometric interpretation of this encoding in the case of the 2-dimensional atlas algorithms considered in this chapter.

13.4. Suppose that u^* denotes a point on the solution manifold to some zero problem and that the columns of the matrix T^* constitute an orthonormal basis of the tangent space to the solution manifold at u^*. Let $\tilde{u} = u^*$ and let the tangent matrix V be obtained by applying an orthonormalization algorithm to the columns of the matrix

$$
\left(\begin{array}{c} \partial_u \Phi(u^* + \theta h T^* \cdot \sigma) \\ T^{*T} \end{array} \right)^{-1} \cdot \left(\begin{array}{c} 0 \\ I_{n-m} \end{array} \right)
$$

for some $\theta \in [0,1]$. Suppose that σ is a unit vector. Show that the direction vector

$$
s = \frac{V^T \cdot T^* \cdot \sigma}{\| V^T \cdot T^* \cdot \sigma \|}
$$

is compatible with continuation from u^* in the direction of the unit vector $T^* \cdot \sigma$.

13.5. Consider the base point $\tilde{u}$, tangent matrix V, and direction vector s obtained in the previous exercise. Suppose that h represents a nominal step size corresponding to the pseudo-arclength predictor $u^* + h T^* \cdot \sigma$. Explain the reason for the modified step size

$$h \| V^T \cdot T^* \cdot \sigma \|$$

and the corresponding linear predictor

$$u^* + h \| V^T \cdot T^* \cdot \sigma \| V \cdot s.$$

13.6. Show that the construction in the previous two exercises is independent of the dimensional deficit of the zero problem and that it reduces to the method implemented in Sect. 12.3 in the case of a 1-dimensional solution manifold.

13.7. Consider the encoding of the `merge` class method in Sect. 13.1. Identify the elements of the encoding that ensure that a direction available for continuation is removed from consideration if the projected distance from the corresponding pseudo-arc-length predictor to the base point of a neighboring chart along the corresponding tangent space is less than the scalar radius R. Apply the `atlas_2d_min` atlas algorithm in this section to the continuation of solutions to the zero problem

$$(u_1 - 1)^2 + u_2^2 + u_3^2 - 1 = 0$$

with the initial solution guess at $u_0 := (2,0,0)$. Explain your observations.

13.8. Consider, again, the atlas algorithm in Sect. 13.1. Provide a complete test suite for the algorithm flow and confirm, by example, the claim regarding redundant coverage made in Sect. 13.1.2.

13.9. What are the principal differences between the encodings of the `merge` class method in Sect. 13.1 and the first version in Sect. 13.2?

13.10. Demonstrate, by example, that the linear and recursive merger implementations of the consolidation phase in Sect. 13.2 are mathematically equivalent. Illustrate their equivalence by a graphical representation of the charts in the atlas prior to consolidation. Do the implementations differ in computational cost?

13.11. Suppose that the scalar radius R is constant across all charts and denote by $\mathcal{N}$ the set of all charts $\{u, T, \cdot, R\}$ that are considered "close" according to the criteria implemented in the `isclose` class method in Sect. 13.3. Show that the set

$$P = \{r \in \mathcal{T}, |r_i| < R + \varepsilon\} \setminus \cup_{\mathcal{N}} \{r \in \mathcal{T}, \|r\| > \|r - \varphi(u)\|\}$$

is a convex polygon in $\mathcal{T}$ consisting of all points that are closer to the origin than to the projection of the base point of any neighboring chart onto $\mathcal{T}$.

13.12. Show that, under the assumptions of the previous exercise, if all neighboring charts have base points distinct from u, then $P \ni 0$. Deduce that $P \cap \{r \in \mathcal{T}, \|r\| < R\} \neq \emptyset$.

13.13. Verify the equivalence

$$\|r\| \leq \|r - \varphi(u')\| \Leftrightarrow r^T \cdot \varphi(u') \leq \frac{1}{2} \|\varphi(u')\|^2.$$

13.14. Prove that the sequence of vertices given by Eq. (13.15) correctly implements the construction implied by Eq. (13.10).

13.15. What is the reason for requiring that the polygon P be exterior to the circle of radius R? Why not the circle of smaller radius $\check{R}$? Demonstrate your analysis with a computational example.

13.16. Verify that the encoding of the `subtract_half_space` class method implements correctly the construction in Eq. (13.15).

13.17. Propose modifications to the `atlas_2d_min` atlas algorithm that allow continuation to proceed even after two consecutive charts fail to qualify as neighbors. What is the reduction of your modified algorithm to the case of a 1-dimensional solution manifold? Apply your algorithm to an implicit parameterization of the immersion of the Klein bottle in $\mathbb{R}^3$ given by

$$\left(\|u\|^2 + 2u_2 - 1 \right) \left(\left(\|u\|^2 - 2u_2 - 1 \right)^2 - 8u_3^2 \right)$$
$$+ 16u_1u_3 \left(\|u\|^2 - 2u_2 - 1 \right) = 0.$$

13.18. Write plotting routines for representing the atlas generated by the atlas algorithm in Sect. 13.3 in terms of

1. a triangular mesh connecting the base points of all atlas charts,

2. a collection of polygons representing all atlas charts,

3. a collection of polygons representing all boundary charts, or

4. a piecewise polygonal path representing boundary edges of all boundary charts.

13.19. Given a matrix T whose columns span the tangent space at some point on the solution manifold of a continuation problem, a tangent matrix V such that $V^T \cdot T$ is invertible, and a unit coordinate vector $\sigma \in \Sigma$, consider continuation along a curve segment in a direction given by $T \cdot \varrho(\sigma)$, where

$$\varrho(\sigma) := \frac{\left(V^T \cdot T \right)^{-1} \cdot \left(T^T \cdot V \right)^{-1} \cdot \sigma}{\left\| \left(V^T \cdot T \right)^{-1} \cdot \left(T^T \cdot V \right)^{-1} \cdot \sigma \right\|}.$$

Make the corresponding modifications to the `predict` class method in Sect. 13.2 and use the modified atlas algorithm to compute a local cover of the solution manifold for the zero problem in Example 13.2. What, if any, differences do you observe in the distribution of base points along the solution manifold?

Chapter 14

Computational Domains

In this chapter, we consider modifications to the atlas algorithms developed in the previous two chapters in order to accommodate volume constraints on the computational domain. Specifically, we wish to perform continuation only if

- the initial chart on the solution manifold lies inside or on the boundary of a computational domain, defined by interval bounds on (a subset of) the continuation parameters, and

- at most, as long as there are charts with nonempty coordinate sets Σ.

Several examples of such interval bounds were already used liberally in illustrating the toolbox functionality in Part II of this text.

Only minimal modifications to the 1- and 2-dimensional atlas algorithms are necessary to support the desired behavior during continuation in a bounded computational domain. By default, and prior to entering the init_atlas state, the COCO core finite-state machine checks the first point located on the solution manifold to ensure that this point lies within the computational domain. When this is not the case, the ep_flag field of the chart in the curr_chart property of the current curve segment is assigned the integer

- 1 if the point lies on the domain boundary (within a given tolerance) or

- 2 if the point lies outside the domain.

In either case, the pt_type field is set equal to 'EX' and the cseg.Status property is set equal to the cseg.CurveSegmentCorrupted integer constant. The transition to the flush state and subsequent call to the AtlasBase.flush class method flushes the content of the cseg.curr_chart structure to storage and terminates the algorithm.

Similarly, default detection of the crossing of a domain boundary by a curve segment on the solution manifold is provided by the COCO core finite-state machine. The algorithm locates a point along the curve segment that lies on the boundary to within a given tolerance and ensures that the corresponding cseg.ptlist property contains a consecutive sequence of charts that terminate on the boundary point. The implementation assigns the integer 1 to the ep_flag field of the final chart in this sequence and sets the corresponding pt_type

367

and `Status` fields equal to `'EP'` and `cseg.CurveSegmentOK`, respectively. In this case, the `AtlasBase.flush` class method assigns the `cseg.BoundaryPoint` integer property to the `cseg.Status` field.

Example 14.1 As an illustration of the default functionality, consider the `atlas_1d_min` class implementation in Sect. 12.3 and the zero function

$$\Phi(u) := (u_1 - 1)^2 + u_2^2 - 1, u \in \mathbb{R}^2, \tag{14.1}$$

encoded in a COCO-compatible function with handle `@circle`. The extract below shows the result of attempting to perform continuation from a point outside the computational domain given by the interval bounds $u_1 \in [2,3]$.

```
>> prob = coco_add_func(coco_prob(), 'circle', @circle, [], ...
   'zero', 'u0', [1.5; 1]);
>> prob = coco_add_pars(prob, '', [1 2], {'x' 'y'});
>> coco(prob, 'run', [], 1, {'x' 'y'}, [2 3]);
```

STEP	DAMPING		NORMS			COMPUTATION	TIMES	
IT SIT	GAMMA	||d||	||f||	||U||	F(x)	DF(x)	SOLVE	
0			2.50e-01	2.55e+00		0.0	0.0	0.0	
1	1 1.00e+00	1.77e-01	1.56e-02	2.46e+00		0.0	0.0	0.0	
2	1 1.00e+00	1.26e-02	7.97e-05	2.45e+00		0.0	0.0	0.0	
3	1 1.00e+00	6.51e-05	2.12e-09	2.45e+00		0.0	0.0	0.0	
4	1 1.00e+00	1.73e-09	0.00e+00	2.45e+00		0.0	0.0	0.0	

```
warning: AtlasBase: initial point is outside computational domain
active boundary or terminal constraints were: EP
```

STEP	TIME	||U||	LABEL	TYPE	x	y
-1	00:00:00	2.4495e+00	1	EX	1.5000e+00	8.6603e-01

In contrast, as shown below, in the case that u_1 is restricted to the interval $[1,2]$, continuation proceeds until the boundary is reached.

```
>> coco(prob, 'run', [], 1, {'x' 'y'}, [1 2]);
```

STEP	DAMPING		NORMS			COMPUTATION	TIMES	
IT SIT	GAMMA	||d||	||f||	||U||	F(x)	DF(x)	SOLVE	
0			2.50e-01	2.55e+00		0.0	0.0	0.0	
1	1 1.00e+00	1.77e-01	1.56e-02	2.46e+00		0.0	0.0	0.0	
2	1 1.00e+00	1.26e-02	7.97e-05	2.45e+00		0.0	0.0	0.0	
3	1 1.00e+00	6.51e-05	2.12e-09	2.45e+00		0.0	0.0	0.0	
4	1 1.00e+00	1.73e-09	0.00e+00	2.45e+00		0.0	0.0	0.0	

STEP	TIME	||U||	LABEL	TYPE	x	y
0	00:00:00	2.4495e+00	1	EP	1.5000e+00	8.6603e-01
10	00:00:00	2.7787e+00	2		1.9303e+00	3.6692e-01
20	00:00:01	2.7975e+00	3		1.9566e+00	-2.9151e-01
30	00:00:01	2.5040e+00	4		1.5675e+00	-8.2336e-01
39	00:00:02	2.0000e+00	5	EP	1.0000e+00	-1.0000e+00

■

14.1   A 1-dimensional atlas algorithm

We now consider further modifications to the 1-dimensional expanding-boundary atlas algorithm developed in Sect. 12.3. Specifically, provided that the initial point on the solution manifold lies inside the computational domain, we seek to have the algorithm perform continuation in each direction along the solution manifold until

- the boundary of the computational domain is reached or

- the maximum number of successfully executed continuation steps is reached.

The desired objective is achieved by straightforward modifications to the `init_atlas` and `flush` class methods, with no changes to the class declaration or the encodings of the other class methods. These modifications ensure that continuation proceeds initially with $s = -1$, and then again from the initial chart in the opposite direction with $s = 1$, until, in both cases, either the boundary of the computational domain is reached or the maximum number of successfully executed continuation steps is reached. In the implementation below, we make use of the `AtlasBase.bddat_set` class method to impose an ordering of charts, by arclength along the solution manifold, in the output of the `coco` entry-point function.

Recall the following encoding of the `init_atlas` class method.

```
function [prob atlas cseg flush] = init_atlas(atlas, prob, cseg)

  chart            = cseg.curr_chart;
  chart.pt         = 0;
  chart.R          = atlas.cont.h;
  chart.s          = [1, -1]*sign(atlas.cont.PtMX);
  atlas.cont.PtMX  = abs(atlas.cont.PtMX);
  chart.pt_type    = 'EP';
  chart.ep_flag    = 1;
  [prob cseg]      = cseg.add_chart(prob, chart);
  flush            = true;

end
```

Here, the assignment to the `s` field of the `chart` structure gave preference to the direction implied by the sign of the `cont.PtMX` property of the `atlas` instance. The atlas boundary would thus continue to expand in this direction for as long as

- the maximum number of successfully executed continuation steps was not reached;

- each application of the corrector algorithm was convergent;

- consecutive charts satisfied the conditions in Eqs. (12.22)–(12.26); and

- the `atlas.boundary` cell array was nonempty.

In this section, we consider the following modified encoding of the `init_atlas` class method.

```
function [prob atlas cseg flush] = init_atlas(atlas, prob, cseg)

  chart            = cseg.curr_chart;
  chart.pt         = 0;
  chart.R          = atlas.cont.h;
  chart.s          = [-1, 1];
  atlas.cont.PtMX  = abs(atlas.cont.PtMX);
  chart.pt_type    = 'EP';
  chart.ep_flag    = 1;
```

```
[prob cseg]     = cseg.add_chart(prob, chart);
flush           = true;
prob = AtlasBase.bddat_set(prob, 'ins_mode', 'prepend');

end
```

As suggested above, the assignment to the s field of the `chart` structure ensures that continuation proceeds initially in the direction given by $s = -1$. Similarly, the `'prepend'` value of the `'ins_mode'` option in the call to the `AtlasBase.bddat_set` class method ensures that the successive points located on the solution manifold with $s = -1$ appear in reverse order in the output from the `coco` entry-point function.

The modified encoding of the `flush` class method, shown below, now provides the required handling of the case that continuation terminates on the boundary of the computational domain.

```
function [prob atlas cseg] = flush(atlas, prob, cseg)

if cseg.Status==cseg.CurveSegmentOK
  [atlas cseg] = atlas.merge(cseg);
end
[prob atlas cseg] = atlas.flush@AtlasBase(prob, cseg);
closed = isempty(atlas.boundary);
if cseg.Status==cseg.CurveSegmentOK
  if closed || (atlas.boundary{1,1}.pt>=atlas.cont.PtMX)
    cseg.Status = cseg.BoundaryPoint;
  end
elseif cseg.Status==cseg.BoundaryPoint && ~closed
  atlas.boundary{1,1}.pt=atlas.cont.PtMX;
  atlas.boundary = atlas.boundary([2:end 1],:);
  prob = AtlasBase.bddat_set(prob, 'ins_mode', 'append');
  if atlas.boundary{1,1}.pt<atlas.cont.PtMX
    cseg.Status = cseg.CurveSegmentOK;
  end
end

end
```

In particular, provided that the `atlas.boundary` property contains two elements and continuation previously proceeded with $s = -1$, continuation now proceeds with $s = 1$ and with successive points on the solution manifold being appended to the output from the `coco` entry-point function.

Example 14.2 The continued extract below shows continuation performed in two stages on the computational domain defined by $1 \leq u_1 \leq 2$.

```
>> prob = coco_add_func(coco_prob(), 'circle', @circle, [], ...
   'zero', 'u0', [1.5; 1]);
>> prob = coco_add_pars(prob, '', [1 2], {'x' 'y'});
>> coco(prob, 'run', [], 1, {'x' 'y'}, [1 2]);
```

STEP		DAMPING		NORMS		COMPUTATION TIMES		
IT	SIT	GAMMA	||d||	||f||	||U||	F(x)	DF(x)	SOLVE
0				2.50e-01	2.55e+00	0.0	0.0	0.0
1	1	1.00e+00	1.77e-01	1.56e-02	2.46e+00	0.0	0.0	0.0
2	1	1.00e+00	1.26e-02	7.97e 05	2.45e+00	0.0	0.0	0.0
3	1	1.00e+00	6.51e-05	2.12e-09	2.45e+00	0.0	0.0	0.0
4	1	1.00e+00	1.73e-09	0.00e+00	2.45e+00	0.0	0.0	0.0

```
STEP      TIME        ||U||  LABEL  TYPE             x            y
   0   00:00:00    2.4495e+00     1  EP    1.5000e+00   8.6603e-01
   8   00:00:01    2.0000e+00     2  EP    1.0000e+00   1.0000e+00
  10   00:00:02    2.7787e+00     3        1.9303e+00   3.6692e-01
  20   00:00:02    2.7975e+00     4        1.9566e+00  -2.9151e-01
  30   00:00:03    2.5040e+00     5        1.5675e+00  -8.2336e-01
  39   00:00:03    2.0000e+00     6  EP    1.0000e+00  -1.0000e+00
```

Although the screen output shows the labeled solution points in the order they are generated, the output from the `coco` entry-point function stored to disk contains the solution points ordered by arclength along the solution manifold. ∎

In order to allow for continuation from an initial point on the domain boundary, we override the default handling through the introduction of the `init_chart` and `init_admissible` class methods, as shown in the following modified declaration of the nonstatic public class methods of the `atlas_1d_min` subclass.

```
methods (Access=public)
    [prob atlas cseg correct] = init_prcond     (atlas, prob, chart)
    [prob atlas cseg]         = init_chart       (atlas, prob, cseg)
    [opts atlas cseg]         = init_admissible(atlas, opts, cseg, S)
    [prob atlas cseg flush]   = init_atlas       (atlas, prob, cseg)
    [prob atlas cseg]         = flush            (atlas, prob, cseg)
    [prob atlas cseg correct] = predict          (atlas, prob, cseg)
    [prob atlas cseg flush]   = add_chart        (atlas, prob, cseg)
end
```

Here, the last input argument of the `init_admissible` class method is a structure whose `ep_flag` field equals

- 0 if the base point of the `cseg.curr_chart` structure is inside the computational domain;

- 1 if the base point lies on the boundary (within a given tolerance); and

- 2 if the base point lies outside the computational domain.

In the case that the base point lies on the boundary, the `dir_flag` field contains a sequence of integers equal in length to the content of the s field of the `cseg.curr_chart` structure. In particular, an entry equal to the `AtlasBase.IsAdmissible` integer constant indicates continuation along a direction pointing into the computational domain, while any other value indicates continuation along a direction tangential to the domain boundary or away from the interior of the domain. The encoding of the `init_admissible` class method, shown below, relies on the default implementation in `AtlasBase`, unless the initial point lies on the boundary of the computational domain.

```
function [opts atlas cseg] = init_admissible(atlas, opts, cseg, S)

flags = (S.dir_flags==atlas.IsAdmissible);
if S.ep_flag==1 && any(flags)
  cseg.curr_chart.s = cseg.curr_chart.s(flags);
else
  [opts atlas cseg] = init_admissible@AtlasBase(atlas, opts, cseg, S);
end

end
```

In the latter case, only those entries of the `s` field corresponding to continuation along directions into the computational domain are retained. The modified encodings of the `init_chart` class method (invoked before `init_admissible`) and the `init_atlas` class method (invoked after `init_admissible`), shown below, now provide the appropriate initialization of chart content before and after the check for admissible continuation directions.

```
function [prob atlas cseg] = init_chart(atlas, prob, cseg)

chart           = cseg.curr_chart;
chart.pt        = 0;
chart.s         = [-1, 1];
atlas.cont.PtMX = abs(atlas.cont.PtMX);
[prob cseg]     = cseg.add_chart(prob, chart);
cseg.curr_chart = cseg.ptlist{1};

end


function [prob atlas cseg flush] = init_atlas(atlas, prob, cseg)

chart           = cseg.curr_chart;
chart.R         = atlas.cont.h;
chart.pt_type   = 'EP';
chart.ep_flag   = 1;
if ~isempty(chart.s) && chart.s(1)<0
  prob = AtlasBase.bddat_set(prob, 'ins_mode', 'prepend');
else
  prob = AtlasBase.bddat_set(prob, 'ins_mode', 'append');
end
cseg.ptlist{1} = chart;
flush          = true;

end
```

Finally, the following minimal modifications to the `merge` class method accommodate the possibility of fewer than two allowable continuation directions in the call to this method from the `flush` class method.

```
function [atlas cseg] = merge(atlas, cseg)

chart = cseg.ptlist{end};
R     = atlas.cont.h;
h     = atlas.cont.Rmarg*R;
nb    = cell(numel(chart.s),4);
for k=1:numel(chart.s)
  sk      = chart.s(k);
  xk      = chart.x+h*(chart.TS*sk);
  nb(k,:) = {chart, xk, sk, h};
end
for i=size(atlas.boundary,1):-1:1
  chart2 = atlas.boundary{i,1};
  if atlas.isneighbor(chart, chart2)
    x2 = atlas.boundary{i,2};
    if norm(chart.TS'*(x2-chart.x))<R
      atlas.boundary(i,:) = [];
    end
    for k=size(nb,1):-1:1
      x1 = nb{k,2};
```

```
            if norm(chart2.TS'*(x1-chart2.x))<R
              nb(k,:) = [];
            end
          end
        end
      end
      atlas.boundary = [nb; atlas.boundary];
      if isempty(atlas.boundary)
        chart.pt_type   = 'EP';
        chart.ep_flag   = 1;
        cseg.ptlist{end} = chart;
      end

    end
```

Example 14.3 We illustrate the handling of initial conditions on the boundary of the computational domain implemented in the init_admissible class method in the continued extract below. In the case of a computational domain given by $1.5 \leq u_1 \leq 2$, continuation requires only one stage of computation.

```
>> coco(prob, 'run', [], 1, {'x' 'y'}, [1.5 2]);
```

	STEP	DAMPING		NORMS		COMPUTATION TIMES		
IT	SIT	GAMMA	||d||	||f||	||U||	F(x)	DF(x)	SOLVE
0				2.50e-01	2.55e+00	0.0	0.0	0.0
1	1	1.00e+00	1.77e-01	1.56e-02	2.46e+00	0.0	0.0	0.0
2	1	1.00e+00	1.26e-02	7.97e-05	2.45e+00	0.0	0.0	0.0
3	1	1.00e+00	6.51e-05	2.12e-09	2.45e+00	0.0	0.0	0.0
4	1	1.00e+00	1.73e-09	0.00e+00	2.45e+00	0.0	0.0	0.0

STEP	TIME	||U||	LABEL	TYPE	x	y
0	00:00:00	2.4495e+00	1	EP	1.5000e+00	8.6603e-01
10	00:00:00	2.7787e+00	2		1.9303e+00	3.6692e-01
20	00:00:01	2.7975e+00	3		1.9566e+00	-2.9151e-01
30	00:00:01	2.5040e+00	4		1.5675e+00	-8.2336e-01
32	00:00:01	2.4495e+00	5	EP	1.5000e+00	-8.6603e-01

Finally, as seen below, continuation on the singular domain $1.5 \leq u_1 \leq 1.5$ results in termination after locating only a single chart on the solution manifold.

```
>> coco(prob, 'run', [], 1, {'x' 'y'}, [1.5 1.5]);
```

	STEP	DAMPING		NORMS		COMPUTATION TIMES		
IT	SIT	GAMMA	||d||	||f||	||U||	F(x)	DF(x)	SOLVE
0				2.50e-01	2.55e+00	0.0	0.0	0.0
1	1	1.00e+00	1.77e-01	1.56e-02	2.46e+00	0.0	0.0	0.0
2	1	1.00e+00	1.26e-02	7.97e-05	2.45e+00	0.0	0.0	0.0
3	1	1.00e+00	6.51e-05	2.12e-09	2.45e+00	0.0	0.0	0.0
4	1	1.00e+00	1.73e-09	0.00e+00	2.45e+00	0.0	0.0	0.0

```
warning: AtlasBase: initial point is not inside computational domain
active boundary or terminal constraints were: EP, EP
```

STEP	TIME	||U||	LABEL	TYPE	x	y
0	00:00:00	2.4495e+00	1		1.5000e+00	8.6603e-01

■

14.2   A 2-dimensional atlas algorithm

We proceed to modify the 2-dimensional atlas algorithm developed in Chap. 13 in order to limit continuation to a given computational domain while preventing premature termination. Following the treatment in the previous section, we first consider the case of an initial chart inside the computational domain and later address the special case of an initial chart on the domain boundary.

The default detection of boundary crossings along curve segments again provides the essential functionality for limiting the covering algorithm to those portions of the solution manifold that lie within a given computational domain. The minimal changes to the subtract_half_space class method, shown in the last two lines of the encoding below, ensure that charts on the domain boundary are designated as belonging to the atlas interior.

```
function chart = subtract_half_space(atlas, chart, test, phi, flag, NB)

k         = find(flag & ~circshift(flag, -1), 1);
flag      = circshift(flag, -k(1));
test      = circshift(test, -k(1));
chart.s   = circshift(chart.s, [0, -k(1)]);
chart.v   = circshift(chart.v, -k(1));
chart.nb  = circshift(chart.nb, [0, -k(1)]);
j         = find(~flag & circshift(flag, -1), 1);
vx1       = chart.v(j)*chart.s(:,j);
vx2       = chart.v(j+1)*chart.s(:,j+1);
nvx1      = vx1-test(j)/((vx2-vx1)'*phi)*(vx2-vx1);
vx1       = chart.v(end)*chart.s(:,end);
vx2       = chart.v(1)*chart.s(:,1);
nvx2      = vx1-test(end)/((vx2-vx1)'*phi)*(vx2-vx1);
chart.s   = [chart.s(:,1:j), nvx1/norm(nvx1), nvx2/norm(nvx2)];
chart.v   = [chart.v(1:j); norm(nvx1); norm(nvx2)];
chart.nb  = [chart.nb(1:j+1), NB];
ep_flag   = (chart.ep_flag && (chart.pt>0));
chart.bv  = find(~ep_flag & (chart.v>atlas.cont.Rmarg*chart.R));

end
```

The following modified encoding of the flush class method now encodes the regular termination of the finite-state machine only if all charts are designated as belonging to the atlas interior or the number of successfully executed continuation steps has reached atlas.cont.PtMX.

```
function [prob atlas cseg] = flush(atlas, prob, cseg)

if cseg.Status==cseg.CurveSegmentOK
  [atlas cseg] = atlas.merge(cseg);
end
[prob atlas cseg] = atlas.flush@AtlasBase(prob, cseg);
if any(cseg.Status==[cseg.CurveSegmentOK cseg.BoundaryPoint])
  if isempty(atlas.boundary) || (atlas.next_pt>atlas.cont.PtMX)
    cseg.Status = cseg.BoundaryPoint;
  else
    cseg.Status = cseg.CurveSegmentOK;
  end
end

end
```

Example 14.4 We illustrate the modified functionality of the `atlas_2d_min` subclass by introducing a nontrivial domain boundary, in terms of interval bounds on the value of a suitably defined monitor function. In particular, consider the zero problem

$$(u_1 - 1)^2 + u_2^2 - 1 = 0, u \in \mathbb{R}^3, \qquad (14.2)$$

and the associated encoding in a COCO-compatible function with handle `@cylinder`. Now let $\Psi : \mathbb{R}^3 \to \mathbb{R}$ be given by $\Psi(u) := \varphi$, where

$$\tan\varphi = \frac{u_3}{u_1 + 0.2}. \qquad (14.3)$$

As shown below, we obtain a bounded portion of the cylindrical solution manifold by restricting the values of Ψ to the interval $[-0.5, 0.5]$.

```
>> prob = coco_add_func(coco_prob(), 'cylinder', @cylinder, [], ...
     'zero', 'u0', [1;0;0]+sqrt([0.5;0.55;0]) );
>> prob = coco_add_func(prob, 'angle', @angle, [], 'active', 'p', ...
     'uidx', (1:3)');
>> prob = coco_add_pars(prob, '', [1, 2, 3], {'x' 'y' 'z'});
>> prob = coco_set(prob, 'cont', 'h', 0.4, 'almax', 20, 'PtMX', 250);
>> coco(prob, 'cylinder', [], 2, {'x' 'z' 'y' 'p'}, ...
     {[], [], [], [-0.5 0.5]});
```

...	NORMS			COMPUTATION TIMES		
...	||f||	||U||	F(x)	DF(x)	SOLVE	
...	5.00e-02	2.63e+00	0.0	0.0	0.0	
...	1.14e-03	2.61e+00	0.0	0.0	0.0	
...	6.44e-07	2.61e+00	0.0	0.0	0.0	
...	2.07e-13	2.61e+00	0.0	0.0	0.0	

...	LABEL	TYPE	x	z	y	p
...	1	EP	1.7071e+00	0.0000e+00	7.0711e-01	0.0000e+00
...	2		1.9268e+00	2.4146e-01	3.7565e-01	1.1305e-01
...	3		1.9688e+00	-4.6349e-01	2.4798e-01	-2.1055e-01
...	4		8.5334e-01	-2.1250e-01	9.8919e-01	-1.9906e-01
...	5		1.9889e+00	2.3724e-01	-1.4830e-01	1.0796e-01
.	.		.	.		
.	.		.	.		
.	.		.	.		
...	55	EP	5.3046e-01	-3.9905e-01	-8.8291e-01	-5.0000e-01
...	56	EP	5.7504e-01	4.2341e-01	-9.0521e-01	5.0000e-01
...	57	EP	7.1245e-01	4.9847e-01	-9.5776e-01	5.0000e-01
...	58	EP	2.4305e-01	-2.4204e-01	-6.5347e-01	-5.0000e-01
...	59	EP	3.0892e-01	2.7802e-01	-7.2278e-01	5.0000e-01

The corresponding manifold cover is illustrated in Fig. 14.1. ∎

As in the case of the 1-dimensional atlas algorithm in the previous section, we handle the case of an initial chart on the boundary of the computational domain by overriding the default `init_chart` and `init_admissible` class methods, and by modifying the `init_atlas` class method, accordingly. Specifically, in the encoding of the class method `init_admissible`, shown below, we remove references to inadmissible continuation directions from the `bv` field of the `cseg.curr_chart` structure. As before, these include direction vectors that correspond to continuation along curve segments with initial tangent vector parallel to the domain boundary or pointing away from the domain interior.

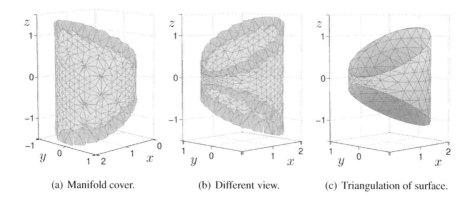

(a) Manifold cover.   (b) Different view.   (c) Triangulation of surface.

Figure 14.1. *Applying the modified 2-dimensional covering algorithm from Sect.* 14.2 *to the constrained cylinder in Example* 14.4 *results in the covering shown in different views in panels* (a) *and* (b). *A triangulation of the surface can be constructed using the neighbor information stored with each chart. A resulting triangulated surface is shown in panel* (c) *in the same view as in panel* (b).

```
function [opts atlas cseg] = init_admissible(atlas, opts, cseg, S)

flags = (S.dir_flags==atlas.IsAdmissible);
if S.ep_flag==1 && any(flags)
   chart            = cseg.curr_chart;
   chart.v(~flags)  = 0.5*(atlas.cont.Rmarg*chart.R);
   chart.bv(~flags) = [];
   cseg.curr_chart  = chart;
else
   [opts atlas cseg] = init_admissible@AtlasBase(atlas, opts, cseg, S);
end

end
```

We accommodate the flow through the finite-state machine by the following encoding of the init_chart class method (invoked before init_admissible) and the init_atlas class method (invoked after init_admissible).

```
function [prob atlas cseg] = init_chart(atlas, prob, cseg)

chart           = cseg.curr_chart;
chart.pt        = atlas.next_pt;
atlas.next_pt   = atlas.next_pt+1;
chart.R         = atlas.cont.h;
chart.s         = atlas.cont.s0;
chart.id        = chart.pt+1;
chart.bv        = atlas.cont.bv0;
chart.nb        = atlas.cont.nb0;
chart.v         = atlas.cont.v0;
[prob cseg]     = cseg.add_chart(prob, chart);
cseg.curr_chart = cseg.ptlist{1};

end
```

```
function [prob atlas cseg flush] = init_atlas(atlas, prob, cseg)

chart          = cseg.curr_chart;
chart.pt_type  = 'EP';
chart.ep_flag  = 1;
cseg.ptlist{1} = chart;
flush          = true;

end
```

The modified class declaration, shown below, is now immediate.

```
classdef atlas_2d_min < AtlasBase

  properties (Access=private)
    boundary = [];
    charts   = {};
    next_pt  = 0;
    cont     = struct();
  end

  methods (Access=private)
    function atlas = atlas_2d_min(prob, cont, dim)
      assert(dim==2, '%s: wrong manifold dimension', mfilename);
      atlas      = atlas@AtlasBase(prob);
      atlas.cont = atlas.get_settings(cont);
    end
  end

  methods (Static)
    function [prob cont atlas] = create(prob, cont, dim)
      atlas = atlas_2d_min(prob, cont, dim);
      prob  = CurveSegment.add_prcond(prob, dim);
      prob  = coco_add_slot(prob, 'atlas', @atlas.save_atlas, ...
        [], 'save_bd');
    end

    function [data res] = save_atlas(prob, data, varargin)
      res.charts   = prob.atlas.charts;
      res.boundary = prob.atlas.boundary;
    end

  end

  methods (Static, Access=private)
    cont = get_settings(cont)
  end

  methods (Access=public)
    [prob atlas cseg correct] = init_prcond      (atlas, prob, chart)
    [prob atlas cseg]         = init_chart        (atlas, prob, cseg)
    [opts atlas cseg]         = init_admissible(atlas, opts, cseg, S)
    [prob atlas cseg flush]   = init_atlas       (atlas, prob, cseg)
    [prob atlas cseg]         = flush            (atlas, prob, cseg)
    [prob atlas cseg correct] = predict          (atlas, prob, cseg)
    [prob atlas cseg flush]   = add_chart        (atlas, prob, cseg)
  end

  methods (Access=private)
    flag = isneighbor(atlas, chart1, chart2)
    flag = isclose(atlas, chart1, chart2)
    [atlas cseg] = merge(atlas, cseg)
```

```
    [atlas chart1 checked] = merge_recursive(atlas, chart1, k, checked)
    chart = subtract_half_space(atlas, chart, test, phi, flag, NB)
  end

end
```

Example 14.5 We continue the analysis of the zero problem in Example 14.4. As shown in the extract below, the default parameterization allows for an initial solution point on a single boundary, or at an intersection of several smooth domain boundaries, provided that at least one element of the s field of the initial chart implies continuation into the interior of the computational domain.

```
>> prob = coco_add_func(coco_prob(), 'cylinder', @cylinder, [], ...
   'zero', 'u0', [1; -1; 0]);
>> prob = coco_add_pars(prob, '', [1 2 3], {'x' 'y' 'z'});
>> prob = coco_set(prob, 'cont', 'Ndirs', 10, 'PtMX', 25);
>> coco(prob, 'cylinder1', [], 2, {'x' 'y' 'z'}, {[] [] [0 1]});
```

...	NORMS		COMPUTATION TIMES		
...	||f||	||U||	F(x)	DF(x)	SOLVE
...	0.00e+00	2.00e+00	0.0	0.0	0.0
...	0.00e+00	2.00e+00	0.0	0.0	0.0

...	LABEL	TYPE	x	y	z
...	1	EP	1.0000e+00	-1.0000e+00	0.0000e+00
...	2	EP	1.0598e+00	-9.9821e-01	0.0000e+00
...	3	EP	1.1222e+00	-9.9250e-01	0.0000e+00
...	4		1.1875e+00	-9.8226e-01	3.9485e-02
...	5	EP	8.9155e-01	-9.9410e-01	0.0000e+00
...	6	EP	1.1861e+00	-9.8253e-01	0.0000e+00
...	7		1.2144e+00	-9.7675e-01	1.0081e-01
...	8	EP	8.9444e-01	-9.9441e-01	2.1055e-01

```
>> coco(prob, 'cylinder2', [], 2, {'x' 'y' 'z'}, {[0 1] [] [0 1]});
```

...	NORMS		COMPUTATION TIMES		
...	||f||	||U||	F(x)	DF(x)	SOLVE
...	0.00e+00	2.00e+00	0.0	0.0	0.0
...	0.00e+00	2.00e+00	0.0	0.0	0.0

...	LABEL	TYPE	x	y	z
...	1	EP	1.0000e+00	-1.0000e+00	0.0000e+00
...	2	EP	1.0000e+00	-1.0000e+00	6.3887e-02
...	3	EP	1.0000e+00	-1.0000e+00	1.2911e-01
...	4	EP	8.9155e-01	-9.9410e-01	0.0000e+00
...	5	EP	1.0000e+00	-1.0000e+00	1.9940e-01
...	6		7.5953e-01	-9.7066e-01	1.3274e-01
...	7	EP	8.2104e-01	-9.8386e-01	0.0000e+00
...	8	EP	1.0000e+00	-1.0000e+00	2.6358e-01
...	9	EP	1.0000e+00	-1.0000e+00	3.2143e-01

The atlases obtained after the first 25 continuation steps are shown in Fig. 14.2.   ∎

14.3   Manifolds of resonant periodic orbits

We conclude this chapter with an application of the 2-dimensional atlas algorithm from the previous section to the continuation of a 2-dimensional family of resonant periodic orbits

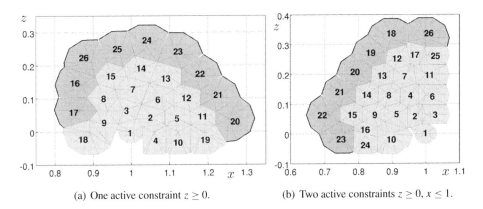

(a) One active constraint $z \geq 0$.   (b) Two active constraints $z \geq 0$, $x \leq 1$.

Figure 14.2. *The 2-dimensional atlas algorithm discussed at the end of Sect.* 14.2 *supports starting on the boundary of the computational domain. This is illustrated with the manifold of Example* 14.5, *where we start on a boundary defined by one active constraint in panel* (a) *and at a corner defined by two active constraints in panel* (b). *In both cases, the initial solution guess is given by* $(x, y, z) = (1, -1, 0)$ *and we show the atlas obtained after* 25 *continuation steps.*

in the Langford dynamical system, given by the vector field

$$f(y, p) := \begin{pmatrix} (y_3 - 0.7) y_1 - \omega y_2 \\ \omega y_1 + (y_3 - 0.7) y_2 \\ 0.6 + y_3 - \frac{1}{3} y_3^3 - \left(y_1^2 + y_2^2\right)(1 + \rho y_3) + \varepsilon y_3 y_1^3 \end{pmatrix}, \qquad (14.4)$$

where $p = (\omega, \rho, \varepsilon)$, introduced previously in Sect. 9.2. As shown there, when $\varepsilon = 0$, this system is equivalent to the system of differential equations

$$\frac{d\theta}{dt} = 0, \qquad (14.5)$$

$$\frac{dr}{dt} = r(z - 0.7), \qquad (14.6)$$

$$\frac{dz}{dt} = 0.6 + z - \frac{1}{3} z^3 - r^2(1 + \rho z), \qquad (14.7)$$

where

$$y = e^{\omega t G} \cdot \begin{pmatrix} r \cos\theta \\ r \sin\theta \\ z \end{pmatrix} \qquad (14.8)$$

and

$$G := \begin{pmatrix} 0 & -1 & 0 \\ 1 & 0 & 0 \\ 0 & 0 & 0 \end{pmatrix}. \qquad (14.9)$$

In this case, a periodic orbit of the transformed system with period T_{po} corresponds to an invariant torus of the equivariant Langford system with parallel flow, consisting of either

- torus-covering quasi-periodic trajectories, in the case that the rotation number

$$\varrho := \frac{2\pi}{\omega T_{po}} \tag{14.10}$$

is irrational, or

- a continuous family of periodic orbits, in the case that ϱ is rational.

The analysis in Sect. 9.2 applied the `'msbvp'` toolbox for continuation of quasi-periodic invariant tori along 1-dimensional arcs in parameter space. In this section, we instead consider the continuation of 2-dimensional families of phase-locked, resonant periodic orbits corresponding to rational rotation numbers.

Consider again the `'coll'`-compatible encodings of the vector field f and its Jacobians $\partial_y f$ and $\partial_p f$ in Sect. 9.2. An approximate periodic orbit for $\omega = 3.5$, $\rho = 0.35$, and $\varepsilon = 0$ is then obtained by a single stage of forward integration, as shown below.

```
>> p0 = [3.5;0.35;0];
>> [t x0] = ode45(@(t,x) lang(x,p0), [0 5.3], [0.3;0;0.4]);
```

We construct a corresponding restricted continuation problem by invoking the toolbox constructor `bvp_isol2seg`, as shown in the following sequence of commands.

```
>> prob = coco_prob();
>> prob = bvp_isol2seg(prob, '', @lang, @lang_DFDX, @lang_DFDP, ...
      t, x0, {'om' 'ro' 'eps'}, p0, @po_bc, @po_bc_DFDX);
```

Here, the `po_bc` function and its Jacobian encode periodicity and a vanishing value for y_2 at $t = 0$. Continuation under simultaneous variations in ρ and ε, where the latter is constrained to the interval $[-0.5, 0.5]$, then yields an embedded 2-dimensional solution manifold. Several projections of this *resonance surface* are shown in Figs. 14.3 and 14.4.

14.4   Conclusions

The default boundary detection relied on in the refined versions of the 1- and 2-dimensional atlas algorithms described in this chapter is a special case of a general event detection and event handling functionality, implemented in the COCO core. The core offers broad support for the introduction of monitor functions with the intent of detecting and continuing special points on the solution manifold associated with critical values of these functions. As event handling is a central feature of any general-purpose tool for numerical continuation, we devote Part IV of this text in its entirety to COCO-compatible toolboxes and atlas algorithms that make nontrivial use of the built-in event handling functionality.

As we close Part III, we reflect briefly on the mathematical assumptions that underpin much of the treatment thus far and, in particular, the theoretical treatment in Chap. 11. Specifically, it is noted that the discussion depended greatly on assumptions of smoothness and application of the implicit function theorem. This theorem enabled the construction of smooth 1-dimensional curve segments on solution manifolds of arbitrary (finite) dimension and the use of linear projection conditions, including the commonly deployed

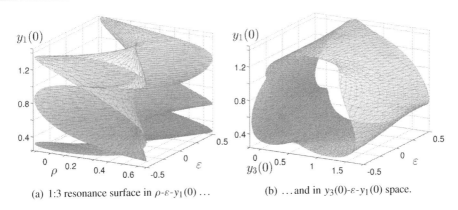

(a) 1:3 resonance surface in ρ-ε-$y_1(0)$ ...   (b) ...and in $y_3(0)$-ε-$y_1(0)$ space.

Figure 14.3. *A so-called* resonance surface *obtained by applying the 2-dimensional covering algorithm from Sect. 14.2 to a boundary-value problem for resonant periodic orbits of the Langford dynamical system, given by the vector field in Eq. (14.4). A projection of this surface onto the ρ-ε-$y_1(0)$ space is shown in panel (a). The striking cyclic triple-S shape is generic for 1:3 resonance surfaces. Similar shapes are observed for all m:n resonance surfaces. The apparent self-intersection is due to projection; the surface is locally isomorphic to a cylinder (b). A more familiar projection, a so-called* Arnol'd tongue, *is shown in Fig. 14.4(a).*

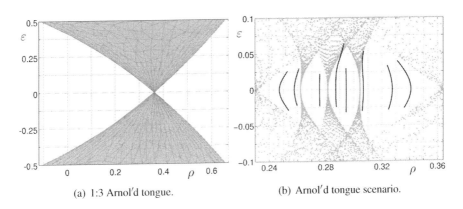

(a) 1:3 Arnol'd tongue.   (b) Arnol'd tongue scenario.

Figure 14.4. *Projecting the resonance surface shown in Fig. 14.3 onto the ρ-ε plane results in an Arnol'd tongue (a). Combining the computation of resonance surfaces with the computation of quasi-periodic arcs, as shown in Sect. 9.2, enables the investigation of the so-called* Arnol'd tongue scenario. *This consists of a countable collection of Arnol'd tongues and a complementary Cantor-like set of quasi-periodic arcs. Panel (b) shows, from left to right, the 1:4, 3:11, 2:7, 3:10 and 1:3 tongues together with quasi-periodic arcs for rotation numbers obtained by a two-level recursive golden-mean subdivision of the interval [1/4, 1/3]. The tongues are represented by a dot for each of the 1,500 resonant orbits computed on each resonance surface. We observe that one of the quasi-periodic arcs enters the 2:7 tongue. This violates theory and is an artifact caused by the absence of control for discretization errors, a task undertaken in Part V.*

pseudo-arclength formalism. Smoothness also formed an essential ingredient in the abstract representation of a curve segment in the `CurveSegment` class, but not in the design of the core finite-state machine or the `AtlasBase` interface. It is not difficult to propose situations where smoothness must be abandoned in favor of less restrictive assumptions that nevertheless support the existence of a solution set of some regularity. The COCO core implementation supports the application of continuation techniques to such contexts, albeit generally not without requiring the implementation of a substitution for the core `CurveSegment` class.

Exercises

14.1. Explain the output from the second call to the `coco` entry-point function in Example 14.1. What elements of the `atlas_1d_min` implementation ensure that the finite-state machine terminates when the boundary is reached?

14.2. What happens if one omits the calls to the `AtlasBase.bddat_set` class method in the `init_atlas` and `flush` class methods in Sect. 14.1?

14.3. Include the assignment

```
atlas.PrintHeadLine = true;
```

after the assignment

```
cseg.Status = cseg.CurveSegmentOK;
```

in the modified `flush` class method in Sect. 14.1. Repeat the analysis in Example 14.2 and comment on the differences in screen output.

14.4. Provide a complete test suite of the conditional flow in the `init_admissible` class method in Sect. 14.1.

14.5. Consider the modification to the normed distances stored in the `chart.v` array by the `init_admissible` class method in Sect. 14.2. What are the consequences to the convexity of the polygon P associated with a boundary chart?

14.6. Consider the modified `atlas_2d_min` class in Sect. 14.2. Is this guaranteed to cover the entire solution manifold within a bounded domain provided that `cont.PtMX` is large enough? Is it possible to have the algorithm terminate even as there remain uncovered patches of the solution manifold adjacent to the domain boundaries?

14.7. Comment on the choice to treat charts on the boundary of the computational domain as belonging to the atlas interior. Can this lead to redundant coverage near the domain boundaries?

14.8. Substitute the `po_isol2orb` constructor for the `bvp_isol2seg` constructor in Sect. 14.3 and repeat the analysis. Comment on your observations of the differences in performance of the atlas algorithm.

14.9. Provide two examples of continuation along 2-dimensional solution manifolds in bounded computational domains defined by curved domain boundaries.

Part IV

Event Handling

Chapter 15

Special Points and Events

In Part III of this text, we considered several implementations of minimal atlas algorithms for covering solution manifolds of restricted continuation problems. In each case, the construction of an atlas relied on a finite-state machine that proceeded to cover the manifold by a repeated process of expansion and consolidation. During expansion, the algorithm would construct a new curve segment from an existing atlas chart. In the subsequent phase of consolidation, this curve segment would be merged into the atlas. In this and the next several chapters, we illustrate the further benefits of this approach in the context of the parallel task to continuation of *event handling*.

Detecting and locating *special points* lies at the heart of applications of continuation. We already saw an illustration of the COCO core *boundary point* detection functionality in Chap. 14. In the context of dynamical systems, special points are most commonly *bifurcation points*, for example, where a solution changes its stability type under variations in system parameters. As we shall see in Part V of this text, special points may also be defined to trigger on *continuation-related events*, e.g., associated with bounds on estimated discretization errors. A well-designed continuation code must contain the essential functionality to locate such special points, to map out branches of special points, and to switch to emerging branches of solutions from special points, when they exist.

15.1 Theoretical framework

Consider again the general case of a zero problem $\Phi = 0$ for $\Phi : \mathbb{R}^n \to \mathbb{R}^m$ for $n > m \geq 1$. Suppose that there exists a curve segment Γ parameterized by h on some interval such that the point $\Gamma_h \in \Gamma$ satisfies the closed continuation problem

$$\begin{pmatrix} \Phi(u) \\ V^T \cdot (u - \tilde{u}) - hs \end{pmatrix} = 0 \qquad (15.1)$$

for some base point $\tilde{u}$, direction vector s, and tangent matrix $V \in \mathbb{R}^{n \times (n-m)}$, where $V^T \cdot V = I_{n-m}$. By construction, Γ coincides with the intersection of the solution manifold with the $(m+1)$-dimensional affine subspace $\tilde{u} + \text{span}\{V \cdot s\} \oplus \text{span}\{V^\perp\}$, where $V^\perp$ is a full-rank matrix such that $V^T \cdot V^\perp = 0$.

15.1.1 Detection and location

Given a continuous function ψ on Γ and some scalar ψ_0, a point Γ_0 on Γ is called a *special point* if $\psi(\Gamma_0) = \psi_0$. A special point associated with a continuous function ψ and a scalar ψ_0 is said to be *detectable* along a curve segment Γ provided that one may find two points $\Gamma_<$ and $\Gamma_>$ on Γ, corresponding to $h = h_<$ and $h = h_>$, respectively, such that

$$\psi(\Gamma_<) - \psi_0 < 0 < \psi(\Gamma_>) - \psi_0. \tag{15.2}$$

The existence of at least one special point along Γ between $\Gamma_<$ and $\Gamma_>$, and for some $h = h_0$, then follows from the intermediate value theorem. For smooth parameterizations $h \mapsto (\psi \circ \Gamma)(h)$, a special point Γ_0 is detectable if

$$\frac{d(\psi \circ \Gamma)}{dh}(h_0) \neq 0. \tag{15.3}$$

We propose three methods for locating Γ_0 or, in the case of the third method, an approximation of Γ_0. As described below, the applicability of each method depends on properties associated with the smoothness of the function ψ and the regularity of the special point on the solution manifold.

Method 1: Smoothness and regularity

Suppose that the matrix

$$\begin{pmatrix} \partial_u \Phi(\Gamma_0) \\ V^T \end{pmatrix} \tag{15.4}$$

is invertible and that the columns of T_0 constitute a basis for the tangent space to the solution manifold at the point Γ_0. As shown in Chap. 11, the vector

$$T_0 \cdot \left(V^T \cdot T_0\right)^{-1} \cdot s \tag{15.5}$$

is then tangent to Γ at the point Γ_0.

Now suppose that ψ is a smooth function of u on a neighborhood of Γ_0 and that

$$\partial_u \psi(\Gamma_0) \cdot T_0 \cdot \left(V^T \cdot T_0\right)^{-1} \cdot s \neq 0, \tag{15.6}$$

i.e., that the rate of change of ψ along the curve segment Γ at Γ_0 is nonzero. It follows that the special point Γ_0 is detectable along Γ. It is now straightforward to show that the point (Γ_0, h_0) is a regular point of the closed continuation problem

$$\begin{pmatrix} \Phi(u) \\ V^T \cdot (u - \tilde{u}) - hs \\ \psi(u) - \psi_0 \end{pmatrix} = 0. \tag{15.7}$$

Without loss of generality, we may suppose that $s_1 \neq 0$. It follows that h may be obtained in terms of u from the expression

$$h = \frac{1}{s_1} \left[V^T \cdot (u - \tilde{u})\right]_1, \tag{15.8}$$

thus resulting in the reduced set of equations

$$
\begin{pmatrix}
\Phi(u) \\
s_1\left[V^T \cdot (u - \tilde{u})\right]_2 - s_2\left[V^T \cdot (u - \tilde{u})\right]_1 \\
\vdots \\
s_1\left[V^T \cdot (u - \tilde{u})\right]_{n-m} - s_{n-m}\left[V^T \cdot (u - \tilde{u})\right]_1 \\
\psi(u) - \psi_0
\end{pmatrix} = 0,
\tag{15.9}
$$

for which Γ_0 is a regular point. The point Γ_0 may thus be located on Γ by applying a Jacobian-based corrector algorithm to this reduced continuation problem, with an initial solution guess chosen somewhere along a suitable interpolant through $\Gamma_<$ and $\Gamma_>$.

Method 2: Regularity

Suppose again that the matrix

$$
\begin{pmatrix}
\partial_u \Phi(\Gamma_0) \\
V^T
\end{pmatrix}
\tag{15.10}
$$

is invertible, but that ψ is not a smooth function of u on a neighborhood of Γ_0. Given two points $\Gamma_<$ and $\Gamma_>$ on either side of Γ_0 along the curve segment Γ, a subdivision algorithm in h may then be applied to the closed continuation problem

$$
\begin{pmatrix}
\Phi(u) \\
V^T \cdot (u - \tilde{u}) - hs
\end{pmatrix} = 0
\tag{15.11}
$$

in order to locate the special point Γ_0. Specifically, we may generate a nested sequence of intervals $(h_{<,i}, h_{>,i})$, each corresponding to a pair of points $\{\Gamma_{<,i}, \Gamma_{>,i}\}$ on Γ, such that

$$
\psi(\Gamma_{<,i}) - \psi_0 < 0 < \psi(\Gamma_{>,i}) - \psi_0
\tag{15.12}
$$

for all i and

$$
\lim_{i \to \infty} \psi(\Gamma_{<,i}) = \lim_{i \to \infty} \psi(\Gamma_{>,i}) = \psi_0.
\tag{15.13}
$$

Naturally, the subdivision algorithm may also be applied in the case that ψ is continuously differentiable. This may be advantageous if the computation of the Jacobian of ψ is prohibitively expensive.

Method 3: Singularity

Finally, suppose that the point Γ_0 on the zero set of $\psi - \psi_0$ coincides with an isolated singular point of the closed continuation problem

$$
\begin{pmatrix}
\Phi(u) \\
V^T \cdot (u - \tilde{u}) - hs
\end{pmatrix} = 0,
\tag{15.14}
$$

i.e., such that the matrix

$$
\begin{pmatrix}
\partial_u \Phi(\Gamma_0) \\
V^T
\end{pmatrix}
\tag{15.15}
$$

is noninvertible. In this case, let $h \mapsto p(h) \in \tilde{u} + \text{span}\{V \cdot s\} \oplus \text{span}\{V^\perp\}$ denote an interpolating function such that

$$p(h_<) = \Gamma_< \tag{15.16}$$

and

$$p(h_>) = \Gamma_>. \tag{15.17}$$

Provided that ψ can be defined along the image of p, a subdivision algorithm in h may now be applied to the equation

$$\psi(p(h)) - \psi_0 = 0. \tag{15.18}$$

This yields, at best, a close approximant to the singular point Γ_0 provided that p is a close approximant to the curve segment Γ. Naturally, this subdivision algorithm may also be applied in the case that Γ_0 corresponds to a regular point on the solution manifold. This may be advantageous if one seeks only an approximate location of the special point for further processing, or if evaluating the Jacobian of the extended continuation problem and/or solving the closed continuation problem is very expensive. Since a subdivision algorithm often involves a large number of steps, locating special points using the algorithm for regular solution points may incur prohibitive computational costs.

15.1.2   Examples

Let $u = (x, \kappa, \lambda)$ and consider the zero problem defined by the zero function

$$\Phi : u \mapsto \kappa - x\left(\lambda - x^2\right), \tag{15.19}$$

whose solutions constitute the set of equilibrium points of the vector field corresponding to the so-called *cusp normal form*. In the absence of monitor functions, all solution points of the restricted continuation problem (which here equals the zero problem) are regular, since

$$\partial_u \Phi = \left(\begin{array}{ccc} 3x^2 - \lambda & 1 & -x \end{array} \right) \tag{15.20}$$

is everywhere rank 1. Indeed, at each point of the corresponding 2-dimensional solution manifold, shown in Figs. 15.1–15.3, the tangent space is spanned by the columns of the matrix

$$T = \left(\begin{array}{cc} 0 & 1 \\ x & \lambda - 3x^2 \\ 1 & 0 \end{array} \right). \tag{15.21}$$

It follows that for every point u on the solution manifold, there exists a matrix $V \in \mathbb{R}^{3 \times 2}$ such that $V^T \cdot V = I_2$ and such that the matrix $V^T \cdot T$ is invertible.

   In the discussion below, we highlight special points associated with geometric features of the solution manifold and describe methods for their detection and location.

Vertical tangencies

The tangent space to the solution manifold contains a vector that is perpendicular to the (κ, λ) plane if and only if

$$\det\left(U^T \cdot T\right) = 0 \tag{15.22}$$

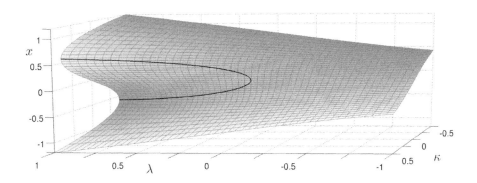

Figure 15.1. *Solution manifold of the zero problem* $\Phi(u) = 0$, *where* Φ *is given in Eq.* (15.19). *The solid curve marks the locus of vertical tangencies, implicitly defined by Eq.* (15.22) *and given by special points associated with the function* ψ *in Eq.* (15.30) *and the value* 0.

for every matrix $U \in \mathbb{R}^{3 \times 2}$ whose columns span the (κ, λ) parameter plane. Clearly, the columns of the matrix

$$U = \begin{pmatrix} 0 & 0 \\ a & b \\ c & d \end{pmatrix} \tag{15.23}$$

span the (κ, λ) plane provided that $ad - bc \neq 0$. In this case,

$$\det\left(U^T \cdot T\right) = (ad - bc)\left(3x^2 - \lambda\right). \tag{15.24}$$

Special points of *vertical tangency* thus occur where $3x^2 - \lambda = 0$, i.e., for $(x, \kappa, \lambda) = \left(x, 2x^3, 3x^2\right)$. Their locus is illustrated by the solid curve in Fig. 15.1.

Let

$$u^* := \begin{pmatrix} x^* \\ 2x^{*3} \\ 3x^{*2} \end{pmatrix} \tag{15.25}$$

represent such a point of vertical tangency and set $T^* := T|_{u=u^*}$. Consider the curve segment Γ through u^* defined by the projection condition

$$V^T \cdot (u - \tilde{u}) - hs = 0, \tag{15.26}$$

with base point $\tilde{u} = u^*$, tangent matrix

$$V = \begin{pmatrix} 0 & 1 \\ \cos \eta & 0 \\ \sin \eta & 0 \end{pmatrix} \tag{15.27}$$

for some $\eta \in [0, 2\pi)$ such that $x^* \cos \eta + \sin \eta \neq 0$, and direction vector

$$s = \begin{pmatrix} 0 \\ 1 \end{pmatrix}. \tag{15.28}$$

It follows by direct computation that the matrix

$$\begin{pmatrix} \partial_u \Phi(u^*) \\ V^T \end{pmatrix} \tag{15.29}$$

is invertible and, equivalently, that $\det\left(V^T \cdot T^*\right) \neq 0$. Moreover, with

$$\psi(u) = 3x^2 - \lambda, \tag{15.30}$$

it follows that

$$\partial_u \psi\left(u^*\right) \cdot T^* \cdot \left(V^T \cdot T^*\right)^{-1} \cdot s \neq 0 \tag{15.31}$$

provided that $x^* \neq 0$. In this case, the point of vertical tangency at $u = u^*$ is detectable along Γ as a special point Γ_0 associated with the function ψ and the scalar 0. By the smoothness properties of ψ and the invertibility of the matrix in Eq. (15.29), it follows that Γ_0 may be located along Γ by applying the first method in Sect. 15.1.1.

As shown above, the function

$$\psi : u \mapsto \det\left(U^T \cdot T\right), \tag{15.32}$$

with T given explicitly in Eq. (15.21), is a smooth function of location along the entire solution manifold. In practice, however, a basis for the tangent space at a particular point on the solution manifold is not available explicitly and, as discussed in Chap. 11, may be obtained numerically subsequent only to the location of such a point. For the purpose of locating vertical tangencies using any of the methods described in Sect. 15.1.1, care must then be taken to ensure even just the continuity of ψ along individual curve segments.

To this end, let Γ again denote a curve segment parameterized by h on some interval $\mathscr{H}$ and suppose that the matrix

$$\begin{pmatrix} \partial_u \Phi(\Gamma_h) \\ V^T \end{pmatrix} \tag{15.33}$$

is a continuous function of h and nonsingular for all $h \in \mathscr{H}$. Now recall the construction of the matrix $T = T_h$, associated with a point Γ_h on Γ, as the unique solution to the equation

$$\begin{pmatrix} \partial_u \Phi(\Gamma_h) \\ V^T \end{pmatrix} \cdot T = \begin{pmatrix} 0 \\ I_{n-m} \end{pmatrix}. \tag{15.34}$$

By the invertibility and continuity of the coefficient matrix, it follows that $T : h \mapsto T_h$ is a continuous function of h on $\mathscr{H}$. Now let

$$\psi : u \mapsto \det\left(U^T \cdot T_h\right) \tag{15.35}$$

and suppose that $\Gamma_0 = \Gamma(h_0)$ corresponds to a special point associated with the function ψ and the scalar 0 for some $h_0 \in \mathscr{H}$. Provided that the special point Γ_0 is detectable along Γ, it follows that it may be located along Γ by applying the second method in Sect. 15.1.1.

Fold points

For $|h| \ll 1$, the curve segment Γ with $\Gamma(0) = u^*$ and $x^* \neq 0$, defined in the previous section, is given explicitly by

$$u(h) - u^* = \begin{pmatrix} h \\ -\dfrac{3h^2 x^* \sin\eta}{x^* \cos\eta + \sin\eta} \\ \dfrac{3h^2 x^* \cos\eta}{x^* \cos\eta + \sin\eta} \end{pmatrix} + \mathcal{O}\left(h^3\right). \tag{15.36}$$

It follows that

$$3x^2(h) - \lambda(h) = 6hx^* + \mathcal{O}\left(h^2\right), \tag{15.37}$$

confirming the transversality of the intersection of Γ with the zero level set of the continuous function $u \mapsto 3x^2 - \lambda$ at $h = 0$.

Now consider some scalar-valued function $g(\kappa, \lambda)$ and denote the value of this function and its derivatives at the point u^* by a $*$ superscript. It follows that $h = 0$ is a simple local extremum of the function

$$h \mapsto g(\kappa(h), \lambda(h)) = g^* + \frac{3h^2 x^*}{x^* \cos\eta + \sin\eta}((\partial_\lambda g)^* \cos\eta - (\partial_\kappa g)^* \sin\eta) + \mathcal{O}(h^3) \quad (15.38)$$

provided that $(\partial_\lambda g)^* \cos\eta - (\partial_\kappa g)^* \sin\eta \neq 0$. In this case, we say that the curve segment exhibits a *fold point* with respect to the function g at u^*.

Consider, for example, the restricted continuation problem $G\left(u, \mu_{\mathbb{J}}\right) = 0$ obtained from the zero function Φ, the family of monitor functions

$$\Psi : u \mapsto \begin{pmatrix} \kappa \\ \lambda \end{pmatrix}, \tag{15.39}$$

and the index array $\mathbb{I} = \{2\}$ (cf. Fig. 15.2). In this case,

$$\partial_{(u,\mu_{\mathbb{J}})} G\left(u, \mu_{\mathbb{J}}\right) = \begin{pmatrix} 3x^2 - \lambda & 1 & -x & 0 \\ 0 & 1 & 0 & -1 \\ 0 & 0 & 1 & 0 \end{pmatrix}, \tag{15.40}$$

whose 1-dimensional nullspace is spanned by the vector

$$\begin{pmatrix} 1 & \lambda - 3x^2 & 0 & \lambda - 3x^2 \end{pmatrix}^T. \tag{15.41}$$

It follows that all points on the 1-dimensional solution manifold to the restricted continuation problem are regular. In this case, and for $\lambda = \lambda^* > 0$, the tangent vector is perpendicular to the κ axis at special points associated with the function

$$\psi : u \mapsto 3x^2 - \lambda \tag{15.42}$$

and the scalar 0, i.e., when $x = x^* = (\lambda^*/3)^{1/2}$ and $\kappa = \kappa^* = 2(\lambda^*/3)^{3/2}$.

Let Γ denote the curve segment through such a vertical tangency defined by the projection condition

$$V^T \cdot \left(\begin{pmatrix} u \\ \mu_{\mathbb{J}} \end{pmatrix} - \begin{pmatrix} \tilde{u} \\ \tilde{\mu}_{\mathbb{J}} \end{pmatrix} \right) - hs = 0 \tag{15.43}$$

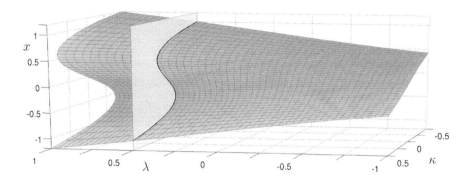

Figure 15.2. *Restricted continuation problem defined by the zero function* Φ *in Eq.* (15.19) *and the monitor functions* Ψ *in Eq.* (15.39), *where* $\mathbb{I} = \{2\}$. *The solution curve is the intersection of the solution manifold of* $\Phi(u) = 0$ *and the hyperplane* $\lambda =$ const. *For* $\lambda = 0.5$, *we find two fold points along the highlighted curve.*

with base point

$$\begin{pmatrix} \tilde{u} \\ \tilde{\mu}_{\mathbb{J}} \end{pmatrix} = \begin{pmatrix} x^* & \kappa^* & \lambda^* & \kappa^* \end{pmatrix}^T, \tag{15.44}$$

tangent matrix

$$V = \begin{pmatrix} 1 & 0 & 0 & 0 \end{pmatrix}^T, \tag{15.45}$$

and direction vector $s = 1$. A projection of Γ onto the (x, λ, κ) space is shown in Fig. 15.2. Along Γ, it follows that

$$3x^2(h) - \lambda(h) = 2\sqrt{3\lambda^*}h + \mathcal{O}\left(h^2\right) \tag{15.46}$$

and

$$\kappa(h) - \kappa^* = -\sqrt{3\lambda^*}h^2 + \mathcal{O}\left(h^3\right), \tag{15.47}$$

i.e., that Γ exhibits a fold with respect to the function $u \mapsto \kappa$ at the detectable vertical tangency associated with the function ψ in Eq. (15.42) and the scalar 0. By the smoothness of ψ and the regularity of Γ_0, this point may be located along Γ by applying the first method in Sect. 15.1.1.

In general, detectable special points associated with folds relative to some monitor function are formulated in terms of zero-crossings of the corresponding component of the associated tangent vector. For a given projection condition, the function ψ may be defined only algorithmically, rather than as a unique function of location on the solution manifold. The general principle of detectability still applies, however, provided that the definition of ψ guarantees its continuity along the curve segment. As an example, with $T = T_h$ given by the solution to the equation

$$\begin{pmatrix} \partial_u \Phi(\Gamma_h) \\ V^T \end{pmatrix} \cdot T = \begin{pmatrix} 0 \\ I_{n-m} \end{pmatrix}, \tag{15.48}$$

it follows that the function $h \mapsto T_h \cdot s$ generates a continuous family of tangent vectors. In this case, a fold is detectable along the corresponding curve segment Γ as a special point associated with the function $\psi(h) := (T_h \cdot s)_i$, for some index i, and the scalar 0. A detectable fold may now be located along the curve segment Γ using the second method in Sect. 15.1.1.

Branch points

Consider the restricted continuation problem $G\left(u, \mu_{\mathbb{J}}\right) = 0$ in $u = (x, \kappa, \lambda)$ obtained from the zero function

$$\Phi : u \mapsto \kappa - x \left(\lambda - x^2\right), \tag{15.49}$$

the family of monitor functions

$$\Psi : u \mapsto \begin{pmatrix} \kappa \\ \lambda \end{pmatrix}, \tag{15.50}$$

and $\mathbb{I} = \{1\}$. In this case,

$$\partial_{(u, \mu_{\mathbb{J}})} G\left(u, \mu_{\mathbb{J}}\right) = \begin{pmatrix} -\lambda + 3x^2 & 1 & -x & 0 \\ 0 & 1 & 0 & 0 \\ 0 & 0 & 1 & -1 \end{pmatrix}, \tag{15.51}$$

whose nullspace is 1-dimensional and spanned by the vector

$$\begin{pmatrix} x & 0 & 3x^2 - \lambda & 3x^2 - \lambda \end{pmatrix}^T, \tag{15.52}$$

as long as either x or $3x^2 - \lambda$ is nonzero. It follows that all points on the corresponding solution manifold away from $x = \lambda = 0$ (and, consequently, $\kappa = 0$) are regular.

It is straightforward to show that the 2-dimensional solution manifold of the original zero problem intersects the $\kappa = 0$ plane along two curves given by $x = 0$ and $\lambda = x^2$, as seen in Fig. 15.3. These curves, in turn, intersect each other at a *branch point* coincident with the singular point at $x = \lambda = 0$ of the restricted continuation problem $G\left(u, \mu_{\mathbb{J}}\right) = 0$ with $\kappa = 0$. By the rank deficiency of the Jacobian in Eq. (15.51), there does not exist any 4×1 matrix V for which the square matrix

$$\begin{pmatrix} \partial_{(u, \mu_{\mathbb{J}})} G\left(u, \mu_{\mathbb{J}}\right) \\ V^T \end{pmatrix}, \tag{15.53}$$

evaluated at the branch point, is invertible.

Now suppose that $\Gamma \in \mathbb{R}^4$ is a curve segment on the solution manifold to the restricted continuation problem that also satisfies the projection condition

$$V^T \cdot \left(\begin{pmatrix} u \\ \mu_{\mathbb{J}} \end{pmatrix} - \begin{pmatrix} \tilde{u} \\ \tilde{\mu}_{\mathbb{J}} \end{pmatrix} \right) - hs = 0 \tag{15.54}$$

with base point corresponding to $\left(\tilde{u}, \tilde{\mu}_{\mathbb{J}}\right) = (0, 0)$, tangent matrix $V = \begin{pmatrix} v_1 & v_2 & v_3 & v_4 \end{pmatrix}^T$, and direction vector $s = 1$. It follows that the function

$$\psi : \left(u, \mu_{\mathbb{J}}\right) \mapsto \det \begin{pmatrix} \partial_{(u, \mu_{\mathbb{J}})} G\left(u, \mu_{\mathbb{J}}\right) \\ V^T \end{pmatrix} \Bigg|_{(u, \mu_{\mathbb{J}}) \in \Gamma} \tag{15.55}$$

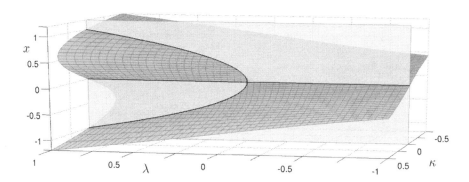

Figure 15.3. *Restricted continuation problem defined by the zero function* Φ *in Eq.* (15.49) *and the monitor functions* Ψ *in Eq.* (15.50), *where* $\mathbb{I} = \{1\}$. *The solution set is the intersection of the solution manifold of* $\Phi(u) = 0$ *and the hyperplane* $\kappa = const.$ *For* $\kappa = 0$, *this is the union of two curves that intersect at* $(0,0,0)$ *in a branch point.*

must vanish at $h = 0$. In the case of the curve segment Γ defined by $x(h) \equiv 0$, it follows that $v_3 + v_4$ must be nonzero, in which case

$$\Gamma : h \mapsto \left(0, 0, \frac{h}{v_3 + v_4}, \frac{h}{v_3 + v_4}\right). \tag{15.56}$$

Substitution then yields

$$\psi\left(\Gamma_h\right) = -h, \tag{15.57}$$

and the branch point is detectable along Γ as a special point associated with the function ψ and the scalar 0. By the noninvertibility of the matrix in Eq. (15.53), the special point may be located approximately near Γ using the third method in Sect. 15.1.1.

In general, let Γ denote a curve segment on the solution manifold of the zero problem $\Phi(u) = 0$ defined by some base point $\tilde{u}$, tangent matrix V, and direction vector s. A detectable branch point along Γ associated with a special point of the continuous function

$$\psi : u \mapsto \det \left(\begin{array}{c} \partial_u \Phi(u) \\ V^T \end{array} \right) \Bigg|_{u \in \Gamma} \tag{15.58}$$

and the scalar 0 may then be located using the third method in Sect. 15.1.1, even in the absence of an explicit expression for Γ.

15.2   A continuation paradigm

For a given continuation problem, let $\widetilde{\Psi}$ denote a collection of scalar-valued continuous monitor functions on $\mathbb{R}^n$, which includes the continuously differentiable *embeddable* functions Ψ introduced in Sect. 2.2. We now associate with each element of $\widetilde{\Psi}$ a (possibly empty) set of numerical values and refer to the corresponding level sets as *event sets*. The intersection of the union of all event sets with the set of solutions to a particular restricted continuation problem is the set of *special points* associated with the continuation problem.

In particular, special points that lie within the intersection of exactly k distinct level sets are said to be of *codimension k*.

We say that an *event* has been detected along a curve segment Γ on the solution manifold when two consecutive points $\Gamma_<, \Gamma_> \in \Gamma$ lie on opposite sides of an event set. As before, by the intermediate value theorem, there then exists at least one special point on the intersection of Γ with this event set. Depending on the nature of the continuation problem at such a special point, one of the *location methods* described above may now be employed to determine the approximate locus of the special point. Notably, k co-detected events do not generally guarantee the existence of a codimension-k special point. When the latter exists, it may, at best, be located recursively by performing continuation along embedded submanifolds corresponding to families of special points of lower codimension.

It is common to be interested only in a subset of the special points associated with a given event function. Special points that fall outside of this set are then referred to as *spurious*. Transitions between spurious and nonspurious solutions constitute special points of higher codimension.

15.2.1   Algorithmic implementation

We recall from Chap. 2 the abstract construction and evaluation of the restriction $G(u, \mu_{\mathbb{J}})$ and its Jacobian in terms of the input-output algorithms Ξ and $\partial \Xi$. Here, given

- the family of zero functions Φ and the Jacobian $\partial_u \Phi$,

- the family of monitor functions Ψ and the Jacobian $\partial_u \Psi$,

- the initial solution guess u_0, and

- the index set $\mathbb{I}$,

it was suggested that a general-purpose constructor

- store access to the functions Φ and Ψ in the data variable Θ such that

$$\Theta(u) = \begin{pmatrix} \Phi(u) \\ \Psi(u) \end{pmatrix}; \tag{15.59}$$

- store access to the Jacobians of the functions Φ and Ψ in the data variable $\partial_u \Theta$ such that

$$\partial_u \Theta(u) = \begin{pmatrix} \partial_u \Phi(u) \\ \partial_u \Psi(u) \end{pmatrix}; \tag{15.60}$$

- store the integer m equal to the number of zero functions;

- store the index set $\mathbb{J} = \{1, \ldots, r\} \setminus \mathbb{I}$, where r equals the number of monitor functions;

- store the column matrix

$$\lambda_0 := \begin{pmatrix} 0 \\ \Psi(u_0) \end{pmatrix} \in \mathbb{R}^{m+r}; \tag{15.61}$$

and

- store the matrix

$$\Lambda_0 := \begin{pmatrix} 0 \\ -I_{\mathbb{J}} \end{pmatrix} \in \mathbb{R}^{(m+r)\times|\mathbb{J}|}, \tag{15.62}$$

where $I_{\mathbb{J}}$ is the submatrix of I_r containing the columns indexed by $\mathbb{J}$.

Given input values for the continuation variables u and the subset $\mu_{\mathbb{J}}$ of the continuation parameters, evaluation of the left-hand side of the restricted continuation problem was then shown to be equivalent to the algorithm Ξ encoded in the three assignments

1. $\lambda \Leftarrow \lambda_0$,

2. $\lambda[m + \mathbb{J}] \Leftarrow \mu_{\mathbb{J}}$, and

3. $\Xi(u, \mu_{\mathbb{J}}) \Leftarrow \Theta(u) - \lambda$.

Similarly, the Jacobian of the left-hand side of the restricted continuation problem, with respect to u and $\mu_{\mathbb{J}}$, was obtained from the algorithm $\partial\Xi$, encoded in the single assignment

1. $\partial\Xi(u, \mu_{\mathbb{J}}) \Leftarrow \begin{pmatrix} \partial_u\Theta(u) & \Lambda_0 \end{pmatrix}$.

The advantage of this formulation was the recognition that the algorithms Ξ and $\partial\Xi$ could be implemented in the core layer independently of the meaning of the variables in u and $\mu_{\mathbb{J}}$, or of the content of Θ and $\partial_u\Theta$, which would be defined elsewhere.

A similar observation allows for a general algorithmic implementation of event detection and location. Here, at the stage of initial construction, it is necessary to provide a computational implementation with

- the family of monitor functions $\widetilde{\Psi}$, which is used to compute the number ν of event functions, and

- a collection $\{\mathcal{N}_i\}_{i=1}^{\nu}$ of (possibly empty) sequences of numerical values, which is used to compute the sequence $\{|\mathcal{N}_i|\}_{i=1}^{\nu}$ of cardinal numbers.

A general-purpose constructor would now

- store access to the functions $\widetilde{\Psi}$ and

- store the 1-dimensional arrays

$$\mathcal{E} := \begin{bmatrix} \cdots & \mathcal{N}_i & \cdots \end{bmatrix} \tag{15.63}$$

and

$$\mathcal{I} := \begin{bmatrix} \cdots & 1_{1,|\mathcal{N}_i|} \otimes \begin{bmatrix} i \end{bmatrix} & \cdots \end{bmatrix} \tag{15.64}$$

for an ordered sequence of all $i \in \{1,\ldots,\nu\}$ such that $|\mathcal{N}_i| \neq 0$.

Given input values for the continuation variables u, evaluation of the monitor functions is then equivalent to the algorithm $\widetilde{\Xi}$, encoded in the single assignment

1. $\widetilde{\Xi}(u) \Leftarrow \widetilde{\Psi}(u)$.

If the output of this algorithm is stored in the 1-dimensional array $\widetilde{\mu}$, then event detection and location can be implemented by monitoring zero-crossings of the array

$$\widetilde{\mu}_{\mathcal{I}} - \mathcal{E} \tag{15.65}$$

and, when necessary and appropriate, fixing the value of one or several components of this array to zero.

15.3   The core interface

The finite-state machine introduced in Chap. 11 agrees with the proposed methodology of covering a solution manifold through the successive construction of curve segments and the merging of each such curve segment with an existing atlas. As suggested above, event detection and location requires at least two consecutive points on such a curve segment on opposite sides of an event set.

15.3.1   Event classification

Two overlapping classifications of events are implemented in COCO: one distinguishing events according to the algorithm used for locating the event, and another according to the action taken subsequent to locating the event. In the latter classification, we distinguish between *boundary*, *terminal*, and *special events*; in the former, we distinguish between *continuation*, *regular*, and *singular events*. In particular, continuation, regular, and singular events are distinguished by the application of the first, second, and third methods, respectively, described in Sect. 15.1.1.

A boundary event occurs if a computational boundary of the solution manifold is crossed during continuation, as described in Chap. 14. An implementation should locate the boundary point, include it in the atlas, and discard points outside the computational domain. A typical example is the restriction to a finite interval of values of a continuation parameter. Boundary events may, for example, be declared implicitly in COCO by including explicit bounds on the continuation parameters in the call to the `coco` entry-point function, as shown in Chap. 14.

A terminal event is an event that indicates that a newly computed point is unacceptable for some reason. An implementation should not locate this event, but include the last computed point in the atlas, mark the corresponding curve segment as invalid, and stop any further continuation beyond this point. A typical example is given by a discretization error estimate exceeding a user-prescribed limit in the absence of adaptation (but see Chap. 20).

A special event is an event that is neither a boundary event nor a terminal event. An implementation should locate the special point, include it in the atlas, and proceed with the continuation. Along each curve segment, the computation of events should proceed in the order indicated: first locate boundary events, then check whether a terminal event occurred, and finally locate all special events.

15.3.2   Function evaluation

The monitor functions $\widetilde{\Psi}$ are evaluated using the algorithm $\widetilde{\Xi}$ during the transition from the init state of the COCO finite-state machine in order to support handling of computational boundaries and during the transition from the add state in order to support event detection and location during continuation. In each case, the output array $\widetilde{\mu}$ of the algorithm and the array $\widetilde{\mu}(\mathcal{I}) - \mathcal{E}$ in Eq. (15.65) are stored in the `chart.p` and `chart.e` fields, respectively, of the corresponding chart structures.

For example, in order to support the determination of whether the initial chart on the solution manifold lies inside a computational domain (defined by interval bounds on a subset of the continuation parameters, as discussed in Chap. 14), the `chart.p` and `chart.e` fields are populated immediately following the execution of the `init_chart` class method. In this case, an admissible initial chart on the interior of the computational domain is one for which the relevant entries of the array stored in `chart.e` are of the appropriate signs.

The monitor functions in $\widetilde{\Psi}$ are also evaluated repeatedly during continuation along the solution manifold, immediately following the execution of the `add_chart` class method, provided that its `flush` output argument is set to `true`. In this case, upon the detection of one or several events along a curve segment, the COCO finite-state machine enters a separate set of states charged with locating the corresponding events, using one of the three methods described in this chapter, and in the order described in the previous section. Control is returned to the ʄluʃɥ state upon the conclusion of this process, with `cseg.Status` equal to the `cseg.CurveSegmentOK` integer constant, even in the case that the algorithm fails to locate an event. In the case of failure to locate a boundary event, the `pt_type` and `ep_flag` fields of the end point of the current curve segment are assigned the values `'MX'` and 2, respectively, for further processing by the atlas algorithm.

In addition to the evaluation of the monitor functions $\widetilde{\Psi}$ during continuation, all embeddable monitor functions Ψ are also evaluated once prior to entering the finite-state machine, in order to initialize values for the continuation parameters $\mu_0 = \Psi(u_0)$, as described in Sect. 2.2.4. This further allows for the imposition of constraints on the subset of continuation parameters referenced by the index array $\mathbb{I}$, associated with holding constant the values of the corresponding monitor functions.

Finally, we note that in the case of the third method described in Sect. 15.1.1, evaluation of the monitor functions $\widetilde{\Psi}$ occurs after each iteration of the subdivision algorithm, without a preceding application of the predictor-corrector algorithm for locating a point on the solution manifold.

15.3.3 Declaring monitor functions and events

Monitor functions may be appended to a continuation problem structure by invoking the `coco_add_func` command, with function type `'inactive'`, `'active'`, `'regular'`, or `'singular'`. The first two of these imply that the corresponding function is added to the set of embeddable monitor functions Ψ. As before, the index of monitor functions with function type `'inactive'` is assigned by default to $\mathbb{I}$, whereas that of monitor functions with function type `'active'` is assigned by default to the complement $\mathbb{J}$. Events associated with monitor functions of function type `'inactive'` or `'active'` are located by means of the first of the three methods mentioned in Sect. 15.1.1. In contrast, events associated with the function types `'regular'` and `'singular'` are located by means of methods two and three, respectively, described in Sect. 15.1.1.

Example 15.1 Let $u \in \mathbb{R}^2$. Let `circ` and `dist` contain COCO-compatible function definitions of the zero function

$$\Phi : u \mapsto u_1^2 + (u_2 - 1)^2 - 1 \tag{15.66}$$

and the monitor function

$$\Psi : u \mapsto \sqrt{u_1^2 + u_2^2}, \tag{15.67}$$

respectively. As before, these functions are added to the continuation problem structure using the assignments

```
>> prob = coco_add_func(coco_prob(), 'circ', @circ, [], 'zero', ...
   'u0', [1; 1.1]);
>> prob = coco_add_func(prob, 'fun2', @dist, [], 'inactive', 'p', ...
   'uidx', [1; 2]);
```

The `coco_add_event` utility may now be used to associate a nonempty set of numerical values with the monitor function and to identify the corresponding events with the point type `'UZ'`, as shown below.

```
>> prob = coco_add_event(prob, 'UZ', 'p', 0.5:0.5:2);
```

In the extract below, the screen output includes special points associated with the point type `'UZ'`, as shown in the `TYPE` column.

```
>> coco(prob, 'run', [], 1, 'p', [0.5, 3]);
```

	STEP	DAMPING		NORMS		COMPUTATION TIMES		
IT	SIT	GAMMA	||d||	||f||	||U||	F(x)	DF(x)	SOLVE
0				1.00e-02	2.10e+00	0.0	0.0	0.0
1	1	1.00e+00	7.43e-03	5.83e-05	2.10e+00	0.0	0.0	0.0
2	1	1.00e+00	2.78e-05	7.85e-10	2.10e+00	0.0	0.0	0.0
3	1	1.00e+00	4.41e-10	0.00e+00	2.10e+00	0.0	0.0	0.0

STEP	TIME	||U||	LABEL	TYPE	p
0	00:00:00	2.1024e+00	1	EP	1.4866e+00
7	00:00:01	1.4142e+00	2	UZ	1.0000e+00
10	00:00:02	9.4446e-01	3		6.6784e-01
12	00:00:02	7.0711e-01	4	UZ	5.0000e-01
12	00:00:02	7.0711e-01	5	EP	5.0000e-01

STEP	TIME	||U||	LABEL	TYPE	p
0	00:00:02	2.1024e+00	6	EP	1.4866e+00
1	00:00:02	2.1213e+00	7	UZ	1.5000e+00
10	00:00:03	2.7040e+00	8		1.9120e+00
20	00:00:03	2.8196e+00	9		1.9937e+00
30	00:00:04	2.5363e+00	10		1.7934e+00
37	00:00:04	2.1213e+00	11	UZ	1.5000e+00
40	00:00:04	1.7539e+00	12		1.2402e+00
43	00:00:05	1.4142e+00	13	UZ	1.0000e+00
48	00:00:05	7.0711e-01	14	EP	5.0000e-01

As shown below, the `coco_bd_read` and `coco_bd_labs` utilities may now be used to extract the solution labels corresponding to these labeled solution points.

```
>> bd = coco_bd_read('run');
>> labs = coco_bd_labs(bd, 'UZ');
```

∎

Previous chapters have discussed the construction of an initializable extended continuation problem in the continuation package COCO through the definition of the functions Φ and Ψ and the associated selection of a subset of elements of Ψ corresponding to a subset $\mathbb{I}$ of the element indices. In COCO, the construction of $\mathbb{I}$ is achieved through an initial assignation at the time of definition of elements of Ψ followed by an optional sequence of additional pairwise exchanges of elements between $\mathbb{I}$ and its complement $\mathbb{J}$. In particular, with the ability to detect special points comes the opportunity to change the content of $\mathbb{I}$ in order to continue families of special points. Pairwise exchanges of function type result from invoking the `coco_xchg_pars` utility as in the assignment

```
prob = coco_xchg_pars(prob, 'p1', 'p2');
```

where the function types of the monitor functions associated with the elements of μ with string identifiers `'p1'` and `'p2'` have been switched. For example, suppose that the corresponding function types were `'inactive'` and `'active'` prior to this call, i.e., that the monitor function associated with `'p1'` was constrained and that associated with `'p2'` was allowed to vary. Then, subsequent to this call, the reverse is true.

It is clear that an arbitrary permutation of function types may be achieved by successive calls to `coco_xchg_pars`, although the total number of monitor functions of each type must remain constant. As a special case, the number of monitor functions of function type `'inactive'`, i.e., the number of elements of $\mathbb{I}$, must equal the cardinality $|\mathbb{I}_0|$ of the initial index array throughout.

Example 15.2 Consider the continuation of two-segment periodic trajectories using the `'hspo'` toolbox defined in Chap. 9, where each segment is governed by the vector field

$$f(y, p; \mathfrak{free}) := \begin{pmatrix} y_2 \\ p_3 \cos y_3 - p_2 y_2 - p_1 y_1 \\ p_4 \end{pmatrix}, \tag{15.68}$$

where the terminal and jump functions for the first segment are given by

$$h(y, p; \mathfrak{impact}) := p_5 - y_1 \tag{15.69}$$

and

$$g(y, p; \mathfrak{bounce}) := \begin{pmatrix} y_1 \\ -p_6 y_2 \\ y_3 \end{pmatrix}, \tag{15.70}$$

respectively; where the terminal and jump functions for the second segment are given by

$$h(y, p; \mathfrak{phase}) := 2\pi - y_3 \tag{15.71}$$

and

$$g(y, p; \mathfrak{phase}) := \begin{pmatrix} y_1 \\ y_2 \\ y_3 - 2\pi \end{pmatrix}, \tag{15.72}$$

respectively; and where $p = (k, c, A, \omega, d, e)$. The `'hspo'`-compatible functions, shown below, encode the corresponding functions.

```
function y = impact(x, p, mode)

x1 = x(1,:);
x2 = x(2,:);
x3 = x(3,:);
p1 = p(1,:);
p2 = p(2,:);
```

```
p3 = p(3,:);
p4 = p(4,:);

switch mode
  case 'free'
    y(1,:) = x2;
    y(2,:) = p3.*cos(x3)-p2.*x2-p1.*x1;
    y(3,:) = p4;
end

end


function y = impact_events(x, p, event)

x1 = x(1,:);
x3 = x(3,:);
p5 = p(5,:);

switch event
  case 'impact'
    y = p5-x1;
  case 'phase'
    y = pi-x3;
end

end


function y = impact_resets(x, p, reset)

p6 = p(6,:);

y = x;
switch reset
  case 'bounce'
    y(2,:) = -p6.*y(2,:);
  case 'phase'
    y(3,:) = y(3,:)-2*pi;
end

end
```

The following sequence of commands then creates a restricted continuation problem that is initializable by a 1-dimensional atlas algorithm. Specifically, the `hspo_isol2segs` constructor is invoked with an initial guess u_0 based on an approximate periodic trajectory generated using two instances of forward integration.

```
>> p0      = [1; 0.1; 1; 1; 1; 0.8];
>> modes  = {'free' 'free'};
>> events = {'impact' 'phase'};
>> resets = {'bounce' 'phase'};
>> f       = @(t, x) impact(x, p0, modes{1});
>> [t1 x1] = ode45(f, [0 3.2], [-0.98; -0.29; -pi]);
>> f       = @(t, x) impact(x, p0, modes{2});
>> [t2 x2] = ode45(f, [0 3.1], [1; -1.36; 0.076]);
>> t0 = {t1  t2};
>> x0 = {x1  x2};
>> hspo_args = {{@impact, @impact_events, @impact_resets}, ...
     modes, events, resets, t0, x0, {'k' 'c' 'A' 'w' 'd' 'e'}, p0};
>> prob = hspo_isol2segs(coco_prob(), '', hspo_args{:});
```

Now suppose that we are interested only in solutions for which $y_2 < 0$ at the initial point of the second segment, i.e., that solutions with $y_2 > 0$ at this point are spurious. To this end, consider the following sequence of commands.

```
>> [data uidx] = coco_get_func_data(prob, 'msbvp.seg2.coll', ...
   'data', 'uidx');
>> prob = coco_add_pars(prob, 'grazing', uidx(data.x0_idx(2)), ...
   'graze', 'active');
>> prob = coco_add_event(prob, 'GR', 'graze', 0);
```

Here, the coco_get_func_data utility is used to extract the toolbox data structure $\mathfrak{D}$ and index array $\mathbb{K}$ associated with the collocation zero problem for the second segment. The coco_add_pars utility is then invoked to introduce an active continuation parameter with string label 'graze' and associated monitor function with function identifier 'grazing' corresponding to the second component of the initial point on this segment. Finally, the coco_add_event utility is invoked to append an event associated with a zero value for this monitor function to the continuation problem structure and to designate this event by the point type 'GR'.

The extract below shows the result of continuation of a 1-dimensional family of two-segment periodic trajectories under variations in the problem parameter A while monitoring the value of the 'graze' continuation parameter.

```
>> prob = coco_set(prob,'cont', 'ItMX', 100);
>> coco(prob, 'run1', [], {'A' 'graze'}, [0.01 1]);
```

STEP	DAMPING		NORMS		COMPUTATION	TIMES	
IT SIT	GAMMA	||d||	||f||	||U||	F(x)	DF(x)	SOLVE
0			6.85e-02	2.24e+01	0.0	0.0	0.0
1 1	1.00e+00	2.13e-01	2.65e-04	2.23e+01	0.0	0.0	0.1
2 1	1.00e+00	1.99e-03	3.62e-09	2.23e+01	0.0	0.0	0.1
3 1	1.00e+00	3.56e-08	8.26e-15	2.23e+01	0.0	0.0	0.1

STEP	TIME	||U||	LABEL	TYPE	A	graze
0	00:00:00	2.2281e+01	1	EP	1.0000e+00	-1.3566e+00
10	00:00:01	2.1383e+01	2		5.5932e-01	-7.5178e-01
20	00:00:02	2.1341e+01	3		2.4156e-01	-3.0237e-01
30	00:00:03	2.1979e+01	4		1.2330e-01	-1.0145e-01
40	00:00:04	2.3197e+01	5		1.0000e-01	-6.7419e-03
41	00:00:04	2.3318e+01	6	GR	1.0000e-01	0.0000e+00
50	00:00:05	2.4813e+01	7		1.1771e-01	8.1226e-02
60	00:00:06	2.6625e+01	8		2.1045e-01	2.4860e-01
70	00:00:07	2.8228e+01	9		4.9045e-01	6.4977e-01
80	00:00:08	2.9925e+01	10		8.8039e-01	1.1864e+00
83	00:00:08	3.0516e+01	11	EP	1.0000e+00	1.3500e+00

We note, in particular, the occurrence of a special point associated with a zero value of the y_2 component at the initial point of the second segment corresponding to a point of grazing contact with the event surface $d - y_1 = 0$. Fig. 15.4 illustrates the projection of an approximation to the solution manifold onto the plane whose coordinates are A and the norm $\| (u, \mu) \|$. Representative orbits are shown in Fig. 15.5.

To continue the family of two-segment trajectories for which $y_2 = 0$ at the initial point of the second segment, we use the msbvp_sol2segs constructor to reconstruct the restricted continuation problem, followed by a pairwise exchange of function type that allows the value of A to vary during continuation while constraining the value of y_2 at the

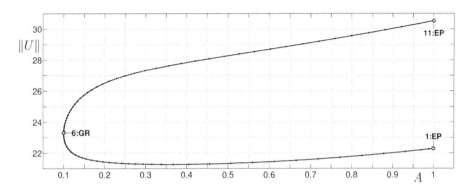

Figure 15.4. *Bifurcation diagram of the impact oscillator from Example* 15.2 *with* $\|U\| := \left\|\left(u, \mu_{\mathbb{J}}\right)\right\|$. *Continuation starts with an impacting orbit at label* 1 *and passes through a grazing point. The part of the curve between labels* 6 *and* 11 *corresponds to nonphysical orbits penetrating the impact surface. Some representative orbits are shown in Fig.* 15.5.

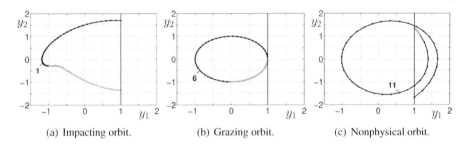

| (a) Impacting orbit. | (b) Grazing orbit. | (c) Nonphysical orbit. |

Figure 15.5. *Some orbits representative of the continuation illustrated in Fig.* 15.4. *Continuation starts with the impacting orbit shown in* (a) *and passes through a grazing point* (b), *after which nonphysical orbits result* (c).

initial point of the second segment. The extract below shows the result of this computation, including the value of the `'graze'` continuation parameter along the solution manifold.

```
>> labgr = coco_bd_labs(bd1, 'GR');
>> prob = msbvp_sol2segs(coco_prob(), '', 'run1', labgr);
>> [data uidx] = coco_get_func_data(prob, 'msbvp.seg2.coll', ...
     'data', 'uidx');
>> prob = coco_add_pars(prob, 'grazing', uidx(data.x0_idx(2)), ...
     'graze', 'active');
>> prob = coco_xchg_pars(prob, 'graze', 'A');
>> prob = coco_set(prob, 'cont', 'ItMX', 100);
>> coco(prob, 'run2', [], {'w' 'A' 'graze'}, {[] [0 1]});

...     NORMS              COMPUTATION TIMES
...     ||f||      ||U||   F(x)   DF(x)   SOLVE
... 8.43e-15   2.33e+01    0.0    0.0     0.0
... 6.56e-15   2.33e+01    0.0    0.0     0.0

... LABEL   TYPE            w              A          graze
...     1   EP      1.0000e+00    1.0000e-01    0.0000e+00
```

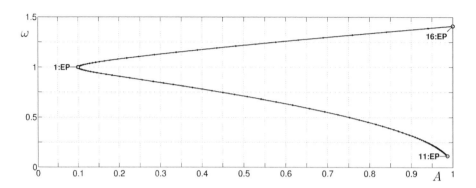

Figure 15.6. *Grazing curve in the (A, ω) parameter plane computed for the impact oscillator in Example 15.2. For the system under consideration, this curve subdivides the parameter plane locally into a region of parameter values for which nonimpacting orbits exist (left-hand side) from parameter values for which orbits have at least one impact (right-hand side). Some representative grazing orbits are shown in Fig. 15.7.*

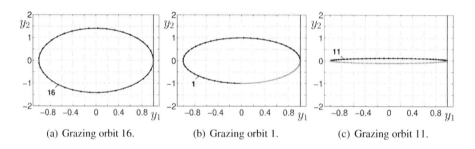

(a) Grazing orbit 16.   (b) Grazing orbit 1.   (c) Grazing orbit 11.

Figure 15.7. *Some grazing orbits representative of the continuation illustrated in Fig. 15.6. All orbits have the same amplitude, but different periods.*

	LABEL	TYPE	w	A	graze
...	2		9.7288e-01	1.1103e-01	0.0000e+00
...	3		9.0507e-01	2.0224e-01	0.0000e+00
...	4		7.4548e-01	4.5047e-01	0.0000e+00
...	5		4.7484e-01	7.7598e-01	0.0000e+00
...	6		3.1292e-01	9.0262e-01	0.0000e+00
...	7		2.3187e-01	9.4652e-01	0.0000e+00
...	8		1.8398e-01	9.6633e-01	0.0000e+00
...	9		1.5245e-01	9.7688e-01	0.0000e+00
...	10		1.3014e-01	9.8315e-01	0.0000e+00
...	11	EP	1.1352e-01	9.8718e-01	0.0000e+00
...	LABEL	TYPE	w	A	graze
...	12	EP	1.0000e+00	1.0000e-01	0.0000e+00
...	13		1.0279e+00	1.1736e-01	0.0000e+00
...	14		1.1052e+00	2.4760e-01	0.0000e+00
...	15		1.3220e+00	7.5925e-01	0.0000e+00
...	16	EP	1.4107e+00	1.0000e+00	0.0000e+00

The resulting *grazing bifurcation curve*, corresponding to the projection of the solution manifold onto the (A, ω) plane, is shown in Fig. 15.6. Representative orbits are shown in Fig. 15.7. ∎

15.4   Conclusions

The dual tasks of event detection and location rely fundamentally on the intermediate value theorem for scalar-valued continuous functions. The natural context in which to consider event handling is therefore during the construction of curve segments on solution manifolds. Here, the intrinsic parameterization in terms of h reduces the task of event location to one of arriving at a value of h associated with a special point.

The challenge of event detection is thus limited to the design of monitor functions that are continuous along individual curve segments, without requiring that these be continuous functions of location along the solution manifold. Continuity as a function of location on the solution manifold might be a valuable property for purposes of characterizing solution points, but should have no effect on the possibility of locating associated special points. This observation allows us to generalize the class of nonembeddable monitor functions to include functions that depend not only on the location on the solution manifold but also on local properties of the curve segment, e.g., the tangent vector for purposes of fold detection and location.

In the next several chapters, we illustrate modifications to some of the template toolboxes developed in Parts I and II that provide support for context-relevant event detection. Chap. 16 further demonstrates the integration of curve-segment-dependent monitor functions into the 1- and 2-dimensional atlas algorithms developed in Part III of this text. In Chap. 17, we illustrate the use of a reverse communication protocol in support of a refined event-detection functionality.

Exercises

15.1. Suppose that the matrix

$$\begin{pmatrix} \partial_u \Phi(\Gamma_0) \\ V^T \end{pmatrix}$$

is invertible and that the columns of T_0 constitute a basis for the tangent space to the solution manifold at the point Γ_0. Show that

- the matrix $V^T \cdot T_0$ is invertible and
- the equivalence

$$\partial_u \Phi(\Gamma_0) \cdot \alpha = 0 \Leftrightarrow \alpha = T_0 \cdot \sigma$$

holds for some σ.

15.2. Suppose that the conditions in the previous exercise apply and, in addition, that

$$\partial_u \psi(\Gamma_0) \cdot T_0 \cdot \left(V^T \cdot T_0 \right)^{-1} \cdot s \neq 0.$$

Show that the $(n+1) \times (n+1)$ matrix

$$\begin{pmatrix} \partial_u \Phi(\Gamma_0) & 0 \\ V^T & -s \\ \partial_u \psi(\Gamma_0) & 0 \end{pmatrix}$$

is invertible.

15.3. Suppose that the conditions in the previous exercises apply and that (Γ_0, h_0) is a solution to the closed continuation problem

$$
\begin{pmatrix}
\Phi(u) \\
V^T \cdot (u - \tilde{u}) - hs \\
\psi(u)
\end{pmatrix} = 0
$$

in the unknowns (u, h). Show that the point (Γ_0, h_0) is a regular point of this continuation problem.

15.4. Suppose that the conditions in the previous exercises apply and that the s_1 component of the direction vector is nonzero. Show that Γ_0 is a regular point of the closed continuation problem

$$
\begin{pmatrix}
\Phi(u) \\
s_1 \left[V^T \cdot (u - \tilde{u}) \right]_2 - s_2 \left[V^T \cdot (u - \tilde{u}) \right]_1 \\
\vdots \\
s_1 \left[V^T \cdot (u - \tilde{u}) \right]_{n-m} - s_{n-m} \left[V^T \cdot (u - \tilde{u}) \right]_1 \\
\psi(u) - \psi_0
\end{pmatrix} = 0.
$$

15.5. Verify that the columns of the matrix

$$
T = \begin{pmatrix}
0 & 1 \\
x & \lambda - 3x^2 \\
1 & 0
\end{pmatrix}
$$

are linearly independent and span the tangent space at every point (x, κ, λ) of the solution manifold to the zero problem

$$
\kappa - x \left(\lambda - x^2 \right) = 0.
$$

15.6. Let $U \in \mathbb{R}^{n \times m}$ and $T \in \mathbb{R}^{n \times m}$ be two arbitrary matrices. Show that span $\{T\}$ contains a vector that is orthogonal to span $\{U\}$, and vice versa, if and only if

$$
\det \left(U^T \cdot T \right) = 0.
$$

15.7. Suppose that $T_\perp : h \mapsto T_{\perp, h}$ is a one-parameter family of matrices whose columns span the tangent space at the corresponding points Γ_h and for which $T_{\perp, h}^T \cdot T_{\perp, h} = I_{n-m}$ for h on some interval. Show that the monitor function $\det \left(U^T \cdot T_{\perp, h} \right)$ is unchanged if we replace $T_{\perp, h}$ by any matrix obtained after a rotation of the column vectors in the tangent plane.

15.8. Under the assumptions of the previous exercise, show that $\det \left(U^T \cdot T_{\perp, h} \right)$ is a continuous function of h on some interval provided that the function $h \mapsto \det \left(V^T \cdot T_{\perp, h} \right)$ is of constant sign on this interval.

15.9. For arbitrary matrices U and T in $\mathbb{R}^{3 \times 2}$, show that

$$
\det \left(U^T \cdot T \right) = (u_1 \times u_2)^T \cdot (t_1 \times t_2),
$$

where u_1 and u_2 denote the first and second columns of U, respectively, and t_1 and t_2 denote the first and second columns of T, respectively. Thus show that a change in sign of the left-hand side along a curve segment for fixed U and continuously varying T implies a change in orientation of the basis of the tangent space given by the columns of T relative to the basis of the parameter plane given by the columns of U.

15.10. Show that every vertical tangency on the solution manifold of the zero problem

$$\kappa - x\left(\lambda - x^2\right) = 0$$

coincides with a point on the curve $(x, \kappa, \lambda) = \left(x, 2x^3, 3x^2\right)$.

15.11. Verify Eq. (15.31) and explain the implications to the intersection between the curve segment Γ and the zero-level set of ψ.

15.12. Let Γ denote a curve segment on the solution manifold to the zero problem

$$\kappa - x\left(\lambda - x^2\right) = 0, u = (x, \kappa, \lambda),$$

corresponding to the projection condition with base point $\tilde{u} = 0$, tangent matrix

$$V = \begin{pmatrix} 0 & 1 \\ \cos\eta & 0 \\ \sin\eta & 0 \end{pmatrix}$$

for $\eta \neq n\pi$, $n \in \mathbb{Z}$, and direction vector $s = \begin{pmatrix} 0 & 1 \end{pmatrix}^T$. Characterize the intersection between Γ and the zero-level set of

$$\psi : u \mapsto 3x^2 - \lambda.$$

Comment on the detectability of the point $\Gamma_0 = 0$ along Γ.

15.13. Verify Eqs. (15.36)–(15.37) and comment on the detectability of the special point associated with the function $\psi : (x, \kappa, \lambda) \mapsto 3x^2 - \lambda$ and the scalar 0.

15.14. Let

$$u(h) := \begin{pmatrix} x(h) \\ \kappa(h) \\ \lambda(h) \end{pmatrix} = u^* + \begin{pmatrix} h \\ -\dfrac{3h^2 x^* \sin\eta}{x^* \cos\eta + \sin\eta} \\ \dfrac{3h^2 x^* \cos\eta}{x^* \cos\eta + \sin\eta} \end{pmatrix} + \mathscr{O}\left(h^3\right)$$

describe the parameterization of a curve segment Γ^* through the point u^* with $x^* \neq 0$. Show that

$$g(\kappa(h), \lambda(h)) = g^* + \frac{3h^2 x^*}{x^* \cos\eta + \sin\eta}\left((\partial_\lambda g)^* \cos\eta - (\partial_\kappa g)^* \sin\eta\right) + \mathscr{O}(h^3)$$

for every scalar-valued function g, where $*$ denotes evaluation at $h = 0$. Describe the local shape of the curve segment relative to the level sets of g in the cases when

• $(\partial_\lambda g)^* \cos\eta - (\partial_\kappa g)^* \sin\eta \neq 0$ and

• $(\partial_\lambda g)^* \cos\eta - (\partial_\kappa g)^* \sin\eta = 0$.

15.15. Consider the restricted continuation problem $G\left(u,\mu_{\mathbb{J}}\right)=0$ in $u=(x,\kappa,\lambda)$ obtained from the zero function

$$\Phi : u \mapsto \kappa - x\left(\lambda - x^2\right),$$

the family of monitor functions

$$\Psi : u \mapsto \begin{pmatrix} \kappa \\ \lambda \end{pmatrix},$$

and $\mathbb{I} = \{1\}$. Find all regular solution points at which the corresponding tangent vector is perpendicular to the λ axis.

15.16. Consider the restricted continuation problem in the previous exercise and a special point given by $u^* = \left(\begin{array}{ccc} x^* & \kappa^* & \lambda^* \end{array}\right)^T$ and $\mu_{\mathbb{J}}^* = \lambda^*$ associated with a vertical tangency of the corresponding tangent vector to the solution manifold. Consider the curve segment Γ through such a vertical tangency defined by the projection condition

$$V^T \cdot \left(\begin{pmatrix} u \\ \mu_{\mathbb{J}} \end{pmatrix} - \begin{pmatrix} \tilde{u} \\ \tilde{\mu}_{\mathbb{J}} \end{pmatrix} \right) - hs = 0$$

with base point $\left(\tilde{u},\tilde{\mu}_{\mathbb{J}}\right) = \left(u^*,\mu_{\mathbb{J}}^*\right)$, tangent matrix

$$V = \left(\begin{array}{cccc} 1 & 0 & 0 & 0 \end{array}\right)^T,$$

and direction vector $s = 1$. Show that, along Γ,

$$3x^2(h) - \lambda(h) = 3\left(4\kappa^*\right)^{1/3}h + \mathcal{O}\left(h^2\right)$$

and

$$\lambda(h) - \lambda^* = 3h^2 + \mathcal{O}\left(h^3\right).$$

Show that the vertical tangency is a fold with respect to the function $\left(u,\mu_{\mathbb{J}}\right) \mapsto \lambda$ along Γ. Comment on the detectability of the vertical tangency and the preferred method for locating this point along Γ.

15.17. Show that the nullspace of the Jacobian of the restricted continuation problem in the previous exercise is 2-dimensional when $x = \kappa = \lambda = 0$. Does this imply that the tangent space to the corresponding solution manifold is 2-dimensional?

15.18. Comment on the detectability and preferred method of locating the special point associated with the function ψ given in Eq. (15.55) and the scalar 0 along the curve segment $\Gamma \in \mathbb{R}^4$ on the solution manifold to the corresponding restricted continuation problem defined by $\lambda = x^2$.

15.19. Consider the restricted continuation problem $G\left(u,\mu_{\mathbb{J}}\right)=0$ for $u=(x,\kappa,\lambda)$ obtained from the zero function

$$\Phi\left(u\right) := \kappa - x\left(\lambda - x^2\right),$$

the family of monitor functions

$$\Psi\left(u\right) = \begin{pmatrix} \kappa \\ \lambda \\ 3x^2 - \lambda \\ x \end{pmatrix},$$

and $\mathbb{I} = \{3\}$. Let $x^* = \kappa^* = 0$ and consider an arbitrary λ^*. Show that the curve segment through the point

$$
\begin{pmatrix} u^* \\ \mu_{\mathbb{J}}^* \end{pmatrix} = \begin{pmatrix} x^* & \kappa^* & \lambda^* & \kappa^* & \lambda^* & x^* \end{pmatrix}^T,
$$

corresponding to the projection condition

$$
v^T \cdot \left(\begin{pmatrix} u \\ \mu_{\mathbb{J}} \end{pmatrix} - \begin{pmatrix} \tilde{u} \\ \tilde{\mu}_{\mathbb{J}} \end{pmatrix} \right) - hs = 0
$$

with base point $\left(\tilde{u}, \tilde{\mu}_{\mathbb{J}}\right) = \left(u^*, \mu_{\mathbb{J}}^*\right)$, tangent matrix

$$
V = \frac{1}{\sqrt{2 + 2\mu_3^{*2}}} \begin{pmatrix} 1 & -\mu_3^* & 0 & -\mu_3^* & 0 & 1 \end{pmatrix}^T,
$$

and direction vector $s = 1$, is given by

$$
x(h) = \frac{h}{\sqrt{2 + 2\lambda^{*2}}} + \mathcal{O}\left(h^3\right),
$$

$$
\kappa(h) = \frac{h\lambda^*}{\sqrt{2 + 2\lambda^{*2}}} + \mathcal{O}\left(h^3\right),
$$

and

$$
\lambda(h) = \lambda^* + \frac{3h^2}{2 + 2\lambda^{*2}} + \mathcal{O}\left(h^3\right).
$$

What is the local shape of this curve segment with respect to the level sets of $x(h)$, $\kappa(h)$, and $\lambda(h)$? Comment on the detectability of the point $(u, \mu_{\mathbb{J}}) = (u^*, \mu_{\mathbb{J}}^*)$ along Γ as a special point with respect to either one of the monitor functions.

15.20. Why is it not possible to conclude the existence of a codimension-k special point from a change in the sign of k monitor functions between consecutive points along a curve segment? Give a counterexample to illustrate your response.

15.21. Why is the event associated with the monitor function in Eq. (15.67) and the value 2 not detected in the continuation run in Example 15.1?

15.22. What happens if you omit the call to the `coco_xchg_pars` utility in the construction of the second continuation problem structure in Example 15.2?

15.23. Use COCO to explore the families of special points in the restricted continuation problems considered in Sect. 15.1.2.

Chapter 16

Atlas Events and Toolbox Integration

The command-line interface to event detection and location in COCO, described in Sect. 15.3, is fully compatible with embedding in COCO-compatible toolboxes. It follows that toolbox-specific monitor functions and special points may be appended to the continuation problem structure concurrently with the embedding of a toolbox instance within this structure. Such toolbox-specific support for event detection and location is consistent with the task-embedding formalism, in which events associated with toolbox-specific properties of points on the solution manifold are defined by the corresponding toolbox. We illustrate this extension of the task-embedding paradigm in the context of the 'alg' and 'po' toolboxes in this and the next chapter.

Until such a time as a closed continuation problem has been constructed, COCO-compatible zero and monitor functions may be added to the continuation problem structure using the coco_add_func utility. By the time the create class method of a subclass of AtlasBase is called, for example, there is still an opportunity to append zero and monitor functions to the continuation problem structure, as the check for the compatibility between the dimensionality of the atlas algorithm and the dimensional deficit occurs only after instantiating an atlas instance.

Obvious candidates for monitor functions, to be added to the continuation problem structure by an atlas algorithm, are those functions that depend on the current projection condition, rather than simply the locus on the solution manifold. The following section demonstrates the use of such monitor functions and the corresponding atlas events in the somewhat trivial context of fold- and branch-point detection and location.

16.1 Event detection in atlas algorithms

Embeddable monitor functions must be expressible in terms of the vector of continuation variables u in order to support their inclusion in an extended continuation problem. This is not required of nonembeddable monitor functions, however, as long as these can be uniquely computed at each point on the solution manifold from properties of the corresponding chart and the current curve segment, notably the current projection condition. Since the projection condition is intimately connected to an atlas algorithm, monitor functions that depend on this condition, e.g., on the tangent matrix V, must be defined only after the restricted continuation problem has been fully assembled. This restriction is accommodated within COCO by its support for delayed construction using the `coco_add_func_after` utility.

An event is said to be an *atlas event* if the associated monitor function depends on the base point $\tilde{u}$, the tangent matrix V, the direction vector s, or any other properties particular to the current curve segment. We append atlas events, and their monitor functions, to the continuation problem structure in the `create` construction method of an `AtlasBase` subclass. In the corresponding call to `coco_add_func`, we assign either the `regular` or `singular` function type and include the `'uidx'` option followed by the string `'all'` to indicate nominal dependence on the entire vector of continuation variables.

Below we consider two examples of atlas events, associated with the detection of fold points and branch points, respectively, along a solution manifold.

16.1.1   Fold points

Let $\Gamma : h \mapsto \Gamma_h$ denote a curve segment obtained from the projection condition

$$V^T \cdot (u - \tilde{u}) - hs = 0 \tag{16.1}$$

in terms of some base point $\tilde{u}$, tangent matrix V, and direction vector s. As shown in the previous chapter, a transversal zero-crossing of the scalar-valued function

$$\psi : h \mapsto (T_h \cdot s)_i , \tag{16.2}$$

where

$$\left(\begin{array}{c} \partial_u \Phi(\Gamma_h) \\ V^T \end{array} \right) \cdot T_h = \left(\begin{array}{c} 0 \\ I_{n-m} \end{array} \right) , \tag{16.3}$$

and for some i, corresponds to a detectable fold point with respect to the ith component of the vector of continuation variables. We avoid the expense of having to solve Eq. (16.3) repeatedly for T_h by locating this special point along Γ using the third method in Sect. 15.1.1. In this case, event location is determined using a suitably interpolated tangent vector along the interpolant $h \mapsto p(h)$, introduced in conjunction with Eq. (15.18).

Notably, the function ψ is only implicitly defined in terms of h and very clearly associated with the curve segment Γ, rather than merely the properties of the chart associated with each point Γ_h along Γ. As suggested above, we integrate the definition of ψ as a nonembedded monitor function within an atlas algorithm and append the corresponding atlas event in the `create` construction method of an `AtlasBase` subclass.

We implement these ideas in a modified version of the `atlas_1d_min` subclass introduced in Sect. 14.1. We begin by making small adjustments to the class methods in order to provide content to the `chart.t` field to represent the tangent vector $T_h \cdot s$. As an example, consider the modified encoding of the `init_atlas` class method, shown below.

```
function [prob atlas cseg flush] = init_atlas(atlas, prob, cseg)

chart          = cseg.curr_chart;
chart.R        = atlas.cont.h;
chart.pt_type  = 'EP';
chart.ep_flag  = 1;
```

```
if ~isempty(chart.s) && chart.s(1)<0
  prob = AtlasBase.bddat_set(prob, 'ins_mode', 'prepend');
  chart.t = -chart.TS;
else
  prob = AtlasBase.bddat_set(prob, 'ins_mode', 'append');
  chart.t = chart.TS;
end
[prob chart]   = cseg.update_p(prob, chart);
cseg.ptlist{1} = chart;
flush          = true;

end
```

Here, the call to the `cseg.update_p` class method ensures that all monitor functions that depend on the `chart.t` field are evaluated following the initial assignment of content to this field. The corresponding changes to the `predict` class method, shown below, guarantee that such an evaluation is associated with the construction of each new curve segment.

```
function [prob atlas cseg correct] = predict(atlas, prob, cseg)

[chart xp s h] = atlas.boundary{1,:};
prcond         = struct('x', chart.x, 'TS', chart.TS, 's', s, 'h', h);
th             = atlas.cont.theta;
if th>=0.5 && th<=1
  xp           = chart.x+(th*h)*(chart.TS*s);
  [prob cseg]  = CurveSegment.create(prob, chart, prcond, xp);
  [prob ch2]   = cseg.update_TS(prob, cseg.curr_chart);
  h            = h*(ch2.TS'*chart.TS);
  xp           = chart.x+h*(ch2.TS*s);
  prcond       = struct('x', chart.x, 'TS', ch2.TS, 's', s, 'h', h);
end
[prob cseg]    = CurveSegment.create(prob, chart, prcond, xp);
[prob chart]   = cseg.update_t(prob, cseg.ptlist{1});
[prob chart]   = cseg.update_p(prob, chart);
cseg.ptlist{1} = chart;
correct        = true;

end
```

Here, the call to the `cseg.update_t` class method updates the content of the `chart.t` field according to the new projection condition.

The changes to the `atlas_1d_min` class declaration, shown below, and in particular the addition of the `add_test_FP` and `test_FP` class methods, provide support for fold-point detection.

```
classdef atlas_1d_min < AtlasBase

  properties (Access=private)
    boundary = {};
    cont     = struct();
  end

  methods (Access=private)
    function atlas = atlas_1d_min(prob, cont, dim)
      assert(dim==1, '%s: wrong manifold dimension', mfilename);
      atlas       = atlas@AtlasBase(prob);
      atlas.cont = atlas.get_settings(cont);
    end
  end
```

```
methods (Static)
  function [prob cont atlas] = create(prob, cont, dim)
    atlas = atlas_1d_min(prob, cont, dim);
    prob  = CurveSegment.add_prcond(prob, dim);
    if atlas.cont.FP
      prob = coco_add_func_after(prob, 'mfunc', ...
        @atlas_1d_min.add_test_FP);
    end
  end

  function prob = add_test_FP(prob)
    p_idx = coco_get_func_data(prob, 'efunc', 'pidx');
    if numel(p_idx)>=1
      fid  = coco_get_id('atlas', 'test', 'FP');
      prob = coco_add_func(prob, fid, @atlas_1d_min.test_FP, [], ...
        'singular', fid, 'uidx', 'all', 'passTangent', 'fdim', 1);
      prob = coco_add_event(prob, 'FP', fid, 0);
    end
  end

  function [data y] = test_FP(prob, data, u, t)
    p_idx = coco_get_func_data(prob, 'efunc', 'pidx');
    y = t(p_idx(1));
  end

end

methods (Static, Access=private)
  cont = get_settings(cont)
end

methods (Access=public)
  [prob atlas cseg correct] = init_prcond     (atlas, prob, chart)
  [prob atlas cseg]         = init_chart       (atlas, prob, cseg)
  [opts atlas cseg]         = init_admissible  (atlas, opts, cseg, S)
  [prob atlas cseg flush]   = init_atlas       (atlas, prob, cseg)
  [prob atlas cseg]         = flush            (atlas, prob, cseg)
  [prob atlas cseg correct] = predict          (atlas, prob, cseg)
  [prob atlas cseg flush]   = add_chart        (atlas, prob, cseg)
end

methods (Access=private)
  flag = isneighbor(atlas, chart1, chart2)
  [atlas cseg] = merge(atlas, cseg)
end

end
```

We append the `test_FP` monitor function to the continuation problem structure using the `add_test_FP` constructor, included above among the public static class methods in the `atlas_1d_min` class declaration. In the `add_test_FP` constructor, as well as in `test_FP`, we note the extraction of the indices of all active continuation parameters in the vector (u, μ) using the `coco_get_func_data` utility with the `'efunc'` and `'pidx'` arguments. Here, in the absence of any active continuation parameters, fold-point detection is disabled.

In a departure from the monitor function syntax introduced previously, the calling syntax of `test_FP` includes the fourth input argument `t`, containing the components of the chart.t field associated with the components of *u* that are included in the third input argument. We enable this argument syntax by including the optional 'passTangent' argument in the call to coco_add_func. The 'fdim' argument, followed by the integer 1, indicates the dimension of the range of the monitor function, thereby eliminating the need to evaluate this function before content has been assigned to the chart.t field.

Finally, we note the need to delay the call to the add_test_FP constructor until after all embeddable monitor functions have been defined. To this end, the create class method makes use of the coco_add_func_after utility and includes in the call to this function the string 'mfunc' and the function handle to the delayed constructor. The modified encoding of the get_settings class method, shown below, provides default content to the atlas.cont.FP property.

```
function cont = get_settings(cont)

defaults.FP     = true;
defaults.h      = 0.1;
defaults.PtMX   = 50;
defaults.theta  = 0.5;
defaults.almax  = 10;
defaults.Rmarg  = 0.95;
cont            = coco_merge(defaults, cont);
cont.almax      = cont.almax*pi/180;

end
```

We illustrate the functionality of this modified atlas algorithm below, following the integration of branch-point detection within its encoding.

16.1.2   Function and chart data

As first described in Chap. 4, the algorithmic design of the COCO core associates unique *function data* with each function added to a continuation problem structure using either coco_add_func or coco_add_slot. The content of the function data structure encodes information used in the function body to guide its execution and parameterizes the internal state of the function. As illustrated in the template toolboxes considered in Part II of this text, function data may be modified upon each call to the corresponding function in order to, for example, provide a record of past calls to the function or to update an internal parameter governing its functionality for later use.

By design, functions added using coco_add_func or coco_add_slot have read and write access to their own function data. In addition, run-time read access to the function data associated with any known function is provided through the coco_get_func_data core utility. In Chap. 5 this access mechanism was used to construct gluing conditions between continuation parameters associated with multiple copies of the problem parameters introduced through repeated calls to the alg_isol2eqn constructor. In Chap. 8, this access mechanism was used to impose additional boundary conditions on the collocation discretization of a solution of a system of ordinary differential equations.

A simple mechanism to enable write access to shared function data was introduced in Sect. 8.2 in the form of a coco_func_data class instance. In this alternative formulation, multiple functions could share a single set of function data by providing to each, at the moment of construction, a unique reference to the class instance. In Sect. 8.2, this mechanism was used to update a toolbox data structure prior to each new continuation step. The use of `coco_func_data` was constrained, not by design but by recommendation, to the function data argument of `coco_add_func` and `coco_add_slot`.

An alternative mechanism for storing and sharing data is afforded in COCO by associating unique *chart data* with each chart on a curve segment, in addition to that already contained within the default chart fields, e.g., `x` and `TS`. Such chart data encodes information specific to this chart and may, for example, be used to provide access for individual toolbox functions to chart-specific properties computed by other functions belonging to the same toolbox or by functions belonging to other toolboxes.

Individual elements of chart data must be instantiated prior to entering the COCO finite-state machine by invoking the `coco_add_chart_data` utility. In the simplest case, a call to this function takes the form

```
prob = coco_add_chart_data(prob, fid);
```

where `fid` is a data identifier to the corresponding element of the chart data. With this calling syntax, the new element of the chart data is initialized to be empty and chart data associated with this entry is set to be automatically inherited by all charts generated from this chart during continuation. This default functionality may be modified by the inclusion of one or two additional arguments in the call to the `coco_add_chart_data` utility. As an example, the command

```
prob = coco_add_chart_data(prob, fid, [], []);
```

instantiates an empty element of the chart data and ensures that this element is again empty in any descendant chart.

Using the `coco_set_chart_data` and `coco_get_chart_data` utilities, read and write access to chart data may be given to any function added to the continuation problem by including the optional argument `'passChart'` in the call to the `coco_add_func` utility. Within such a function, a particular element of the chart data may then be accessed and modified provided that its data identifier is known to the function, either because it is stored in the continuation problem structure or because it has been stored by a toolbox constructor in the function's own data. An example of the former construction will be given in the next section. The case of toolbox-specific chart data will be expanded upon in the next chapter.

16.1.3 Branch point detection

In Sect. 15.1, we proposed using a subdivision method for locating special points coincident with simple branch points of the closed continuation problem

$$\begin{pmatrix} \Phi(u) \\ V^T \cdot (u - \tilde{u}) - hs \end{pmatrix} = 0, \tag{16.4}$$

i.e., points for which the matrix

$$\begin{pmatrix} \partial_u \Phi(u) \\ V^T \end{pmatrix} \tag{16.5}$$

is noninvertible and at which the corresponding determinant goes through zero continuously and transversally. We consider, in this section, the use of chart data in support of the detection and approximate location of such special points associated with the vanishing of a suitably scaled version of the determinant of this matrix. As the matrix is defined not only by the function Φ of some extended continuation problem but also by the tangent matrix V of the projection condition, we seek to implement the corresponding monitor function within an atlas algorithm.

We make the following two observations. As noted earlier, the matrix in Eq. (16.5) depends not only on the vector u of continuation variables but also on the tangent matrix V of the current curve segment. It follows that a call to the `coco_add_func` utility, to add a suitable monitor function to the continuation problem, would need to include the `'passChart'` optional argument, since otherwise the tangent matrix information could not be made available to the monitor function.

More importantly, we note that the evaluation of the determinant requires the computation of the Jacobian $\partial_u \Phi(u)$. Not only may this be computationally costly, but it also seems unnecessary, since this Jacobian is already evaluated during the execution of the `cseg.add_chart` class method, in advance of a subsequent evaluation of $\widetilde{\Psi}$. It would, of course, be ideal if the Jacobian $\partial_u \Phi(u)$, computed by this latter function or, even better, the desired determinant could be stored with the chart and later extracted by the monitor function designed for locating a branch point.

The desired functionality is, in fact, implemented in the default linear solver algorithm used by the `cseg.add_chart` class method to solve the equation

$$\begin{pmatrix} \partial_u \Phi(u) \\ V^T \end{pmatrix} \cdot T = \begin{pmatrix} 0 \\ I_{n-m} \end{pmatrix} \tag{16.6}$$

for the tangent space matrix T. Provided that the command

```
prob = coco_set(prob, 'lsol', 'det', true);
```

is executed prior to entering the finite-state machine, the linear solver algorithm stores a scaled version of the determinant of the coefficient matrix in the `det` field of the chart data element associated with the data identifier `'lsol'`.

We proceed to consider modifications to the atlas algorithm presented in Sect. 16.1.1. The encoding of the `atlas_1d_min` class declaration, shown below, provides support for branch-point detection by the method suggested above.

```
classdef atlas_1d_min < AtlasBase

  properties (Access=private)
    boundary = {};
    cont     = struct();
  end

  methods (Access=private)
    function atlas = atlas_1d_min(prob, cont, dim)
      assert(dim==1, '%s: wrong manifold dimension', mfilename);
      atlas      = atlas@AtlasBase(prob);
      atlas.cont = atlas.get_settings(cont);
    end
  end

  methods (Static)
    function [prob cont atlas] = create(prob, cont, dim)
      atlas = atlas_1d_min(prob, cont, dim);
```

```
      prob  = CurveSegment.add_prcond(prob, dim);
      if atlas.cont.FP
        prob = coco_add_func_after(prob, 'mfunc', ...
          @atlas_1d_min.add_test_FP);
      end
      if atlas.cont.BP
        prob = coco_set(prob, 'lsol', 'det', true);
        fid  = coco_get_id('atlas', 'test', 'BP');
        prob = coco_add_func(prob, fid, @atlas.test_BP, [], ...
          'singular', fid, 'uidx', 'all', 'passChart', ...
          'returnsProb', 'fdim', 1);
        prob = coco_add_event(prob, 'BP', fid, 0);
      end
    end

    function prob = add_test_FP(prob)
      p_idx = coco_get_func_data(prob, 'efunc', 'pidx');
      if numel(p_idx)>=1
        fid  = coco_get_id('atlas', 'test', 'FP');
        prob = coco_add_func(prob, fid, @atlas_1d_min.test_FP, [], ...
          'singular', fid, 'uidx', 'all', 'passTangent', 'fdim', 1);
        prob = coco_add_event(prob, 'FP', fid, 0);
      end
    end

    function [data y] = test_FP(prob, data, u, t)
      p_idx = coco_get_func_data(prob, 'efunc', 'pidx');
      y = t(p_idx(1));
    end

    function [prob data chart y] = test_BP(prob, data, chart, u)
      cdata = coco_get_chart_data(chart, 'lsol');
      if ~isfield(cdata, 'det')
        [prob chart] = prob.cseg.update_det(prob, chart);
        cdata = coco_get_chart_data(chart, 'lsol');
      end
      y = cdata.det;
    end

  end

  methods (Static, Access=private)
    cont = get_settings(cont)
  end

  methods (Access=public)
    [prob atlas cseg correct] = init_prcond    (atlas, prob, chart)
    [prob atlas cseg]         = init_chart      (atlas, prob, cseg)
    [opts atlas cseg]         = init_admissible (atlas, opts, cseg, S)
    [prob atlas cseg flush]   = init_atlas      (atlas, prob, cseg)
    [prob atlas cseg]         = flush           (atlas, prob, cseg)
    [prob atlas cseg correct] = predict         (atlas, prob, cseg)
    [prob atlas cseg flush]   = add_chart       (atlas, prob, cseg)
  end

  methods (Access=private)
    flag = isneighbor(atlas, chart1, chart2)
    [atlas cseg] = merge(atlas, cseg)
  end

end
```

Here, the encoding of the `test_BP` method adheres to the standard practice of reusing chart data whenever possible and appropriate. As expected, the call to the COCO core utility `coco_get_chart_data` assigns the chart data element associated with the `'lsol'` data identifier to `cdata`. To accommodate a possible change to the `prob` continuation problem structure, `prob` is included as the first output argument of the `test_BP` function. We enable this modified argument syntax by including the `'returnsProb'` flag among the arguments in the corresponding call to `coco_add_func` in the `create` class method.

The modifications to the `get_settings` class method, shown below, enable branch-point detection and location, unless otherwise indicated by an explicit assignment of the boolean value `false` to the `'BP'` setting of the `'cont'` toolbox.

```
function cont = get_settings(cont)

defaults.FP    = true;
defaults.BP    = true;
defaults.h     = 0.1;
defaults.PtMX  = 50;
defaults.theta = 0.5;
defaults.almax = 10;
defaults.Rmarg = 0.95;
cont           = coco_merge(defaults, cont);
cont.almax     = cont.almax*pi/180;

end
```

16.1.4 Examples

We demonstrate the application of the modified `atlas_1d_min` atlas algorithm to the zero problem given by the zero function

$$\Phi(u) = (u_1^2 + u_2^2)^2 + u_2^2 - u_1^2 \tag{16.7}$$

and the COCO-compatible encoding in the `lemniscate` function. As described in Chap. 12, we identify the atlas algorithm to the COCO finite-state machine by making the appropriate modifications to the local project settings in `coco_project_opts`.

Example 16.1 The extract below shows the successful detection of the branch point at $u = (0,0)$ and the fold points at $u = (\pm 1, 0)$.

```
>> prob = coco_add_func(coco_prob(), 'lemniscate', @lemniscate, ...
     [], 'zero', 'u0', [0.6; 0.4]);
>> prob = coco_add_pars(prob, '', [1 2], {'x' 'y'});
>> prob = coco_set(prob,'cont', 'almax', 15);
>> cont_pars = {'x' 'y' 'atlas.test.FP' 'atlas.test.BP'};
>> coco(prob, 'lemniscate', [], 1, cont_pars);
```

...	NORMS		COMPUTATION TIMES		
...	$\|\|f\|\|$	$\|\|U\|\|$	F(x)	DF(x)	SOLVE
...	7.04e-02	1.02e+00	0.0	0.0	0.0
...	4.86e-03	9.87e-01	0.0	0.0	0.0
...	2.95e-05	9.85e-01	0.0	0.0	0.0
...	1.11e-09	9.85e-01	0.0	0.0	0.0
...	0.00e+00	9.85e-01	0.0	0.0	0.0

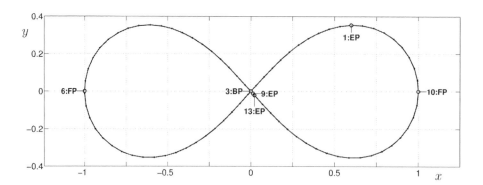

Figure 16.1. *Computation of the lemniscate curve in Example* 16.1 *with detection of atlas events. Two fold points at labels* 6 *and* 10 *and a branch point at label* 3 *are located. The covering closes at the end points* 9 *and* 13.

```
... LABEL  TYPE           x            y atlas.test.FP atlas.test.BP
...    1   EP     6.0000e-01   3.5339e-01   -7.0687e-01   -1.0000e+00
...    2          3.0589e-02   3.0532e-02   -5.0140e-01   -2.6843e-01
...    3   BP    -2.7142e-08  -2.7626e-08   -5.0000e-01    4.3787e-07
...    4         -5.1434e-01  -3.4368e-01   -6.9352e-01    1.0000e+00
...    5         -9.9976e-01  -1.2717e-02   -2.6975e-02    1.0000e+00
...    6   FP    -1.0000e+00   4.0174e-05    2.0914e-08    1.0000e+00
...    7         -5.3945e-01   3.4805e-01    6.9936e-01    1.0000e+00
...    8   BP     3.9544e-08  -4.0114e-08    5.0000e-01   -6.3677e-07
...    9   EP     1.2513e-02  -1.2509e-02    5.0023e-01   -1.3368e-01
...   10   FP     1.0000e+00   2.9602e-05   -3.4454e-08   -1.0000e+00
...   11          9.9195e-01  -7.2837e-02   -1.5423e-01   -1.0000e+00
...   12          4.5544e-01  -3.2857e-01   -6.7466e-01   -1.0000e+00
...   13   EP     2.1378e-02  -2.1359e-02   -5.0068e-01   -2.0470e-01
```

Notably, the method chosen for locating the singular point finds only an approximant to this point along a suitably designed interpolant through two points on either side of the event set. The solution manifold and the associated atlas events are shown in Fig. 16.1. ∎

As a second example, we consider the continuation along a family of periodic responses of period $2\pi/\omega$ of the harmonically excited *Duffing equation*, given by

$$\frac{d^2x}{dt^2} + \lambda\frac{dx}{dt} + \alpha x + \epsilon x^3 = A\cos\omega t, \tag{16.8}$$

under variations in the excitation frequency ω. Notably, if $x(t)$ denotes one such periodic orbit, then $x(t+\tau)$ is a solution to the identical differential equation, albeit with the *phase-shifted excitation* $A\cos\omega(t+\tau)$.

We obtain a 'po'-compatible formulation by replacing the explicitly nonautonomous excitation $\cos\omega t$ with the sinusoidal steady-state output $r\cos\theta$ of the nonlinear oscillator

$$\frac{dr}{dt} = r(1-r), \frac{d\theta}{dt} = \omega. \tag{16.9}$$

In rectangular coordinates, and coupled to a first-order formulation of Eq. (16.8), we obtain the corresponding four-dimensional vector field

$$
f(y,p) := \begin{pmatrix} y_2 \\ p_4 y_4 - p_1 y_2 - p_2 y_1 - p_3 y_1^3 \\ y_3 + p_5 y_4 - y_3 \left(y_3^2 + y_4^2 \right) \\ -p_5 y_3 + y_4 - y_4 \left(y_3^2 + y_4^2 \right) \end{pmatrix},
\tag{16.10}
$$

where $p := (\lambda, \alpha, \epsilon, A, \omega)$.

Example 16.2 A *frequency-response curve*, illustrating the variations in the norm $\| (u, \mu) \|$ under variations in ω along the 1-dimensional solution manifold obtained for fixed values of λ, α, ϵ, and A, is shown in Fig. 16.2. The extract below shows the sequence of assignments used to construct and initialize the corresponding restricted continuation problem, given a 'coll'-compatible encoding of the vector field in Eq. (16.10) in the function duff and executing a continuation for $\omega \in [1/2, 7/2]$.

```
>> p0 = [0.2; 1; 1; 2.5; 1];
>> x0 = [0; 0; 1; 0];
>> [t0 x0] = ode45(@(t,x) duff(x,p0), [0 30], x0);
>> [t0 x0] = ode45(@(t,x) duff(x,p0), [0 2*pi], x0(end,:)');
>> prob = coco_set(coco_prob(), 'coll', 'NTST', 30);
>> prob = po_isol2orb(prob, '', @duff, t0, x0, ...
       {'la' 'al' 'eps' 'A' 'om'}, p0);
>> prob = coco_set(prob, 'cont', 'PtMX', 500, 'h', 0.5, 'almax', 30);
>> coco(prob, 'duffing', [], 1, 'om', [0.5 3.5]);
```

As seen in Fig. 16.2, the solution manifold includes four fold points and two branch points of periodic orbits. Representative orbits are shown in Fig. 16.3. ∎

16.2 A toolbox template

A notable shortcoming of the mechanism for fold detection and location integrated within the atlas_1d_min subclass is its dependence on properties of the current curve segment, rather than the location on the solution manifold. Accordingly, the test_FP monitor function was appended to the continuation problem structure as an element of the collection of nonembeddable monitor functions, eliminating the possibility of imposing constraints on this function by deactivating the corresponding continuation parameter. While this restriction applies to the general continuation problem, particular classes of continuation problems, and particular families of fold points, support more specialized implementations, with full support for embeddable monitor functions.

As an example, we proceed to consider the development of a toolbox that adds to the extended continuation problem constructed by the 'alg' toolbox in Chap. 4 those additional elements of Ψ required to detect, locate, and trace folds with respect to individual problem parameters. This toolbox serves as a template for a basic bifurcation analysis tool for equilibria of continuous-time dynamical systems, as illustrated further in the next chapter, and may be easily modified to apply to the case of fixed points of maps and, consequently, multisegment periodic trajectories.

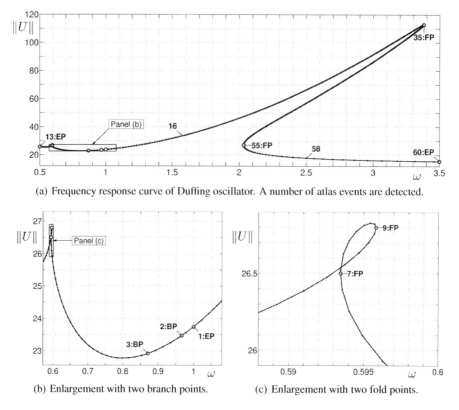

(a) Frequency response curve of Duffing oscillator. A number of atlas events are detected.

(b) Enlargement with two branch points.   (c) Enlargement with two fold points.

Figure 16.2. *Frequency response curve of the Duffing oscillator given by the vector field in Eq.* (16.10) *for $A = 2.5$, $\lambda = 0.2$, and $\alpha = \varepsilon = 1$, as computed in Example* 16.2 *with $\|U\| := \left\| \left(u, \mu_{\mathbb{J}} \right) \right\|$. A number of atlas events are detected along this curve, as shown in panel* (a) *and the enlargements* (b) *and* (c). *For sample solutions at labels* 16, 35, *and* 58 *along the curve we observe a characteristic phase shift, as illustrated in Fig.* 16.3.

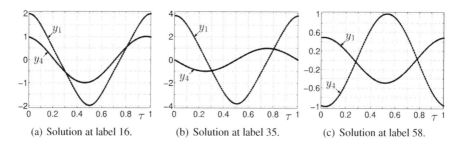

(a) Solution at label 16.   (b) Solution at label 35.   (c) Solution at label 58.

Figure 16.3. *For oscillations obtained at both sides of the main resonance peak of the frequency response curve shown in Fig.* 16.2, *we observe a characteristic phase shift between the forcing $y_4(t)$ and the response $y_1(t)$, here shown over the normalized time $\tau = t/T$. To the left, both oscillations are in phase* (a), *while to the right we observe an antiphase response* (c). *The transition occurs close to the fold point at label* 35 (b).

16.2.1   Problem formulation

Suppose, as in Chap. 4, that there is a natural decomposition of the vector of continuation variables:

$$u := (x, p) \in \mathbb{R}^m \times \mathbb{R}^{n-m}, \tag{16.11}$$

where x denotes problem variables, p denotes problem parameters, and $n > m \geq 1$. Let $\Phi : \mathbb{R}^m \times \mathbb{R}^{n-m} \to \mathbb{R}^m$ denote the corresponding family of zero functions and let the default extended continuation problem be defined by $F(u, \mu) = 0$, where

$$F(u, \mu) = \begin{pmatrix} \Phi(x, p) \\ p - \mu \end{pmatrix} \tag{16.12}$$

and $\mu \in \mathbb{R}^{n-m}$. An initializable restricted continuation problem then follows with the assignment $\mathbb{I} = \{1, \ldots, n - m\}$.

As in the previous chapter, folds with respect to individual problem parameters (while keeping all other problem parameters fixed) coincide with special points on the solution manifold associated with the vanishing of the function

$$\psi : u \mapsto \det\left(U^T \cdot T\right), \tag{16.13}$$

where $U^T = \begin{pmatrix} 0 & I_{n-m} \end{pmatrix}$ and the columns of $T \in \mathbb{R}^{n \times (n-m)}$ form a continuous basis of the tangent space at each point of the solution manifold. It follows that folds with respect to individual problem parameters occur only at points where the matrix $\partial_x \Phi$ is singular.

Let $u^* := (x^*, p^*)$ denote a fold that coincides with a special point associated with the determinant $\det(\partial_x \Phi)$ and the scalar 0 and assume that this point is detectable along a curve segment Γ. Following Govaerts and Kuznetsov, now suppose that b and c have been chosen such that the *bordered matrix*

$$M := \begin{pmatrix} \partial_x \Phi & b \\ c^T & 0 \end{pmatrix} \tag{16.14}$$

is invertible along Γ. Let v, w, h, and g be uniquely defined by the equations

$$M \cdot \begin{pmatrix} v \\ -h \end{pmatrix} = \begin{pmatrix} 0 \\ 1 \end{pmatrix} \tag{16.15}$$

and

$$\begin{pmatrix} w^T & -g \end{pmatrix} \cdot M = \begin{pmatrix} 0 & 1 \end{pmatrix}. \tag{16.16}$$

By Cramer's rule, it follows that

$$h = g = -\frac{\det \partial_x \Phi}{\det M} \tag{16.17}$$

and thus that the fold is a detectable special point along Γ associated with the function $(u, \mu_{\mathbb{J}}) \mapsto h$ and the scalar 0. Implicit differentiation of Eq. (16.15) and premultiplication by $\begin{pmatrix} w^T & -g \end{pmatrix}$ further yields

$$\partial_x h = w^T \cdot \partial_{xx} \Phi[v, \cdot] \tag{16.18}$$

and

$$\partial_p h = w^T \cdot \partial_{xp} \Phi[v, \cdot] \tag{16.19}$$

for the Jacobians with respect to the problem variables and the problem parameters. Here, the equalities

$$\partial_{xx} \Phi(x, p)[t_x, \cdot] = \frac{d}{d\epsilon} \partial_x \Phi(x + \epsilon t_x, p)|_{\epsilon=0} \tag{16.20}$$

and

$$\partial_{xp} \Phi(x, p)[t_x, \cdot] = \frac{d}{d\epsilon} \partial_p \Phi(x + \epsilon t_x, p)\big|_{\epsilon=0} \tag{16.21}$$

may be used in order to approximate $\partial_x h$ and $\partial_p h$ by suitable finite-difference approximations.

It remains to determine whether vectors b and c can be found to ensure the invertibility of the matrix M in Eq. (16.14) throughout continuation. To this end, suppose that Γ consists only of regular points and let b and c be the left and right singular vectors of unit length corresponding to the smallest singular value h of $\partial_x \Phi$ at some point along Γ, i.e., such that

$$\begin{pmatrix} b^T & -h \end{pmatrix} \cdot \begin{pmatrix} \partial_x \Phi & b \\ c^T & 0 \end{pmatrix} = \begin{pmatrix} 0 & 1 \end{pmatrix} \tag{16.22}$$

and

$$\begin{pmatrix} \partial_x \Phi & b \\ c^T & 0 \end{pmatrix} \cdot \begin{pmatrix} c \\ -h \end{pmatrix} = \begin{pmatrix} 0 \\ 1 \end{pmatrix} \tag{16.23}$$

at this point. The equation

$$\begin{pmatrix} \partial_x \Phi & b \\ c^T & 0 \end{pmatrix} \cdot \begin{pmatrix} \alpha \\ \beta \end{pmatrix} = 0 \tag{16.24}$$

implies that

$$\partial_x \Phi \cdot \alpha + \beta b = 0, \quad c^T \cdot \alpha = 0, \tag{16.25}$$

and, consequently, that

$$\partial_x \Phi \cdot (h\alpha + \beta c) = 0. \tag{16.26}$$

For nonsingular $\partial_x \Phi$, it follows that $h \neq 0$ and $h\alpha + \beta c = 0$. The second condition in Eq. (16.25) then implies that $\alpha = \beta = 0$. Suppose, instead, that $\partial_x \Phi$ has a rank loss of 1 and recall that the image of $\partial_x \Phi$ coincides with the orthogonal complement to the kernel of $(\partial_x \Phi)^T$. Since the kernel of $(\partial_x \Phi)^T$ equals span$\{b\}$, this immediately implies that β must equal 0 and, consequently, that $\alpha = 0$, since the kernel of $\partial_x \Phi$ equals span$\{c\}$. It follows by continuity that the matrix in Eq. (16.14) is invertible on some portion of the curve segment containing the chosen point.

While the computational effort associated with determining the singular vectors of $\partial_x \Phi$ may be reasonable during the initialization of the toolbox data structure, it is desirable to substitute a less operationally intensive algorithm for subsequent updates. To this end, suppose that b_0 and c_0 are two unit vectors such that

$$\begin{pmatrix} \partial_x \Phi & b_0 \\ c_0^T & 0 \end{pmatrix} \tag{16.27}$$

is invertible at a previously located point on the solution manifold. Let b_1 be defined from the unique solution to the equation

$$\begin{pmatrix} b_1^T & h_1 \end{pmatrix} \cdot \begin{pmatrix} \partial_x \Phi & b_0 \\ c_0^T & 0 \end{pmatrix} = \begin{pmatrix} 0 & 1 \end{pmatrix} \tag{16.28}$$

or, equivalently,

$$b_1^T \cdot \partial_x \Phi + h_1 c_0^T = 0, \, b_1^T \cdot b_0 = 1. \tag{16.29}$$

Now suppose that c_0 is close to the eigenspace of $(\partial_x \Phi)^T \cdot \partial_x \Phi$ corresponding to the smallest eigenvalue of this matrix and that this eigenvalue is simple. It follows that the matrix

$$\begin{pmatrix} \partial_x \Phi & b_1 \\ c_0^T & 0 \end{pmatrix} \tag{16.30}$$

is invertible, since

$$\partial_x \Phi \cdot \alpha + \beta b_1 = 0, \, c_0^T \cdot \alpha = 0, \tag{16.31}$$

then implies that $\alpha = \beta = 0$. The update of the *bordering vectors* b and c is now accomplished by defining c_1 in terms of the unique solution to the equation

$$\begin{pmatrix} \partial_x \Phi & b_1 \\ c_0^T & 0 \end{pmatrix} \cdot \begin{pmatrix} c_1 \\ g_1 \end{pmatrix} = \begin{pmatrix} 0 \\ 1 \end{pmatrix}. \tag{16.32}$$

16.2.2 Encoding

We proceed to modify the last version of the `'alg'` toolbox from Chap. 5 in order to support detection and location of fold points, as well as the continuation along submanifolds of fold points on the solution manifold to the extended continuation problem $F(u, \mu) = 0$. We leave as is the `alg_isol2eqn` and `alg_sol2eqn` toolbox constructors, the `alg_arg_check` error checking function, the `alg_read_solution` toolbox extractor, the zero-function wrapper `alg_F`, and the `alg_bddat` slot function. As we have reason to evaluate the Jacobian $\partial_x \Phi$ also outside of the Jacobian of the zero-function wrapper, we introduce the functions `alg_fhan_DFDX` and `alg_fhan_DFDP`, shown below.

```
function Jx = alg_fhan_DFDX(data, x, p)

if isempty(data.dfdxhan)
  Jx = coco_ezDFDX('f(x,p)', data.fhan, x, p);
else
  Jx = data.dfdxhan(x, p);
end

end
```

```
function Jp = alg_fhan_DFDP(data, x, p)

if isempty(data.dfdphan)
  Jp = coco_ezDFDP('f(x,p)', data.fhan, x, p);
else
  Jp = data.dfdphan(x, p);
end

end
```

We modify the encoding of the full Jacobian of the zero function as follows.

```
function [data J] = alg_DFDU(prob, data, u)

x  = u(data.x_idx);
p  = u(data.p_idx);

J1 = alg_fhan_DFDX(data, x, p);
J2 = alg_fhan_DFDP(data, x, p);
J  = sparse([J1 J2]);

end
```

The function `alg_fold`, shown below, now encodes the embeddable monitor function ψ for detecting and locating fold points, defined in the previous section.

```
function [data y] = alg_fold(prob, data, u)

x  = u(data.x_idx);
p  = u(data.p_idx);
Jx = alg_fhan_DFDX(data, x, p);
v  = [Jx data.b; data.c' 0]\data.rhs;
y  = v(end);

end
```

Here, the `u` input argument contains the components of the vector of continuation variables corresponding to x and p, collectively. The evaluation of the Jacobian of the monitor function with respect to these components is encoded in the `alg_fold_DFDU` function, shown below.

```
function [data J] = alg_fold_DFDU(prob, data, u)

x  = u(data.x_idx);
p  = u(data.p_idx);
Jx = alg_fhan_DFDX(data, x, p);
M  = [Jx data.b; data.c' 0];
v  = M\data.rhs;
w  = data.rhs'/M;

h  = 1.0e-4*(1+norm(x));
J0 = alg_fhan_DFDX(data, x-h*v(data.x_idx), p);
J1 = alg_fhan_DFDX(data, x+h*v(data.x_idx), p);
hx = -w(data.x_idx)*(0.5/h)*(J1-J0);

J0 = alg_fhan_DFDP(data, x-h*v(data.x_idx), p);
J1 = alg_fhan_DFDP(data, x+h*v(data.x_idx), p);
hp = -w(data.x_idx)*(0.5/h)*(J1-J0);

J  = [hx hp];

end
```

We initialize the bordering vectors b and c by invoking the MATLAB svds command to extract the singular value of the Jacobian of the zero problem that is closest to 0 together with its left and right singular vectors, as shown in the following modified encoding of the alg_init_data function.

```
function data = alg_init_data(data, x0, p0)

xdim        = numel(x0);
pdim        = numel(p0);
data.x_idx = (1:xdim)';
data.p_idx = xdim+(1:pdim)';

Jx          = alg_fhan_DFDX(data, x0, p0);
[data.b, ~, data.c] = svds(Jx,1,0);
data.rhs   = [zeros(xdim,1); 1];

end
```

Finally, from the theoretical treatment in the previous section, we recognize the need to update the content of the function data structure during continuation in order to ensure the invertibility of the coefficient matrices in Eq. (16.14). Following the methodology introduced in Sect. 8.2, we therefore assume that the function data structure for the second event function is cast as a coco_func_data class instance. The alg_update function, shown below, makes these modifications to the toolbox data structure prior to each new continuation step.

```
function data = alg_update(prob, data, cseg, varargin)

chart   = cseg.src_chart;
uidx    = coco_get_func_data(prob, data.tbid, 'uidx');
u       = chart.x(uidx);
x       = u(data.x_idx);
p       = u(data.p_idx);
Jx      = alg_fhan_DFDX(data, x, p);
w       = data.rhs'/[Jx data.b; data.c' 0];
data.b = w(data.x_idx)';
data.b = data.b/norm(data.b);
v       = [Jx data.b; data.c' 0]\data.rhs;
data.c = v(data.x_idx);
data.c = data.c/norm(data.c);

end
```

We complete the encoding of the modified 'alg' toolbox by appending the alg_fold monitor function to the extended continuation problem in the alg_construct_eqn constructor, as shown below.

```
function prob = alg_construct_eqn(prob, tbid, data, sol)

prob = coco_add_func(prob, tbid, @alg_F, @alg_DFDU, data, 'zero', ...
  'u0', sol.u);
uidx = coco_get_func_data(prob, tbid, 'uidx');
if ~isempty(data.pnames)
   fid  = coco_get_id(tbid, 'pars');
   prob = coco_add_pars(prob, fid, uidx(data.p_idx), data.pnames);
end
prob = coco_add_slot(prob, tbid, @coco_save_data, data, 'save_full');
```

```
if data.alg.norm
  data.tbid = tbid;
  prob = coco_add_slot(prob, tbid, @alg_bddat, data, 'bddat');
end

switch data.alg.FO
  case {'regular', 'active'}
    fid_FO = coco_get_id(tbid, 'test', 'FO');
    data.tbid = tbid;
    data = coco_func_data(data);
    prob = coco_add_func(prob, fid_FO, @alg_fold, ...
      @alg_fold_DFDU, data, data.alg.FO, fid_FO, ...
      'uidx', uidx, 'fdim', 1);
    prob = coco_add_slot(prob, tbid, @alg_update, data, 'update');
    prob = coco_add_event(prob, 'FO', fid_FO, 0);
  otherwise
end

end
```

Here, we note the use of the `'FO'` setting of the `'alg'` toolbox to distinguish between the use of the `alg_fold` as a monitor function of function type `'regular'` for fold detection and location and of function type `'active'` for fold (detection, location, and) continuation, respectively. A default value of the `'FO'` setting is then provided by the modified encoding of the `alg_get_settings` function, shown below.

```
function data = alg_get_settings(prob, tbid, data)

defaults.norm = false;
defaults.FO   = 'regular';
data.alg = coco_merge(defaults, coco_get(prob, tbid));
assert(islogical(data.alg.norm), ...
  '%s: input for ''norm'' option is not boolean', tbid);
assert(ischar(data.alg.FO), ...
  '%s: input for ''FO'' option is not a string', tbid);

end
```

16.2.3 The cusp normal form

We return to the analysis of equilibria of the cusp normal form corresponding to the zero function

$$\Phi(u) = \kappa - x\left(\lambda - x^2\right), \tag{16.33}$$

where $u = (x, \kappa, \lambda)$ and κ and λ are considered problem parameters. Let the functions cusp, cusp_DFDX, and cusp_DFDP contain `'alg'`-compatible encodings of Φ and its Jacobians $\partial_x \Phi$ and $\partial_p \Phi$. In the extract below, we first continue along a 1-dimensional branch of solutions to the zero problem $\Phi(u) = 0$, for fixed $\lambda = 0.5$, with event location using the second method in Sect. 15.1.1.

```
>> alg_args = {@cusp, @cusp_DFDX, @cusp_DFDP, 0, {'ka' 'la'}, ...
     [0; 0.5]};
>> prob = alg_isol2eqn(coco_prob(), '', alg_args{:});
>> coco(prob, 'regular', [], 1, {'ka' 'la' 'alg.test.FO'}, [-0.5 0.5]);
```

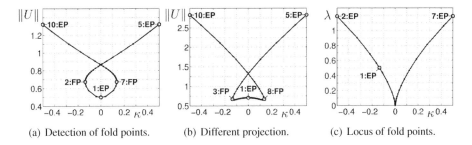

(a) Detection of fold points. (b) Different projection. (c) Locus of fold points.

Figure 16.4. *Fold-point detection and continuation for the cusp normal form considered in Sect. 16.2.3 using the monitor function defined in Sect. 16.2.1 embedded in the* alg *toolbox. Panels* (a) *and* (b) *show the result of computing the solution to the restricted continuation problem illustrated in Fig. 15.2 with the monitor function added as either* regular *or* active, *respectively, and with* $\|U\| := \left\| (u, \mu_{\mathbb{J}}) \right\|$. *The difference in appearance is due to the exclusion or inclusion of the value of the monitor function in the vector of continuation parameters* μ. *Two fold points are located in each case. Restarting at one of these fold points while restricting the monitor function to* 0 *results in a covering of the locus of fold points* (c); *see also Fig. 15.1.*

```
...     NORMS                COMPUTATION TIMES
...     ||f||       ||U||    F(x)   DF(x)   SOLVE
... 0.00e+00   5.00e-01      0.0    0.0     0.0
... 0.00e+00   5.00e-01      0.0    0.0     0.0

... LABEL  TYPE           ka            la   alg.test.FO
...     1  EP      0.0000e+00   5.0000e-01   -5.0000e-01
...     2  FO     -1.3608e-01   5.0000e-01    4.8773e-08
...     3         -1.2073e-01   5.0000e-01    2.9755e-01
...     4          1.0362e-01   5.0000e-01    1.3915e+00
...     5  EP      5.0000e-01   5.0000e-01    2.5000e+00

... LABEL  TYPE           ka            la   alg.test.FO
...     6  EP      0.0000e+00   5.0000e-01   -5.0000e-01
...     7  FO      1.3608e-01   5.0000e-01    4.8773e-08
...     8          1.2073e-01   5.0000e-01    2.9755e-01
...     9         -1.0362e-01   5.0000e-01    1.3915e+00
...    10  EP     -5.0000e-01   5.0000e-01    2.5000e+00
```

We graph a projection of the solution manifold onto the plane whose coordinates are κ and $\|(u, \mu)\|$ in Fig. 16.4(a). The following extract shows the differences in screen output that result from embedding the fold-detection monitor function in the extended continuation problem, in which case event location is implemented using the first method in Sect. 15.1.1.

```
>> alg_args = {@cusp, @cusp_DFDX, @cusp_DFDP, 0, {'ka' 'la'}, ...
     [0; 0.5]};
>> prob = coco_set(coco_prob(), 'alg', 'FO', 'active');
>> prob = alg_isol2eqn(prob, '', alg_args{:});
>> coco(prob, 'active', [], 1, {'ka' 'la' 'alg.test.FO'}, [-0.5 0.5]);
```

```
...     NORMS                COMPUTATION TIMES
...     ||f||       ||U||    F(x)   DF(x)   SOLVE
```

```
... 0.00e+00  7.07e-01    0.0    0.0    0.0
... 0.00e+00  7.07e-01    0.0    0.0    0.0

... LABEL  TYPE              ka          la  alg.test.FO
...     1  EP      0.0000e+00  5.0000e-01  -5.0000e-01
...     2          -9.9771e-02  5.0000e-01  -3.5323e-01
...     3  FO      -1.3608e-01  5.0000e-01   3.3307e-16
...     4          -5.3301e-02  5.0000e-01   7.5253e-01
...     5  EP      5.0000e-01  5.0000e-01   2.5000e+00

... LABEL  TYPE              ka          la  alg.test.FO
...     6  EP      0.0000e+00  5.0000e-01  -5.0000e-01
...     7           9.9771e-02  5.0000e-01  -3.5323e-01
...     8  FO       1.3608e-01  5.0000e-01   3.3307e-16
...     9           5.3301e-02  5.0000e-01   7.5253e-01
...    10  EP      -5.0000e-01  5.0000e-01   2.5000e+00
```

The corresponding projection onto the plane whose coordinates are κ and $\|(u, \mu)\|$ is shown in Fig. 16.4(b). The differences in projection result from the inclusion of an additional element in the vector of continuation parameters.

We may proceed to reconstruct the continuation problem structure from one of the fold points detected during the first run, and to continue a branch of such fold points, for varying κ and λ. The continued extract below demonstrates the use of the `coco_xchg_pars` utility in order to switch the assignment of integer indices for the `'ka'` and `'alg.test.FO'` continuation parameters between $\mathbb{I}$ and $\mathbb{J}$.

```
>> prob = coco_set(coco_prob(), 'alg', 'FO', 'active');
>> labs = coco_bd_labs(bd1, 'FO');
>> prob = alg_sol2eqn(prob, '', 'regular', labs(1));
>> prob = coco_xchg_pars(prob, 'la', 'alg.test.FO');
>> coco(prob, 'cusp', [], 1, {'ka' 'la' 'alg.test.FO'}, [-0.5 0.5]);
```

```
...     NORMS              COMPUTATION TIMES
...    .||f||      ||U||    F(x)   DF(x)  SOLVE
... 2.78e-17  8.39e-01    0.0    0.0    0.0
... 2.78e-17  8.39e-01    0.0    0.0    0.0

... LABEL  TYPE              ka          la  alg.test.FO
...     1  EP      -1.3608e-01  5.0000e-01   4.8773e-08
...     2  EP      -5.0000e-01  1.1906e+00   4.8773e-08

... LABEL  TYPE              ka          la  alg.test.FO
...     3  EP      -1.3608e-01  5.0000e-01   4.8773e-08
...     4          -1.1000e-04  4.3387e-03   4.8773e-08
...     5           5.4742e-04  1.2647e-02   4.8773e-08
...     6           3.4574e-02  2.0057e-01   4.8773e-08
...     7  EP       5.0000e-01  1.1906e+00   4.8773e-08
```

The corresponding solution manifold is shown in Fig. 16.4(c). We obtain an alternative discrete approximation of the branch of fold points by deploying fold-point detection and location during continuation with the final version of the 2-dimensional atlas algorithm in Chap. 14, under variations in κ and λ, as shown in the extract below.

```
>> alg_args = {@cusp, @cusp_DFDX, @cusp_DFDP, 0, {'ka' 'la'}, ...
       [0; 0.5]};
>> prob = alg_isol2eqn(coco_prob(), '', alg_args{:});
>> prob = coco_set(prob, 'cont', 'atlas', @atlas_2d_min.create);
>> prob = coco_set(prob, 'cont', 'h', .075, 'almax', 35);
```

```
>> prob = coco_set(prob, 'cont', 'NPR', 100, 'PtMX', 2000);
>> coco(prob, 'cuspsurface', [], 2, {'ka' 'la' 'alg.test.FO'}, ...
      {[-0.5 0.5], [-1 1]});
```

```
...   NORMS              COMPUTATION TIMES
...   ||f||      ||U||    F(x)  DF(x)  SOLVE
... 0.00e+00  7.07e-01    0.0    0.0    0.0
... 0.00e+00  7.07e-01    0.0    0.0    0.0

... LABEL  TYPE           ka           la  alg.test.FO
...     1  EP     0.0000e+00   5.0000e-01  -5.0000e-01
...     2  FO     8.6948e-02   3.7092e-01   1.5002e-07
...     3        -9.0680e-02   3.8655e-01  -8.5092e-02
...     4  FO     1.1050e-01   4.3518e-01   6.7116e-08
...     5  FO     5.0597e-02   2.5853e-01   3.8570e-08
...     6  FO    -5.9422e-02   2.8778e-01   1.2572e-08
...     7  FO    -7.4921e-02   3.3587e-01   1.1958e-08
...     8  FO    -7.5608e-02   3.3792e-01   1.3632e-08
...     9  FO    -1.1201e-01   4.3914e-01   3.2571e-08
...    10  FO    -1.3872e-01   5.0644e-01   2.8053e-08
.        .          .           .
.        .          .           .
.        .          .           .
...   180  EP     3.0807e-01  -1.0000e+00   1.2436e+00
...   181  EP     5.0000e-01  -9.6386e-01   1.5288e+00
...   182  EP    -3.6638e-01  -1.0000e+00   1.3274e+00
...   183  EP    -5.0000e-01  -9.2340e-01   1.5192e+00
...   184  EP    -2.5520e-01  -1.0000e+00   1.1745e+00
...   185  EP    -3.0167e-01  -1.0000e+00   1.2348e+00
...   186  EP    -1.9919e-01  -1.0000e+00   1.1107e+00
...   187  EP    -4.4849e-01  -1.0000e+00   1.4550e+00
...   188  EP    -5.0000e-01  -9.7031e-01   1.5305e+00
...   189  EP    -4.9546e-01  -1.0000e+00   1.5315e+00
```

The fold points detected during continuation are overlaid on the 2-dimensional solution manifold to the original zero problem in Fig. 16.5. In contrast to the continuation along the branch of fold points, here detection relies on the construction of curve segments that are transversal to this branch. The sparsity of detected fold points in Fig. 16.5, along parts of the 2-dimensional solution manifold, indicates the construction of a number of curve segments adjacent, and close to parallel, to the branch of fold points.

16.3 An alternative constructor

In the case that the `alg_fold` event function is included among the monitor functions, a fold point located using this event function may be continued in parameter space by an exchange of continuation parameters between $\mathbb{I}$ and $\mathbb{J}$ so as to constrain the value of `alg_fold` to zero.

As an alternative formulation that enables the continuation of a branch of fold points, consider the extended *Moore–Spence* continuation problem $\tilde{F}(\tilde{u}, \mu) = 0$ for $\tilde{u} := (x, p, v) \in \mathbb{R}^m \times \mathbb{R}^{n-m} \times \mathbb{R}^m$, where

$$\tilde{F}(\tilde{u}, \mu) := \begin{pmatrix} \Phi(x, p) \\ p - \mu \\ \partial_x \Phi(x, p) \cdot v \\ v^T \cdot v - 1 \end{pmatrix}. \tag{16.34}$$

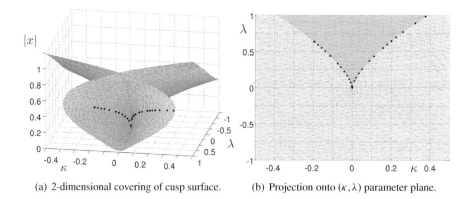

(a) 2-dimensional covering of cusp surface.   (b) Projection onto (κ, λ) parameter plane.

Figure 16.5. *The covering of the cusp surface obtained in the last run of Sect. 16.2.3; fold points are marked as black dots. The 2-dimensional atlas algorithm locates a fold point whenever a curve segment intersects an event surface transversally. In panel* (a), *one can see that events between neighboring charts may remain undetected if no connecting curve segment was constructed by the covering algorithm. Projecting the manifold onto the (κ, λ) parameter plane again reveals the fold curve as the border curve between different shades of gray* (b).

It follows that v is a right unit eigenvector of $\partial_x \Phi(x, p)$ corresponding to a zero eigenvalue. An initializable restricted continuation problem, whose solution corresponds to a fold point, then follows with the assignment $\mathbb{I} = \{1, \ldots, i-1, i+1, \ldots, n-m\}$ for some $1 \leq i \leq n-m$. Furthermore, continuation along a 1-dimensional family of folds is enabled by removing one additional index from $\mathbb{I}$.

The following suite of functions provides complete support for the construction of a corresponding restricted continuation problem with dimensional deficit -1. In particular, the implementation supports continuation of solutions to the extended continuation problem in Eq. (16.34) starting from a fold point located using either of the event functions considered in the previous sections.

```
function prob = alg_FO2FO(prob, oid, varargin)

tbid = coco_get_id(oid, 'alg');
str  = coco_stream(varargin{:});
run  = str.get;
if ischar(str.peek)
  soid = str.get;
else
  soid = oid;
end
lab = str.get;

[sol data] = alg_read_solution(soid, run, lab);
prob       = coco_set(prob, tbid, 'FO', false);
data       = alg_get_settings(prob, tbid, data);
prob       = alg_construct_eqn(prob, tbid, data, sol);

[data uidx] = coco_get_func_data(prob, tbid, 'data', 'uidx');
prob        = alg_create_FO(prob, data, uidx, sol);

end
```

```
function prob = alg_create_FO(prob, data, uidx, sol)

Jx         = alg_fhan_DFDX(data, sol.x, sol.p);
[v0, ~]    = eigs(Jx, 1, 0);
data.v_idx = data.p_idx(end)+(1:numel(v0));

fid  = coco_get_id(data.tbid, 'fold_cond');
prob = coco_add_func(prob, fid, @alg_FO, @alg_FO_DFDU, ...
  data, 'zero', 'uidx', [uidx(data.x_idx); uidx(data.p_idx)], ...
  'u0', v0);

end

function [data y] = alg_FO(prob, data, u)

Jx = alg_fhan_DFDX(data, u(data.x_idx), u(data.p_idx));
v  = u(data.v_idx);
y  = [Jx*v; v'*v-1];

end

function [data J] = alg_FO_DFDU(prob, data, u)

x   = u(data.x_idx);
p   = u(data.p_idx);
Jx  = alg_fhan_DFDX(data, x, p);
v   = u(data.v_idx);

h   = 1.0e-4*(1+norm(x));
J0  = alg_fhan_DFDX(data, x-h*v, p);
J1  = alg_fhan_DFDX(data, x+h*v, p);
Jxx = (0.5/h)*(J1-J0);

J0  = alg_fhan_DFDP(data, x-h*v, p);
J1  = alg_fhan_DFDP(data, x+h*v, p);
Jpx = (0.5/h)*(J1-J0);

J = [Jxx Jpx Jx; zeros(1,numel(data.x_idx)+numel(data.p_idx)) 2*v'];

end
```

We note, in the encoding of `alg_FO_DFDU`, the use of the finite-difference approximations in Eqs. (16.20)–(16.21) for computing second-order directional derivatives.

16.4 Conclusions

Whereas atlas events are associated with localized features of the solution manifold, events embedded in COCO-compatible toolboxes provide for the detection and further processing of isolated solutions with context-dependent properties, unique to a subset of solutions of a particular continuation object. It is a natural step, following the definition of a problem class, to pursue the formulation of suitable monitor functions that may be used to identify such special solutions, as well as to design the toolbox data structure to support the introduction of problem-specific events in parent objects. The modifications to the 'alg' toolbox in this chapter, and to the 'po' toolbox in the next chapter, only scratch the surface of this process of problem and toolbox design. Nevertheless, they provide the general principles upon which all subsequent development rests.

The use of the `coco_xchg_pars` utility, described in the previous chapter, and the associated pairwise switching of indices between the $\mathbb{I}$ and $\mathbb{J}$ index arrays offer a simple, but often very convenient, mechanism for activating and deactivating constraints, following the detection and location of associated events. In the case of nonembedded monitor functions, however, continuation of submanifolds of special points requires the formulation of augmented extended continuation problems, as suggested in Sect. 16.3. In this context, the literature on the formulation of so-called *minimally augmented continuation problems* provides both theoretical depth and opportunities for further practice in algorithm design and implementation.

Exercises

16.1. Consider the two-point boundary-value problem obtained from the ordinary differential equation
$$v''(t) + p_2 v(t) + p_1 e^{v(t)} = 0$$
and the homogeneous *Dirichlet* boundary conditions
$$v(0) = v(1) = 0$$
for some parameter vector p. Show that $v(t) \equiv 0$ is a solution for all p_2 provided that $p_1 = 0$. Do solutions exist for all $p_1 \neq 0$?

16.2. Discretize the solution $v(t)$ to the boundary-value problem in the previous exercise on the uniform mesh $t_0 < t_1 < \cdots < t_{N-1} < t_N$, where $t_i = i/N$, such that $v_i = v(t_i)$. Moreover, for $1 \leq i \leq N-1$, discretize the differential equation by the following finite-difference approximation:
$$v''(t_i) = N^2 (v_{i+1} - 2v_i + v_{i-1}).$$

Use the `'alg'` toolbox to construct a corresponding continuation problem with initial solution guess given by $v_i = 0$, for $i = 0, \ldots, N$, when $p_1 = 0$. Perform continuation for different values of the discretization order N using the modified atlas algorithm described in Sect. 16.1 and comment on the existence of fold points along the solution manifold of the original boundary-value problem.

16.3. Let the vector x consist of the values v_i for $1 \leq i \leq N-1$ from the previous exercise and suppose that
$$e^x = \begin{pmatrix} e^{v_1} \\ \vdots \\ e^{v_{N-1}} \end{pmatrix}.$$

Show that the discretized zero function then reads as
$$\Phi(x, p) := D \cdot x + p_2 x + p_1 e^x,$$
where
$$D := N^2 \begin{pmatrix} -2 & 1 & & & \\ 1 & -2 & 1 & & \\ & \ddots & \ddots & \ddots & \\ & & 1 & -2 & 1 \\ & & & 1 & -2 \end{pmatrix}$$

is an $(N-1) \times (N-1)$ matrix. Consider the trivial solution $v_i = 0$, for $i = 0$, $\ldots, N$, when $p_1 = 0$ and identify values of p_2 for which the rectangular matrix $\left(\ \partial_x \Phi(x,p) \quad \partial_{p_2} \Phi(x,p) \ \right)$ is not full rank. How do these values change with N? Can you relate these values to the eigenvalues of the linear boundary-value problem

$$v''(t) + \lambda v(t) = 0, \qquad v(0) = v(1) = 0?$$

16.4. Demonstrate the application of the 'alg' toolbox from Sect. 16.2.2 to the detection, location, and continuation of fold points along the solution manifold associated with the zero function from the previous exercise. Use the modified atlas algorithm described in Sect. 16.1 to detect branch points along curves of fold points and show the coincidence of such a branch point with one of the singularities observed in the previous exercise.

16.5. Consider the restricted continuation problem $G\left(u, \mu_{\mathbb{J}}\right) = 0$, where

$$G\left(u, \mu_{\mathbb{J}}\right) := \left. \left(\begin{array}{c} \Phi(x,p) \\ p - \mu \end{array} \right) \right|_{\mu_{\mathbb{I}} = \mu_{\mathbb{I}}^*},$$

with $\mathbb{I} = \{1, \ldots, n - m - 1\}$ and $p = (p_1, p_2) \in \mathbb{R}^{n-m-1} \times \mathbb{R}$. Show that regular points $u^* := (x^*, p^*)$ on the solution set of this restricted problem are those points for which

- the Jacobian $\partial_x \Phi(x^*, p^*)$ is invertible or

- $\partial_x \Phi(x^*, p^*)$ has a rank loss of 1 and span $\{\partial_{p_2} \Phi(x^*, p^*)\}$ does not lie in the range of $\partial_x \Phi(x^*, p^*)$.

Show that a necessary condition for the existence of a fold with respect to $u \mapsto p_2$ is the existence of a nullvector of the Jacobian $\partial_x \Phi(x^*, p^*)$.

16.6. Consider the zero function $\Phi : \mathbb{R}^n \to \mathbb{R}^m$ in terms of the vector of continuation variables $u := (x, p) \in \mathbb{R}^m \times \mathbb{R}^{n-m}$. Let

$$U := \left(\begin{array}{c} 0 \\ I_{n-m} \end{array} \right).$$

Show that

$$\left(\begin{array}{c} \partial_u \Phi \\ U^T \end{array} \right)$$

is singular if and only if the matrix $\partial_x \Phi$ is singular.

16.7. Show that a necessary condition for the existence of a fold with respect to a problem parameter in the algebraic continuation problem considered in Sect. 16.2 is the vanishing of an eigenvalue of the Jacobian $\partial_x \Phi(x^*, p^*)$. Show that the determinant

$$\det \partial_x \Phi(x, p)$$

and, equivalently, the product of the eigenvalues of $\partial_x \Phi(x, p)$ each constitute viable monitor functions for detecting and locating fold points along the solution manifold.

16.8. Verify Eqs. (16.18)–(16.19).

16.9. Review the discussion of generalized turning points in [Griewank, A. and Reddien, G.W., "Characterization and computation of generalized turning points," *SIAM Journal on Numerical Analysis*, 21(1), pp. 176–185, 1984] and propose suitable modifications to the 'alg' toolbox for the detection and location of such points.

16.10. Suppose that the curve segment Γ on the solution manifold of the zero problem $\Phi(u) = 0$ for $u = (x, p)$ consists only of regular points. Let b and c be the left and right singular vectors of unit length corresponding to the smallest singular value h of $\partial_x \Phi(x, p)$ at some point $u^* := (x^*, p^*)$ along Γ. Show that the bordered matrix

$$\begin{pmatrix} \partial_x \Phi & b \\ c^T & 0 \end{pmatrix}$$

is invertible on some portion of Γ containing u^*.

16.11. Suppose that c_0 is close to the eigenspace of $(\partial_x \Phi)^T \cdot \partial_x \Phi$ corresponding to the smallest eigenvalue of this matrix and that this eigenvalue is simple. Show that

$$\partial_x \Phi \cdot \alpha + \beta b_1 = 0, \quad c_0^T \cdot \alpha = 0$$

implies that $\alpha = \beta = 0$.

16.12. Let λ be a scalar and consider the scaling

$$\lambda \mapsto \tilde{\lambda} := \frac{2\lambda}{\max(1, |\lambda|) + |\lambda|}.$$

Show that $\tilde{\lambda} = \lambda/|\lambda|$ for $|\lambda| \geq 1$ and $|\lambda| < |\tilde{\lambda}| < 2|\lambda|$ for $|\lambda| < 1$. Now let Λ consist of the eigenvalues of the Jacobian $\partial_x \Phi(x, p)$ and denote by $\tilde{\Lambda}$ the collection $\{\tilde{\lambda}, \lambda \in \Lambda\}$. Show that the product

$$\prod_{\tilde{\lambda} \in \tilde{\Lambda}} \tilde{\lambda}$$

changes sign when $\det \partial_x \Phi(x, p)$ changes sign. Implement a monitor function that evaluates to this product and illustrate its use in fold detection and location. Can you think of a reason for introducing the scaling transformation, rather than operating directly on the eigenvalues of $\partial_x \Phi(x, p)$?

16.13. Implement a monitor function that evaluates the determinant $\det \partial_x \Phi(x, p)$ and illustrate its use in fold detection and location. How do you expect this algorithm to perform for large systems? Test your algorithm on the boundary-value problem in Exercises 16.1–16.2 and comment on your observations.

16.14. Implement a monitor function that evaluates to the real part of the eigenvalue of $\partial_x \Phi(x, p)$ with smallest-in-magnitude real part, if this is smaller than 1, and 1 otherwise. Illustrate its use in fold detection and location. How does this compare with the other monitor functions considered in this chapter?

16.15. Can the alg_sol2eqn constructor of the modified 'alg' toolbox in Sect. 16.2 be used to initialize an extended continuation problem, corresponding to continuation along a branch of fold points, from an approximate fold point located using the modified 1-dimensional atlas algorithm in Sect. 16.1.1? Show this by example or explain how you would modify the constructor to enable such an initialization.

16.16. Review [Schwetlick, H. and Schnabel, U., "Iterative computation of the smallest singular value and the corresponding singular vectors of a matrix," *Linear Algebra and Its Applications*, 371, pp. 1–30, 2003] and comment on the implications to the algorithm for updating the bordering vectors b and c in Sect. 16.2.1.

16.17. Include the condition number of the matrix

$$\begin{pmatrix} \partial_x \Phi & b \\ c^T & 0 \end{pmatrix}$$

in the output of the monitor function `alg_fold` in the modified encoding of the `'alg'` toolbox in Sect. 16.2 and repeat the numerical analysis of the examples in this section. Comment on the effect of the algorithm for updating the bordering vectors b and c on the numerical invertibility of this matrix.

16.18. Repeat the analysis of the previous exercise, but first replace the update algorithm for the bordering vectors b and c with one, where c_1 is obtained as the unique solution to the equation

$$\begin{pmatrix} \partial_x \Phi & b_0 \\ c_0^T & 0 \end{pmatrix} \cdot \begin{pmatrix} c_1 \\ g_1 \end{pmatrix} = \begin{pmatrix} 0 \\ 1 \end{pmatrix}.$$

16.19. Can you quantify the difference in computational expense between using the MATLAB `svds` function or the update algorithm in Sect. 16.2.1 to compute the bordering vectors b and c?

16.20. Let h be obtained as in Sect. 16.2.1. Show that h^2 is the smallest eigenvalue of the symmetric matrices $(\partial_x \Phi)^T \cdot \partial_x \Phi$ and $\partial_x \Phi \cdot (\partial_x \Phi)^T$ with corresponding eigenvectors c and b, respectively.

16.21. Consider again the coupled differential equations

$$x'' - \frac{p_4^2}{p_1} \left((p_6 + 1)x - x^2 y - p_3 \frac{e^{\frac{p_4}{\sqrt{p_5}} z} + e^{\frac{p_4}{\sqrt{p_5}}(1-z)}}{1 + e^{\frac{p_4}{\sqrt{p_5}}}} \right) = 0$$

and

$$y'' - \frac{p_4^2}{p_2} \left(x^2 y - p_6 x \right) = 0$$

for the unknown functions $x(z)$ and $y(z)$, together with the boundary conditions

$$x(0) = x(1) = p_3, \qquad y(0) = y(1) = \frac{p_6}{p_3}$$

introduced in Chap. 4 of [Govaerts, W.J.F., *Numerical Methods for Bifurcations of Dynamical Equilibria*, SIAM, Philadelphia, 2000] as a model of a 1-dimensional medium of a Belusov–Zhabotinsky-type chemical oscillator and previously investigated in Exercise 3.24. For $p_4 = 0$, a solution is given by $x(z) = p_3$ and $y(z) = p_6/p_3$ for all $z \in [0,1]$. Use your finite-difference approximation from Exercise 3.24 to perform continuation of this solution under variations in p_4 with the `atlas_1d_min` atlas algorithm from Sect. 16.1. Explore the presence of fold points and branch points along the solution manifold for different orders of discretization.

Chapter 17

Event Handlers and Branch Switching

The classification of events, introduced in Chap. 15, establishes the core response to event detection. The distinction between continuation, regular, and singular events is reflected in the use of a core implementation for event location that is appropriate to the smoothness properties of the corresponding monitor function and the regularity of the special point on the solution manifold. Similarly, the classification of special points encountered during continuation as boundary, terminal, and special events ensures a well-defined behavior of the finite-state machine. Thus, in the case of boundary events, the finite-state machine terminates the current curve segment at the boundary point, assigns the `'EP'` point type to the corresponding chart, and sets the `chart.ep_flag` field to 1. We relied on this functionality in Chap. 14. Similarly, in the case of a terminal event, the finite-state machine sets `cseg.Status` to `cseg.TerminalEventDetected`, resulting in the termination of any of the atlas algorithms developed in this text.

In the case of special events, and to a degree in the case of boundary and terminal events, toolbox- or user-defined *event handlers* may be introduced to override or refine the response to event detection. Thus, the point type assigned to a special event might be modified by properties of the special point, other than its immediate relationship to the corresponding monitor function. Indeed, event handlers could even be relied upon to establish refined conditions under which event location should be skipped. Finally, event handlers may be used to append additional data to charts associated with special points, which could be used for further processing subsequent to the current execution of the `coco` entry-point function.

Such functionality could be useful for characterizing bifurcation points in dynamical systems in terms of features of the local dynamics that do not follow immediately from the existence of a special point. As an example, toolbox-specific chart data could be used to store an initial solution guess for a distinct restricted continuation problem corresponding to a solution manifold branching off of the original solution manifold at the bifurcation point.

We illustrate below the theoretical framework and core interface for incorporating event handlers within COCO-compatible toolboxes and atlas algorithms. The treatment opens up toolbox development to a vast enterprise of problem-class-specific monitor functions, and associated events, with opportunities for flexible event handling and automation.

439

17.1   Toolbox event handlers

17.1.1   Hopf bifurcations

A *Hopf bifurcation* of an equilibrium y^* of the nonlinear dynamical system

$$\frac{dy}{dt} = f(y, p), \qquad (17.1)$$

for $(y, p) \in \mathbb{R}^n \times \mathbb{R}^p$, is said to occur along a curve segment through a point p^* in parameter space provided that

- there is a simple nonzero pair, $\pm i\omega$, of eigenvalues of the Jacobian

$$A = \partial_y f\left(y^*, p^*\right); \qquad (17.2)$$

- there are no other eigenvalues with zero real part; and

- the real parts of these imaginary eigenvalues have nonzero directional derivatives along the tangent vector to the curve segment.

From dynamical systems theory we recall that such a bifurcation point is associated with a change in the stability of the equilibrium, as well as the local coexistence with the equilibrium of a family of periodic orbits emanating from the equilibrium y^* at p^* (see also Sect. 8.2).

We may detect and locate Hopf bifurcations by associating them with special points of the function ψ_{HB}, whose value equals the product of the sums of all unique unordered pairs of eigenvalues. Since one such pair is $\{i\omega, -i\omega\}$, it follows that this function vanishes at the Hopf bifurcation. The converse is not true, however, since a pair $\{\lambda, -\lambda\}$ for $\lambda \in \mathbb{R}$ would also result in a zero sum. A special point characterized by such an eigenvalue combination is called a *neutral saddle* and is not related to a system bifurcation. Branches of neutral saddles connect smoothly with branches of Hopf bifurcations. A transition from a branch of neutral saddles to a branch of Hopf bifurcations occurs at a codimension-2 special point known as a *Bogdanov–Takens point*.

Given the above observation, there appears to be value in being able to choose whether to locate an event associated with the nonembedded monitor function ψ_{HB} in the case that it corresponds to a neutral saddle, rather than a Hopf bifurcation point. As the monitor function itself contains no information on the character of the special point, it is clear that additional functionality must be provided by a toolbox or user-defined function to enable this determination.

In COCO, the finite-state machine associated with event detection supports this type of event supervision through the implementation of a *reverse communication* protocol and the use of *event handlers*. We illustrate this paradigm first by extending the implementation of the 'alg' toolbox to include the detection and location of Hopf bifurcations and the optional location of neutral saddles.

To this end, consider the encoding below of the monitor function ψ_{HB}.

```
function [data y] = alg_hopf(prob, data, u)

x   = u(data.x_idx);
p   = u(data.p_idx);
```

```
Jx = alg_fhan_DFDX(data, x, p);
la = eig(Jx);
la = la(data.la_idx1)+la(data.la_idx2);
sc = abs(la);
y  = real(prod((2*la)./(max(1,sc)+sc)));

end
```

The encoding assumes that the `la_idx1` and `la_idx2` fields of the `data` argument contain the first and second parts of all pairs of the form (i, j), where $1 \le i < j \le n$. We assign these fields in the modified `alg_init_data` function, shown below.

```
function data = alg_init_data(data, x0, p0)

xdim        = numel(x0);
pdim        = numel(p0);
data.x_idx = (1:xdim)';
data.p_idx = xdim+(1:pdim)';

Jx              = alg_fhan_DFDX(data, x0, p0);
[data.b, ~, data.c] = svds(Jx,1,0);
data.rhs        = [zeros(xdim,1); 1];
I               = triu(true(xdim),1);
A               = repmat((1:xdim)', [1 xdim]);
data.la_idx1 = A(I);
A               = A';
data.la_idx2 = A(I);

end
```

The modified constructor now takes the form shown below.

```
function prob = alg_construct_eqn(prob, tbid, data, sol)

prob = coco_add_func(prob, tbid, @alg_F, @alg_DFDU, data, 'zero', ...
  'u0', sol.u);
uidx = coco_get_func_data(prob, tbid, 'uidx');
if ~isempty(data.pnames)
  fid  = coco_get_id(tbid, 'pars');
  prob = coco_add_pars(prob, fid, uidx(data.p_idx), data.pnames);
end
prob = coco_add_slot(prob, tbid, @coco_save_data, data, 'save_full');
if data.alg.norm
  data.tbid = tbid;
  prob = coco_add_slot(prob, tbid, @alg_bddat, data, 'bddat');
end

switch data.alg.FO
  case {'regular', 'active'}
    fid_FO = coco_get_id(tbid, 'test', 'FO');
    data.tbid = tbid;
    data = coco_func_data(data);
    prob = coco_add_func(prob, fid_FO, @alg_fold, ...
      @alg_fold_DFDU, data, data.alg.FO, fid_FO, ...
      'uidx', uidx, 'fdim', 1);
    prob = coco_add_slot(prob, tbid, @alg_update, data, 'update');
    prob = coco_add_event(prob, 'FO', fid_FO, 0);
end

if data.alg.HB
  fid_HB = coco_get_id(tbid, 'test', 'HB');
```

```
   prob = coco_add_func(prob, fid_HB, @alg_hopf, data, ...
      'regular', fid_HB, 'uidx', uidx);
   prob = coco_add_event(prob, 'HB', fid_HB, 0);
end

end
```

Here, the 'HB' setting is assigned default content in the modified get_settings function, shown below.

```
function data = alg_get_settings(prob, tbid, data)

defaults.norm = false;
defaults.FO   = 'regular';
defaults.HB   = true;
data.alg = coco_merge(defaults, coco_get(prob, tbid));
assert(islogical(data.alg.norm), ...
   '%s: input for ''norm'' option is not boolean', tbid);
assert(ischar(data.alg.FO), ...
   '%s: input for ''FO'' option is not a string', tbid);
assert(islogical(data.alg.HB), ...
   '%s: input for ''HB'' option is not boolean', tbid);

end
```

Example 17.1 We illustrate the application of the modified 'alg' toolbox by seeking to detect and locate Hopf bifurcations of equilibria in a model of population dynamics governed by the vector field

$$f(y, p) = \begin{pmatrix} y_1 - y_1 y_2/(1 + p_1 y_1) \\ y_1 y_2/(1 + p_1 y_1) - y_2 - p_2 y_2^2 \end{pmatrix}, \tag{17.3}$$

for $p_1, p_2 > 0$, under variations in the problem parameters. It is straightforward to show that ψ_{HB} vanishes along a branch of equilibria obtained when

$$p_1 = \frac{1 \pm \sqrt{1 - 4p_2} + 2p_2}{2(2 + p_2)} \tag{17.4}$$

for $0 < p_2 < \frac{1}{4}$. By a more detailed analysis, the eigenvalues of the corresponding Jacobians are purely imaginary provided that $0 < p_1 < \sqrt{2} - 1$, and real and opposite in sign for $\sqrt{2} - 1 < p_1 < 1/2$. Both eigenvalues vanish at the codimension-2 Bogdanov–Takens point

$$(p_1, p_2) = \left(\sqrt{2} - 1, \frac{1}{2}\left(\sqrt{2} - 1\right)\right). \tag{17.5}$$

We apply the modified 'alg' toolbox to the encoding of the vector field f in the popul function, as shown in the extract below.

```
>> alg_args = {@popul, [1.76; 1.52], {'p1' 'p2'}, [0.3; 0.1]};
>> prob = alg_isol2eqn(coco_prob(), '', alg_args{:});
>> prob = coco_add_pars(prob, 'pars', [1 2], {'x', 'y'});
>> prob = coco_set(prob, 'cont', 'h', 0.15, 'PtMX', 1000);
>> coco(prob, 'run', [], 2, {'p1' 'p2' 'x' 'y' }, ...
      {[0 0.5], [0 0.25], [0 10], [0 10]});
```

```
...    NORMS                  COMPUTATION TIMES
...     ||f||        ||U||     F(x)   DF(x)   SOLVE
... 9.22e-03   3.32e+00       0.0    0.0     0.0
... 2.04e-05   3.32e+00       0.0    0.0     0.0
... 1.70e-10   3.32e+00       0.0    0.0     0.0
... 4.36e-16   3.32e+00       0.0    0.0     0.0

... LABEL  TYPE            p1             p2            x             y
...    1    EP     3.0000e-01     9.8415e-02    1.7564e+00    1.5269e+00
...    2    HB     2.2316e-01     2.0474e-01    1.8061e+00    1.4031e+00
...    3           3.5834e-01     1.6520e-02    1.6228e+00    1.5815e+00
...    4    HB     2.3710e-01     2.1420e-01    1.9019e+00    1.4509e+00
...    5    EP     2.2315e-01     2.5000e-01    1.9501e+00    1.4352e+00
...    6    HB     2.1046e-01     1.9550e-01    1.7269e+00    1.3634e+00
...    7    EP     2.0349e-01     2.5000e-01    1.8502e+00    1.3765e+00
...    8           2.5800e-01     6.6677e-02    1.5220e+00    1.3927e+00
...    9    EP     3.7752e-01     0.0000e+00    1.6065e+00    1.6065e+00
...   10    EP     3.9982e-01     0.0000e+00    1.6662e+00    1.6662e+00
 .     .      .            .              .
 .     .      .            .              .
 .     .      .            .              .
...  343    EP     4.5638e-01     1.4331e-01    1.0000e+01    5.5638e+00
...  344    EP     4.4649e-01     1.5185e-01    1.0000e+01    5.4649e+00
...  345    EP     5.0000e-01     1.1250e-01    9.7006e+00    5.8503e+00
...  346    EP     5.0000e-01     1.1235e-01    9.7328e+00    5.8664e+00
...  347    EP     4.6560e-01     1.3579e-01    1.0000e+01    5.6560e+00
...  348    EP     5.0000e-01     1.1190e-01    9.8302e+00    5.9151e+00
...  349    EP     4.7683e-01     1.2718e-01    1.0000e+01    5.7683e+00
...  350    EP     5.0000e-01     1.1147e-01    9.9216e+00    5.9608e+00
...  351    EP     4.9755e-01     1.1271e-01    1.0000e+01    5.9755e+00
...  352    EP     4.8937e-01     1.1822e-01    1.0000e+01    5.8937e+00
```

Continuation is performed here, using the 2-dimensional atlas algorithm developed in Sect. 14.2, on a computational domain defined by $p_1 \in [0, 0.5]$, $p_2 \in [0, 0.25]$, $y_1 \in [0, 10]$, and $y_2 \in [0, 10]$. The corresponding 2-dimensional solution manifold is shown in Fig. 17.1(a). As seen in Fig. 17.1(b), the projection of the zero-level set of ψ_{HB} onto the (p_1, p_2) plane is the curve given in Eq. (17.4). As the algorithm is unable to discriminate between Hopf bifurcations and neutral saddles, they are both detected as events of point type 'HB'. ∎

17.1.2 A reverse communication protocol

We proceed to construct an event handler, capable of ascertaining the nature of the event associated with a detected zero crossing of the monitor function ψ_{HB} introduced in the previous section, and of instructing the COCO finite-state machine appropriately. In COCO, event handlers must be encoded in a MATLAB-compatible function with the following function syntax.

```
function [data cseg msg] = evhan(prob, data, cseg, cmd, msg)
```

The first input argument, prob, contains a copy of the continuation problem structure used by COCO to store all properties of the continuation problem and any associated algorithms during execution. The content of this variable may be accessed in the function body, but changes made to the copy do not survive execution of the function. As the second input argument, data, is identical to the first output argument, its content may be accessed and

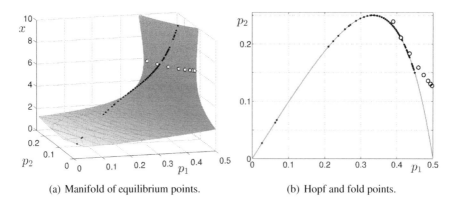

(a) Manifold of equilibrium points.   (b) Hopf and fold points.

Figure 17.1. *Covering of a manifold of equilibrium points of Example* 17.1. *The black dots mark events associated with* ψ_{HB} *and the value* 0, *which are either Hopf points or neutral saddles. The circles mark fold points. From panel* (a) *it is evident that the loci of Hopf and fold points intersect each other transversally. When projected onto the* (p_1, p_2) *parameter plane, these curves have a tangency* (b). *The gray curve in panel* (b) *is the locus of zeros of* ψ_{HB} *given by Eq.* (17.4). *The distinction between different events associated with the same monitor function is afforded by event handlers; see Fig.* 17.3.

permanently modified within the function body for reference in a subsequent function call or by the COCO finite-state machine. This argument contains the event-handler function data structure and provides a parameterization of the implementation of the event handler. Finally, as before, the `cseg` input argument contains an instance of the `CurveSegment` class corresponding to the current curve segment. Changes made to this variable survive execution of the event handler, although neither of the examples considered below takes advantage of this facility.

The schematic flow diagram in Fig. 17.2 illustrates the reverse communication protocol that defines the use of event handlers in COCO. Here, the `cmd` input distinguishes between actions taken by the event handler immediately following the detection of an associated event and immediately following the location of such an event, respectively. In particular, `cmd` equals `'init'` unless a special point has been located, in which case it equals `'check'`.

Upon detection of an event corresponding to a function ψ along a curve segment Γ, the finite-state machine assigns default content to the `msg` structure contained in the final input argument and returned by the event handler in the third output argument. This includes

- the `u0` field, containing the point $\Gamma_<$;

- the `u1` field, containing the point $\Gamma_>$;

- the `e0` field, containing the scalar $\psi(\Gamma_<)$;

- the `e1` field, containing the scalar $\psi(\Gamma_>)$;

- the `evidx` field, containing the integer index of the event in the list of possible events;

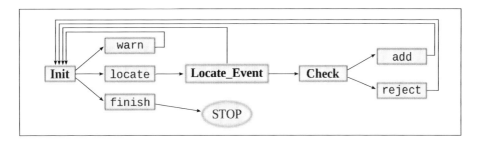

Figure 17.2. *The reverse communication protocol implemented in* COCO *is a three-state finite-state machine. An event handler triggers a state transition from either* **Init** *or* **Check** *and determines the subsequent flow through the finite-state machine by setting the* action *field of the* msg *message structure to one of the values following an outgoing edge of the current state. Since the lifetime of the message structure spans from* **Init** *until* STOP, *transitions triggered by an event handler can depend on (parts of) the complete history of state transitions. As a typical case, this allows an implementation to determine the state prior to* **Init**, *which in turn supports a distinct response depending on whether or not* **Locate_Event** *was successful.*

- the pidx field, containing the integer index of the monitor function in the list of monitor functions; and

- the pars field, containing the string identifier for the monitor function.

The event handler, in turn, communicates its instructions back to the finite-state machine by assigning content to the action, point_type, and idx fields, as well as the optional wmsg and callMX fields. In particular, the msg.action field must be assigned

- one of the strings 'locate', 'warn', or 'finish' when cmd equals 'init' and

- either of the strings 'add' or 'reject' when cmd equals 'check'.

In the case that the msg.action field equals 'locate', the msg.idx field must contain the integer index for the desired event in the list of possible events and the msg.point_type field must contain the label to be assigned to this special point. In this case, the finite-state machine seeks to locate the event using the method associated with the function type of the encoding of ψ. If, instead, the msg.action field equals 'warn', the finite-state machine issues a warning message. In lieu of a default message, an event-specific warning message may be assigned to the msg.wmsg field. Finally, in the case that the msg.action field equals 'finish', the finite-state machine terminates the reverse communication protocol and either proceeds with handling the next event or transitions to the flush state.

In the case that the finite-state machine is instructed to locate the event and does so successfully, control is returned to the event handler with cmd equal to 'check'. If the msg.action field equals 'add' at the end of execution of the event handler, then the finite-state machine is instructed to append the special point to the current curve segment (and, possibly, to discard charts that lie beyond this special point in the case that it designates a computational boundary). If desired, the content of the point_type field may be changed by the event handler when cmd equals 'check'. In contrast, if msg.action is assigned the string 'reject', then the special point is ignored by the finite-state machine. Either way, control returns again to the event handler, this time with cmd equal to 'init'.

Finally, we note that the first time that cmd equals 'init' following the detection of an event, the msg.callMX field may be set to an integer representing the maximum number of allowed consecutive calls to the event handler for locating a particular special point. In the absence of an explicit value, a default value of 100 is used by the finite-state machine.

The COCO core utility function coco_add_event is used to associate a COCO-compatible event handler with a particular special point. As an example, the following command appends the event handler evhan with function data structure data to the continuation problem and associates it with the detection of a special event corresponding to a zero-crossing of the value of the monitor function with function identifier fid.

```
prob = coco_add_event(prob, @evhan, data, 'SP', fid, 0);
```

17.1.3 Encoding

We proceed to encode the alg_evhan_HB event handler so as to instruct the COCO finite-state machine to

- locate every Hopf bifurcation associated with ψ_{HB} and assign the 'HB' point type to the corresponding chart and

- optionally locate every neutral saddle associated with ψ_{HB} and assign the 'NSad' point type to the corresponding chart.

To this end, we first modify the alg_hopf encoding as follows.

```
function [data chart y] = alg_hopf(prob, data, chart, u)

cdata = coco_get_chart_data(chart, data.tfid);
if ~isempty(cdata) && isfield(cdata, 'la')
  la = cdata.la;
else
  x  = u(data.x_idx);
  p  = u(data.p_idx);
  Jx = alg_fhan_DFDX(data, x, p);
  la = eig(Jx);
  chart = coco_set_chart_data(chart, data.tfid, struct('la', la));
end
la = la(data.la_idx1)+la(data.la_idx2);
sc = abs(la);
y  = real(prod((2*la)./(max(1,sc)+sc)));

end
```

We note here the use of 'alg'-specific chart data to store an array of the eigenvalues of the Jacobian $\partial_x \Phi$. We adhere to the practice of first seeking to extract chart data associated with the data.tfid identifier, if previously computed, and performing only the necessary computations if this chart data is missing. The desired event-handling functionality is now implemented in the encoding shown below.

```
function [data cseg msg] = alg_evhan_HB(prob, data, cseg, cmd, msg)

switch cmd
```

```
case 'init'
  if isfield(msg, 'finish') || strcmp(msg.action, 'warn')
    msg.action = 'finish';
  elseif strcmp(msg.action, 'locate')
    msg.action = 'warn';
  else
    cdata = coco_get_chart_data(cseg.ptlist{1}, data.tfid);
    la0 = cdata.la;
    cdata = coco_get_chart_data(cseg.ptlist{end}, data.tfid);
    la1 = cdata.la;
    switch abs(sum(sign(real(la0)))-sum(sign(real(la1))))
      case 4
        msg.point_type = 'HB';
        msg.action     = 'locate';
      case 0
        msg.point_type = 'NSad';
        if data.alg.NSad
          msg.action   = 'locate';
        else
          msg.action   = 'finish';
        end
      otherwise
        msg.point_type = 'HB';
        msg.action     = 'warn';
        msg.wmsg       = 'could not determine type of event';
    end
    msg.idx = 1;
  end
case 'check'
  msg.action = 'add';
  msg.finish = true;
end

end
```

Here, we note that a Hopf bifurcation is associated with a change of 4 in the difference between the number of eigenvalues with positive and negative real part, respectively, whereas no change in this number would occur in the case of a neutral saddle.

The modified `alg_construct_eqn` constructor now reads as follows.

```
function prob = alg_construct_eqn(prob, tbid, data, sol)

prob = coco_add_func(prob, tbid, @alg_F, @alg_DFDU, data, 'zero', ...
  'u0', sol.u);
uidx = coco_get_func_data(prob, tbid, 'uidx');
if ~isempty(data.pnames)
  fid  = coco_get_id(tbid, 'pars');
  prob = coco_add_pars(prob, fid, uidx(data.p_idx), data.pnames);
end
prob = coco_add_slot(prob, tbid, @coco_save_data, data, 'save_full');
if data.alg.norm
  data.tbid = tbid;
  prob = coco_add_slot(prob, tbid, @alg_bddat, data, 'bddat');
end

switch data.alg.FO
  case {'regular', 'active'}
    fid_FO = coco_get_id(tbid, 'test', 'FO');
    data.tbid = tbid;
    data = coco_func_data(data);
    prob = coco_add_func(prob, fid_FO, @alg_fold, ...
```

```
      @alg_fold_DFDU, data, data.alg.FO, fid_FO, ...
      'uidx', uidx, 'fdim', 1);
    prob = coco_add_slot(prob, tbid, @alg_update, data, 'update');
    prob = coco_add_event(prob, 'FO', fid_FO, 0);
  end

  if data.alg.HB
    fid_HB = coco_get_id(tbid, 'test', 'HB');
    data.tfid = fid_HB;
    prob = coco_add_chart_data(prob, fid_HB, [], []);
    prob = coco_add_func(prob, fid_HB, @alg_hopf, data, ...
      'regular', fid_HB, 'uidx', uidx, 'passChart');
    prob = coco_add_event(prob, @alg_evhan_HB, data, 'SP', fid_HB, 0);
  end

end
```

Finally, we provide default content for the `'HB'` and `'NSad'` toolbox settings in the modified `alg_get_settings` function, shown below.

```
function data = alg_get_settings(prob, tbid, data)

defaults.norm = false;
defaults.FO   = 'regular';
defaults.HB   = true;
defaults.NSad = false;
data.alg = coco_merge(defaults, coco_get(prob, tbid));
assert(islogical(data.alg.norm), ...
  '%s: input for ''norm'' option is not boolean', tbid);
assert(ischar(data.alg.FO), ...
  '%s: input for ''FO'' option is not a string', tbid);
assert(islogical(data.alg.HB), ...
  '%s: input for ''HB'' option is not boolean', tbid);
assert(islogical(data.alg.NSad), ...
  '%s: input for ''NSad'' option is not boolean', tbid);

end
```

Example 17.2 We repeat the analysis from Example 17.1 using the modified `'alg'` toolbox, with the `'NSad'` setting of the `'alg'` toolbox assigned each of the boolean values `false` (by default) and `true`, respectively. The result of the analysis is shown in the two panels in Fig. 17.3. We note in particular the transition between a branch of Hopf bifurcations and a branch of neutral saddles at a Bogdanov–Takens point, coincident with a codimension-2 point along the branch of fold points also shown in the figure.   ■

17.2   Bifurcations of periodic orbits

We follow the model set forth by the modifications to the `'alg'` toolbox in this and the previous chapter by modifying the `'po'` and `'hspo'` toolboxes to allow for detection and location of bifurcations of single- and multisegment periodic orbits of autonomous dynamical systems associated with critical combinations of the corresponding Floquet multipliers. In particular, we seek to implement detection and location for

- *saddle-node bifurcations*, associated with a single Floquet multiplier crossing the unit circle in the complex plane through the real number 1;

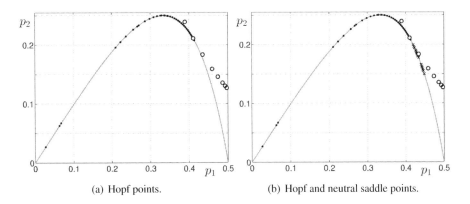

(a) Hopf points. (b) Hopf and neutral saddle points.

Figure 17.3. *Repeating the computation of Example* 17.1 *using an event handler, as shown in Sect.* 17.1.3, *allows for a distinction between Hopf and neutral saddle points; compare with Fig.* 17.1(b). *Panel* (a) *shows only Hopf and fold points, while panel* (b) *also shows neutral saddle points marked with* ×. *The gray curve is the locus of zeros of* ψ_{HB} *given by Eq.* (17.4).

- *period-doubling bifurcations*, associated with a single Floquet multiplier crossing the unit circle in the complex plane through the real number -1;

- *Neimark–Sacker bifurcations*, associated with a pair of complex conjugate Floquet multipliers crossing the unit circle in the complex plane (at points away from a strong resonance, corresponding to an nth root of 1 for $n = 1, \ldots, 4$);

- optionally, *neutral saddle points*, associated with a pair of real Floquet multipliers whose product crosses 1.

As in the case of Hopf bifurcations of equilibria, we find it convenient to rely on a monitor function that provides for the detection of Neimark–Sacker bifurcations, as well as neutral saddle points, and to rely on an event handler to discriminate between these alternatives.

To this end, we consider the following four monitor functions:

$$\psi_1 : u \mapsto \prod_{i=1}^{n} (\lambda_i - 1), \tag{17.6}$$

$$\psi_2 : u \mapsto \prod_{i=1}^{n} (\lambda_i + 1), \tag{17.7}$$

$$\psi_3 : u \mapsto \begin{cases} 1 & \text{when } n = 1, \\ \prod_{1 \le i < j \le n} (\lambda_i \lambda_j - 1) & \text{when } n > 1, \end{cases} \tag{17.8}$$

$$\psi_4 : u \mapsto \sum_{i \in [1,n], |\lambda_i| > 1} 1. \tag{17.9}$$

Here, zero-crossings of ψ_1, ψ_2, and ψ_3 correspond to saddle-node bifurcations, period-doubling bifurcations, and (possibly) Neimark–Sacker bifurcations, respectively. The ψ_4

monitor function, on the other hand, provides a discrete count of the number of Floquet multipliers with magnitude greater than 1. Any integer value greater than 0 thus corresponds to an unstable periodic orbit.

17.2.1   Single-segment orbits

We implement the monitor functions $\psi_1, \ldots, \psi_4$ from Eqs. (17.6)–(17.9) in the po_TF function, shown below.

```
function [data chart y] = po_TF(prob, data, chart, u)

cdata = coco_get_chart_data(chart, data.tfid);
if ~isempty(cdata) && isfield(cdata, 'la')
  la = cdata.la;
else
  fdata     = coco_get_func_data(prob, data.var_id, 'data');
  M         = fdata.M(fdata.M1_idx,:);
  la        = eig(M);
  [vv idx]  = sort(abs(la-1));
  la        = la(idx(2:end));
  chart     = coco_set_chart_data(chart, data.tfid, struct('la', la));
end
y(1,1) = prod(la-1);
y(2,1) = prod(la+1);
if numel(la)>1
  NS_TF  = la(data.la_idx1).*la(data.la_idx2);
  y(3,1) = prod(NS_TF(:)-1);
else
  y(3,1) = 1;
end
y(4,1) = sum(abs(la)>1);

end
```

As in the case of the modified encoding of alg_hopf in the previous section, we use chart data to store a sorted list of the Floquet multipliers, omitting the trivial eigenvalue at 1. The data.var_id field is here assumed to contain a toolbox identifier associated with an instance of the version of the 'var_coll' toolbox from Sect. 10.1.3, appended to the continuation problem structure in order to store the fundamental solution matrix in the function data structure.

   We initialize the la.idx1 and la.idx2 index arrays in the modified po_init_data function, shown below.

```
function data = po_init_data(prob, tbid, data)

stbid     = coco_get_id(tbid, 'seg.coll');
[fdata u0] = coco_get_func_data(prob, stbid, 'data', 'u0');

dim  = fdata.dim;
NTST = fdata.coll.NTST;
NCOL = fdata.coll.NCOL;
rows = [1:dim 1:dim];
cols = [fdata.x0_idx fdata.x1_idx];
vals = [ones(1,dim) -ones(1,dim)];
J    = sparse(rows, cols, vals, dim, dim*NTST*(NCOL+1));

data.x0_idx = fdata.x0_idx;
```

```
data.x1_idx = fdata.x1_idx;
data.intfac = fdata.Wp'*fdata.wts2*fdata.W;
data.xp0    = u0(fdata.xbp_idx)'*data.intfac;
data.J      = [J; data.xp0];

I           = triu(true(dim-1),1);
A           = repmat((1:dim-1)', [1 dim-1]);
data.la_idx1 = A(I);
A           = A';
data.la_idx2 = A(I);

end
```

The `po_evhan_NS` event handler, shown below, is now a straightforward modification of `alg_evhan_HB` from the previous section.

```
function [data cseg msg] = po_evhan_NS(prob, data, cseg, cmd, msg)

switch cmd
  case 'init'
    if isfield(msg, 'finish') || strcmp(msg.action, 'warn')
      msg.action = 'finish';
    elseif strcmp(msg.action, 'locate')
      msg.action = 'warn';
    else
      cdata = coco_get_chart_data(cseg.ptlist{1}, data.tfid);
      la0 = cdata.la;
      cdata = coco_get_chart_data(cseg.ptlist{end}, data.tfid);
      la1 = cdata.la;
      switch abs(sum(sign(abs(la0)-1))-sum(sign(abs(la1)-1)))
        case 4
          msg.point_type = 'NS';
          msg.action     = 'locate';
        case 0
          msg.point_type = 'NSad';
          if data.alg.NSad
            msg.action     = 'locate';
          else
            msg.action     = 'finish';
          end
        otherwise
          msg.point_type = 'NS';
          msg.action     = 'warn';
          msg.wmsg       = 'could not determine type of event';
      end
      msg.idx = 1;
    end
  case 'check'
    msg.action = 'add';
    msg.finish = true;
end

end
```

The modified `po_close_orb` toolbox closer, shown below, incorporates the corresponding call to the `var_coll_add` constructor from Sect. 10.1.3.

```
function prob = po_close_orb(prob, tbid, data)

data.tbid = tbid;
data = coco_func_data(data);
```

```
prob = coco_add_slot(prob, tbid, @po_update, data, 'update');
segtbid     = coco_get_id(tbid, 'seg.coll');
[fdata uidx] = coco_get_func_data(prob, segtbid, 'data', 'uidx');
prob = coco_add_func(prob, tbid, @po_F, @po_DFDU, data, 'zero', ...
  'uidx', uidx(fdata.xbp_idx));
fid  = coco_get_id(tbid, 'period');
prob = coco_add_pars(prob, fid, uidx(fdata.T_idx), fid, 'active');
prob = coco_add_slot(prob, tbid, @coco_save_data, data, 'save_full');

if data.po.bifus
  segoid = coco_get_id(tbid, 'seg');
  prob = var_coll_add(prob, segoid);
  data.var_id = coco_get_id(segoid, 'var');
  tfid = coco_get_id(tbid, 'test');
  data.tfid = tfid;
  tfps = coco_get_id(tfid, {'SN' 'PD' 'NS' 'stab'});
  prob = coco_add_chart_data(prob, tfid, [], []);
  prob = coco_add_func(prob, tfid, @po_TF, data, ...
    'regular', tfps, 'requires', data.var_id, 'passChart');
  prob = coco_add_event(prob, 'SN', tfps{1}, 0);
  prob = coco_add_event(prob, 'PD', tfps{2}, 0);
  prob = coco_add_event(prob, @po_evhan_NS, data, tfps{3}, 0);
end

end
```

The call to `coco_add_func`, which appends `po_TF` to the continuation problem structure, makes use of the `'requires'` flag followed by an array of function identifiers for monitor functions that must be evaluated prior to an evaluation of `po_TF`. This ensures that the content of the `'var_coll'` toolbox data structure is updated prior to evaluation of the `po_TF` monitor functions. The modifications to the `po_get_settings` function, shown below, provide default values for the `'bifus'` and `'NSad'` toolbox settings.

```
function data = po_get_settings(prob, tbid, data)

defaults.bifus = true;
defaults.NSad  = false;
data.po = coco_merge(defaults, coco_get(prob, tbid));
assert(islogical(data.po.bifus), ...
  '%s: input for ''bifus'' option is not boolean', tbid);
assert(islogical(data.po.NSad), ...
  '%s: input for ''NSad'' option is not boolean', tbid);

end
```

Finally, the following modified encodings of the `po_isol2orb` and `po_sol2orb` toolbox constructors ensure that `po_get_settings` is invoked appropriately during construction.

```
function prob = po_isol2orb(prob, oid, varargin)

tbid   = coco_get_id(oid, 'po');
str    = coco_stream(varargin{:});
segoid = coco_get_id(tbid, 'seg');
prob   = coll_isol2seg(prob, segoid, str);

data = po_get_settings(prob, tbid, struct());
data = po_init_data(prob, tbid, data);
prob = po_close_orb(prob, tbid, data);

end
```

```
function prob = po_sol2orb(prob, oid, varargin)

ttbid = coco_get_id(oid, 'po');
str   = coco_stream(varargin{:});
run   = str.get;
if ischar(str.peek)
  stbid = coco_get_id(str.get, 'po');
else
  stbid = ttbid;
end
lab = str.get;

toid = coco_get_id(ttbid, 'seg');
soid = coco_get_id(stbid, 'seg');
prob = coll_sol2seg(prob, toid, run, soid, lab);
data = coco_read_solution(stbid, run, lab);
data = po_get_settings(prob, tbid, data);
data = po_init_data(prob, ttbid, data);
prob = po_close_orb(prob, ttbid, data);

end
```

17.2.2 Multisegment orbits

We generalize the treatment from the previous section to the detection and location of bifurcations of multisegment periodic orbits of hybrid dynamical systems, again associated with the critical combinations of the corresponding Floquet multipliers, shown at the beginning of Sect. 17.2. The hspo_TF function, shown below, mirrors the implementation in po_TF.

```
function [data chart y] = hspo_TF(prob, data, chart, u)

cdata = coco_get_chart_data(chart, data.tfid);
if ~isempty(cdata) && isfield(cdata, 'la')
  la = cdata.la;
else
  P = hspo_P(prob, data, u);
  M = P{1};
  for i=2:data.nsegs
    M = P{i}*M;
  end
  la = eig(M);
  chart = coco_set_chart_data(chart, data.tfid, struct('la', la));
end
y(1,1) = prod(la-1);
y(2,1) = prod(la+1);
if numel(la)>1
  NS_TF = la(data.la_idx1).*la(data.la_idx2);
  y(3,1) = prod(NS_TF(:)-1);
else
  y(3,1) = 1;
end
y(4,1) = sum(abs(la)>1);

end
```

Here, the auxiliary function hspo_P is identical to the encoding in Sect. 10.1.5. The modifications to the hspo_init_data function, shown below, are again immediate.

```matlab
function data = hspo_init_data(data, fhan, dfdxhan, dfdphan, ...
  modes, events, resets, x0, p0)

data.hhan     = fhan{2};
data.dhdxhan  = dfdxhan{2};
data.dhdphan  = dfdphan{2};
data.ghan     = fhan{3};
data.dgdxhan  = dfdxhan{3};
data.dgdphan  = dfdphan{3};
data.modes    = modes;
data.events   = events;
data.resets   = resets;

nsegs         = numel(events);
data.nsegs    = nsegs;
data.x0_idx   = cell(1,nsegs);
data.x1_idx   = cell(1,nsegs);
cdim          = 0;
dim           = zeros(1,nsegs);
data.dim      = dim;
for i=1:nsegs
  dim(i)        = size(x0{i},2);
  data.dim(i)   = dim(i);
  data.x0_idx{i} = cdim+(1:dim(i))';
  data.x1_idx{i} = cdim+(1:dim(i))';
  cdim          = cdim+dim(i);
end
data.cdim = cdim;
rows = [];
cols = [];
off  = 0;
pdim = numel(p0);
data.pdim = pdim;
for i=1:nsegs
  rows = [rows; repmat(off+1, [dim(i)+pdim 1])];
  rows = [rows; repmat(off+1+(1:dim(mod(i,data.nsegs)+1))', ...
    [1+dim(i)+pdim 1])];
  cols = [cols; nsegs+cdim+data.x1_idx{i}; nsegs+2*cdim+(1:pdim)'];
  c2   = repmat(nsegs+cdim+data.x1_idx{i}', ...
    [dim(mod(i,data.nsegs)+1) 1]);
  c3   = repmat(nsegs+2*cdim+(1:pdim), [dim(mod(i,data.nsegs)+1) 1]);
  cols = [cols; nsegs+data.x0_idx{mod(i,data.nsegs)+1}; c2(:); c3(:)];
  off  = off+dim(mod(i,data.nsegs)+1)+1;
end
data.rows = rows;
data.cols = cols;

I             = triu(true(dim(1)),1);
A             = repmat((1:dim(1))', 1, dim(1));
data.la_idx1  = A(I);
A             = A';
data.la_idx2  = A(I);

end
```

The additional `data` input argument is compatible with the modified encoding of the constructor `hspo_isol2segs` shown below.

```
function prob = hspo_isol2segs(prob, oid, varargin)

tbid = coco_get_id(oid, 'hspo');
str  = coco_stream(varargin{:});
fhan = str.get;
dfdxhan = cell(1,3);
dfdphan = cell(1,3);
if is_empty_or_func(str.peek)
  dfdxhan = str.get;
  if is_empty_or_func(str.peek)
    dfdphan = str.get;
  end
end
modes  = str.get('cell');
events = str.get('cell');
resets = str.get('cell');
t0     = str.get('cell');
x0     = str.get('cell');
pnames = {};
if iscellstr(str.peek('cell'))
  pnames = str.get('cell');
end
p0 = str.get;

hspo_arg_check(tbid, fhan, dfdxhan, dfdphan, ...
  modes, events, resets, t0, x0, p0, pnames);
coll = {};
for i=1:numel(modes)
  coll = [coll, {...
    coll_func(fhan{1},    modes{i}), ...
    coll_func(dfdxhan{1}, modes{i}), ...
    coll_func(dfdphan{1}, modes{i}), ...
    t0{i}, x0{i}, p0}];
end
hspo_bc_data = hspo_get_settings(prob, tbid);
hspo_bc_data = hspo_init_data(hspo_bc_data, fhan, dfdxhan, dfdphan, ...
  modes, events, resets, x0, p0);
prob = msbvp_isol2segs(prob, oid, coll{:}, pnames, ...
  @hspo_bc, @hspo_bc_DFDX, hspo_bc_data);
if hspo_bc_data.hspo.bifus
  prob = hspo_add_bifus(prob, oid, tbid, hspo_bc_data);
end

end
```

The `hspo_TF` monitor function is appended to the continuation problem structure by a suitable call to the `coco_add_func` constructor in the `hspo_add_bifus` function, shown below.

```
function prob = hspo_add_bifus(prob, oid, tbid, data)

cids = {};
vids = {};
msid = coco_get_id(oid, 'msbvp');
for i=1:numel(data.modes)
  soid = coco_get_id(msid,sprintf('seg%d', i));
  prob = var_coll_add(prob, soid);
  cids = [cids {coco_get_id(soid, 'coll')}];
  vids = [vids {coco_get_id(soid, 'var')}];
end
```

```
data.msid = msid;
data.cids = cids;
data.vids = vids;
[fdata uidx] = coco_get_func_data(prob, msid, 'data', 'uidx');
data.p_idx = numel(fdata.x1_idx)+(1:numel(fdata.p_idx));
tfid = coco_get_id(tbid, 'test');
data.tfid = tfid;
tfps = coco_get_id(tfid, {'SN' 'PD' 'NS' 'stab'});
prob = coco_add_chart_data(prob, tfid, [], []);
prob = coco_add_func(prob, tfid, @hspo_TF, data, 'regular', tfps, ...
  'uidx', [uidx(fdata.x1_idx); uidx(fdata.p_idx)], ...
  'requires', vids, 'passChart');
prob = coco_add_event(prob, 'SN', tfps{1}, 0);
prob = coco_add_event(prob, 'PD', tfps{2}, 0);
prob = coco_add_event(prob, @hspo_NS_han, data, 'SP', tfps{3}, 0);

end
```

Here, the `hspo_NS_han` event handler is identical, except for the obvious difference in name, to the encoding of `po_NS_han`. Finally, we provide default values for the `'bifus'` and `'NSad'` toolbox settings in the `hspo_get_settings` function, in turn obtained from `po_get_settings` by the text substitution `po → hspo`.

Example 17.3 In Example 16.2, we accommodated the case of harmonic excitation of a nonlinear oscillator through the introduction of an additional pair of state variables with the appropriate steady-state behavior. Here, we instead consider the case of a periodic bang-bang excitation strategy, in which the excitation term is piecewise constant and intermittently equal to A and $-A$, respectively, over half a period. In this case, a periodic response with the period of the excitation may be naturally separated into a two-segment orbit, where each segment corresponds to a given constant value of the excitation.

Consider, in particular, the Duffing equation with bang-bang forcing, described by the vector fields

$$f(y, p; \mathfrak{pos}) := \begin{pmatrix} y_2 \\ A - \lambda y_2 - \alpha y_1 - \varepsilon y_1^3 \\ 1 \end{pmatrix} \tag{17.10}$$

and

$$f(y, p; \mathfrak{neg}) := \begin{pmatrix} y_2 \\ -A - \lambda y_2 - \alpha y_1 - \varepsilon y_1^3 \\ 1 \end{pmatrix}, \tag{17.11}$$

the event function

$$h(y, p; \mathfrak{phase}) := \pi/\omega - y_3, \tag{17.12}$$

and the reset function

$$g(y, p; \mathfrak{phase}) := \begin{pmatrix} y_1 \\ y_2 \\ 0 \end{pmatrix}, \tag{17.13}$$

where $y \in \mathbb{R}^3$ and $p = (\lambda, \alpha, \varepsilon, A, \omega)$. A periodic response with the period of the excitation $(2\pi/\omega)$ then corresponds to a two-segment periodic orbit with signature

$$\{(\mathfrak{neg}, \mathfrak{phase}, \mathfrak{phase}), (\mathfrak{pos}, \mathfrak{phase}, \mathfrak{phase})\}.$$

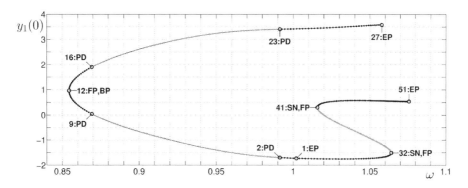

Figure 17.4. *Frequency response curve of the Duffing oscillator from Example* 17.3 *for forcing amplitude* $A = 26$ *under variations in the forcing frequency* ω, *while the remaining parameters are set to* $\lambda = 0.2$ *and* $\alpha = \varepsilon = 1$. *Here, black dots correspond to periodic orbits with all Floquet multipliers within the unit circle, whereas gray dots correspond to periodic orbits with at least one Floquet multiplier outside of the unit circle. A number of toolbox and atlas events are detected along this curve. As a special case, the branch point at label* 12 *is a pitchfork bifurcation point that marks a characteristic transition from nonsymmetric responses through a symmetric one to symmetry-conjugate responses. This transition is illustrated with sample solutions at labels* 9, 12 *and* 16 *in Fig.* 17.5.

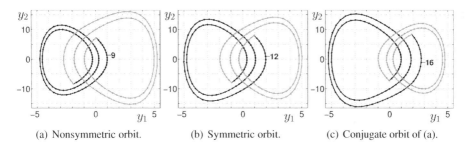

    (a) Nonsymmetric orbit.   (b) Symmetric orbit.   (c) Conjugate orbit of (a).

Figure 17.5. *The Duffing equation with bang-bang forcing in Example* 17.3 *is equivariant under rotations of the* (y_1, y_2) *plane by* π, *combined with a phase shift by* π. *As a consequence, we expect to observe symmetric as well as nonsymmetric oscillations, where existence of a nonsymmetric oscillation implies existence of the symmetry-conjugate oscillation. In the present case, continuation starts at a family of nonsymmetric oscillations, passes through a symmetry-increasing bifurcation point, and continues to produce symmetry-conjugate oscillations, as shown in the sequence of plots in panels (a)–(c). The solutions in panels (a) and (c) are obtained for the same value of* ω, *but on different sides of the branch point. Clearly, one can transform (a) into (c) by a rotation by* π *and exchanging black for gray.*

Fig. 17.4 shows the result of continuation of a 1-dimensional family of such two-segment periodic orbits under variations in ω with the atlas algorithm developed in Sect. 16.1. The analysis reveals a number of local bifurcations, as well as atlas events, including a branch-point singularity at a *pitchfork bifurcation point*. Representative orbits are shown in Fig. 17.5. ∎

17.3   Branch switching

As final examples of the utility of event handlers, we consider their use in enabling switching between intersecting manifolds in the solution set to a restricted continuation problem, or between solution manifolds to distinct continuation problems associated with a given dynamical system. We begin by demonstrating automated branch switching of the first type within a modified 1-dimensional atlas algorithm. As an example of branch switching of the second type, we consider the construction of a restricted continuation problem for continuation along a branch of periodic orbits of period $2T$ emanating from a primary branch of periodic orbits of period T at a period-doubling bifurcation point.

17.3.1   Branch points

We propose below modifying the 1-dimensional expanding-boundary atlas algorithm in Sect. 16.1 to enable continuation along a network of 1-dimensional solution branches. In particular, we associate branch-point detection with an event handler that stores a tangent matrix, appropriate for continuation along the emerging branch, in the chart data. Finally, we modify the `merge` class method in order to include the corresponding predictors in the `boundary` cell array. Here, key use is again made of the default linear solver for computing the nullvector of the matrix

$$\begin{pmatrix} \partial_u \Phi(u) \\ V^T \end{pmatrix} \tag{17.14}$$

at the branch point.

The modifications to the `atlas_1d_min` class declaration are shown below.

```
classdef atlas_1d_min < AtlasBase

  properties (Access=private)
    boundary = {};
    cont     = struct();
    cdid     = '';
  end

  methods (Access=private)
    function atlas = atlas_1d_min(prob, cont, dim)
      assert(dim==1, '%s: wrong manifold dimension', mfilename);
      atlas      = atlas@AtlasBase(prob);
      atlas.cont = atlas.get_settings(cont);
    end
  end

  methods (Static)
    function [prob cont atlas] = create(prob, cont, dim)
      atlas = atlas_1d_min(prob, cont, dim);
      prob  = CurveSegment.add_prcond(prob, dim);
      atlas.cdid = coco_get_id('atlas', 'BP');
      prob = coco_add_chart_data(prob, atlas.cdid, [], []);
      if atlas.cont.FP
        prob = coco_add_func_after(prob, 'mfunc', ...
          @atlas_1d_min.add_test_FP);
      end
```

```
  if atlas.cont.BP
    prob = coco_set(prob, 'lsol', 'det', true);
    fid  = coco_get_id('atlas', 'test', 'BP');
    prob = coco_add_func(prob, fid, @atlas.test_BP, [], ...
      'singular', fid, 'uidx', 'all', 'passChart', ...
      'returnsProb', 'fdim', 1);
    data = struct('cdid', atlas.cdid);
    prob = coco_add_event(prob, @atlas.evhan_BP, data, fid, 0);
  end
end

function prob = add_test_FP(prob)
  p_idx = coco_get_func_data(prob, 'efunc', 'pidx');
  if numel(p_idx)>=1
    fid  = coco_get_id('atlas', 'test', 'FP');
    prob = coco_add_func(prob, fid, @atlas_1d_min.test_FP, [], ...
      'singular', fid, 'uidx', 'all', 'passTangent', 'fdim', 1);
    prob = coco_add_event(prob, 'FP', fid, 0);
  end
end

function [data y] = test_FP(prob, data, u, t)
  p_idx = coco_get_func_data(prob, 'efunc', 'pidx');
  y = t(p_idx(1));
end

function [prob data chart y] = test_BP(prob, data, chart, u)
  cdata = coco_get_chart_data(chart, 'lsol');
  if ~isfield(cdata, 'det')
    [prob chart] = prob.cseg.update_det(prob, chart);
    cdata = coco_get_chart_data(chart, 'lsol');
  end
  y = cdata.det;
end

function [data cseg msg] = evhan_BP(prob, data, cseg, cmd, msg)

  switch cmd
    case 'init'
      if isfield(msg, 'finish') || strcmp(msg.action, 'warn')
        msg.action = 'finish';
      elseif strcmp(msg.action, 'locate')
        msg.action = 'warn';
      else
        msg.point_type = 'BP';
        msg.action     = 'locate';
        msg.idx = 1;
      end
    case 'check'
      msg.action = 'add';
      msg.finish = true;
      chart = cseg.curr_chart;
      cdata = coco_get_chart_data(chart, 'lsol');
      chart = coco_set_chart_data(chart, data.cdid, ...
        struct('TS', cdata.v, 'ign', msg.evidx));
      cseg.curr_chart = chart;
  end

  end
end
```

```
methods (Static, Access=private)
  cont = get_settings(cont)
end

methods (Access=public)
  [prob atlas cseg correct] = init_prcond      (atlas, prob, chart)
  [prob atlas cseg]         = init_chart       (atlas, prob, cseg)
  [opts atlas cseg]         = init_admissible  (atlas, opts, cseg, S)
  [prob atlas cseg flush]   = init_atlas       (atlas, prob, cseg)
  [prob atlas cseg]         = flush            (atlas, prob, cseg)
  [prob atlas cseg correct] = predict          (atlas, prob, cseg)
  [prob atlas cseg flush]   = add_chart        (atlas, prob, cseg)
end

methods (Access=private)
  flag = isneighbor(atlas, chart1, chart2)
  [atlas prob cseg] = merge(atlas, prob, cseg)
end

end
```

These include a call to `coco_add_event`, in the `create` class method, with reference to the `evhan_BP` event handler, as well as the initialization of, and subsequent assignment of content to, chart data with identifier `'atlas.BP'`. Here, the `cdata.v` field contains a normalized eigenvector $\tilde{v}$ of the matrix in Eq. (17.14) corresponding to the smallest singular value. At a branch point $u^\dagger$, it follows that $\tilde{v}$ lies in the intersection of $\mathcal{N}\left[\partial_u \Phi(u^\dagger)\right]$ and the orthogonal complement to span$\{V\}$. Now recall that the tangent vector to every curve segment on the solution manifold must lie in the nullspace $\mathcal{N}\left[\partial_u \Phi(u^\dagger)\right]$. Thus, if the vector V is tangential to one of the two branches intersecting at $u^\dagger$, and if their intersection is transversal, then the tangent vector to the second branch has a nonzero projection onto $\tilde{v}$ and there exists a locally unique solution along this branch to the closed continuation problem

$$\left(\begin{array}{c} \Phi(u) \\ \tilde{v}^T \cdot \left(u - u^\dagger\right) - h \end{array} \right) = 0 \qquad (17.15)$$

for sufficiently small, but nonzero, h.

Provided that the `'atlas.BP'` element of the corresponding chart data is nonempty and includes a `TS` field, the modified encoding of the `merge` class method, shown below, grows the `boundary` array accordingly.

```
function [atlas prob cseg] = merge(atlas, prob, cseg)

chart = cseg.ptlist{end};
R     = atlas.cont.h;
h     = atlas.cont.Rmarg*R;
nb    = cell(numel(chart.s),4);
for k=1:numel(chart.s)
  sk      = chart.s(k);
  xk      = chart.x+h*(chart.TS*sk);
  nb(k,:) = {chart, xk, sk, h};
end
for i=size(atlas.boundary,1):-1:1
  chart2 = atlas.boundary{i,1};
```

```
    if atlas.isneighbor(chart, chart2)
      x2 = atlas.boundary{i,2};
      if norm(chart.TS'*(x2-chart.x))<R
        atlas.boundary(i,:) = [];
      end
      for k=size(nb,1):-1:1
        x1 = nb{k,2};
        if norm(chart2.TS'*(x1-chart2.x))<R
          nb(k,:) = [];
        end
      end
    end
  end
end
atlas.boundary = [nb; atlas.boundary];
for j=2:numel(cseg.ptlist)-1
  cdata = coco_get_chart_data(cseg.ptlist{j}, atlas.cdid);
  if ~isempty(cdata) && isfield(cdata, 'TS')
    chart    = cseg.ptlist{j};
    chart.t = cdata.TS;
    [prob cseg2] = CurveSegment.create_initial(prob, chart);
    chart    = cseg2.curr_chart;
    chart.TS = cseg2.prcond.TS;
    chart.pt = 0;
    chart.ignore_evs = cdata.ign;
    nb = [{chart, chart.x+h*chart.TS, 1, h}; ...
      {chart, chart.x-h*chart.TS, -1, h}];
    for i=size(atlas.boundary,1):-1:1
      chart2 = atlas.boundary{i,1};
      if atlas.isneighbor(chart, chart2)
        x2 = atlas.boundary{i,2};
        if norm(chart.TS'*(x2-chart.x))<R
          atlas.boundary(i,:) = [];
        end
        for k=size(nb,1):-1:1
          x1 = nb{k,2};
          if norm(chart2.TS'*(x1-chart2.x))<R
            nb(k,:) = [];
          end
        end
      end
    end
    atlas.boundary = [nb; atlas.boundary];
  end
end
if isempty(atlas.boundary)
  chart.pt_type    = 'EP';
  chart.ep_flag    = 1;
  cseg.ptlist{end} = chart;
end

end
```

Example 17.4 We continue the analysis of the Duffing equation with bang-bang excitation from Example 17.3. Fig. 17.6 shows the result of continuation with the 1-dimensional atlas algorithm from Sect. 16.1 for forcing frequency $\omega = 1$ and varying forcing amplitude A. Representative orbits are shown in Fig. 17.7. The result of continuation with the modified atlas algorithm is shown in Fig. 17.8. Here, continuation along additional branches on the solution set emanating from singular branch points is automated using the event-handling functionality. Representative orbits are shown in Fig. 17.9. ∎

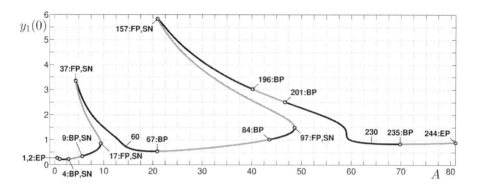

Figure 17.6. *Amplitude response curve of the Duffing oscillator from Example 17.4 for $\omega = 1$ computed with the atlas code developed in Sect. 16.1. Again, $\lambda = 0.2$ and $\alpha = \varepsilon = 1$. Along this curve we indicate the location of atlas and toolbox events. In addition, we encode stability information provided by the toolbox monitor function. Here, black corresponds to stable and gray to unstable oscillations. We observe two resonance peaks, where passage through each peak corresponds to the addition of an oscillation during one forcing half cycle, as illustrated with sample solutions at labels 1, 60 and 230 in Fig. 17.7. Note that all solutions along this response curve are symmetric. The analysis of this example continues in Fig. 17.8.*

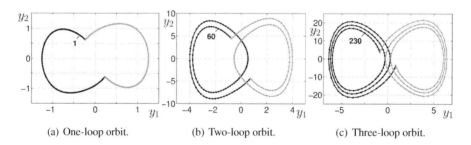

(a) One-loop orbit.   (b) Two-loop orbit.   (c) Three-loop orbit.

Figure 17.7. *After each passage through a resonance peak along the amplitude response curve shown in Fig. 17.6, we observe the addition of an oscillation during one half cycle of the forcing.*

17.3.2   Period-doubling bifurcations of single-segment orbits

We conclude the development of event handling in the 'po' toolbox by implementing an event handler associated with the detection of a period-doubling bifurcation. In particular, we seek to provide support for continuation along the branch of periodic orbits of twice the original period that emanates from the period-doubling bifurcation. To this end, the encoding of the event handler, shown below, stores an approximation of the period-doubled periodic orbit in the chart data of the chart associated with the special point.

```
function [data cseg msg] = po_evhan_PD(prob, data, cseg, cmd, msg)

switch cmd
```

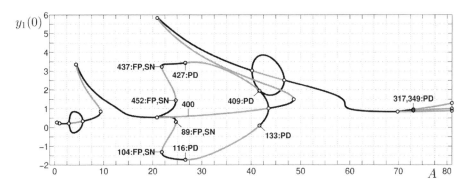

Figure 17.8. *Amplitude response curve of the Duffing oscillator from Example 17.4 for ω = 1 computed with the atlas code developed in Sect. 17.3, which implements automatic branch switching at branch points. Again, λ = 0.2 and α = ε = 1. In addition to the results shown in Fig. 17.6, we obtain a number of families of nonsymmetric oscillations emerging from the branch points located during continuation; see also Fig. 17.9. Three of these form closed families, and the others terminate at the computational boundary. Along the closed family in the center of the figure, a number of toolbox events are located, including period-doubling bifurcation points. The analysis of this example continues in Fig. 17.10. The overlap of two curves at the period-doubling point 409 is an artifact of projection.*

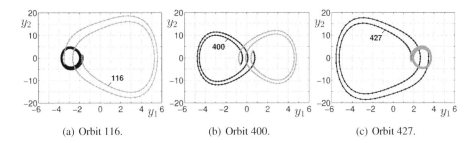

(a) Orbit 116.   (b) Orbit 400.   (c) Orbit 427.

Figure 17.9. *A sample of periodic orbits for A ≈ 27 on the family of symmetric (b), and along the closed family of nonsymmetric (a) and (c), two-loop oscillations located in the center of Fig. 17.8. The latter two orbits are again symmetry conjugate and lie at period-doubling points.*

```
case 'init'
  if isfield(msg, 'finish') || strcmp(msg.action, 'warn')
    msg.action = 'finish';
  elseif strcmp(msg.action, 'locate')
    msg.action = 'warn';
  else
    msg.point_type = 'PD';
    msg.action     = 'locate';
    msg.idx = 1;
  end
case 'check'
```

```
          msg.action = 'add';
          msg.finish = true;
          vdata = coco_get_func_data(prob, data.var_id, 'data');
          [uidx fdata] = coco_get_func_data(prob, vdata.tbid, ...
            'uidx', 'data');
          chart = cseg.curr_chart;
          u = chart.x(uidx);
          x = u(fdata.xbp_idx);
          T = u(fdata.T_idx);
          p = u(fdata.p_idx);

          M     = vdata.M;
          M1    = M(vdata.M1_idx,:);
          [v d] = eig(M1);
          [m i] = min(diag(d)+1);
          xp1   = reshape(x+0.01*M*v(:,i), fdata.xbp_shp)';
          xp1   = xp1(fdata.tbp_idx,:);
          t1    = fdata.tbp(fdata.tbp_idx)*T;
          xp2   = reshape(x-0.01*M*v(:,i), fdata.xbp_shp)';
          xp2   = xp2(fdata.tbp_idx,:);
          t2    = fdata.tbp(fdata.tbp_idx)*T;
          x0    = [xp1; xp2(2:end,:)];
          t0    = [t1; T+t2(2:end)];
          chart = coco_set_chart_data(chart, data.efid, ...
            struct('pd_x0', x0, 'pd_t0', t0, 'pd_p', p));
          cseg.curr_chart = chart;
    end

    end
```

Here, the eigenvector of the fundamental matrix corresponding to the eigenvalue -1 is used to generate two perturbations to the original periodic orbit, which are subsequently concatenated as an approximation of a period-doubled periodic orbit. The initial solution guess is then stored among the chart data with the `data.efid` data identifier for later processing.

The `po_evhan_PD` event handler is included with the construction of the periodic orbit object by modifying the `po_close_orb` closer, as shown below.

```
function prob = po_close_orb(prob, tbid, data)

data.tbid = tbid;
data = coco_func_data(data);
prob = coco_add_slot(prob, tbid, @po_update, data, 'update');
segtbid     = coco_get_id(tbid, 'seg.coll');
[fdata uidx] = coco_get_func_data(prob, segtbid, 'data', 'uidx');
prob = coco_add_func(prob, tbid, @po_F, @po_DFDU, data, 'zero', ...
  'uidx', uidx(fdata.xbp_idx));
fid  = coco_get_id(tbid, 'period');
prob = coco_add_pars(prob, fid, uidx(fdata.T_idx), fid, 'active');
prob = coco_add_slot(prob, tbid, @coco_save_data, data, 'save_full');

if data.po.bifus
  segoid = coco_get_id(tbid, 'seg');
  prob = var_coll_add(prob, segoid);
  data.var_id = coco_get_id(segoid, 'var');
  tfid = coco_get_id(tbid, 'test');
  data.tfid = tfid;
  tfps = coco_get_id(tfid, {'SN' 'PD' 'NS' 'stab'});
  prob = coco_add_chart_data(prob, tfid, [], []);
  prob = coco_add_func(prob, tfid, @po_TF, data, ...
```

```
            'regular', tfps, 'requires', data.var_id, 'passChart');
    prob = coco_add_event(prob, 'SN', tfps{1}, 0);
    data.efid = coco_get_id(tbid, 'PD');
    prob = coco_add_chart_data(prob, data.efid, [], []);
    prob = coco_add_event(prob, @po_evhan_PD, data, tfps{2}, 0);
    prob = coco_add_event(prob, @po_evhan_NS, data, tfps{3}, 0);
  end

end
```

17.3.3 Period-doubling bifurcations of multisegment orbits

We generalize the treatment of branch switching at period-doubling bifurcations to the case of multisegment periodic orbits of hybrid dynamical systems. The `hspo_PD_han` event handler, shown below, is consistent with the theoretical results in Sect. 10.1.5, in particular Eq. (10.57).

```
function [data cseg msg] = hspo_PD_han(prob, data, cseg, cmd, msg)

switch cmd
  case 'init'
    if isfield(msg, 'finish') || strcmp(msg.action, 'warn')
      msg.action = 'finish';
    elseif strcmp(msg.action, 'locate')
      msg.action = 'warn';
    else
      msg.point_type = 'PD';
      msg.action     = 'locate';
      msg.idx = 1;
    end
  case 'check'
    msg.action = 'add';
    msg.finish = true;

    chart   = cseg.curr_chart;
    cdata   = coco_get_chart_data(chart, data.tfid);
    M       = cdata.M;
    [v d]   = eig(M);
    [m idx] = min(diag(d)+1);
    v       = 0.01*v(:,idx);

    t0   = {};
    x0   = {};
    for i=1:data.nsegs
      vdata = coco_get_func_data(prob, data.vids{i}, 'data');
      [uidx fdata] = coco_get_func_data(prob, vdata.tbid, ...
        'uidx', 'data');
      u = chart.x(uidx);
      x = u(fdata.xbp_idx);
      T = u(fdata.T_idx);
      p = u(fdata.p_idx);

      x = reshape(x+vdata.M*v, fdata.xbp_shp)';
      t0 = [t0 {fdata.tbp(fdata.tbp_idx)*T}];
      x0 = [x0 {x(fdata.tbp_idx,:)}];
      v = cdata.P{i}*v;
    end
    for i=1:data.nsegs
      vdata = coco_get_func_data(prob, data.vids{i}, 'data');
```

```
        [uidx fdata] = coco_get_func_data(prob, vdata.tbid, ...
          'uidx', 'data');
        u  = chart.x(uidx);
        x  = u(fdata.xbp_idx);
        T  = u(fdata.T_idx);
        p  = u(fdata.p_idx);

        x  = reshape(x+vdata.M*v, fdata.xbp_shp)';
        t0 = [t0 {fdata.tbp(fdata.tbp_idx)*T}];
        x0 = [x0 {x(fdata.tbp_idx,:)}];
        v  = cdata.P{i}*v;
      end

      pd         = struct('t0', {t0}, 'x0', {x0}, 'p', {p});
      pd.modes   = [data.modes  data.modes];
      pd.events  = [data.events data.events];
      pd.resets  = [data.resets data.resets];
      chart = coco_set_chart_data(chart, data.efid, pd);
      cseg.curr_chart = chart;
  end

  end
```

The following modified encoding of the `hspo_TF` monitor function ensures that the composite sensitivity matrix and the sequence of segment-specific products, corresponding to individual values of j in Eq. (10.57), are stored as chart data.

```
function [data chart y] = hspo_TF(prob, data, chart, u)

cdata = coco_get_chart_data(chart, data.tfid);
if ~isempty(cdata) && isfield(cdata, 'la')
  la = cdata.la;
else
  P = hspo_P(prob, data, u);
  M = P{1};
  for i=2:data.nsegs
    M = P{i}*M;
  end
  la = eig(M);
  chart = coco_set_chart_data(chart, data.tfid, ...
    struct('M', M, 'la', la, 'P', {P}));
end
y(1,1) = prod(la-1);
y(2,1) = prod(la+1);
if numel(la)>1
  NS_TF = la(data.la_idx1).*la(data.la_idx2);
  y(3,1) = prod(NS_TF(:)-1);
else
  y(3,1) = 1;
end
y(4,1) = sum(abs(la)>1);

end
```

The necessary modifications to the `hspo_add_bifus` constructor are now immediate and left as an exercise.

Example 17.5 We illustrate the expanded branch-switching functionality by applying the modified 'hspo' toolbox to the continued analysis of the Duffing equation with bang-bang excitation from Examples 17.3 and 17.4. The bifurcation diagram in Fig. 17.10 builds on

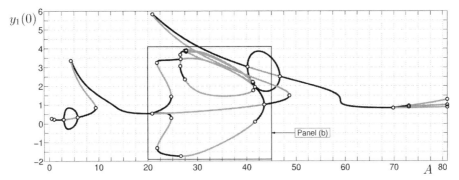

(a) Bifurcation diagram of Duffing oscillator with families of nonsymmetric and period-doubled solutions. A sequence of enlargements of details is shown in panels (b) to (d).

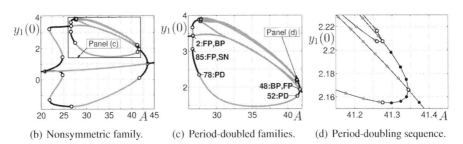

(b) Nonsymmetric family.   (c) Period-doubled families.   (d) Period-doubling sequence.

Figure 17.10. *Continuation of the analysis from Fig.* 17.8. *Using the algorithm for branch switching at period-doubling points developed in Sect.* 17.3.3, *we can further complete the bifurcation diagram* (a). *Panel* (b) *shows a zoom into the closed family, along which period-doubling points were detected previously. A continuation of the emerging period-doubled orbits results in another closed family, along which period-doubling points are again detected* (c). *Repetition of this procedure results in evidence of a sequence of period-doubling bifurcations. The onset of this sequence is shown in the enlargement* (d), *which includes local families of period-2, -4, and -8 orbits; see also Fig.* 17.11. *Note that the families of period-doubled solutions were computed with the atlas algorithm without automatic branch switching to prevent redundant coverage.*

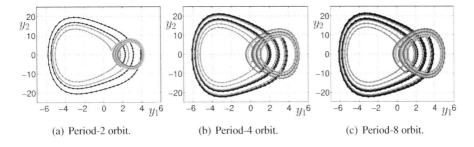

(a) Period-2 orbit.   (b) Period-4 orbit.   (c) Period-8 orbit.

Figure 17.11. *A sequence of period-doubled orbits along the families shown in Fig.* 17.10(d).

the results illustrated in Fig. 17.8 and includes additional solution branches corresponding to families of multisegment periodic orbits of twice, four times, and eight times the period along the original solution manifold. Representative orbits are shown in Fig. 17.11. ∎

17.4 Conclusions

It is difficult to imagine the practical utility of numerical continuation without core support for event detection and location. The nature of the solutions to a mathematical problem, molded to fit the continuation paradigm, is often illuminated precisely through the structure imposed by special points and events on the solution manifold. In the context of dynamical systems, for example, bifurcations of equilibria, fixed points, or periodic orbits imply changes in the local (and, sometimes, global) character of the flow that, in turn, reveal hitherto unknown features of the system behavior.

The response to the detection and location of an event offers problem-specific opportunities for deploying continuation within a framework of intelligent exploration. Such a framework takes advantage of new information gleaned from event-specific features in order to further uncover properties of the overall system. The use of event handlers and the branch-switching paradigm explored in this chapter provide rudimentary, but powerful, ingredients of such a framework.

A particularly powerful context for event handling is in the introduction of adaptive changes to finite-dimensional discretizations of infinite-dimensional mathematical problems during continuation. Here, events are introduced in order to impose run-time modifications to the mathematical formulation of the extended continuation problem in order to maintain acceptable bounds on discretization errors. We devote Part V of this text to several examples of such adaptive implementations.

Exercises

17.1. Explain the construction of the `data.la_idx1` and `data.la_idx2` fields in the `alg_construct_eqn` constructor in Sect. 17.1.

17.2. Consider the vector field

$$f(y, p) = \begin{pmatrix} y_1 - y_1 y_2 / (1 + p_1 y_1) \\ y_1 y_2 / (1 + p_1 y_1) - y_2 - p_2 y_2^2 \end{pmatrix}$$

for $p_1, p_2 > 0$, modified slightly from Exercise 9 in Chap. 10 of [Kuznetsov, Yu.A., *Elements of Applied Bifurcation Theory*, Springer-Verlag, New York, 1998]. Find all equilibria and examine their stability properties. Show that ψ_{HB} vanishes along a branch of equilibria provided that

$$p_1 = \frac{1 \pm \sqrt{1 - 4p_2} + 2p_2}{2(2 + p_2)}$$

for $0 < p_2 < \frac{1}{4}$.

17.3. Show that there exists a Bogdanov–Takens point of the vector field in the previous exercise when

$$(p_1, p_2) = \left(\sqrt{2} - 1, \frac{1}{2} \left(\sqrt{2} - 1 \right) \right).$$

17.4. Find an expression for the relationship between the problem parameters p_1 and p_2 in the previous two exercises along the curve of fold points found in Example 17.1.

17.5. Illustrate the reverse communication protocol in the form of a flow diagram for the `alg_evhan_HB` event handler in Sect. 17.1. Verify the conditions on the difference

```
abs(sum(sign(real(la0)))-sum(sign(real(la1))))
```

and give examples of situations that might render this quantity different from 0 or 4.

17.6. A Hopf bifurcation is said to be *supercritical* if the periodic orbits, emanating from the bifurcation point under parameter variations, are stable limit cycles and *subcritical* otherwise. Review the discussion in Chaps. 5 and 10 of [Kuznetsov, Yu.A., *Elements of Applied Bifurcation Theory*, Springer-Verlag, New York, 1998] as pertains to the computation of the *first Lyapunov exponent* and its relationship to the nature of the Hopf bifurcation. Modify the `alg_evhan_HB` event handler to include a computation of the first Lyapunov coefficient in the `'check'` state of the event handler and the storing of the result as chart data.

17.7. Review the use of the bialternate matrix product and a suitable bordered matrix approach to detect, locate, and continue families of Hopf bifurcations of equilibria in [Kuznetsov, Yu.A., *Elements of Applied Bifurcation Theory*, Springer-Verlag, New York, 1998]. Implement a corresponding embeddable monitor function in the `'alg'` toolbox and demonstrate its use on the vector field in Example 17.1.

17.8. Explain the need for the `'requires'` flag in the second call to `coco_add_func` in the encoding of `po_close_orb` in Sect. 17.2.1.

17.9. Embed the `'alg'` toolbox in a toolbox designed for the continuation and analysis of fixed points of smooth maps $f : \mathbb{R}^n \to \mathbb{R}^n$. Implement a suite of monitor functions and event handlers appropriate for the detection, location, and continuation of bifurcations associated with such fixed points.

17.10. Include with the toolbox developed in the previous exercise an event handler associated with a period-doubling bifurcation that stores an approximation of the period-2 orbit in the chart data of the corresponding chart. Use this to demonstrate branch switching at a period-doubling bifurcation point and illustrate your implementation by analyzing the Hénon map

$$f : u \mapsto \left(\begin{array}{c} 1 - \alpha u_1^2 + u_2 \\ \beta u_1 \end{array} \right).$$

17.11. Modify the `'hspo'` toolbox in Sect. 17.2.2 to support detection, location, and continuation of bifurcation points associated with orbital degeneracies at the terminal points of individual segments, for example, points of grazing contact or points on the intersection of two codimension-1 surfaces.

17.12. What changes must be made to the `'hspo'` toolbox in Sect. 17.2.2 to support restarting continuation from a previously stored solution point? Why was this not necessary in Chap. 9?

17.13. Use COCO to reproduce the numerical results illustrated in Figs. 17.4–17.5.

17.14. Suppose that $\Phi : \mathbb{R}^n \to \mathbb{R}^{n-1}$ and consider the nullvector $\tilde{v}$ of the square matrix

$$\begin{pmatrix} \partial_u \Phi(u) \\ V^T \end{pmatrix}$$

at a simple branch point $u^\dagger$ corresponding to the intersection of two 1-dimensional branches of solutions to the zero problem $\Phi(u) = 0$. Suppose that V is tangential to one of the two branches. Explain why there exists a locally unique solution along the second branch to the closed continuation problem

$$\begin{pmatrix} \Phi(u) \\ \tilde{v}^T \cdot (u - u^\dagger) - h \end{pmatrix} = 0$$

for sufficiently small, but nonzero, h.

17.15. How does the boundary array grow in the automated branch-switching atlas algorithm described in Sect. 17.3.1? Can you guarantee that the algorithm will not produce a redundant cover of portions of the solution set?

17.16. What is the significance of the `ign` field of the `'atlas.BP'` element of the chart data that is assigned in the `evhan_BP` event handler in Sect. 17.3.1? Illustrate the result of applying this atlas algorithm when omitting this field and explain your observations.

17.17. Propose a modification to the branch-switching atlas algorithm in Sect. 17.3.1 that would label each branch with a unique integer index. Suggest a use for this functionality.

17.18. Propose a modification to the branch-switching atlas algorithm in the previous exercise that would limit the number of switches performed during a single run.

17.19. As an alternative to the computation of the vector $\tilde{v}$ in Sect. 17.3.1, review [Hughes, J. and Friedman, M., "A bisection-like algorithm for branch switching at a simple branch point," *Journal of Scientific Computing*, 41(1), pp. 62–69, 2009] and propose a COCO-compatible implementation.

17.20. As an alternative to the computation of the vector $\tilde{v}$ in Sect. 17.3.1, explore the use of the *algebraic branching equation*, described in [Beyn, W.-J., Champneys, A., Doedel, E.J., Kuznetsov, Yu.A., Sandstede, B., and Govaerts, W., "Numerical continuation and computation of normal forms," in Fiedler, B. (ed.) *Handbook of Dynamical Systems* III: *Towards Applications*. Elsevier, Amsterdam (2001)] for obtaining a suitable tangent matrix for branch switching at a simple singular point.

17.21. Use COCO to reproduce the numerical results illustrated in Figs. 17.6–17.9.

17.22. Explain the computation of an approximate period-doubled periodic orbit in the `po_evhan_PD` and `hspo_evhan_PD` event handlers. Modify the implementations so as to replace the numerical constant 0.01 with a quantity scaled by the size of the original periodic orbit.

17.23. Use COCO to reproduce the numerical results illustrated in Figs. 17.10–17.11.

17.24. The vector field

$$f : (y, p) \mapsto \begin{pmatrix} \frac{1}{p_4}\left(-(p_1 + p_2)y_1 + p_2 y_2 - p_5 y_1^3 + p_6(y_2 - y_1)^3\right) \\ p_2 y_1 - (p_2 + p_3)y_2 - y_3 - p_6(y_2 - y_1)^3 \\ y_2 \end{pmatrix}$$

is analyzed in great detail using the AUTO software package (see [Doedel, E. and Oldeman, B., *AUTO-07P: Continuation and Bifurcation Software for Ordinary Differential Equations*, manual, 2012]). Use the modified 'po' toolbox together with the branch-switching atlas algorithm from Sect. 17.3 to detect simple branch points along families of periodic orbits of the dynamical system corresponding to this vector field and demonstrate the branch-switching functionality.

17.25. Review [Weber, H., "Multigrid bifurcation iteration," *SIAM Journal on Numerical Analysis*, 22(2), pp. 262–279, 1985] and apply the branch-switching atlas algorithm in Sect. 17.3 to the determination of branches of nontrivial solutions, emanating from critical values of a problem parameter λ, in the example problems discussed therein.

Part V

Adaptation

Chapter 18

Pointwise Adaptation and Comoving Meshes

In this chapter, we begin a study of integrating *adaptivity*, in the formulation of discretized zero problems, with continuation along a solution manifold. We are concerned here with changes to the number of continuation variables or their meaning, as well as with (accompanying) changes to the number of zero functions. We anticipate that such changes would be imposed adaptively in order to maintain bounds on discretization errors during continuation.

Four distinct methods of adaptation during continuation are considered in this part of the text. In the first method implemented in this chapter, continuation terminates when a suitably formulated discretization error estimate exceeds a critical value. Following such an event, we modify the discretization and reconstruct, from scratch, a restricted continuation problem with an accompanying initial solution guess. Continuation then proceeds until termination is again triggered by an excessive discretization error. We term such a method one of *pointwise adaptation*, since no information about the local geometry of the solution manifold is provided to the atlas algorithm.

In the second method demonstrated in this chapter, the discretization parameters are included among the continuation variables and solved for accordingly. We obtain an augmented extended continuation problem by appending additional conditions to the continuation zero problem in order to allow the discretization to change concurrently with the discretized solution during continuation. In the example shown in this chapter, the resultant *comoving mesh* seeks to provide an equidistant discretization of the state-space representation of the solution trajectory through a suitably modified collocation zero problem.

In Chap. 19, we implement a *mesh-preserving* discretization method that supports adaptive changes to the discretization order, without a change in the meaning or the number of continuation variables. In this case, the solution to the discretized continuation problem parameterizes an approximant through a transformation that is independent of adaptive changes to the discretization scheme. We conclude in Chap. 20 with examples of methods of adaptation in which both the meaning and the number of continuation variables may change during continuation. Such *moving-mesh* schemes are here accommodated through minor modifications to the atlas algorithms and by associating a remeshing function with each zero function that depends on the discretization.

18.1   A brute-force approach

18.1.1   Discretization orders, parameters, and errors

Recall the introductory treatment of the *catenary* problem from the theory of calculus of variations in Chap. 1. There, we sought a smooth curve $y = f(x) > 0$ on the interval $[0, 1]$ such that $f(0) = 1$ and $f(1) = Y$, for some known nonnegative constant Y, and such that the integral functional

$$J(f) = 2\pi \int_0^1 f(x)\sqrt{1 + (f'(x))^2}\, dx \qquad (18.1)$$

attained a stationary value at the chosen curve.

By definition, solutions to the catenary problem are infinite-dimensional mathematical objects. In order to support a computational implementation, they must be approximated by members of some class of finitely parameterized approximants. To this end, we proposed two distinct discretization schemes in Chap. 1. In the first case, polynomial approximants p_1 and p_2 of degree m for f and its derivative f' were formulated in terms of $2m + 2$ unknown coefficients. A corresponding set of $2m + 2$ zero functions was then obtained by requiring that

- the polynomial approximants satisfy the Euler–Lagrange equations

$$\frac{d}{dx}\begin{pmatrix} f \\ f' \end{pmatrix} = \begin{pmatrix} f' \\ (1 + f'^2)/f \end{pmatrix} \qquad (18.2)$$

  at a finite set of collocation nodes $\{x_i\}_{i=1}^m$ on the interval $[0, 1]$ and

- p_1 satisfy the boundary conditions at $x = 0$ and $x = 1$.

In the second case, a single polynomial approximant p of degree m for f was formulated in terms of $m + 1$ unknown coefficients. This polynomial was substituted into a quadrature discretization of the integral (18.1) in terms of some set of quadrature nodes and weights. A corresponding set of $m + 1$ zero functions was then obtained using the method of Lagrange multipliers by requiring that the resultant function of the polynomial coefficients attain a local extremum along the constraint manifold imposed by the boundary conditions on p. In both cases, choices were made regarding the *orders of discretization*, e.g., the degree of the approximating polynomials or the number of quadrature nodes, as well as regarding the *discretization parameters*, e.g., the locations of the collocation nodes or quadrature nodes and the values of the quadrature weights. With each such choice came a source of *discretization error* that would need to be considered carefully in order to gain confidence in the obtained solutions.

Notably, in all previous chapters, it was (sometimes tacitly) assumed that neither the orders of discretization nor the values of the discretization parameters would change during continuation. Discretization orders such as NTST (N) and NCOL (m), introduced in the encoding of the 'coll' toolbox in Chap. 7, were assigned values at the outset of the construction of a segment object. No changes were made to these numbers during the subsequent continuation. The implementation was further restricted to a uniform mesh

$$0 = t_1 \leq \cdots \leq t_j = \frac{j-1}{N} T \leq \cdots \leq t_{N+1} = T \tag{18.3}$$

in the independent variable t of the differential equation in y, choosing to approximate y on each of the equal-sized intervals $[t_j, t_{j+1}]$, for $j = 1, \ldots, N$, by a linear combination of Lagrange polynomials. On each such interval, the corresponding Lagrange polynomials in the independent variable σ were, in turn, generated using a uniform mesh

$$-1 = \sigma_1 \leq \cdots \leq \sigma_j = 2\frac{j-1}{m} - 1 \leq \cdots \leq \sigma_{m+1} = 1 \tag{18.4}$$

that remained fixed during continuation. Finally, the collocation/quadrature nodes z_l and the corresponding quadrature weights w_l, for $l = 1, \ldots, m$ (used, for example, when discretizing the integral phase condition in the po toolbox in Sect. 8.2), were precomputed at construction and kept fixed during continuation.

It is easy to imagine cases where different choices of integers for the discretization orders, and numerical values for the discretization parameters, would result in approximants within the same estimated error bounds, but at vastly different computational expense. Optimal choices would be expected to reduce the computational expense, without violating the desired bounds on the discretization errors. Such considerations are important even when seeking a single approximant for some given set of values of the problem parameters, e.g., Y in the catenary problem. During continuation, moreover, the optimal choices for discretization orders and parameters would typically vary with changes to the values of the problem parameters.

In the absence of adaptive changes to the discretization orders or parameters, discretization errors may vary greatly along the solution manifold. In this case, very conservative choices may be required in order to allow continuation to proceed along the desired portion of the solution manifold, without violating error bounds. There appears to be great value in supporting adaptivity during continuation in order to maintain a suitable balance between error tolerances and computational efficiency. In the remainder of this part of the book, we proceed to explore mechanisms for implementing such functionality in COCO-compatible toolboxes and atlas algorithm.

Example 18.1 Consider the continuation of periodic orbits $y(t) \in \mathbb{R}^2$ of the dynamical system given by the vector field

$$f(y, p) := \begin{pmatrix} y_2 \\ -y_1 + \varepsilon y_2 \left(\frac{1}{2} - y_2^2 \right) \end{pmatrix}, \tag{18.5}$$

where $p = (\varepsilon)$. The extract below shows the result of applying the 'po' toolbox with the default choices $N = 10$ and $m = 4$ for the discretization orders NTST and NCOL, respectively.

```
>> eps0 = 0.1;
>> t0    = linspace(0, 2*pi, 100)';
>> x0    = [sin(t0) cos(t0)];
>> prob  = coco_set(coco_prob(), 'cont', 'ItMX', [0 200]);
>> prob1 = po_isol2orb(prob, '', @pneta, t0, x0, 'eps', eps0);
```

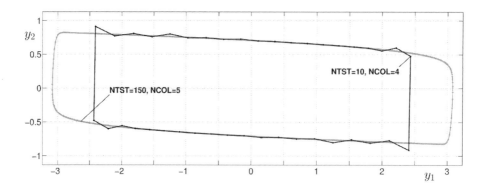

Figure 18.1. *Comparison of approximate periodic orbits of the dynamical system with vector field given by Eq. (18.5) at $\varepsilon = 20$, obtained in Example 18.1 using continuation for different uniform meshes. For the default settings with $N = 10$ and $m = 4$, we obtain a poor approximation. Increasing the discretization parameters to $N = 150$ and $m = 5$ results in a solution curve that seems acceptable, at least by visual inspection. A large subset of the additional mesh points contributes little to the improvement, however; only a small number is allocated along the problematic vertical part of the orbit.*

```
>> coco(prob1, 'run1', [], 1, 'eps', [0.1 20]);
```

STEP		DAMPING		NORMS			COMPUTATION	TIMES	
IT	SIT	GAMMA	||d||	||f||	||U||	F(x)	DF(x)	SOLVE	
0				5.01e-02	1.14e+01	0.0	0.0	0.0	
1	1	1.00e+00	1.01e+00	6.47e-03	1.08e+01	0.0	0.0	0.0	
2	1	1.00e+00	2.68e-01	4.03e-04	1.06e+01	0.0	0.0	0.0	
3	1	1.00e+00	1.91e-02	1.98e-06	1.06e+01	0.0	0.0	0.0	
4	1	1.00e+00	9.46e-05	4.87e-11	1.06e+01	0.0	0.0	0.0	
5	1	1.00e+00	2.33e-09	1.56e-15	1.06e+01	0.0	0.0	0.0	

STEP	TIME	||U||	LABEL	TYPE	eps
0	00:00:00	1.0599e+01	1	EP	1.0000e-01
10	00:00:00	1.2398e+01	2		2.8405e+00
20	00:00:01	1.6505e+01	3		5.7782e+00
30	00:00:02	2.1101e+01	4		8.5855e+00
40	00:00:02	2.5870e+01	5		1.1326e+01
50	00:00:03	3.0687e+01	6		1.4111e+01
60	00:00:03	3.4338e+01	7		1.7177e+01
69	00:00:04	3.7009e+01	8	EP	2.0000e+01

In this case, the dimension of the domain of the corresponding extended continuation problem is 102. A state-space representation of the approximate solution obtained for $\varepsilon = 20$ is shown in Fig. 18.1. In comparison, the continued extract below shows the result of applying the 'po' toolbox with $N = 150$ and $m = 5$.

```
>> prob2 = coco_set(prob, 'coll', 'NTST', 150, 'NCOL', 5);
>> prob2 = po_isol2orb(prob2, '', @pneta, t0, x0, 'eps', eps0);
>> coco(prob2, 'run2', [], 1, 'eps', [0.1 20]);
```

STEP		DAMPING		NORMS			COMPUTATION	TIMES	
IT	SIT	GAMMA	||d||	||f||	||U||	F(x)	DF(x)	SOLVE	
0				1.80e-02	3.13e+01	0.0	0.0	0.0	

```
        1   1   1.00e+00   4.28e+00   1.87e-03   2.72e+01    0.0      0.0      0.0
        2   1   1.00e+00   1.14e+00   1.16e-04   2.61e+01    0.0      0.0      0.0
        3   1   1.00e+00   8.08e-02   5.71e-07   2.61e+01    0.0      0.1      0.0
        4   1   1.00e+00   4.01e-04   1.40e-11   2.61e+01    0.0      0.1      0.0
        5   1   1.00e+00   9.84e-09   1.31e-14   2.61e+01    0.0      0.1      0.1

    STEP        TIME          ||U||   LABEL   TYPE              eps
       0    00:00:00     2.6060e+01       1   EP       1.0000e-01
      10    00:00:02     2.7174e+01       2            1.8461e+00
      20    00:00:03     3.0543e+01       3            3.9170e+00
      30    00:00:05     3.4500e+01       4            5.7939e+00
      40    00:00:06     3.8713e+01       5            7.6142e+00
      50    00:00:08     4.3088e+01       6            9.4046e+00
      60    00:00:09     4.7580e+01       7            1.1175e+01
      70    00:00:10     5.2160e+01       8            1.2929e+01
      80    00:00:12     5.6806e+01       9            1.4672e+01
      90    00:00:13     6.1506e+01      10            1.6405e+01
     100    00:00:14     6.6247e+01      11            1.8130e+01
     110    00:00:16     7.1022e+01      12            1.9849e+01
     111    00:00:16     7.1444e+01      13   EP       2.0000e+01
```

This greater resolution is accompanied by a significant increase in the dimension of the domain of the extended continuation problem to 1802. The corresponding state-space trajectory for $\varepsilon = 20$ is also shown in Fig. 18.1. From the figure, it is clear that the added resolution afforded by the denser mesh serves little purpose for the majority of the solution trajectory. On the other hand, the lower-resolution discretization is clearly unable to resolve the rapid variation in y_2. ∎

18.1.2 Adaptive changes to the discretization order

A first attempt at integrating adaptivity into the continuation paradigm developed thus far is afforded by the following brute-force approach. Let the monitor function ψ evaluate to an estimate of the discretization error and associate a terminal event with the value of this function exceeding a critical tolerance. Upon the triggering of such an event, changes made outside of the atlas algorithm to the discretization orders and parameters, as well as a recomputation of the initial solution guess, should then allow continuation to proceed until the next terminal event. We illustrate below this paradigm in the context of the 'coll' toolbox, encoded in Chap. 7.

Recall the coll_sol2seg generalized constructor, repeated below for convenience.

```
function prob = coll_sol2seg(prob, oid, varargin)

tbid = coco_get_id(oid, 'coll');
str  = coco_stream(varargin{:});
run  = str.get;
if ischar(str.peek)
  soid = str.get;
else
  soid = oid;
end
lab = str.get;

[sol data] = coll_read_solution(soid, run, lab);
data       = coll_get_settings(prob, tbid, data);
data       = coll_init_data(data, sol.x, sol.p);
```

```
sol        = coll_init_sol(data, sol.t, sol.x, sol.p);
prob       = coll_construct_seg(prob, tbid, data, sol);

end
```

We note, in particular, the call to the `coll_get_settings` function, shown below.

```
function data = coll_get_settings(prob, tbid, data)

defaults.NTST = 10;
defaults.NCOL = 4;
if ~isfield(data, 'coll')
  data.coll = [];
end
data.coll = coco_merge(defaults, coco_merge(data.coll, ...
  coco_get(prob, tbid)));
NTST = data.coll.NTST;
assert(numel(NTST)==1 && isnumeric(NTST) && mod(NTST,1)==0, ...
  '%s: input for option ''NTST'' is not an integer', tbid);
NCOL = data.coll.NCOL;
assert(numel(NCOL)==1 && isnumeric(NCOL) && mod(NCOL,1)==0, ...
  '%s: input for option ''NCOL'' is not an integer', tbid);

end
```

It is clear from this implementation that the values of the discretization orders NTST and NCOL, stored with an existing solution file, may be overwritten by assigning different values to the optional settings associated with the `'coll'` toolbox, using the `coco_set` utility. A compatible initial solution guess is then obtained by the call to the `coll_init_sol` function, shown below.

```
function sol = coll_init_sol(data, t0, x0, p0)

t0 = t0(:);
T0 = t0(end)-t0(1);
t0 = (t0-t0(1))/T0;
x0 = interp1(t0, x0, data.tbp)';

sol.u = [x0(:); T0; p0];

end
```

Here, the `tbp` field of the `data` structure contains the t_{bp} array in Eq. (7.13), recomputed with the new values of N and m (i.e., NTST and NCOL) in the preceding call to `coll_init_data`.

We proceed to introduce a monitor function whose value corresponds to an estimate of the discretization error for a given discretization order. Consider, for a moment, the approximation of a scalar-valued function y on the interval $[\sigma_1, \sigma_{m+1}]$ by a polynomial p_m of degree m that interpolates the values of y at a sequence $\{\sigma_i\}_{i=1}^{m+1}$ of mesh nodes. The pointwise interpolation error $y(\sigma) - p_m(\sigma)$ is then given in the *Lagrange form* by the expression

$$\frac{1}{(m+1)!} y^{(m+1)}(\tilde{\sigma}) \prod_{i=1}^{m+1} (\sigma - \sigma_i) \tag{18.6}$$

for some $\sigma_1 \leq \tilde{\sigma} \leq \sigma_{m+1}$.

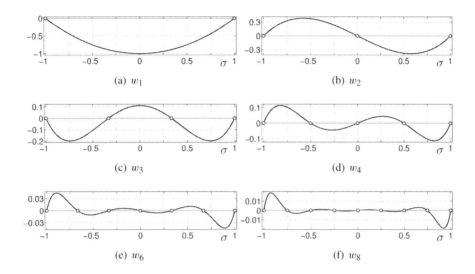

Figure 18.2. *Graphs of the function $w_m(\sigma) := \prod_{i=1}^{m+1}(\sigma - \sigma_i)$ for typical choices of $m = 1,2,3,4,6,8$. We observe that, for all these functions, $|w_m|$ assumes its global maximum between the first two zeros located at σ_1 and σ_2, as claimed in Example 18.2.*

Example 18.2 We obtain an upper bound for the product

$$\prod_{i=1}^{m+1} |\sigma - \sigma_i| \tag{18.7}$$

by replacing σ by the value in the interval $[\sigma_1, \sigma_{m+1}]$ that yields a maximum for the product. As suggested by Fig. 18.2, in the case of a uniform mesh

$$\sigma_i = 2\frac{i-1}{m} - 1, i = 1,\dots,m+1, \tag{18.8}$$

this maximum is obtained for a value of σ in the interval $[\sigma_1, \sigma_2]$. ∎

If the function y is unknown, except for its values at the mesh nodes σ_i, $i = 1,\dots,m+1$, then the interpolation error in Eq. (18.6) must be estimated by replacing the derivative $y^{(m+1)}(\tilde{\sigma})$ with a quantity computable in terms of the polynomial interpolant p_m. Of course, since the latter is of degree m, its $(m+1)$th derivative must vanish. We rely instead on the following *spectral assumption*:

> There exists an integer M such that the coefficient in front of σ^m in the polynomial interpolant p_k, for every $k \geq m > M$, is bounded from above by the coefficient in front of σ^M in the polynomial interpolant p_M.

For sufficiently large m, it follows that an upper bound on the derivative $y^{(m+1)}(\tilde{\sigma})$ may be obtained by the product $(m+1)!c_m$, where c_m is the coefficient in front of σ^m in the polynomial p_m. An upper bound for the discretization error in Eq. (18.6) is then given by

$$|c_m| \max_\sigma \prod_{i=1}^{m+1} |\sigma - \sigma_i|. \tag{18.9}$$

In the case of a family of polynomial interpolants, one for each component of the function $y : \mathbb{R} \to \mathbb{R}^n$, let c_m denote the vector of the corresponding coefficients of the mth order power of σ in each of the interpolating polynomials. We then propose the following estimate for the discretization error:

$$\|c_m\| \max_\sigma \prod_{i=1}^{m+1} |\sigma - \sigma_i| \tag{18.10}$$

in terms of the Euclidean norm $\| \cdot \|$.

In the collocation continuation problem, we obtain the coefficients of the mth order power of σ in the polynomial interpolants $g_j(\sigma)$ from the formula

$$\sum_{i=1}^{m+1} \upsilon_{(m+1)(j-1)+i} \prod_{k=1, k \neq i}^{m+1} \frac{1}{\sigma_i - \sigma_k} \tag{18.11}$$

for $j = 1, \ldots, N$. In terms of the array υ_{bp}, it follows that the array of coefficients may be obtained from a product $W^{(m)} \cdot \upsilon_{bp}$, where

$$W^{(m)} = I_N \otimes \left(L^{(m)} \otimes I_n \right) \tag{18.12}$$

for some $1 \times (m+1)$ row matrix $L^{(m)}$.

Example 18.3 Recall the notation P_k for the mapping

$$P_k(A)(i_1, \ldots, i_{k-1}, i_{k+1}, \ldots, i_n) := \prod_{i_k} A(i_1, \ldots, i_n) \tag{18.13}$$

from the space of n-dimensional arrays (with $n \geq k$) to the space of $(n-1)$-dimensional arrays. Furthermore, let t_1 denote the 2-dimensional array whose (j, k) entry is given by

$$t_1(j, k) = \begin{cases} \left(\sigma_j - \sigma_k \right)^{-1}, & k \neq j, \\ 1, & k = j. \end{cases} \tag{18.14}$$

It follows that the jth entry of the matrix $L^{(m)}$ in Eq. (18.12) is given by

$$P_2(t_1). \tag{18.15}$$

The function `coll_Lm` shown below computes the matrix $L^{(m)}$.

```
function A = coll_Lm(ts)

p = numel(ts);

sj = repmat(reshape(ts, [1 p]), [p 1]);
sk = repmat(reshape(ts, [p 1]), [1 p]);
t1 = sj-sk;
idx = abs(t1)<=eps;
t1(idx) = 1;

A = 1./t1;
A = prod(A, 2)';

end
```

Here, the transpose in the last assignment ensures that the output argument `A` is a row matrix. ∎

The inclusion of the following assignments in a modified encoding of `coll_init_data` provides for the computation of $W^{(m)}$ and the upper bound in Example 18.2.

```
mmap       = coll_Lm(tm);
rows       = reshape(1:dim*NTST, [dim NTST]);
rows       = repmat(rows, [bpdim 1]);
cols       = repmat(1:xbpdim, [dim 1]);
Wm         = repmat(kron(mmap, eye(dim)), [1 NTST]);
data.Wm    = sparse(rows, cols, Wm);
x          = linspace(tm(1), tm(2), 51);
y          = arrayfun(@(x) prod(x-tm), x);
data.wn    = max(abs(y));
```

The discretization error estimate and its scaling by the error tolerance given in `data.coll.` TOL are now encoded in the `coll_err` monitor function, shown below.

```
function [data y] = coll_err(prob, data, u)

cp = reshape(data.Wm*u, [data.dim data.coll.NTST]);
y  = data.wn*max(sqrt(sum(cp.^2,1)));
y  = [y; y/data.coll.TOL];

end
```

We append this monitor function to the segment object, and define a terminal event triggered by the scaled discretization error estimate exceeding 1, in the modified encoding of the constructor `coll_construct_seg`, shown below.

```
function prob = coll_construct_seg(prob, tbid, data, sol)

prob = coco_add_func(prob, tbid, @coll_F, @coll_DFDU, data, 'zero', ...
  'u0', sol.u);
uidx = coco_get_func_data(prob, tbid, 'uidx');
if ~isempty(data.pnames)
  fid  = coco_get_id(tbid, 'pars');
  prob = coco_add_pars(prob, fid, uidx(data.p_idx), data.pnames);
end
```

```
prob = coco_add_slot(prob, tbid, @coco_save_data, data, 'save_full');
efid = coco_get_id(tbid, {'err' 'err_TF'});
prob = coco_add_func(prob, efid{1}, @coll_err, data, ...
  'regular', efid, 'uidx', uidx(data.xbp_idx));
prob = coco_add_event(prob, 'MXCL', 'MX', efid{2}, '>', 1);

end
```

Default content for the error tolerance is encoded in the modified `coll_get_settings` function, shown below.

```
function data = coll_get_settings(prob, tbid, data)

defaults.NTST = 10;
defaults.NCOL = 4;
if ~isfield(data, 'coll')
  data.coll = [];
end
data.coll = coco_merge(defaults, coco_merge(data.coll, ...
  coco_get(prob, tbid)));
if ~coco_exist('TOL', 'class_prop', prob, tbid, '-no-inherit-all')
  data.coll.TOL = coco_get(prob, 'corr', 'TOL')^(2/3);
end
NTST = data.coll.NTST;
assert(numel(NTST)==1 && isnumeric(NTST) && mod(NTST,1)==0, ...
  '%s: input for option ''NTST'' is not an integer', tbid);
NCOL = data.coll.NCOL;
assert(numel(NCOL)==1 && isnumeric(NCOL) && mod(NCOL,1)==0, ...
  '%s: input for option ''NCOL'' is not an integer', tbid);

end
```

Here, we invoke the `coco_exist` utility in order to assign a default interpolation error tolerance, in the event that this setting has not been explicitly assigned in the construction of the continuation problem structure. In particular, let δ_{corr} denote the error tolerance of the nonlinear corrector, stored in the `'TOL'` setting of the `'corr'` toolbox. Similarly, let δ_{coll} denote the error tolerance of the collocation scheme, stored in the `'TOL'` setting of the `'coll'` toolbox. The assignment

$$\delta_{\mathrm{coll}} = \delta_{\mathrm{corr}}^{2/3} \tag{18.16}$$

seeks to ensure that the number of significant digits in the value of the interpolating polynomial at any point in the interval $[-1, 1]$ is a factor of 2/3 of the corresponding number of significant digits in the values of the continuation variables obtained after correction. Here, the remaining digits are used as a guard against propagation of truncation errors.

Example 18.4 The extracts below show a continued analysis of the family of periodic orbits considered in Example 18.1. In the first case, with $N = 10$ and $m = 4$, continuation terminates at $\varepsilon = 1.8936$, after seven successfully completed continuation steps.

```
>> prob  = coco_set(prob, 'coll', 'TOL', 1.0e-3);
>> prob2 = po_isol2orb(prob, '', @pneta, t0, x0, 'eps', 0.1);
>> coco(prob2, 'run1', [], 1, {'eps' 'po.seg.coll.err'}, [0.1 20]);
```

```
         STEP   DAMPING                  NORMS              COMPUTATION TIMES
    IT  SIT      GAMMA      ||d||       ||f||      ||U||    F(x)   DF(x)  SOLVE
     0                                5.01e-02  1.14e+01    0.0    0.0    0.0
     1   1    1.00e+00   1.01e+00   6.47e-03  1.08e+01    0.0    0.0    0.0
     2   1    1.00e+00   2.68e-01   4.03e-04  1.06e+01    0.0    0.0    0.0
     3   1    1.00e+00   1.91e-02   1.98e-06  1.06e+01    0.0    0.0    0.0
     4   1    1.00e+00   9.46e-05   4.87e-11  1.06e+01    0.0    0.0    0.0
     5   1    1.00e+00   2.33e-09   1.56e-15  1.06e+01    0.0    0.0    0.0

    STEP     TIME        ||U||    LABEL  TYPE          eps po.seg.coll.err
     0    00:00:00   1.0599e+01      1   EP       1.0000e-01      5.0591e-05
     7    00:00:00   1.1440e+01      2   MXCL     1.8936e+00      8.9142e-04
     8    00:00:00   1.1733e+01          MXCL     2.2147e+00      1.4916e-03
```

As shown below, restarting continuation from this solution, after a change in discretization order to $N = 20$ and $m = 4$, allows continuation to proceed, within the established bound on the discretization error estimate, until $\varepsilon = 4.619$.

```
>> prob  = coco_set(prob, 'coll', 'NTST', 20, 'NCOL', 4);
>> prob2 = po_sol2orb(prob, '', 'run1', 2);
>> coco(prob2, 'run2', [], 1, {'eps' 'po.seg.coll.err'}, [0.1 20]);
```

```
         STEP   DAMPING                  NORMS              COMPUTATION TIMES
    IT  SIT      GAMMA      ||d||       ||f||      ||U||    F(x)   DF(x)  SOLVE
     0                                1.12e-01  1.29e+01    0.0    0.0    0.0
     1   1    1.00e+00   2.17e-02   5.78e-05  1.29e+01    0.0    0.0    0.0
     2   1    1.00e+00   1.05e-04   8.51e-10  1.29e+01    0.0    0.0    0.0
     3   1    1.00e+00   2.17e-09   2.98e-15  1.29e+01    0.0    0.0    0.0

    STEP     TIME        ||U||    LABEL  TYPE          eps po.seg.coll.err
     0    00:00:00   1.2915e+01      1   EP       1.8936e+00      9.7915e-05
    10    00:00:00   1.5868e+01      2            4.3386e+00      4.2909e-04
    11    00:00:01   1.6279e+01      3   MXCL     4.6190e+00      7.6882e-04
    12    00:00:01   1.6697e+01          MXCL     4.8977e+00      1.2590e-03
```

Finally, with a further change of discretization order to $N = 150$ and $m = 5$, continuation proceeds from $\varepsilon = 4.619$ to $\varepsilon = 20$, while the discretization error estimate remains within the desired tolerance.

```
>> prob  = coco_set(prob, 'coll', 'NTST', 150, 'NCOL', 5);
>> prob2 = po_sol2orb(prob, '', 'run2', 3);
>> coco(prob2, 'run3', [], 1, {'eps' 'po.seg.coll.err'}, [0.1 20]);
```

```
         STEP   DAMPING                  NORMS              COMPUTATION TIMES
    IT  SIT      GAMMA      ||d||       ||f||      ||U||    F(x)   DF(x)  SOLVE
     0                                3.54e-02  3.19e+01    0.0    0.0    0.0
     1   1    1.00e+00   4.58e-02   4.04e-05  3.20e+01    0.0    0.0    0.0
     2   1    1.00e+00   1.90e-04   8.17e-10  3.20e+01    0.0    0.1    0.0
     3   1    1.00e+00   4.13e-09   1.64e-14  3.20e+01    0.0    0.1    0.0

    STEP     TIME        ||U||    LABEL  TYPE          eps po.seg.coll.err
     0    00:00:00   3.1961e+01      1   EP       4.6190e+00      3.0971e-08
    10    00:00:02   3.5321e+01      2            6.1588e+00      1.7635e-07
    20    00:00:03   3.9571e+01      3            7.9720e+00      9.3032e-07
    30    00:00:05   4.3973e+01      4            9.7578e+00      3.2115e-06
    40    00:00:07   4.8484e+01      5            1.1524e+01      1.0550e-05
    50    00:00:09   5.3078e+01      6            1.3276e+01      4.4581e-05
```

```
60   00:00:11   5.7737e+01    7          1.5017e+01   6.9714e-05
70   00:00:13   6.2445e+01    8          1.6749e+01   8.9607e-05
80   00:00:16   6.7194e+01    9          1.8473e+01   3.7918e-04
89   00:00:18   7.1444e+01   10   EP     2.0000e+01   4.7651e-04
```

The results of this analysis are illustrated in Fig. 18.3.   ∎

18.2   (Co)moving meshes

In this section, we generalize the use of adaptive changes to the discretization of solutions to boundary-value problems, by enabling changes to the locus of individual mesh points, in addition to the number of mesh intervals.

Consider, as in Chap. 6, a solution $y(t) \in \mathbb{R}^n$ on the interval $[0, T]$, for some yet-to-be-determined scalar T, to the ordinary differential equation

$$\frac{dy}{dt} = f(y, p),$$
(18.17)

where the vector field $f : \mathbb{R}^n \times \mathbb{R}^q \to \mathbb{R}^n$ is parameterized by a vector of problem parameters $p \in \mathbb{R}^q$. We seek below to approximate the unknown function y on the interval $[0, T]$ in terms of a continuous function of t, expressed on each of N intervals as a polynomial of degree m and parameterized by the unknown values at $m + 1$ base points.

In contrast to Chap. 6, we no longer require equal-sized intervals in the independent variable t. Instead, consider the nonlinear transformation

$$t = t(\tau) = T \int_0^\tau \kappa(s)ds, \tau \in [0, 1],$$
(18.18)

for some positive function $\kappa : [0, 1] \to \mathbb{R}$ such that

$$\int_0^1 \kappa(s)ds = 1.$$
(18.19)

It follows that $\upsilon(\tau) := y(t(\tau))$ satisfies the equation

$$\frac{d\upsilon}{d\tau} = T\kappa f(\upsilon, p)$$
(18.20)

on the interval $[0, 1]$. We proceed to approximate the unknown function υ on the interval $[0, 1]$ in terms of a continuous function of τ, expressed on each of N *equal-sized* intervals as a polynomial of degree m and parameterized by the unknown values at $m + 1$ base points.

To this end, given the positive integer N, consider again the uniform partition

$$0 = \tau_1 < \cdots < \tau_j := \frac{j-1}{N} < \cdots < \tau_{N+1} = 1$$
(18.21)

and the linear transformation

$$\tau = \tau^{(j)}(\sigma) := \tau_j + \frac{(1+\sigma)}{2}\left(\tau_{j+1} - \tau_j\right), \sigma \in [-1, 1],$$
(18.22)

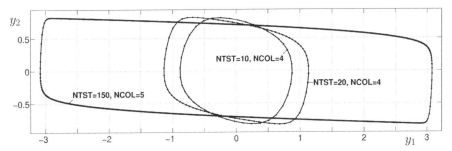

(a) Terminal solutions produced in each of the three runs in Example 18.4. The orbit with NTST equal to 150 is at $\varepsilon = 20$.

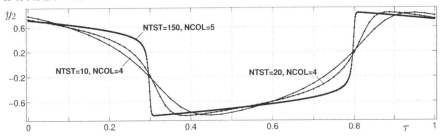

(b) Time profiles of y_2. The two sharp fronts give rise to a large, but localized, error.

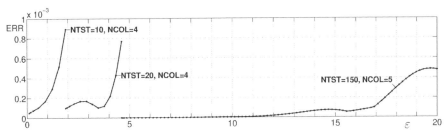

(c) The estimated error for different meshes plotted against variations in ε. The tolerance is 10^{-3}, and the first two runs, with NTST equal to 10 and 20, eventually exceed this tolerance and terminate. The last run, with NTST equal to 150 and NCOL equal to 5, is successful up to $\varepsilon = 20$.

Figure 18.3. *Continuation of periodic orbits of the dynamical system given by the vector field in Eq. (18.5) using pointwise adaptive changes to the discretization order, as described in Sect. 18.1.2. Here, a terminal event is associated with a monitor function that computes an estimate of the discretization error. Continuation stops automatically whenever the estimated error exceeds a predefined limit. As shown in Example 18.4, this approach allows one to compute an approximate solution family within a desired tolerance. Here, we need to restart twice with a finer mesh at terminal solution points, represented by the corresponding approximate periodic orbits in panel (a) together with the solution for $\varepsilon = 20$. The reason for the failure to remain within the desired tolerance with low discretization order is explained by the sharp fronts evident in the corresponding time profiles in panel (b). Panel (c) shows the estimated error plotted against variations in ε.*

on the interval $\left[\tau_j, \tau_{j+1}\right]$ for $j \in \{1, \ldots, N\}$. Let $\upsilon^{(j)}(\sigma) := \upsilon\left(\tau^{(j)}(\sigma)\right)$ and $\kappa^{(j)}(\sigma) = \kappa\left(\tau^{(j)}(\sigma)\right)$. It follows that

$$\frac{d\upsilon^{(j)}}{d\sigma} = \frac{1}{2N} T \kappa^{(j)} f\left(\upsilon^{(j)}, p\right) \tag{18.23}$$

for all σ on the interval $[-1, 1]$. As before,

$$\tau^{(j)}(1) = \tau_j = \tau^{(j+1)}(-1) \tag{18.24}$$

for every $1 \le j < N$. Continuity of the original solution across the interval $[0, T]$ then implies that

$$\upsilon^{(j)}(1) = \upsilon^{(j+1)}(-1) \tag{18.25}$$

for $1 \le j < N$.

In accordance with the stated objective, we proceed to approximate $\upsilon^{(j)}$ by a polynomial of degree m. For simplicity, we consider the case where $\kappa^{(j)}(\sigma) := \kappa_j$ for all σ on $[-1, 1]$. It follows that

$$\sum_{j=1}^{N} \kappa_j = N \tag{18.26}$$

and

$$\tau \in \left[\tau_j, \tau_{j+1}\right] \Rightarrow t \in \left[\frac{T}{N}\sum_{i=1}^{j-1}\kappa_i, \frac{T}{N}\sum_{i=1}^{j}\kappa_i\right] \tag{18.27}$$

for $1 \le j \le N$. On this interval, we now obtain

$$y(t) = \upsilon^{(j)}\left(\frac{2N}{\kappa_j}\left(\frac{t}{T} - \frac{1}{N}\sum_{i=1}^{j-1}\kappa_i\right) - 1\right). \tag{18.28}$$

Now consider approximating $\upsilon^{(j)}$ on $[-1, 1]$ by the polynomial

$$g_j(\sigma) := \sum_{i=1}^{m+1} \mathcal{L}_i(\sigma) \upsilon_{(m+1)(j-1)+i} \tag{18.29}$$

expressed in terms of the Lagrange polynomials $\mathcal{L}_i$, $i = 1, \ldots, m+1$, for some partition

$$-1 = \sigma_1 < \cdots < \sigma_i < \cdots < \sigma_{m+1} = 1. \tag{18.30}$$

As before,

$$g_j(\sigma_i) = \upsilon_{(m+1)(j-1)+i} \tag{18.31}$$

for $i \in \{1, \ldots, m+1\}$. The *collocation zero problem* is now given by

- the collocation conditions

$$0 = \frac{dg_j}{d\sigma}(z_l) - \frac{T\kappa_j}{2N} f\left(g_j(z_l), p\right), \tag{18.32}$$

where

$$\frac{dg_j}{d\sigma}(z_l) = \sum_{i=1}^{m+1} \mathcal{L}'_i(z_l) \, \upsilon_{(m+1)(j-1)+i} \tag{18.33}$$

and

$$\frac{T\kappa_j}{2N} f\left(g_j(z_l), p\right) = \frac{T\kappa_j}{2N} f\left(\sum_{i=1}^{m+1} \mathcal{L}_i(z_l) \, \upsilon_{(m+1)(j-1)+i}, p\right) \tag{18.34}$$

for $j = 1, \dots, N$ and $l = 1, \dots, m$, and

- the continuity conditions

$$0 = g_{j+1}(\sigma_1) - g_j(\sigma_{m+1}) = \upsilon_{(m+1)j+1} - \upsilon_{(m+1)(j-1)+m+1} \tag{18.35}$$

for $j = 1, \dots, N-1$.

Here, z_l, $l = 1, \dots, m$, denotes a set of m collocation nodes on the interval $[-1, 1]$.

A vectorized formulation of the collocation zero problem may be obtained by introducing the 1-dimensional array κ whose jth entry is given by the scalar κ_j for $j = 1, \dots, N$. Specifically, denote by

- κ_f the $n \times Nm$, 2-dimensional array whose ith entry equals the ith entry of the 1-dimensional array $\kappa \otimes 1_{nm,1}$,

- $\kappa_{\partial_y f}$ the $n \times n \times Nm$, 3-dimensional array whose ith entry equals the ith entry of the 1-dimensional array $\kappa \otimes 1_{n^2m,1}$, and

- $\kappa_{\partial_p f}$ the $n \times q \times Nm$, 3-dimensional array whose ith entry equals the ith entry of the 1-dimensional array $\kappa \otimes 1_{nmq,1}$.

Let a $*$ denote element-by-element multiplication. The collocation conditions are then obtained from the zero function

$$\left(\upsilon_{bp}, T, p\right) \mapsto \frac{T}{2N} \mathfrak{vec}\left(\kappa_f * f\left(\mathfrak{vec}_n\left(W \cdot \upsilon_{bp}\right), 1_{1,Nm} \otimes p\right)\right) - W' \cdot \upsilon_{bp} \tag{18.36}$$

in terms of the known matrices W and W' (introduced in Chap. 6). Its Jacobians with respect to υ_{bp}, T, and p then equal

$$\frac{T}{2N} \mathfrak{diag}\left(\kappa_{\partial_y f} * \partial_y f\left(\mathfrak{vec}_n\left(W \cdot \upsilon_{bp}\right), 1_{1,Nm} \otimes p\right)\right) \cdot W - W', \tag{18.37}$$

$$\frac{1}{2N} \mathfrak{vec}\left(\kappa_f * f\left(\mathfrak{vec}_n\left(W \cdot \upsilon_{bp}\right), 1_{1,Nm} \otimes p\right)\right), \tag{18.38}$$

and

$$\frac{T}{2N} \mathfrak{transp}\left(\kappa_{\partial_p f} * \partial_p f\left(\mathfrak{vec}_n\left(W \cdot \upsilon_{bp}\right), 1_{1,Nm} \otimes p\right)\right), \tag{18.39}$$

respectively.

As in Chap. 6, the continuity conditions are obtained from the zero function

$$Q \cdot \upsilon_{bp} \tag{18.40}$$

in terms of the known matrix Q. The corresponding Jacobians with respect to υ_{bp}, T, and p then equal Q, 0, and 0, respectively.

18.3   A comoving-mesh algorithm

Changes to the locus of the mesh nodes may be imposed during continuation, either

- through the application of a transformation in terms of the vector of continuation variables and the current discretization, following the successful location of each point on the solution manifold, or

- through the simultaneous solving for the continuation variables and the discretization parameters, as part of the continuation problem.

In the former case, we speak of an adaptive moving-mesh algorithm. It is clear that such an approach is necessary if the discretization order were to change adaptively. In the latter case, we speak of an adaptive comoving-mesh algorithm, as the mesh "moves" jointly with the solution during continuation. By integrating the discretization parameters in the continuation problem, such a formulation relies on the existing atlas algorithm for ensuring a desirable coverage of the solution manifold.

We consider moving-mesh algorithms in Chap. 20 and focus the remainder of this section on an example of a comoving-mesh algorithm.

18.3.1   Equidistributed arclength

In the comoving-mesh algorithm, considered below, the discretization parameters κ_j, for $j = 1, \ldots, N$, are included among the continuation variables. By definition, it follows that the function

$$\kappa \mapsto \sum_{j=1}^{N} \kappa_j - N \tag{18.41}$$

must be included among the zero functions of the augmented extended continuation problem. We obtain an additional set of conditions on the elements of κ by coupling their values to the other continuation variables, most notably v_{bp}.

To this end, let Γ denote a curve segment on the solution manifold, parameterized by h. Suppose that the elements of κ vary along Γ according to the differential equation

$$\frac{d\kappa_j}{dh} = -\left(l_j - \frac{1}{N} \sum_{s=1}^{N} l_s \right), \tag{18.42}$$

where l_j denotes the arclength, in state space, along the graph of the function v on the interval $[\tau_j, \tau_{j+1}]$. It follows, by definition, that the instantaneous rate of change of κ is zero at points along Γ where

$$l_j = \frac{1}{N} \sum_{s=1}^{N} l_s \tag{18.43}$$

for all $j = 1, \ldots, N$. In this case, the mesh is *equidistributed* by arclength along the graph of v. Moreover, for every mesh,

$$\frac{d}{dh} \sum_{j=1}^{N} \kappa_j = 0; \tag{18.44}$$

i.e., the sum $\sum_{j=1}^{N} \kappa_j$ is conserved along Γ.

Example 18.5 Consider the alternative evolution equation

$$\frac{d\kappa_j}{dh} = -\left(l_j - \frac{1}{N}\sum_{s=1}^{N} l_s\right) - \lambda\left(\kappa_j - 1\right) \tag{18.45}$$

for some $\lambda > 0$. It follows that

$$\frac{d}{dh}\sum_{j=1}^{N}\kappa_j = -\lambda\left(\sum_{j=1}^{N}\kappa_j - N\right), \tag{18.46}$$

i.e., that the manifold defined by

$$\sum_{j=1}^{N}\kappa_j = N \tag{18.47}$$

is asymptotically attractive along Γ. If this equality holds for the initial conditions, it will hold for all h. ∎

We proceed to replace υ on the interval $[\tau_j, \tau_{j+1}]$ by the piecewise-linear interpolant through the points $\upsilon^{(j)}(\sigma_i)$, for $i = 1, \ldots, m+1$, and to discretize the evolution equations in Eq. (18.45) with respect to a finite step size h. There is no guarantee that the linear invariant $\sum_{j=1}^{N}\kappa_j$ remains conserved by the discretized scheme. Moreover, even if it is conserved, there is no guarantee that it remains equal to N in the presence of round-off errors.

Example 18.6 The explicit and the implicit Euler schemes both conserve linear invariants. When applied to the modified evolution equations in Eq. (18.45), they yield

$$\sum_{j=1}^{N}\kappa_j^{(k+1)} - N = (1 - h\lambda)\left(\sum_{j=1}^{N}\kappa_j^{(k)} - N\right) \tag{18.48}$$

and

$$\sum_{j=1}^{N}\kappa_j^{(k+1)} - N = \frac{1}{1+h\lambda}\left(\sum_{j=1}^{N}\kappa_j^{(k)} - N\right), \tag{18.49}$$

respectively. The manifold defined by

$$\sum_{j=1}^{N}\kappa_j = N \tag{18.50}$$

is thus asymptotically attractive provided that $|1 - h\lambda| < 1$ for the explicit Euler scheme, and for all h for the implicit Euler scheme. ∎

In contrast to the observations in Example 18.6, the conservation of the linear invariant $\sum_{j=1}^{N} \kappa_j$ is not guaranteed for the mixed Euler scheme

$$\frac{\kappa_j^{(k+1)} - \kappa_j^{(k)}}{h} = -\left(l_j^{(k+1)} - \frac{1}{N} \sum_{s=1}^{N} l_s^{(k)} \right) - \lambda \left(\kappa_j^{(k)} - 1 \right). \tag{18.51}$$

Indeed, even if $\sum_{j=1}^{N} \kappa_j^{(k)} = N$, no value of λ will guarantee that $\sum_{j=1}^{N} \kappa_j^{(k+1)} = N$. Consider, instead, the discretization

$$\frac{\kappa_j^{(k+1)} - \kappa_j^{(k)}}{h} = -\left(l_j^{(k+1)} - \frac{1}{N} \sum_{s=1}^{N} l_s^{(k)} \right) - \lambda \kappa_j^{(k)}, \tag{18.52}$$

where λ is chosen such that $\sum_{j=1}^{N} \kappa_j^{(k+1)} = N$. We expect that, for a nearly equidistributed mesh with respect to arclength along the graph of v, there exists a $\lambda \approx 0$ that satisfies this condition.

We now arrive at an augmented zero problem, in terms of the vector of continuation variables $u := (v_{bp}, T, p, \kappa, \lambda)$, given by the collocation zero problem, the *conservation condition*

$$\sum_{j=1}^{N} \kappa_j = N, \tag{18.53}$$

and the *mesh conditions*

$$\kappa_j - \lambda \kappa_j^* + h \left(l_j - \frac{1}{N} \sum_{s=1}^{N} l_s^* \right) = 0 \tag{18.54}$$

for $j = 1, \dots, N$. Here, κ^* and l_j^* are reference values that are updated before each new continuation step to agree with the values for κ and l_j obtained for the corresponding base chart.

Example 18.7 The arclength l_j along the piecewise-linear interpolant through the points $v^{(j)}(\sigma_i)$, for $i = 1, \dots, m+1$, may be obtained for all $j = 1, \dots, N$ by suitable manipulation of the v_{bp} array. Specifically, let v denote the $n \times (m+1) \times N$, 3-dimensional array whose (k, i, j) entry equals $v_{(m+1)(j-1)+i,k}$, where the second subscript refers to a vector component. In this case, the 3-dimensional array

$$\Delta v = v\big(:, \{1, \dots, m\}, : \big) - v\big(:, \{2, \dots, m+1\}, : \big) \tag{18.55}$$

contains all the separations between consecutive mesh values along $v^{(j)}$. The arclength l_j is now obtained from the jth entry of the 1-dimensional array l, in turn obtained by eliminating all singleton dimensions from the 3-dimensional array

$$S_2 \left(\sqrt{S_1 (\Delta v * \Delta v)} \right), \tag{18.56}$$

where $\sqrt{}$ denotes a vectorized application of the square-root operator. ∎

The choice of a mixed Euler scheme is a compromise between a desire for a stable discretization and a zero problem with a banded structure of the Jacobian of the mesh conditions with respect to v_{bp}. Let

$$\delta_{i,j} := \sqrt{\sum_{k=1}^{n} \left(v_{(m+1)(j-1)+i,k} - v_{(m+1)(j-1)+i+1,k} \right)^2} \tag{18.57}$$

for $i = 1,\ldots,m$ and $j = 1,\ldots,N$ such that

$$l_j = \sum_{i=1}^{m} \delta_{i,j}. \tag{18.58}$$

It follows that

$$\frac{\partial l_j}{\partial v_{(m+1)(j-1)+1,b}} = \frac{v_{(m+1)(j-1)+1,b} - v_{(m+1)(j-1)+2,b}}{\delta_{1,j}} \tag{18.59}$$

and

$$\frac{\partial l_j}{\partial v_{(m+1)j,b}} = -\frac{v_{(m+1)j-1,b} - v_{(m+1)j,b}}{\delta_{m,j}} \tag{18.60}$$

for $b = 1,\ldots,n$. Finally, for $a = 2,\ldots,m$ and $b = 1,\ldots,n$, we obtain

$$\frac{\partial l_j}{\partial v_{(m+1)(j-1)+a,b}} = -\frac{v_{(m+1)(j-1)+a-1,b} - v_{(m+1)(j-1)+a,b}}{\delta_{a-1,j}}$$
$$+ \frac{v_{(m+1)(j-1)+a,b} - v_{(m+1)(j-1)+a+1,b}}{\delta_{a,j}}. \tag{18.61}$$

18.3.2 Initialization

Given an initial partition

$$0 = t_1 \leq \cdots \leq t_{M+1} = T \tag{18.62}$$

of the interval $[0,T]$, we seek to initialize the κ array such that

$$\sum_{j=1}^{N} \kappa_j = N. \tag{18.63}$$

To this end, let $\hat{t}$ denote a continuous interpolant such that

$$\hat{t}\left((i-1)\frac{N}{M} + 1 \right) = t_i, i = 1,\ldots,M+1. \tag{18.64}$$

In particular, $\hat{t}(1) = 0$ and $\hat{t}(N+1) = T$. It follows that the 1-dimensional array κ, for which

$$\kappa_j = N \frac{\hat{t}(j+1) - \hat{t}(j)}{\hat{t}(N+1)}, j = 1,\ldots,N, \tag{18.65}$$

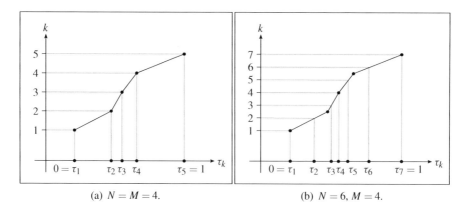

(a) $N = M = 4$.   (b) $N = 6$, $M = 4$.

Figure 18.4. *Initialization of the κ array, as shown in Example* 18.8. *The piecewise-linear interpolant $\hat{t}^{-1}$, here shown as a function of normalized time $\tau = t/T$, is constructed such that $\hat{t}\left((k-1)\frac{N}{M}+1\right) = \tau_k$ for $k = 1,\ldots,M+1$. As is evident from this construction visualized in* (a), *when $N = M$, the inverse of the sequence $\{1,\ldots,N+1\}$ under $\hat{t}$ is the sequence $\{\tau_1,\ldots,\tau_{N+1}\}$ itself. An initialization following the outlined procedure thus preserves the initial time distribution of mesh points when $N = M$. If the number of mesh points is changed, the new mesh will have a time distribution that incorporates adaptation information provided with an initial solution* (b).

trivially satisfies the condition in Eq. (18.63). Moreover,

$$\frac{T}{N}\sum_{i=1}^{j}\kappa_i = \hat{t}(j+1), \quad j = 1,\ldots,N, \tag{18.66}$$

i.e.,

$$\tau \in \left[\tau_j, \tau_{j+1}\right] \Rightarrow t \in \left[\hat{t}(j), \hat{t}(j+1)\right] \tag{18.67}$$

for $1 \le j \le N$.

Example 18.8 A natural choice for $\hat{t}$ is given by the piecewise-linear interpolant

$$s \mapsto \frac{iN-(s-1)M}{N}t_i + \frac{(s-1)M-(i-1)N}{N}t_{i+1} \tag{18.68}$$

for

$$s \in \left[(i-1)\frac{N}{M}+1, i\frac{N}{M}+1\right] \tag{18.69}$$

and $i = 1,\ldots,M$. Fig. 18.4 illustrates the relationship between the sequences t_i for $i = 1,\ldots,M+1$ and $\hat{t}(j)$ for $j = 1,\ldots,N+1$. ∎

In the special case that $M = N$, it follows that $\hat{t}(j) = t_j$ and

$$\kappa_j = \frac{N}{T}\left(t_{j+1}-t_j\right), \quad j = 1,\ldots,N. \tag{18.70}$$

The a priori construction in Eq. (18.65) thus conserves the partition in t when the desired number of mesh intervals agrees with the number of mesh intervals in the initial partition.

18.3.3 Encoding

Only relatively small changes to the `'coll'` toolbox are necessary in order to support the adaptive changes to the problem mesh implied by the comoving-mesh formulation. In order to eliminate redundant construction of toolbox data, subsequent to adaptive changes to the discretization parameters, as well as to anticipate adaptive changes to the discretization order N, consider a differentiation of the content of the toolbox data structure into the subfields

- `int`, containing all properties of the problem formulation that are independent of the discretization order and the discretization parameters;

- `maps`, containing all properties of the problem formulation that depend on the discretization order but not on the discretization parameters; and

- `mesh`, containing all properties of the problem formulation that depend on the discretization parameters.

It follows that the content of `data.int` remains unchanged during continuation, whereas that of `data.mesh` reflects changes to the discretization during continuation.

Consider, as an example, the following modified encoding of the `coll_F` zero function.

```
function [data y] = coll_F(prob, data, u)

maps = data.maps;

x  = u(maps.xbp_idx);
T  = u(maps.T_idx);
p  = u(maps.p_idx);
ka = u(maps.ka_idx);
la = u(maps.la_idx);

fka = ka(maps.fka_idx);
xx  = reshape(maps.W*x, maps.x_shp);
pp  = repmat(p, maps.p_rep);

ode = fka.*data.fhan(xx, pp);
ode = (0.5*T/maps.NTST)*ode(:)-maps.Wp*x;
cnt = maps.Q*x;

v   = reshape(x, maps.v_shp);
msh = v(:,1:end-1,:)-v(:,2:end,:);
msh = squeeze(sum(sqrt(sum(msh.*msh,1)),2));
msh = [ka-la*data.ka+data.h*(msh-data.mean_msh); sum(ka)-maps.NTST];

y = [ode; cnt; msh];

end
```

Here, `ka` contains the 1-dimensional array κ and `ka(maps.ka_idx)` the 2-dimensional array κ_f. We note further the use of the MATLAB `squeeze` command to eliminate singleton dimensions from its argument, following the description in Example 18.7. Here, the `data.ka`, `data.h`, and `data.mean_msh` fields are assumed to contain the κ^* array, the step size h, and the scalar average $N^{-1} \sum_{j=1}^{N} l_j^*$, respectively.

The modified encoding of the `coll_DFDU` function, shown below, makes similar use of the `maps.dxka_idx` and `maps.dpka_idx` fields of the toolbox data structure in order to assign the $\kappa_{\partial_x f}$ and $\kappa_{\partial_p f}$ arrays to the `dxka` and `dpka` variables, respectively.

```
function [data J] = coll_DFDU(prob, data, u)

maps = data.maps;
int  = data.int;

x   = u(maps.xbp_idx);
T   = u(maps.T_idx);
p   = u(maps.p_idx);
ka  = u(maps.ka_idx);

fka  = ka(maps.fka_idx);
dxka = ka(maps.dxka_idx);
dpka = ka(maps.dpka_idx);

xx = reshape(maps.W*x, maps.x_shp);
pp = repmat(p, maps.p_rep);

if isempty(data.dfdxhan)
  dxode = coco_ezDFDX('f(x,p)v', data.fhan, xx, pp);
else
  dxode = data.dfdxhan(xx, pp);
end
dxode = dxka.*dxode;
dxode = sparse(maps.dxrows, maps.dxcols, dxode(:));
dxode = (0.5*T/maps.NTST)*dxode*maps.W-maps.Wp;

ode   = data.fhan(xx, pp);
dTode = fka.*ode;
dTode = (0.5/maps.NTST)*dTode(:);
dkaode = (0.5*T/maps.NTST)*ode;
dkaode = sparse(maps.karows, maps.fka_idx, dkaode(:));

if isempty(data.dfdphan)
  dpode = coco_ezDFDP('f(x,p)v', data.fhan, xx, pp);
else
  dpode = data.dfdphan(xx, pp);
end
dpode = dpka.*dpode;
dpode = sparse(maps.dprows, maps.dpcols, dpode(:));
dpode = (0.5*T/maps.NTST)*dpode;

J1 = [dxode dTode dpode dkaode; ...
  maps.Q sparse(maps.Qnum, 1+maps.pdim+maps.NTST)];

v   = reshape(x, maps.v_shp);
du  = v(:,1:end-1,:)-v(:,2:end,:);
dsq = 1./sqrt(sum(du.*du,1));
dsq = du.*repmat(dsq, [int.dim 1 1]);

df1 = dsq(:,1,:);
df2 = -dsq(:,1:end-1,:)+dsq(:,2:end,:);
df3 = -dsq(:,end,:);
df  = data.h*cat(2,df1,df2,df3);
```

```
rows = reshape(1:maps.NTST, [1 1 maps.NTST]);
rows = repmat(rows, [int.dim int.NCOL+1 1]);
cols = reshape(maps.xbp_idx, maps.v_shp);
vals = df;

rows = [rows(:); (1:maps.NTST)'];
cols = [cols(:); maps.ka_idx];
vals = [vals(:); ones(maps.NTST,1)];

rows = [rows(:); (1:maps.NTST)'];
cols = [cols(:); maps.la_idx*ones(maps.NTST,1)];
vals = [vals(:); -data.ka];

rows = [rows(:); (maps.NTST+1)*ones(maps.NTST,1)];
cols = [cols(:); maps.ka_idx];
vals = [vals(:); ones(maps.NTST,1)];

J3 = sparse(rows, cols, vals);

J = sparse([J1 zeros(int.dim*(int.NCOL+1)*maps.NTST-int.dim,1); J3]);

end
```

We further accommodate the differentiated toolbox data structure through the following modified encoding of the `coll_err` function.

```
function [data y] = coll_err(prob, data, u)

int  = data.int;
maps = data.maps;

cp = reshape(maps.Wm*u, [int.dim maps.NTST]);
y  = maps.wn*max(sqrt(sum(cp.^2,1)));
y  = [y; y/data.coll.TOL];

end
```

As before, the content of the toolbox data structure is initialized through a call to the `coll_init_data` function, shown below.

```
function data = coll_init_data(data, t0, x0, p0)

NCOL = data.coll.NCOL;
dim  = size(x0,2);
data.int = coll_interval(NCOL, dim);

NTST = data.coll.NTST;
pdim = numel(p0);
data.maps = coll_maps(data.int, NTST, pdim);

t  = linspace(0, NTST, numel(t0));
tt = interp1(t, t0, 0:NTST, 'linear');
tt = tt*(NTST/tt(end));
data.mesh = coll_mesh(data.int, data.maps, tt);

data.h        = 0;
data.mean_msh = 0;
data.ka       = data.mesh.ka;

end
```

Here, the construction of the `tt` variable follows the formalism in Example 18.8 in the previous section. The assignment of initial content to `data.h`, `data.ka`, and `data.mean_msh` enables the initial call to the `coll_F` zero function.

The call to the function `coll_interval`, shown below, initializes all mesh-independent properties of the formulation.

```
function int = coll_interval(NCOL, dim)

int.NCOL = NCOL;
int.dim  = dim;

[int.tc int.wt] = coll_nodes(NCOL);
int.tm = linspace(-1, 1, NCOL+1)';
pmap    = coll_L(int.tm, int.tc);
dmap    = coll_Lp(int.tm, int.tc);
mmap    = coll_Lm(int.tm);
int.W  = kron(pmap, eye(dim));
int.Wp = kron(dmap, eye(dim));
int.Wm = kron(mmap, eye(dim));

end
```

We omit the functions `coll_nodes`, `coll_L`, `coll_Lp`, and `coll_Lm`, as their encodings are unaffected by the change to the toolbox data structure. The following encoding of the `coll_maps` function now initializes all properties of the problem formulation that depend on the discretization order N, but not on the discretization parameters contained in κ.

```
function maps = coll_maps(int, NTST, pdim)

NCOL = int.NCOL;
dim  = int.dim;

maps.NTST = NTST;
maps.pdim = pdim;

bpnum  = NCOL+1;
bpdim  = dim*(NCOL+1);
xbpnum = (NCOL+1)*NTST;
xbpdim = dim*(NCOL+1)*NTST;
cndim  = dim*NCOL;
xcnnum = NCOL*NTST;
xcndim = dim*NCOL*NTST;
cntnum = NTST-1;
cntdim = dim*(NTST-1);

maps.xbp_idx  = (1:xbpdim)';
maps.T_idx    = xbpdim+1;
maps.p_idx    = xbpdim+1+(1:pdim)';
maps.ka_idx   = xbpdim+1+pdim+(1:NTST)';
maps.la_idx   = xbpdim+1+pdim+NTST+1;
maps.Tp_idx   = [maps.T_idx; maps.p_idx];
maps.fka_idx  = kron(1:NTST, ones(dim, NCOL));
maps.dxka_idx = reshape(kron(maps.fka_idx, ones(1, dim)), ...
  [dim dim NCOL*NTST]);
maps.dpka_idx = reshape(kron(maps.fka_idx, ones(1,pdim)), ...
  [dim pdim NCOL*NTST]);
maps.tbp_idx  = setdiff(1:xbpnum, 1+bpnum*(1:cntnum))';
```

```
maps.x_shp    = [dim xcnnum];
maps.xbp_shp  = [dim xbpnum];
maps.v_shp    = [int.dim int.NCOL+1 maps.NTST];
maps.p_rep    = [1 xcnnum];

rows          = reshape(1:xcndim, [cndim NTST]);
rows          = repmat(rows, [bpdim 1]);
cols          = repmat(1:xbpdim, [cndim 1]);
W             = repmat(int.W, [1 NTST]);
Wp            = repmat(int.Wp, [1 NTST]);
maps.W        = sparse(rows, cols, W);
maps.Wp       = sparse(rows, cols, Wp);

temp          = reshape(1:xbpdim, [bpdim NTST]);
Qrows         = [1:cntdim 1:cntdim];
Qcols         = [temp(1:dim, 2:end) temp(cndim+1:end, 1:end-1)];
Qvals         = [ones(cntdim,1) -ones(cntdim,1)];
maps.Q        = sparse(Qrows, Qcols, Qvals, cntdim, xbpdim);
maps.Qnum     = cntdim;

maps.dxrows   = repmat(reshape(1:xcndim, [dim xcnnum]), [dim 1]);
maps.dxcols   = repmat(1:xcndim, [dim 1]);
maps.dprows   = repmat(reshape(1:xcndim, [dim xcnnum]), [pdim 1]);
maps.dpcols   = repmat(1:pdim, [dim xcnnum]);
maps.karows   = 1:xcndim;

maps.x0_idx   = (1:dim)';
maps.x1_idx   = xbpdim-dim+(1:dim)';

rows          = reshape(1:dim*NTST, [dim NTST]);
rows          = repmat(rows, [bpdim 1]);
cols          = repmat(1:xbpdim, [dim 1]);
Wm            = repmat(int.Wm, [1 NTST]);
maps.Wm       = sparse(rows, cols, Wm);
x             = linspace(int.tm(1), int.tm(2), 51);
y             = arrayfun(@(x) prod(x-int.tm), x);
maps.wn       = max(abs(y));

end
```

Finally, all properties of the problem formulation that depend on the discretization order N and on the discretization parameters contained in κ are initialized in the `coll_mesh` function, shown below.

```
function mesh = coll_mesh(int, maps, tmi)

mesh.tmi = tmi;

dim  = int.dim;
NCOL = int.NCOL;
NTST = maps.NTST;

bpnum  = NCOL+1;
xcndim = dim*NCOL*NTST;

ka       = diff(tmi);
mesh.ka  = ka';

wts       = repmat(int.wt, [dim NTST]);
kas       = kron(ka,ones(dim,NCOL));
mesh.wts1 = wts(1,:);
```

```
mesh.kas1 = kas(1,:);
mesh.wts2 = spdiags(wts(:), 0, xcndim, xcndim);
mesh.kas2 = spdiags(kas(:), 0, xcndim, xcndim);

t  = repmat(tmi(1:end-1)/NTST, [bpnum 1]);
tt = repmat((0.5/NTST)*(int.tm+1), [1 NTST]);
tt = t+repmat(ka, [bpnum 1]).*tt;
mesh.tbp = tt(:)/tt(end);

end
```

The use of the MATLAB `diff` function completes the initial construction of κ, described in the previous section. Here, the `kas1` and `kas2` fields are constructed for possible use by zero or monitor functions implementing integration in t.

The `coll_update` function, shown below, updates κ^*, $N^{-1}\sum_{j=1}^{N} l_j^*$, and other elements of the toolbox data structure that depend on the discretization parameters, once prior to continuation and then again before each new continuation step.

```
function data = coll_update(prob, data, cseg, varargin)

maps = data.maps;
int  = data.int;

data.h = cseg.prcond.h;

base_chart = cseg.src_chart;
uidx       = coco_get_func_data(prob, data.tbid, 'uidx');
u          = base_chart.x(uidx);
data.ka    = u(maps.ka_idx);
tmi        = [0 cumsum(data.ka')];
data.mesh  = coll_mesh(data.int, data.maps, tmi);

v   = reshape(u(maps.xbp_idx), [int.dim int.NCOL+1 maps.NTST]);
msh = v(:,1:end-1,:)-v(:,2:end,:);
msh = squeeze(sum(sqrt(sum(msh.*msh,1)),2));
data.mean_msh = mean(msh);

end
```

The assignment

```
data.h = cseg.prcond.h;
```

ensures that the step size h used in the mesh conditions equals the step size associated with the current projection condition. The encoding of the `'update_h'`-compatible slot function `coll_update_h`, shown below, ensures that this remains the case also during event location.

```
function data = coll_update_h(prob, data, h, varargin)

data.h = h;

end
```

The following modified encoding of the `coll_construct_seg` constructor now invokes `coco_add_slot` accordingly.

```
function prob = coll_construct_seg(prob, tbid, data, sol)

data.tbid = tbid;
data = coco_func_data(data);
prob = coco_add_func(prob, tbid, @coll_F, @coll_DFDU, data, 'zero', ...
  'u0', sol.u);
uidx = coco_get_func_data(prob, tbid, 'uidx');
if ~isempty(data.pnames)
   fid  = coco_get_id(tbid, 'pars');
   prob = coco_add_pars(prob, fid, uidx(data.maps.p_idx), data.pnames);
end
prob = coco_add_slot(prob, tbid, @coco_save_data, data, 'save_full');
efid = coco_get_id(tbid, {'err' 'err_TF'});
prob = coco_add_func(prob, efid{1}, @coll_err, data, ...
   'regular', efid, 'uidx', uidx(data.maps.xbp_idx));
prob = coco_add_event(prob, 'MXCL', 'MX', efid{2}, '>', 1);
prob = coco_add_slot(prob, tbid, @coll_update, data, 'update');
prob = coco_add_slot(prob, tbid, @coll_update_h, data, 'update_h');

end
```

The generalized toolbox constructors `coll_isol2seg` and `coll_sol2seg` are now modified in order to include the additional input arguments `t0` and `sol.t`, respectively, in the calls to `coll_init_data`.

```
function prob = coll_isol2seg(prob, oid, varargin)

tbid = coco_get_id(oid, 'coll');
str  = coco_stream(varargin{:});
data.fhan = str.get;
data.dfdxhan  = [];
data.dfdphan  = [];
if is_empty_or_func(str.peek)
  data.dfdxhan = str.get;
  if is_empty_or_func(str.peek)
    data.dfdphan = str.get;
  end
end
t0 = str.get;
x0 = str.get;
data.pnames = {};
if iscellstr(str.peek('cell'))
  data.pnames = str.get('cell');
end
p0 = str.get;

coll_arg_check(tbid, data, t0, x0, p0);
data = coll_get_settings(prob, tbid, data);
data = coll_init_data(data, t0, x0, p0);
sol  = coll_init_sol(data, t0, x0, p0);
prob = coll_construct_seg(prob, tbid, data, sol);

end


function prob = coll_sol2seg(prob, oid, varargin)

tbid = coco_get_id(oid, 'coll');
str  = coco_stream(varargin{:});
run  = str.get;
if ischar(str.peek)
  soid = str.get;
```

```
else
  soid = oid;
end
lab = str.get;

[sol data] = coll_read_solution(soid, run, lab);
data       = coll_get_settings(prob, tbid, data);
data       = coll_init_data(data, sol.t, sol.x, sol.p);
sol        = coll_init_sol(data, sol.t, sol.x, sol.p);
prob       = coll_construct_seg(prob, tbid, data, sol);

end
```

We finally account for the introduction of the `maps` and `mesh` fields in the following modified encodings of the `coll_init_sol` and `coll_read_solution` functions.

```
function sol = coll_init_sol(data, t0, x0, p0)

t0 = t0(:);
T0 = t0(end)-t0(1);
if abs(T0)>eps
  t0 = (t0-t0(1))/T0;
  x0 = interp1(t0, x0, data.mesh.tbp)';
else
  x0 = repmat(x0(1,:), size(data.mesh.tbp))';
end
ka = data.mesh.ka;
la = 1;

sol.u = [x0(:); T0; p0; ka; la];

end
```

```
function [sol data] = coll_read_solution(oid, run, lab)

tbid         = coco_get_id(oid, 'coll');
[data chart] = coco_read_solution(tbid, run, lab);
maps         = data.maps;
mesh         = data.mesh;

sol.t = mesh.tbp(maps.tbp_idx)*chart.x(maps.T_idx);
xbp   = reshape(chart.x(maps.xbp_idx), maps.xbp_shp)';
sol.x = xbp(maps.tbp_idx,:);
sol.p = chart.x(maps.p_idx);

end
```

Fig. 18.5 shows the results of applying the comoving-mesh adaptation algorithm to the continuation of periodic orbits considered in Examples 18.1 and 18.4. As seen in Fig. 18.5(c), the estimated discretization error stays within the desired tolerance, even with a significant decrease in the discretization order N, relative to the brute-force method.

18.4   Conclusions

With the introduction of adaptive changes to the discretization orders and/or parameters, one is made acutely aware of the distinction between the numerical output of a computational algorithm and the original mathematical object that is being sought and, hopefully,

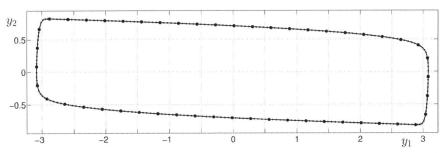

(a) Approximate periodic orbit for $\varepsilon = 20$. The fat dots mark the end points of mesh intervals. The arclengths of all mesh intervals are approximately equal, while points inside a mesh interval are equidistributed with respect to time.

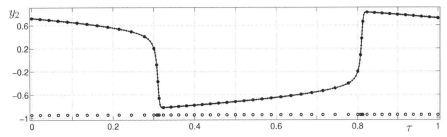

(b) Time profile of y_2. The circles at the bottom mark the end points of the mesh intervals.

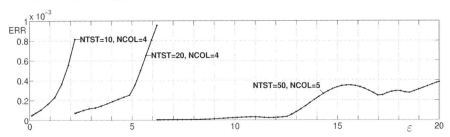

(c) The estimated error for different meshes plotted against variations in ε. The first two runs, with NTST equal to 10 and 20, respectively, eventually exceed the desired tolerance of 10^{-3} and terminate. The subsequent run, with NTST and NCOL equal to 50 and 5, respectively, is successful up to $\varepsilon = 20$.

Figure 18.5. *Continuation of periodic orbits of the dynamical system given by the vector field in Eq. (18.5) using comoving-mesh adaptation aimed at equalizing the arclength across all collocation intervals in state space, as described in Sect. 18.3.1. Here, the distribution of mesh points is uniquely determined along the solution manifold by appending a family of zero functions to the continuation problem, corresponding to a mixed Euler method applied to a gradient vector field. In contrast to Examples 18.1 and 18.4, the comoving-mesh algorithm allows one to compute the solution family with the same desired tolerance as in Fig. 18.3, but with only 1/3 of the number of mesh points. Again, we need to restart twice with a finer mesh at terminal solution points. The solution obtained for $\varepsilon = 20$ is shown in panel (a). Panel (b) shows the corresponding time profile and illustrates the adaptive time mesh. The estimated error plotted against variations in ε is shown in panel (c).*

approximated by this output. A number of observations bring this home. Among these is the realization that changes to the discretization orders amount to a change to the dimension of the domain (and range) of the associated zero problem. Thus, although the dimension of the solution manifold obtained for each choice of discretization order is fixed throughout continuation, the dimension of the space within which it is embedded may change significantly.

A consequence of such changes to the dimension of the domain is the challenge associated with making comparisons between points found on solution manifolds for different discretization orders. It is clear that only by making reference to properties of the illusive infinite-dimensional mathematical object, and to the approximation of these properties by finite-dimensional approximants, may viable comparisons be made.

A related challenge is to the detection of special points on the solution manifold associated with critical values of suitably formulated monitor functions. In the pointwise adaptation method shown in this chapter, this is of no concern, since changes to the discretization are imposed only in between individual runs of the `coco` entry-point function. With the integration of adaptivity into an atlas algorithm, however, comes the need to support detection of special points only along individual curve segments on the solution manifold, corresponding to a fixed discretization and a given projection condition. We return to such an integrated paradigm in Chap. 20.

Exercises

18.1. Consider the product

$$\prod_{i=1}^{m+1} |\sigma - \sigma_i|,$$

where $\sigma \in [-1, 1]$ and

$$-1 = \sigma_1 \leq \cdots \leq \sigma_{m+1} = 1.$$

Determine the value of σ for which the product attains its maximal value in the case of

- a uniform mesh

$$\sigma_i = 2\frac{i-1}{m} - 1$$

for $i = 1, \ldots, m+1$ and

- a *Chebyshev* mesh

$$\sigma_i = \cos \pi \frac{m-i+1}{m}$$

for $i = 1, \ldots, m+1$.

18.2. Verify Eq. (18.15) for the entries of the matrix $L^{(m)}$ in Eq. (18.12).

18.3. Verify the encoding of the function `coll_Lm` in Example 18.3.

18.4. Justify the constraint in Eq. (18.19) on the function κ.

18.5. Derive Eq. (18.26) from the definition of κ_j and the integral condition in Eq. (18.19).

18.6. Let κ_j and $\upsilon^{(j)}$ be defined as in Sect. 18.2. Show that

$$y(t) = \upsilon^{(j)} \left(\frac{2N}{\kappa_j} \left(\frac{t}{T} - \frac{1}{N} \sum_{i=1}^{j-1} \kappa_i \right) - 1 \right)$$

for t in the interval

$$\left[\frac{T}{N} \sum_{i=1}^{j-1} \kappa_i, \frac{T}{N} \sum_{i=1}^{j} \kappa_i \right].$$

18.7. Verify Eqs. (18.36)–(18.39) and explain the reason for the $*$ operation.

18.8. Given a 1-dimensional array κ whose jth entry is given by the scalar κ_j for $j = 1, \ldots, N$, implement MATLAB commands for computing

- the $n \times Nm$, 2-dimensional array κ_f whose ith entry equals the ith entry of the 1-dimensional array $\kappa \otimes 1_{nm,1}$.
- the $n \times n \times Nm$, 3-dimensional array $\kappa_{\partial_y f}$ whose ith entry equals the ith entry of the 1-dimensional array $\kappa \otimes 1_{n^2 m,1}$.
- the $n \times q \times Nm$, 3-dimensional array $\kappa_{\partial_p f}$ whose ith entry equals the ith entry of the 1-dimensional array $\kappa \otimes 1_{nmq,1}$.

18.9. Consider the system of differential equations

$$\frac{d\kappa}{dh} = - \left(l(\kappa) - \frac{1_{N,1}}{N} \sum_{s=1}^{N} l_s(\kappa) \right),$$

in the unknown function $\kappa : \mathbb{R} \to \mathbb{R}^N$ and for $l : \mathbb{R}^N \to \mathbb{R}^N$. Show that the sum $\sum_{j=1}^{N} \kappa_j$ is a constant of motion.

18.10. Let κ and l be defined as in the previous exercise. Consider the system of differential equations

$$\frac{d\kappa}{dh} = - \left(l(\kappa) - \frac{1_{N,1}}{N} \sum_{s=1}^{N} l_s(\kappa) \right) - \lambda \left(\kappa - 1_{N,1} \right),$$

for some $\lambda > 0$. Explain why the manifold defined by

$$\sum_{j=1}^{N} \kappa_j = N$$

is asymptotically attractive as $h \to \infty$.

18.11. Derive Eqs. (18.48) and (18.49) and explain why they guarantee that the manifold defined by

$$\sum_{j=1}^{N} \kappa_j = N$$

is asymptotically attractive provided that $|1 - h\lambda| < 1$ for the explicit Euler scheme, and for all h for the implicit Euler scheme.

18.12. Consider the mixed Euler scheme

$$\frac{\kappa_j^{(k+1)} - \kappa_j^{(k)}}{h} = -\left(l_j^{(k+1)} - \frac{1}{N}\sum_{s=1}^{N} l_s^{(k)}\right) - \lambda\left(\kappa_j^{(k)} - 1\right).$$

Show that there does not exist a value of λ which guarantees that

$$\sum_{j=1}^{N} \kappa_j^{(k)} = N \Rightarrow \sum_{j=1}^{N} \kappa_j^{(k+1)} = N.$$

What about in the case of the following difference equation?

$$\frac{\kappa_j^{(k+1)} - \kappa_j^{(k)}}{h} = -\left(l_j^{(k+1)} - \frac{1}{N}\sum_{s=1}^{N} l_s^{(k)}\right) - \lambda\kappa_j^{(k)}.$$

Under what conditions on h is the manifold $\sum_{j=1}^{N} \kappa_j = N$ asymptotically attractive if λ in the previous equation is replaced by

$$-\frac{1}{N}\sum_{j=1}^{N}\left(l_j^{(k+1)} - l_j^{(k)}\right)?$$

18.13. What is the relationship between the variable λ in Eq. (18.52) and the continuation variable λ in Eq. (18.54)?

18.14. Verify that the jth entry of the 1-dimensional array, obtained by eliminating all singleton dimensions form the array in Eq. (18.56), equals the arclength l_j along the piecewise-linear interpolant through the points $v^{(j)}(\sigma_i)$ for $i = 1,\ldots,m+1$.

18.15. Verify Eqs. (18.59)–(18.61) and identify the corresponding lines of code in the implementation of coll_DFDU in Sect. 18.3.3.

18.16. Given a partition

$$0 = t_1 \leq \cdots \leq t_{M+1} = T,$$

let $\hat{t}$ denote a continuous interpolant such that

$$\hat{t}\left((i-1)\frac{N}{M}+1\right) = t_i, i = 1,\ldots,M+1.$$

Show that the 1-dimensional array κ given by

$$\kappa_j = N\frac{\hat{t}(j+1) - \hat{t}(j)}{\hat{t}(N+1)}, j = 1\ldots,N,$$

satisfies the condition

$$\sum_{j=1}^{N}\kappa_j = N.$$

18.17. Consider the 1-dimensional array κ defined in the previous exercise. Show that

$$\kappa_j = \frac{N}{T}\left(t_{j+1} - t_j\right)$$

for $j = 1,\dots,N$, when $N = M$.

18.18. Identify the elements of the encoding of the modified `'coll'` toolbox that implement the initial construction of κ shown in Sect. 18.3.2.

18.19. Why is it necessary to call `coll_mesh` from within the `coll_update` function?

18.20. What changes are necessary to the `'po'` toolbox to accommodate the differentiation of content of the `'coll'` toolbox data structure?

18.21. Apply the modified encoding of the `'coll'` and `'po'` toolboxes from Sect. 18.3.3 and from the previous exercise to the problem considered in Example 18.1 and reproduce the results in Fig. 18.5. Provide a comprehensive study that explores the sensitivity of the algorithm to the choice of the error tolerance and the (fixed) discretization order.

18.22. Replace the mixed Euler scheme in the adaptive `'coll'` toolbox with a fully implicit formulation, as described in Example 18.6.

18.23. Propose an alternative implementation of a comoving mesh-adaptive collocation algorithm that equidistributes a combination of arclength and curvature along the piecewise-polynomial interpolant.

Chapter 19

A Spectral Toolbox

The concern in this chapter is again on periodic orbits of dynamical systems given by nonlinear vector fields on some n-dimensional state space. In place of the orthogonal collocation scheme employed in Sect. 8.2, we consider a *Fourier spectral scheme*, in which the parameterization of solution approximants is achieved through the amplitudes of various Fourier modes. We support adaptation by allowing for variations in the number of nonzero Fourier modes, within predefined bounds, during continuation.

The encoding of the *spectral continuation problem* in the 'dft' toolbox follows the vectorization paradigm and design structure of the toolboxes introduced in Part II of this text, most notably 'coll' and 'po'.

19.1   The spectral continuation problem

19.1.1   The discrete Fourier transform

Let $h : [0,1] \to \mathbb{R}$ denote a known function and consider the finite sample $\{h_j\}_{j=1}^N$, where

$$h_j = h(t_j) \tag{19.1}$$

and

$$t_j = \frac{j-1}{N}. \tag{19.2}$$

Let i denote the imaginary unit and consider the infinite sequence $\{H_l\}_{l=-\infty}^{\infty}$, where

$$H_l := \sum_{j=1}^N h_j \exp\left(-2\pi i \frac{(l-1)(j-1)}{N}\right). \tag{19.3}$$

It follows directly from the definition that this sequence is periodic in l with period N, i.e., that

$$H_{l+N} = H_l. \tag{19.4}$$

Moreover, if a bar denotes a complex conjugate, then

$$\overline{H_l} = H_{N+2-l}. \tag{19.5}$$

509

It thus suffices to compute H_l for $1 \le l \le N/2 + 1$, since all other values can be obtained through conjugation and periodicity. Notably, Eq. (19.5) implies that

$$\overline{H_1} = H_1, \tag{19.6}$$

i.e., that H_1 is real. Similarly, when N is even, it follows that

$$\overline{H_{N/2+1}} = H_{N/2+1}, \tag{19.7}$$

i.e., that $H_{N/2+1}$ is real. We define the *discrete Fourier transform* of $\{h_j\}_{j=1}^N$ as the finite sequence $\{H_l\}_{l=1}^N$.

Example 19.1 In MATLAB, the discrete Fourier transform of a sample $h(t_j)$, for t_j given in Eq. (19.2), may be obtained using the `fft` function, as shown in the extract below.

```
>> fft(2*rand(3,1)-1)

ans =

  -0.4767
   0.0787 + 0.3501i
   0.0787 - 0.3501i

>> fft(2*rand(6,1)-1)

ans =

   1.9155
   0.1887 - 0.7314i
  -0.4925 + 0.7104i
  -1.1786
  -0.4925 - 0.7104i
   0.1887 + 0.7314i
```

Here, periodicity is implicit, while the conjugation property in Eq. (19.5) is immediately evident. ∎

Now let c_m, $m = -M, \ldots, M$, be a set of $2M + 1$ complex *Fourier coefficients* such that $c_{-m} = \overline{c_m}$ and consider the real-valued periodic function

$$h(t) := \sum_{m=-M}^{M} c_m \exp(2\pi i m t) = c_0 + 2\Re\left(\sum_{m=1}^{M} c_m \exp(2\pi i m t)\right), \tag{19.8}$$

where $\Re(\cdot)$ denotes the real part of the argument. Suppose that h is sampled at the discrete points t_j in Eq. (19.2) for some even integer N larger than $2M$. The lth element of the discrete Fourier transform of the corresponding sequence $\{h_j\}_{j=1}^N$ then equals

$$H_l = \sum_{m=-M}^{M} c_m \cdot \begin{cases} N, & \mathrm{mod}(m - l + 1, N) = 0, \\ 0, & \mathrm{mod}(m - l + 1, N) \ne 0. \end{cases} \tag{19.9}$$

In particular, for $1 \le l \le N/2 + 1$,

$$H_l = \begin{cases} N c_{l-1}, & 1 \le l \le M + 1, \\ 0, & M + 2 \le l \le N/2 + 1. \end{cases} \tag{19.10}$$

From

$$h'(t) = \sum_{m=-M}^{M} 2\pi i m c_m \exp(2\pi i m t), \qquad (19.11)$$

it further follows that the discrete Fourier transform of the sequence $\{h'_j\}_{j=1}^{N}$, obtained by sampling h' at the discrete points t_j, is given by the sequence $\{H'_l\}_{l=1}^{N}$, where

$$H'_l = \begin{cases} 2\pi i (l-1) N c_{l-1}, & 1 \leq l \leq M+1, \\ 0, & M+2 \leq l \leq N/2+1. \end{cases} \qquad (19.12)$$

Finally, given a scalar σ, let

$$h_\sigma(t) := h(t+\sigma) = \sum_{m=-M}^{M} c_m \exp(2\pi i m \sigma) \exp(2\pi i m t). \qquad (19.13)$$

The discrete Fourier transform of the sequence $\{h_{\sigma,j}\}_{j=1}^{N}$, obtained by sampling h_σ at the discrete points t_j, is then given by the sequence $\{H_{\sigma,l}\}_{l=1}^{N}$, where

$$H_{\sigma,l} = \begin{cases} \exp(2\pi i (l-1)\sigma) N c_{l-1}, & 1 \leq l \leq M+1, \\ 0, & M+2 \leq l \leq N/2+1. \end{cases} \qquad (19.14)$$

Example 19.2 It follows directly from substitution of Eq. (19.8) that the lth Fourier coefficient may be obtained from the Fourier integral formula

$$c_l = \int_0^1 h(t) \exp(-2\pi i l t) \, dt. \qquad (19.15)$$

The corresponding Riemann sum

$$\sum_{j=1}^{N} h(t_j) \exp(-2\pi i l t_j)(t_{j+1} - t_j) \qquad (19.16)$$

then equals $1/N$ times the lth element, H_l, of the corresponding discrete Fourier transform. For a function of the form in Eq. (19.8), but with $M \to \infty$, it follows that $1/N$ times the discrete Fourier transform limits on the corresponding sequence of Fourier coefficients as $N \to \infty$. ∎

19.1.2 Matrix representations

With a slight abuse of notation, for an arbitrary sequence $\{h_j\}_{j=1}^{N}$ of sampled data, let h denote the N-dimensional row matrix whose jth entry equals h_j. Similarly, let H denote the N-dimensional row matrix whose lth entry equals the lth element H_l of the corresponding discrete Fourier transform. Then

$$H = h \cdot F, \qquad (19.17)$$

where F denotes the symmetric square matrix whose (j,l) entry equals

$$\exp\left(-2\pi i\frac{(j-1)(l-1)}{N}\right). \tag{19.18}$$

Moreover,

$$h = H \cdot F^{-1}, \tag{19.19}$$

where the (j,l) entry of the symmetric matrix F^{-1} is given by

$$\frac{1}{N}\exp\left(2\pi i\frac{(j-1)(l-1)}{N}\right). \tag{19.20}$$

The existence of F^{-1} guarantees a one-to-one relationship between the sequence $\{h_j\}_{j=1}^N$ and its discrete Fourier transform or, equivalently, between the sequence $\{H_l\}_{l=1}^N$ and its *inverse discrete Fourier transform* $\{h_j\}_{j=1}^N$.

Example 19.3 Let $h^{(i)}$ denote a sequence of sampled data $\{\delta_{ij}\}_{j=1}^N$, where δ denotes the Kronecker delta. The corresponding discrete Fourier transform $h^{(i)} \cdot F$ is then given by the ith row of the matrix F. Since the MATLAB fft routine is vectorized, it follows that we may compute F by applying the fft routine to the identity matrix I_N, as in

    fft(eye(6))

in the case that $N = 6$. Similar observations apply to the computation of the matrix F^{-1} using the MATLAB ifft routine. ∎

The matrix notation extends naturally to rectangular matrices H and h, where each row corresponds to a different component of a vector-valued function. Specifically, consider the family of real-valued periodic functions

$$h^{(i)}(t) := \sum_{m=-M}^{M} c_m^{(i)}\exp(2\pi imt), i = 1,\ldots,n, \tag{19.21}$$

for some integer n such that $c_{-m}^{(i)} = \overline{c_m^{(i)}}$. Suppose that N is an even integer that is larger than $2M$ and let h, h', and h_σ denote the $n \times N$ matrices whose (i,j) entries equal $h^{(i)}(t_j)$, $h^{(i)\prime}(t_j)$, and $h^{(i)}(t_j+\sigma)$, respectively, where the t_j are given in Eq. (19.2). Then, the lth column of the discrete Fourier transform $H := h \cdot F$ is given by

$$H_l = \begin{cases} Nc_{l-1} & \text{for } 1 \le l \le M+1, \\ 0 & \text{for } M+2 \le l \le N-M, \\ N\overline{c_{N+1-l}} & \text{for } N-M+1 \le l \le N, \end{cases} \tag{19.22}$$

where c_l, for $l = 0,\ldots,M$, denotes the $n \times 1$ *Fourier coefficient matrix* whose ith element equals $c_l^{(i)}$. From Eqs. (19.12) and (19.14), it further follows that the $n \times N$ matrices $H' := h' \cdot F$ and $H_\sigma := h_\sigma \cdot F$ satisfy the matrix relations

$$H' = H \cdot D, \quad H_\sigma = H \cdot S, \tag{19.23}$$

where D and S are diagonal $N \times N$ matrices whose lth diagonal elements equal

$$D_{ll} := \begin{cases} 2\pi i(l-1) & \text{for } 1 \le l \le M+1, \\ 0 & \text{for } M+2 \le l \le N-M, \\ -2\pi i(N+1-l) & \text{for } N-M+1 \le l \le N \end{cases} \tag{19.24}$$

and

$$S_{ll} := \begin{cases} \exp(2\pi i(l-1)\sigma) & \text{for } 1 \le l \le M+1, \\ 0 & \text{for } M+2 \le l \le N-M, \\ \exp(-2\pi i(N+1-l)\sigma) & \text{for } N-M+1 \le l \le N. \end{cases} \tag{19.25}$$

Naturally, it again follows that $h = H \cdot F^{-1}$, $h' = H' \cdot F^{-1}$, and $h_\sigma = H_\sigma \cdot F^{-1}$. In particular, $h' = h \cdot \tilde{D}$ and $h_\sigma = h \cdot \tilde{S}$, where

$$\tilde{D} := F \cdot D \cdot F^{-1}, \tilde{S} := F \cdot S \cdot F^{-1}. \tag{19.26}$$

Finally, we note that, provided h is obtained from Eq. (19.21),

$$\left. \frac{dH_\sigma}{d\sigma} \right|_{\sigma=0} = H \cdot \left. \frac{dS}{d\sigma} \right|_{\sigma=0} = H \cdot D = H'. \tag{19.27}$$

19.1.3   The spectral zero problem

Consider again the family of real-valued periodic functions in Eq. (19.21). Suppose that there exists a vector-valued function $f : \mathbb{R}^n \times \mathbb{R}^q \to \mathbb{R}^n$ such that

$$\frac{d}{dt} h^{(i)}(t) = T f^{(i)} \left(\left(h^{(1)}(t) \quad \dots \quad h^{(n)}(t) \right)^T, p \right) \tag{19.28}$$

for all $t \in [0,1]$ and some scalar T. Here, $f^{(i)}$ denotes the ith element of f and p is a vector of problem parameters. By the above treatment, this equality implies that

$$H \cdot D - T f \left(H \cdot F^{-1}, 1_{1,N} \otimes p \right) \cdot F = 0, \tag{19.29}$$

where the notation $f \left(H \cdot F^{-1}, 1_{1,N} \otimes p \right)$ implies a vectorized application of the function f to the columns of its matrix arguments.

Given a $k \times l$ matrix A, we define the $nl \times nk$ matrix $A^* = A^T \otimes I_n$. It is then straightforward to show that

$$\mathfrak{vec}_n(w) \cdot A = \mathfrak{vec}_n \left(A^* \cdot w \right) \tag{19.30}$$

for every nk-dimensional vector w. Let $\eta := \mathfrak{vec}(H)$. It follows that Eq. (19.29) is equivalent to the equation

$$D^* \cdot \eta - T F^* \cdot \mathfrak{vec} \left(f \left(\mathfrak{vec}_n \left(\left(F^{-1} \right)^* \cdot \eta \right), 1_{1,N} \otimes p \right) \right) = 0. \tag{19.31}$$

The autonomous nature of the vector field f in Eq. (19.28) implies that this equation is also satisfied by the family $h_\sigma^{(i)}$, $i = 1, \dots, n$, for arbitrary values of σ. Let $\eta_\sigma := \mathfrak{vec}(H_\sigma) = S^* \cdot \eta$. Eq. (19.31) then implies that

$$D^* \cdot S^* \cdot \eta - T F^* \cdot \mathfrak{vec} \left(f \left(\mathfrak{vec}_n \left(\left(F^{-1} \right)^* \cdot S^* \cdot \eta \right), 1_{1,N} \otimes p \right) \right) = 0 \tag{19.32}$$

for every σ. Differentiation with respect to σ and evaluation at $\sigma = 0$ now yield

$$D^* \cdot D^* \cdot \eta - T F^* \cdot \mathfrak{diag}\left(\partial_y f\left(\mathfrak{vec}_n\left(\left(F^{-1}\right)^* \cdot \eta\right), 1_{1,N} \otimes p\right)\right) \cdot \left(F^{-1}\right)^* \cdot D^* \cdot \eta = 0,$$

(19.33)

i.e., that $D^* \cdot \eta$ is a nullvector of the Jacobian

$$D^* - T F^* \cdot \mathfrak{diag}\left(\partial_y f\left(\mathfrak{vec}_n\left(\left(F^{-1}\right)^* \cdot \eta\right), 1_{1,N} \otimes p\right)\right) \cdot \left(F^{-1}\right)^*$$

(19.34)

of the left-hand side of Eq. (19.31) with respect to the components of η.

Suppose, instead, that we are given a system of n differential equations

$$\frac{d}{dt} h^{(i)} = T f^{(i)}\left(h^{(1)}, \ldots, h^{(n)}, p\right), \, i = 1, \ldots, n.$$

(19.35)

For a periodic solution to these equations of period 1, we may choose to approximate its components by finite Fourier expansions as in Eq. (19.21), with $2M$ smaller than some even integer N. Consider the $n(2M+1)$-dimensional column matrix

$$x := \begin{pmatrix} c_0 \\ \Re(c_1) \\ \Im(c_1) \\ \vdots \\ \Re(c_M) \\ \Im(c_M) \end{pmatrix},$$

(19.36)

where $\Im(\cdot)$ denotes the imaginary part of the argument. It follows that

$$\eta = \mathfrak{vec}(H) = W \cdot x,$$

(19.37)

where the $nN \times n(2M+1)$ matrix

$$W := N \begin{pmatrix} 1 & 0 \\ 0 & \tilde{I}_M \otimes \begin{pmatrix} 1 & i \end{pmatrix} \\ 0 & 0 \\ 0 & \tilde{I}_M \otimes \begin{pmatrix} 1 & -i \end{pmatrix} \end{pmatrix} \otimes I_n,$$

(19.38)

and the (i, j) entry of the matrix $\tilde{I}_M$ equals 1 if $i + j = M$, and 0 otherwise. Now let $D^*_{n(M+1)}$ and $F^*_{n(M+1)}$ denote the matrices obtained from the first $n(M+1)$ rows of D^* and F^*. The *spectral zero problem* in the components of x is then given by the vanishing of the function

$$D^*_{n(M+1)} \cdot W \cdot x - T F^*_{n(M+1)} \cdot \mathfrak{vec}\left(f\left(\mathfrak{vec}_n\left(\left(F^{-1}\right)^* \cdot W \cdot x\right), 1_{1,N} \otimes p\right)\right)$$

(19.39)

obtained from the left-hand side of Eq. (19.31).

19.1.4 A phase condition

From the existence of a nullvector of the Jacobian in Eq. (19.34), it follows that every solution x to the spectral zero problem corresponds to a nullvector

$$
\begin{pmatrix}
0 \\
-\Im(c_1) \\
\Re(c_1) \\
\vdots \\
-M\Im(c_M) \\
M\Re(c_M)
\end{pmatrix}
\tag{19.40}
$$

of the corresponding Jacobian

$$
D^*_{n(M+1)} \cdot W - T F^*_{n(M+1)} \cdot \mathfrak{diag}\left(\partial_y f\left(\mathfrak{vec}_n\left(\left(F^{-1}\right)^* \cdot W \cdot x\right), 1_{1,N} \otimes p\right)\right) \cdot \left(F^{-1}\right)^* \cdot W.
\tag{19.41}
$$

For fixed T and p, the spectral zero problem is thus singular at every point on its solution manifold.

We proceed to regularize the continuation problem by including T in the vector of continuation variables and appending an additional zero function. Specifically, in analogy with the introduction of an integral phase condition in Sect. 8.2, consider a family of reference functions

$$
h^{*(i)}(t) := \sum_{m=-M}^{M} c_m^{*(i)} \exp(2\pi \mathrm{i} m t), i = 1, \ldots, n,
\tag{19.42}
$$

such that $c_{-m}^{*(i)} = \overline{c_m^{*(i)}}$. We now require that the unknown periodic solutions of Eq. (19.35) of the form in Eq. (19.21) satisfy the integral condition

$$
\int_0^1 \sum_{i=1}^n \left(\frac{d}{dt} h^{*(i)}(t)\right) h^{(i)}(t) dt = 0.
\tag{19.43}
$$

It follows that the Fourier coefficient matrices c_m, $m = 1, \ldots, M$, must satisfy the *phase condition*

$$
\sum_{m=1}^{M} m\left(\Re\left(c_m^*\right)^T \cdot \Im(c_m) - \Im\left(c_m^*\right)^T \cdot \Re(c_m)\right) = 0.
\tag{19.44}
$$

The combined zero problem obtained from the (real and, as appropriate, imaginary parts of the) spectral zero problem, together with the discretized integral phase condition, has a nominal dimensional deficit of q, i.e., equal to the number of problem parameters.

Let

$$
x^* := \begin{pmatrix}
c_0^* \\
\Re\left(c_1^*\right) \\
\Im\left(c_1^*\right) \\
\vdots \\
\Re\left(c_M^*\right) \\
\Im\left(c_M^*\right)
\end{pmatrix}.
\tag{19.45}
$$

The phase condition in Eq. (19.44) is automatically satisfied by $x = x^*$. It is reasonable, therefore, to initialize x^* with the initial solution guess x_0 and, subsequently, to update x^* before each new continuation step with the value of x for some previously located point on the solution manifold.

19.1.5 Adaptivity and convergence

We proceed to augment the continuation problem so as to support an adaptive change to M, as long as $2M + 2 \leq N$. To this end, when $N > 2M + 2$, let

$$d := \begin{pmatrix} \Re(c_{M+1}) \\ \Im(c_{M+1}) \\ \vdots \\ \Re\left(c_{N/2-1}\right) \\ \Im\left(c_{N/2-1}\right) \end{pmatrix} \tag{19.46}$$

for a sequence of unknown complex numbers $\{c_i\}_{i=M+1}^{\frac{N}{2}-1}$. We append the array d to the vector of continuation variables and include the *trivial zero problem*

$$d = 0 \tag{19.47}$$

with the combined continuation problem defined in the previous sections. The nominal dimensional deficit then remains equal to q.

For a given integer M and vector x, consider next the matrix

$$\mathfrak{vec}_n\left(F^* \cdot \mathfrak{vec}\left(f\left(\mathfrak{vec}_n\left(\left(F^{-1}\right)^* \cdot W \cdot x\right), 1_{1,N} \otimes p\right)\right)\right). \tag{19.48}$$

Provided that the solution to the differential equations in Eq. (19.35) is of the form in Eq. (19.21), columns $M + 2$ through $\frac{N}{2} + 1$ of this matrix must equal zero. We may thus estimate the *discretization error* (cf. Fig. 19.1), incurred by approximating the actual periodic solution by a function family of the form Eq. (19.21), by the square root of the sum of the squared amplitudes of the elements contained in this submatrix.

To this end, we first recognize the equivalent spectral zero problem obtained with the substitutions

$$F \mapsto \frac{1}{N}F, \qquad F^{-1} \mapsto NF^{-1}, \qquad W \mapsto \frac{1}{N}W, \tag{19.49}$$

for which

- the residual in Eq. (19.39) is independent of N (cf. Example 19.2) and

- the sequence of error estimates obtained for increasing N is expected to be convergent as $N \to \infty$.

We make use of this equivalent formulation in the implementation in the next section.

During continuation, we may now choose to increase M (within the available bound) when the discretization error exceeds an upper critical bound and to decrease M when the discretization error falls below a lower critical bound. Notably, an increase in M implies that some of the continuation variables that were previously elements of d now get

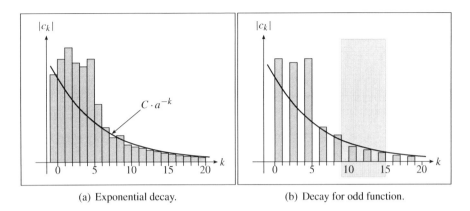

(a) Exponential decay.   (b) Decay for odd function.

Figure 19.1. *For real-analytical 2π-periodic functions, the sequence $\{|c_k|\}_{k=0}^{\infty}$ of the absolute values of the Fourier coefficients decays asymptotically with an exponential rate. There thus exist constants $C > 0$ and $a > 1$, and a mode number $K \geq 0$, such that $|c_k| \leq C \cdot a^{-k}$ holds for all $k \geq K$. If one interprets the sequence $\{|c_k|\}_{k=0}^{\infty}$ as a piecewise-constant function, as indicated in panel (a), one obtains the estimate $S_{K+1} := \sum_{k=K+1}^{\infty} |c_k| \leq \int_K^{\infty} C \cdot a^{-x} dx = C \cdot a^{-K}/\ln a$. Making the simplifying assumptions that $|c_K| \approx C \cdot a^{-K}$ and $\ln a = 1$ gives $S_{K+1} \leq |c_K|$. The tail S_{K+1} can thus be estimated by the first term omitted from the sum S_K. Some caution is necessary when using this idea for estimating an approximation error. For example, for odd functions it holds that $c_{2k} = 0$ for $k = 0, 1, 2, \ldots$. A robust error estimation should therefore use the values of $|c_k|$ from a window $k_0 \leq k \leq k_1$ of size $k_1 - k_0 \geq 1$, as indicated in panel (b) with $k_0 = 10$ and $k_1 = 15$. A related idea of error estimation is implemented in Sect. 19.1.5 in the context of the spectral zero problem.*

reassigned to x. Provided that the discretization error is kept below a reasonable upper bound, the initial solution guess afforded by letting these elements equal zero is expected to still be close to the corresponding solution manifold. Similarly, in the case that M is decreased, some of the continuation variables that were previously elements of x would get reassigned to d. By the linearity of the trivial zero problem, their values would become zero within one iteration of the corrector algorithm. We note that changes in M must be accompanied by related changes to the matrices D and W.

19.2 Encoding

We provide below a listing of toolbox functions that encode the construction of the spectral continuation problem in $u := (x, d, T, p) \in \mathbb{R}^{n(2M+1)} \times \mathbb{R}^{n(N-2M-2)} \times \mathbb{R} \times \mathbb{R}^q$ obtained from

- the family of zero functions Φ corresponding to the $n(2M+1)$ real and imaginary parts of the spectral zero problem, the $n(N-2M-2)$ components of the trivial zero problem, and the phase condition;

- the family of monitor functions $\Psi : u \mapsto \begin{pmatrix} p & T \end{pmatrix}^T$; and

- an initial solution guess $u_0 := (x_0, 0, T_0, p_0)$.

We include with the continuation problem a nonembedded monitor function that computes the estimated discretization error, and provide for a mechanism to vary adaptively the discretization order M, accordingly.

19.2.1 Common functions

The following encoding of the `'dft'` toolbox mirrors the design of the `'coll'` and `'po'` toolboxes in previous chapters. The `dft_isol2orb` constructor, shown below together with the `is_empty_or_func` subfunction, is identical to `coll_isol2seg` except for the changes in function names and the ordering of the calls to `dft_init_sol` and `dft_init_data`.

```
function prob = dft_isol2orb(prob, oid, varargin)

tbid = coco_get_id(oid, 'dft');
str  = coco_stream(varargin{:});
data.fhan = str.get;
data.dfdxhan   = [];
data.dfdphan   = [];
if is_empty_or_func(str.peek)
  data.dfdxhan = str.get;
  if is_empty_or_func(str.peek)
    data.dfdphan = str.get;
  end
end
t0 = str.get;
x0 = str.get;
data.pnames = {};
if iscellstr(str.peek('cell'))
  data.pnames = str.get('cell');
end
p0 = str.get;
if strcmpi(str.peek, 'end-dft')
  str.skip;
end

dft_arg_check(tbid, data, t0, x0, p0);
data = dft_get_settings(prob, tbid, data);
sol  = dft_init_sol(data, t0, x0, p0);
data = dft_init_data(data, sol);
prob = dft_construct_orb(prob, tbid, data, sol);

end

function flag = is_empty_or_func(x)
  flag = isempty(x) || isa(x, 'function_handle');
end
```

The same observation applies to the `dft_sol2orb` constructor shown below.

```
function prob = dft_sol2orb(prob, oid, varargin)

tbid = coco_get_id(oid, 'dft');
str  = coco_stream(varargin{:});
run  = str.get;
if ischar(str.peek)
  soid = str.get;
  lab  = str.get;
```

```
else
  soid = oid;
  lab  = str.get;
end
if strcmp(str.peek, 'end-dft')
  str.skip;
end

[sol data] = dft_read_solution(soid, run, lab);
data       = dft_get_settings(prob, tbid, data);
sol        = dft_init_sol(data, sol.t, sol.x, sol.p);
data       = dft_init_data(data, sol);
prob       = dft_construct_orb(prob, tbid, data, sol);

end
```

The `dft_init_sol` function, shown below, computes the values of N and M from the integers stored in the `dft.NMAX` and `dft.NMOD` fields of the `data` input argument. The MATLAB `fft` function is then used to compute the discrete Fourier transform of the sampled interpolant.

```
function sol = dft_init_sol(data, t0, x0, p0)

N    = 2*data.dft.NMAX+2;
NMOD = data.dft.NMOD;

t0 = t0(:);
T0 = t0(end)-t0(1);
t0 = (t0-t0(1))/T0;
x0 = interp1(t0, x0, (0:N-1)'/N)';

x0 = fft(x0.').';
x0 = x0(:,1:NMOD+1)/N;
xh = [real(x0(:,2:end)); imag(x0(:,2:end))];
x0 = [x0(:,1) reshape(xh, [size(x0,1) 2*NMOD])];

sol.x0 = x0;
sol.T0 = T0;
sol.p0 = p0;
sol.u0 = [x0(:); zeros(size(x0,1)*(N-2*NMOD-2),1); T0; p0];

end
```

In a similar way, the `dft_read_solution` toolbox extractor, shown below, computes the 1-dimensional array of discrete samples

$$\mathfrak{vec}_n\left(\left(F^{-1}\right)^* \cdot W \cdot x\right) \tag{19.50}$$

corresponding to the sequence $t_j := T(j-1)/N$, $j = 1, \ldots, N$. It then appends a copy of the starting point to the end of the array in order to obtain a periodic sample.

```
function [sol data] = dft_read_solution(oid, run, lab)

tbid        = coco_get_id(oid, 'dft');
[data chart] = coco_read_solution(tbid, run, lab);

dim = data.dim;
N   = 2*data.dft.NMAX+2;
```

```
sol.t = chart.x(data.T_idx)*(0:N)/N;
sol.c = reshape(chart.x(data.xf_idx), [dim data.dft.NMOD*2+1]);
sol.x = reshape(real(data.FinvsW*chart.x(data.xf_idx)), [dim N]);
sol.x = [sol.x sol.x(:,1)]';
sol.p = chart.x(data.p_idx);
sol.T = chart.x(data.T_idx);

end
```

Finally, other than the function name, the `dft_arg_check` function is lifted verbatim from the `'coll'` toolbox.

19.2.2 The zero problem

The following encoding of the `dft_F` function implements the family of zero functions Φ.

```
function [data y] = dft_F(prob, data, u)

xf = u(data.xf_idx);
xd = u(data.xd_idx);
T  = u(data.T_idx);
p  = u(data.p_idx);

dim  = data.dim;

pp = repmat(p, data.p_rep);
xp = reshape(data.FinvsW*xf, data.x_shp);
f  = data.fhan(xp, pp);
f  = T*data.Fs*f(:)-data.Wp*xf;
f  = [real(f(1:dim)); real(f(dim+1:end)); imag(f(dim+1:end))];
y  = [f; xd; data.phs*xf];

end
```

Here, the toolbox data structure contains the fields

- `xf_idx` equal to the array $[\![\ 1 \ \cdots \ n(2M+1) \ ]\!]$;

- `xd_idx` equal to the array $[\![\ n(2M+1)+1 \ \cdots \ n(N-1) \ ]\!]$;

- `T_idx` equal to $n(N-1)+1$;

- `p_idx` equal to the array $[\![\ n(N-1)+2 \ \cdots \ n(N-1)+1+q \ ]\!]$;

- `dim` equal to n;

- `p_rep` equal to the array $[\![\ 1 \ \ N \ ]\!]$;

- `FinvsW` equal to the matrix $\left(F^{-1}\right)^* \cdot W$;

- `x_shp` equal to the array $[\![\ n \ \ N \ ]\!]$;

- `fhan` equal to a function handle to a vectorized encoding of the vector field f;

- `Fs` equal to the matrix $F^*_{n(M+1)}$;

- `Wp` equal to the matrix $D^*_{n(M+1)} \cdot W$; and

- phs equal to the matrix

$$x'^* := \begin{pmatrix} 0 & -\Im\left(c_1^{*T}\right) & \Re\left(c_1^{*T}\right) & \cdots & -M\Im\left(c_M^{*T}\right) & M\Re\left(c_M^{*T}\right) \end{pmatrix}. \quad (19.51)$$

The Jacobian of the family of zero functions is encoded in the dft_DFDU function, shown below.

```
function [data J] = dft_DFDU(prob, data, u)

xf = u(data.xf_idx);
T  = u(data.T_idx);
p  = u(data.p_idx);

dim  = data.dim;
pdim = data.pdim;
mdim = data.mdim;
ddim = data.ddim;

pp = repmat(p, data.p_rep);
xp = reshape(data.FinvsW*xf, data.x_shp);
if isempty(data.dfdxhan)
  df = coco_ezDFDX('f(x,p)v', data.fhan, xp, pp);
else
  df = data.dfdxhan(xp, pp);
end
df = sparse(data.dxrows, data.dxcols, df(:));
df = T*data.Fs*df*data.FinvsW - data.Wp;
df = [real(df(1:dim,:)); real(df(dim+1:end,:)); imag(df(dim+1:end,:))];

f = data.fhan(xp, pp);
f = data.Fs*f(:);
f = [real(f(1:dim)); real(f(dim+1:end)); imag(f(dim+1:end))];

J1 = [df, sparse(mdim, ddim), f; data.jac];

if isempty(data.dfdphan)
  df = coco_ezDFDP('f(x,p)v', data.fhan, xp, pp);
else
  df = data.dfdphan(xp, pp);
end
df = sparse(data.dprows, data.dpcols, df(:));
df = T*data.Fs*df;
df = [real(df(1:dim,:)); real(df(dim+1:end,:)); imag(df(dim+1:end,:))];

if pdim>0
  dfcont = sparse(ddim+1, pdim);
else
  dfcont = [];
end

J2 = [df; dfcont];

J = sparse([J1 J2]);

end
```

Here, it is assumed that the toolbox data structure contains the additional fields

- pdim equal to q;

- `mdim` equal to $n(2M+1)$;

- `ddim` equal to $n(N-2M-2)$;

- `dfdxhan` equal to a function handle to a vectorized encoding of the Jacobian $\partial_y f$;

- `dfdphan` equal to a function handle to a vectorized encoding of the Jacobian $\partial_p f$;

- `dxrows` equal to the array $1_{n,1} \otimes \mathfrak{vec}_n\left(\begin{bmatrix} 1 & \cdots & nN \end{bmatrix}\right)$;

- `dxcols` equal to the array $1_{n,1} \otimes \begin{bmatrix} 1 & \cdots & nN \end{bmatrix}$;

- `jac` equal to the $n(N-2M-2)+1 \times n(N-1)+1$ matrix

$$\begin{pmatrix} 0 & I_{n(N-2M-2)} & 0 \\ x'^* & 0 & 0 \end{pmatrix};$$
(19.52)

- `dprows` equal to the array $1_{q,1} \otimes \mathfrak{vec}_n\left(\begin{bmatrix} 1 & \cdots & nN \end{bmatrix}\right)$; and

- `dpcols` equal to the array $1_{n,N} \otimes \begin{bmatrix} 1 & \cdots & q \end{bmatrix}$.

The `dft_init_data` and `dft_init_modes` functions, shown below, are used to populate the fields of the toolbox data structure.

```
function data = dft_init_data(data, sol)

N = 2*data.dft.NMAX+2;

dim  = size(sol.x0,1);
pdim = numel(sol.p0);
data.dim   = dim;
data.pdim  = pdim;
data.T_idx = dim*(N-1)+1;
data.p_idx = dim*(N-1)+1+(1:pdim)';
data.p_rep = [1 N];
data.x_shp = [dim N];

data.Forig  = (1/N)*fft(eye(N));
Finv        = N*ifft(eye(N));
data.Fsorig = kron(data.Forig, speye(dim));
data.Finvs  = kron(Finv, speye(dim));

data.dxrows = repmat(reshape(1:dim*N, [dim N]), [dim 1]);
data.dxcols = repmat(1:dim*N, [dim 1]);
data.dprows = repmat(reshape(1:dim*N, [dim N]), [pdim 1]);
data.dpcols = repmat(1:pdim, [dim N]);

data = dft_init_modes(data, sol);

end


function data = dft_init_modes(data, sol)

N    = 2*data.dft.NMAX+2;
NMOD = data.dft.NMOD;
dim  = data.dim;
```

```
mdim = dim*(2*NMOD+1);
ddim = dim*(N-2*NMOD-2);
data.mdim   = mdim;
data.ddim   = ddim;
data.xf_idx = 1:data.mdim;
data.xd_idx = data.mdim+(1:data.ddim)';

phs = repmat(1:NMOD, [dim 1]);
phs = [-phs; phs].*[sol.x0(:,3:2:end); sol.x0(:,2:2:end)];
data.phs = [zeros(dim,1); phs(:)]';
row = sparse(1:ddim, 1+mdim:ddim+mdim, ones(1,ddim), ddim, data.T_idx);
data.jac = [row; data.phs, sparse(1,ddim+1)];

D         = 2*pi*1i*diag([0:NMOD zeros(1,N-2*NMOD-2+1) -NMOD:-1]);
Ds        = kron(D, speye(dim));
W         = kron([1, zeros(1,2*NMOD);
   zeros(N-1,1), [kron(speye(NMOD), [1 1i]);
   zeros(N-2*NMOD-1,2*NMOD);
   kron(fliplr(speye(NMOD)), [1 -1i])]], speye(dim));
data.Wp   = Ds(1:dim*(NMOD+1),:)*W;
data.F    = data.Forig(:,NMOD+2:N/2+1);
data.Fs   = data.Fsorig(1:dim*(NMOD+1),:);
data.FinvsW = real(data.Finvs*W);

end
```

The division of the initialization of the `data` toolbox data structure into the two functions `dft_init_data` and `dft_init_modes` supports adaptive changes to M, without having to recompute quantities that depend only on n and N.

19.3   Adaptivity

We propose next to integrate adaptivity in the `'dft'` toolbox. Specifically, we include a `dft_update` slot function, whose execution is triggered by the `'update'` signal, emitted, for example, by the `CurveSegment.create` function during the execution of the `predict` class method of an instance of a subclass of `AtlasBase`. The encoding, shown below, estimates the discretization error and increases or decreases M depending on the relation of this error to the window provided by upper and lower bounds stored in the `dft.TOLINC` and `dft.TOLDEC` fields of the toolbox data structure, as well as the relation of the value of M to its upper and lower bounds, stored in the `dft.NMAX` and `dft.NMIN` fields, respectively.

```
function data = dft_update(prob, data, cseg, varargin)

uidx = coco_get_func_data(prob, data.tbid, 'uidx');
u    = cseg.src_chart.x(uidx);
xf   = u(data.xf_idx);
p    = u(data.p_idx);

pp   = repmat(p, data.p_rep);
xp   = reshape(data.FinvsW*xf, data.x_shp);
err  = data.fhan(xp, pp)*data.F;
err  = real(err.*conj(err));
err  = sqrt(sum(err(:)));

dft  = data.dft;
NMOD = dft.NMOD;
```

```
NMAX = dft.NMAX;
NMIN = dft.NMIN;
dim  = data.dim;
sol.x0 = reshape(xf, [dim 2*NMOD+1]);
NMODi = min(ceil(NMOD*1.1+1),   NMAX);
NMODd = max(ceil(NMOD/1.025-1), NMIN);
if err>dft.TOLINC && NMOD~=NMODi
  data.dft.NMOD = NMODi;
  sol.x0 = [sol.x0 zeros(dim, 2*(NMODi-NMOD))];
  data = dft_init_modes(data, sol);
elseif err<dft.TOLDEC && NMOD~=NMODd
  data.dft.NMOD = NMODd;
  sol.x0 = sol.x0(:,1:2*NMODd+1);
  data = dft_init_modes(data, sol);
end

end
```

Here, the upper bound NMAX on NMOD is consistent with the requirement that M be less than or equal to $N/2 - 1$.

Default values for the maximum, minimum, and initial discretization order, the error tolerance, and the upper and lower bounds of the adaptation window are set by the dft_get_settings function, shown below. As in the case of the comoving-mesh algorithm implemented in the 'coll' toolbox in Chap. 18, the default error tolerance is related to the error tolerance of the nonlinear corrector.

```
function data = dft_get_settings(prob, tbid, data)

defaults.NMAX = 8;
defaults.NMIN = 3;
defaults.NMOD = 3;
if ~isfield(data, 'dft')
  data.dft = [];
end
data.dft = coco_merge(defaults, coco_merge(data.dft, ...
  coco_get(prob, tbid)));

if ~coco_exist('TOL', 'class_prop', prob, tbid, '-no-inherit-all')
  data.dft.TOL = coco_get(prob, 'corr', 'TOL')^(2/3);
end
defaults.TOLINC = data.dft.TOL/5;
defaults.TOLDEC = data.dft.TOL/20;
data.dft = coco_merge(defaults, data.dft);

NMOD = data.dft.NMOD;
assert(numel(NMOD)==1 && isnumeric(NMOD) && mod(NMOD,1)==0, ...
  '%s: input for option ''NMOD'' is not an integer', tbid);
NMAX = data.dft.NMAX;
assert(numel(NMAX)==1 && isnumeric(NMAX) && mod(NMAX,1)==0, ...
  '%s: input for option ''NMAX'' is not an integer', tbid);
NMIN = data.dft.NMIN;
assert(numel(NMIN)==1 && isnumeric(NMIN) && mod(NMIN,1)==0, ...
  '%s: input for option ''NMIN'' is not an integer', tbid);
assert(NMIN<=NMOD && NMOD<=NMAX, ...
  '%s: input violates ''NMIN<=NMOD<=NMAX''', tbid);
end
```

Finally, the following encoding of the dft_error function implements three monitor functions whose values equal the discretization error, the discretization error normalized by the tolerance, and the discretization order M, respectively.

```
function [data y] = dft_error(prob, data, u)

xf   = u(data.xf_idx);
p    = u(data.p_idx);

pp   = repmat(p, data.p_rep);
xp   = reshape(data.FinvsW*xf, data.x_shp);
err  = data.fhan(xp, pp)*data.F;
err  = real(err.*conj(err));
err  = sqrt(sum(err(:)));

dft = data.dft;
y   = [err; err/dft.TOL; dft.NMOD];

end
```

The `dft_construct_orb` constructor, shown below, appends the spectral continuation problem to the continuation problem structure `prob`.

```
function prob = dft_construct_orb(prob, tbid, data, sol)

data.tbid = tbid;
data = coco_func_data(data);
prob = coco_add_func(prob, tbid, @dft_F, @dft_DFDU, data, ...
  'zero', 'u0', sol.u0);
uidx = coco_get_func_data(prob, tbid, 'uidx');
fid  = coco_get_id(tbid, 'period');
prob = coco_add_pars(prob, fid, uidx(data.T_idx), fid, 'active');
if ~isempty(data.pnames)
  fid  = coco_get_id(tbid, 'pars');
  prob = coco_add_pars(prob, fid, uidx(data.p_idx), data.pnames);
end
prob = coco_add_slot(prob, tbid, @coco_save_data, data, 'save_full');
prob = coco_add_slot(prob, tbid, @dft_update, data, 'update');

efid = coco_get_id(tbid, {'err' 'err_TF' 'NMOD'});
prob = coco_add_func(prob, efid{1}, @dft_error, data, ...
  'regular', efid, 'uidx', uidx);
prob = coco_add_event(prob, 'MXCL', 'MX', efid{2}, '>', 1);

end
```

Example 19.4 The results in Fig. 19.2 illustrate the application of the adaptive 'dft' toolbox to the continuation of periodic orbits of the dynamical system given by the vector field in Eq. (18.5), as shown in the extract below.

```
>> eps0 = 0.1;
>> t0 = linspace(0, 2*pi, 100)';
>> x0 = [sin(t0) cos(t0)];
>> prob = coco_set(coco_prob(), 'cont', 'ItMX', 100);
>> prob = coco_set(prob, 'dft', 'TOL', 1e-3);
>> prob2 = coco_set(prob, 'dft', 'NMAX', 250, 'NMOD', 20);
>> prob2 = dft_isol2orb(prob2, '', @pneta, t0, x0, 'eps', eps0);
>> coco(prob2, 'run1', [], 1, {'eps' 'dft.err' 'dft.NMOD'}, [0.1 20]);

...     NORMS              COMPUTATION TIMES
...     ||f||       ||U||   F(x)  DF(x)  SOLVE
```

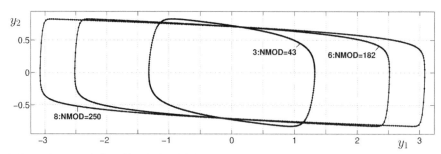

(a) Approximate periodic orbits obtained during continuation for increasing ε. The number of modes used for each approximation is indicated by the label.

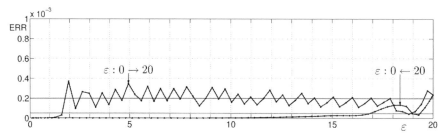

(b) Estimated error during continuation for increasing and decreasing values of ε. The tolerance is 10^{-3}, and the adaptation window is indicated with two horizontal lines.

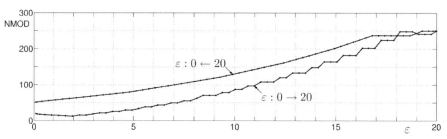

(c) Number of Fourier modes during continuation with 'NMAX' equal to 250.

Figure 19.2. *Continuation of periodic orbits of the dynamical system given by the vector field in Eq. (18.5) using the adaptive spectral toolbox, as implemented in Sects. 19.2 and 19.3. The rapid decay of Fourier coefficients illustrated in Fig. 19.1 enables adaptation by exchanging equations that either fix a variable at 0 or make it equal to a Fourier mode. In contrast to the results shown in Figs. 18.3 and 18.5, continuation here computes the complete family in a single run. Panel* (a) *shows sample orbits. The effects of adaptation are illustrated in panels* (b) *and* (c).

```
... 1.11e-01  8.91e+00    0.0    0.0    0.0
... 1.45e-02  8.91e+00    0.0    0.0    0.0
... 9.00e-04  8.91e+00    0.0    0.1    0.0
... 4.43e-06  8.91e+00    0.0    0.1    0.0
... 1.09e-10  8.91e+00    0.0    0.1    0.0
... 3.82e-15  8.91e+00    0.0    0.1    0.0
```

```
...   LABEL   TYPE              eps        dft.err      dft.NMOD
...       1   EP       1.0000e-01    2.2998e-16     1.9000e+01
...       2            2.9090e+00    2.1485e-04     1.8000e+01
...       3            5.9593e+00    1.0184e-04     4.3000e+01
...       4            8.8821e+00    2.3325e-04     7.0000e+01
...       5            1.1752e+01    2.3347e-04     1.0800e+02
...       6            1.4593e+01    1.3619e-04     1.6400e+02
...       7            1.7416e+01    1.4677e-04     2.2400e+02
...       8   EP       2.0000e+01    2.3723e-04     2.5000e+02
>> prob2 = dft_sol2orb(prob, '', 'run1', 8);
>> coco(prob2, 'run2', [], 1, {'eps' 'dft.err' 'dft.NMOD'}, [0.1 20]);

...     NORMS                 COMPUTATION TIMES
...     ||f||        ||U||     F(x)   DF(x)   SOLVE
... 1.15e-12    3.91e+01       0.0     0.0     0.0
... 1.87e-13    3.91e+01       0.0     2.4     0.5

...   LABEL   TYPE              eps        dft.err      dft.NMOD
...       1   EP       2.0000e+01    2.3723e-04     2.5000e+02
...       2            1.7667e+01    1.1382e-04     2.3700e+02
...       3            1.4846e+01    3.2362e-05     1.9600e+02
...       4            1.2007e+01    1.3229e-05     1.4900e+02
...       5            9.1400e+00    2.6729e-06     1.1200e+02
...       6            6.2236e+00    9.2490e-08     8.2000e+01
...       7            3.1917e+00    3.2295e-12     6.1000e+01
...       8   EP       1.0000e-01    2.2040e-16     4.1000e+01
```

Selected approximate orbits, obtained during continuation for increasing ε, are shown in Fig. 19.2(a). As seen in Fig. 19.2(b), the estimated error remains bounded close to the adaptation window during continuation for both increasing and decreasing values of ε. Finally, we note that although the number of oscillatory Fourier modes (M) varies during continuation, as seen in Fig. 19.2(c), the number of mesh points (N) remains constant. ∎

19.4 Conclusions

The use of the `'update'` signal, relied upon in the implementation of adaptivity in the `'dft'` toolbox in this chapter, requires care and deliberation. As a rule of thumb, we restrict its use to cases in which the base chart of the curve segment provides a local cover of the solution manifold, even after a corresponding change to a toolbox data structure during continuation. In other cases, where this cannot be guaranteed, adaptivity must be introduced using the brute-force techniques described in the previous chapter, or the more sophisticated integration with an atlas algorithm, as introduced in the next chapter.

Consider, as an example, the `po_update` function from the `'po'` toolbox from Chap. 8, repeated below for convenience.

```
function data = po_update(prob, data, cseg, varargin)

fid          = coco_get_id(data.tbid, 'seg.coll');
[fdata uidx] = coco_get_func_data(prob, fid, 'data', 'uidx');
u            = cseg.src_chart.x;
data.xp0     = u(uidx(fdata.xbp_idx))'*data.intfac;
data.J(end,:) = data.xp0;

end
```

Here, the content of the `cseg.src_chart.x` field is used to update the phase condition, as described in Sect. 8.2. Although the base point of the first, and only, chart of the `cseg.ptlist` array remains a numerically exact solution (within the established tolerance) to the modified zero problem, the columns of the matrix stored in the `TS` field constitute only an approximate basis for the tangent space to the solution manifold. The subsequent call, in the `predict` class method, to `cseg.update_t` similarly assigns only an approximate tangent vector to the `cseg.ptlist{1}.t` field. Similar observations apply to the effects of the `per_bc_update` function on the continuation problem considered in Sect. 8.3.

It is less immediate that the adaptive change to the discretization order M introduced in the `'dft'` toolbox implies only small changes to the solution manifold, in support of the continued use of a previously computed chart as a local approximant. The assumption that this is indeed the case is implicit in the reliance on the `dft_update` slot function, but calls for retrospection and study.

Exercises

19.1. Consider the infinite sequence $\{H_l\}_{l=-\infty}^{\infty}$, where

$$H_l := \sum_{j=1}^{N} h_j \exp\left(-2\pi i \frac{(l-1)(j-1)}{N}\right)$$

in terms of a sample $\{h_j\}_{j=1}^{N}$ of a real-valued function h. Show that

$$H_{l+N} = H_l$$

and

$$\overline{H_l} = H_{N+2-l},$$

where $\overline{a+ib} = a - ib$ for $a, b \in \mathbb{R}$.

19.2. Consider the infinite sequence $\{H_l\}_{l=-\infty}^{\infty}$, where

$$H_l := \sum_{j=1}^{N} h_j \exp\left(-2\pi i \frac{(l-1)(j-1)}{N}\right)$$

in terms of a sample $\{h_j\}_{j=1}^{N}$ of a real-valued function h. Use the results of the previous exercise to show that H_1 and, if N is even, $H_{N/2+1}$ are both real.

19.3. Consider the real-valued periodic function

$$h(t) := \sum_{m=-M}^{M} c_m \exp(2\pi i m t) = c_0 + 2\Re\left(\sum_{m=1}^{M} c_m \exp(2\pi i m t)\right),$$

where $c_{-m} = \overline{c_m}$. Write this function in terms of a finite sum of cosine and sine terms.

19.4. Show that if k is a multiple of N, then

$$\sum_{j=1}^{N} \exp\left(2\pi i \frac{(j-1)k}{N}\right) = N.$$

19.5. Suppose that N is even. Show that if the integer k is not a multiple of N, then

$$\sum_{j=1}^{N} \exp\left(2\pi i \frac{(j-1)k}{N}\right) = 0.$$

Hint: show that the sum

$$\left[1 - \exp\left(2\pi i \frac{k}{N}\right)\right] \sum_{j=1}^{N} \exp\left(2\pi i \frac{(j-1)k}{N}\right)$$

equals

$$\exp\left(2\pi i \frac{(l+1)k}{N}\right)\left[1 - \exp(2\pi i k)\right].$$

19.6. Consider the real-valued periodic function

$$h(t) = c_0 + 2\Re\left(\sum_{m=1}^{M} c_m \exp(2\pi i m t)\right),$$

sampled at an even number of discrete points $t_j = \frac{j-1}{N}$, $j = 1,\ldots,N$, and let $\{H_l\}_{l=1}^{N}$ denote the corresponding discrete Fourier transform. Let $c_{-m} = \overline{c_m}$. Use the result of the previous two exercises to show that

$$H_l = \sum_{m=-M}^{M} c_m \cdot \begin{cases} N, & \mathrm{mod}(m-l+1, N) = 0, \\ 0, & \mathrm{mod}(m-l+1, N) \neq 0. \end{cases}$$

19.7. Suppose that N is even and greater than $2M$. Show that, for $1 \leq l \leq N/2 + 1$, the sum

$$\sum_{m=-M}^{M} c_m \cdot \begin{cases} N, & \mathrm{mod}(m-l+1, N) = 0, \\ 0, & \mathrm{mod}(m-l+1, N) \neq 0, \end{cases}$$

equals

$$\begin{cases} N c_{l-1}, & 1 \leq l \leq M+1, \\ 0, & M+2 \leq l \leq N/2 + 1. \end{cases}$$

19.8. Consider the real-valued periodic function

$$h(t) := \sum_{m=-M}^{M} c_m \exp(2\pi i m t) = c_0 + 2\Re\left(\sum_{m=1}^{M} c_m \exp(2\pi i m t)\right),$$

where $c_{-m} = \overline{c_m}$. Show that there is a one-to-one relationship between the Fourier coefficients of h and the first $M+1$ elements of the corresponding discrete Fourier transform $\{H_l\}_{l=1}^{N}$ provided that N is even and greater than $2M$.

19.9. How do the conclusions in the previous several exercises change if $N \leq 2M$?

19.10. Suppose that N is even and let F denote the symmetric square matrix whose (j,l) entry equals

$$\exp\left(-2\pi i \frac{(j-1)(l-1)}{N}\right).$$

Use the results of the previous exercises to show that the (j,l) entry of the inverse F^{-1} is given by

$$\frac{1}{N}\exp\left(2\pi i \frac{(j-1)(l-1)}{N}\right).$$

19.11. Verify that the matrix F^{-1} in the previous exercise may be obtained by applying the MATLAB ifft routine to the identity matrix I_N.

19.12. Consider the family of vector-valued periodic functions

$$h(t) := \sum_{m=-M}^{M} c_m \exp(2\pi i m t),$$

where $c_{-m}^{(i)} = \overline{c_m^{(i)}}$. Let H, H', and H_σ denote the discrete Fourier transforms of $h(t_j)$, $h'(t_j)$, and $h(t_j+\sigma)$, where t_j is given in Eq. (19.2) and N is an even integer that is larger than $2M$. Show that

$$H' = H \cdot D, \; H_\sigma = H \cdot S,$$

where D and S are diagonal matrices whose diagonal elements are given in Eqs. (19.24) and (19.25), respectively. Is this still true when $N \leq 2M$?

19.13. Consider the family of vector-valued periodic functions

$$h(t) := \sum_{m=-M}^{M} c_m \exp(2\pi i m t),$$

where $c_{-m}^{(i)} = \overline{c_m^{(i)}}$. Let H denote the discrete Fourier transforms of $h(t_j)$, where t_j is given in Eq. (19.2) and N is an even integer that is larger than $2M$. Suppose that there exists a vector-valued function $f : \mathbb{R}^n \times \mathbb{R}^q \to \mathbb{R}^n$ such that

$$h'(t) = T f(h(t), p)$$

for all $t \in [0,1]$ and some scalar T. Show that

$$H \cdot D - T f\left(H \cdot F^{-1}, 1_{1,N} \otimes p\right) \cdot F = 0.$$

19.14. Let A denote an arbitrary $k \times l$ matrix. It follows that

$$\mathfrak{vec}_n(w) \cdot A = \mathfrak{vec}_n\left(\left(A^T \otimes I_n\right) \cdot w\right)$$

for every nk-dimensional vector w. Let W be an $n \times k$ matrix. Show that

$$\mathfrak{vec}(W \cdot A) = \left(A^T \otimes I_n\right) \cdot \mathfrak{vec}(W).$$

19.15. Use the result of the previous exercise to show that the equation

$$H \cdot D - Tf\left(H \cdot F^{-1}, 1_{1,N} \otimes p\right) \cdot F = 0$$

is equivalent to the equation

$$D^* \cdot \eta - TF^* \cdot \mathfrak{vec}\left(f\left(\mathfrak{vec}_n\left(\left(F^{-1}\right)^* \cdot \eta\right), 1_{1,N} \otimes p\right)\right) = 0,$$

where $\eta := \mathfrak{vec}(H)$.

19.16. Let S be the diagonal matrix whose diagonal elements are given in Eq. (19.25) for an even integer N that is larger than $2M$. Show that differentiation of the matrix

$$F^* \cdot \mathfrak{vec}\left(f\left(\mathfrak{vec}_n\left(\left(F^{-1}\right)^* \cdot S^* \cdot \eta\right), 1_{1,N} \otimes p\right)\right),$$

with respect to σ, and evaluation at $\sigma = 0$ yield

$$F^* \cdot \mathfrak{diag}\left(\partial_y f\left(\mathfrak{vec}_n\left(\left(F^{-1}\right)^* \cdot \eta\right), 1_{1,N} \otimes p\right)\right) \cdot \left(F^{-1}\right)^* \cdot D^* \cdot \eta.$$

19.17. Consider the vector-valued functions

$$h(t) := \sum_{m=-M}^{M} c_m \exp(2\pi \mathrm{i} m t)$$

and

$$h^*(t) := \sum_{m=-M}^{M} c_m^* \exp(2\pi \mathrm{i} m t)$$

such that $c_{-m} = \overline{c_m}$ and $c_{-m}^* = \overline{c_m^*}$. Show that the integral

$$\int_0^1 \sum_{i=1}^{n} \left(\frac{d}{dt} h^{*(i)}(t)\right) h^{(i)}(t) dt$$

is given by the sum

$$\sum_{m=1}^{M} m\left(\Re\left(c_m^*\right)^T \cdot \Im(c_m) - \Im\left(c_m^*\right)^T \cdot \Re(c_m)\right).$$

19.18. Verify the equivalence of the spectral zero problem given by the vanishing of the expression in Eq. (19.39) with the version obtained with the substitutions in Eq. (19.49). Show that the residual of the latter is independent of N and that the sequence of corresponding error estimates is convergent as $N \to \infty$. Why is this not the case for the original formulation in terms of the discrete Fourier transform?

19.19. Revisit the formulation of the zero problem in Sect. 19.1.3 with $N = 2M$ in terms of the number of unknowns and the discretization of the governing equations of motion. Hint: be careful with real and imaginary parts.

19.20. The zero problem formulated in Sect. 19.1.3 imposes conditions on the unknown solution vector in the frequency domain. When $N = 2M$, it is equally valid to impose conditions in the time domain. Develop the corresponding mathematical formulation and implement this in COCO. Consider using the fast Fourier transform and its inverse as implemented in MATLAB.

19.21. Implement an adaptive toolbox using the zero problem formulated in the previous exercise with a predefined maximum value for N.

19.22. Consider the selection criterion for the discretization order of a spectral continuation problem discussed in [Grolet, A. and Thouverez, F., "On a new harmonic selection technique for harmonic balance method," *Mechanical Systems and Signal Processing*, 30, pp. 43–60, 2012]. Modify the 'dft' toolbox to rely on this criterion.

19.23. Why does the discretization order M change so slowly during continuation with decreasing ε in Example 19.4, even as the estimated discretization error remains well below the desired tolerance? How might one modify the dft_update function to make adaptive modifications to the discretization order that account for rapid variations in the discretization error along the solution manifold?

Chapter 20

Integrating Adaptation in Atlas Algorithms

The method of pointwise adaptation deployed in Chap. 18, albeit functional, lacks the beauty of automation. Its demands on manual intervention after the triggering of a terminal event, associated with an excessive discretization error, make it suitable primarily for development and testing. The `dft` toolbox proposed in the previous chapter, on the other hand, provides for an integrated approach to adaptive discretization, albeit within a framework (decomposition into Fourier modes) that may require unnecessarily high discretization order in the event of highly localized contributions to the discretization error.

In this chapter, we consider as an alternative an adaptive implementation of the `coll` toolbox that relies on the integration of support for adaptation in an atlas algorithm. As a consequence of such integration, we go beyond the pointwise focus of the brute-force methodology in Chap. 18 to a scheme that concerns itself also with changes to the local shape of the solution manifold.

20.1 Adaptive mesh refinements

We frame the problem of adaptive mesh refinements initially in the context of the parameter-dependent construction of an approximately equidistributed partition of an increasing scalar-valued function on $\mathbb{R}$.

20.1.1 Equidistributed partitions

An order-N partition

$$-1 = t_1 \leq \cdots \leq t_N = 1 \tag{20.1}$$

of the interval $[-1, 1]$ is said to be *equidistributed* with respect to an increasing function $f : \mathbb{R} \to \mathbb{R}$ if

$$f\left(t_j\right) = f\left(t_1\right) + \frac{j-1}{N-1}\left(f\left(t_N\right) - f\left(t_1\right)\right) \tag{20.2}$$

for $j = 1, \ldots, N$.

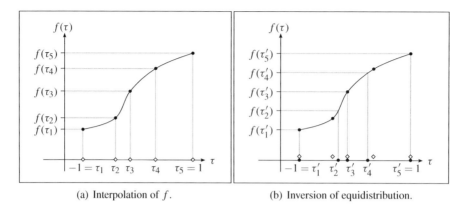

(a) Interpolation of f.   (b) Inversion of equidistribution.

Figure 20.1. *The action of the remeshing map* Υ *defined in Eq.* (20.6) *is computed in a two-step algorithm. In a first step, an interpolation of an increasing function of τ is constructed for a given sequence* $\{\tau_k\}_{k=1}^N$, *as indicated with diamonds in panel* (a). *In a second step, one computes the inverse of an equidistributed sequence of f-values to obtain the sequence* $\{\tau_k'\}_{k=1}^N$, *which is equidistributed with respect to f and represented by the black dots in panel* (b).

Example 20.1 In the case of the function

$$t \mapsto \frac{\tanh pt}{\tanh p}, \tag{20.3}$$

the partition in Eq. (20.1) is equidistributed provided that

$$t_j = \frac{1}{p} \operatorname{arctanh}\left(\left(-1 + 2\frac{j-1}{N-1}\right)\tanh p\right). \tag{20.4}$$

Notably, for fixed order N, and as $|p|$ increases, the interior mesh points cluster around $t = 0$. Although the partition equidistributes the function values on the interval $[f(-1), f(1)]$ for all values of p, the mesh points thus provide an increasingly nonuniform cover of the interval $[-1, 1]$ as $|p| \to \infty$. ∎

Let $f : \mathbb{R} \to \mathbb{R}$ again denote an increasing function on the interval $[-1, 1]$ and consider the construction in Fig. 20.1. Suppose that we are given an interpolation scheme that results in an increasing function υ, for each order-N partition of the interval $[-1, 1]$, such that

$$\upsilon\left(t_j\right) = -1 + 2\frac{f\left(t_j\right) - f\left(t_1\right)}{f\left(t_N\right) - f\left(t_1\right)}. \tag{20.5}$$

It follows, in particular, that $\upsilon : [-1, 1] \to [-1, 1]$ is one-to-one and onto. Let Υ be the map on the collection of all such order-N partitions, defined by

$$\left(\Upsilon\left(\{t_k\}_{k=1}^N\right)\right)_j := \upsilon^{-1}\left(-1 + 2\frac{j-1}{N-1}\right), \tag{20.6}$$

where υ^{-1} denotes the increasing inverse of the function υ. It follows immediately from Eq. (20.2) that

$$t_j = \left(\Upsilon \left(\{t_k\}_{k=1}^N \right) \right)_j \tag{20.7}$$

if and only if the partition in Eq. (20.1) is equidistributed with respect to f, independently of the interpolation scheme.

Example 20.2 Let the sequence $\{t_j^*\}_{j=1}^N$ correspond to an order-N partition of the interval $[-1,1]$, which is equidistributed with respect to an increasing function $f : \mathbb{R} \to \mathbb{R}$. As stated above, this partition is a fixed point of the map Υ, independently of the choice of the interpolating function υ.

Now consider a perturbation $t_k = t_k^* + \delta$ to one of the interior mesh points, i.e., for some $k = 2,\ldots,N-1$, and let $t_j = t_j^*$ for $j \neq k$. Suppose, furthermore, that f is convex on the interval $[t_{k-1}, t_{k+1}]$. Denote by υ the piecewise-linear interpolant, for which

$$\upsilon\left(t_j\right) = -1 + 2 \frac{f\left(t_j\right) - f\left(t_1\right)}{f\left(t_N\right) - f\left(t_1\right)}, \tag{20.8}$$

for $j = 1,\ldots,N$. Since the original partition is equidistributed with respect to f, it follows that

$$\upsilon\left(t_j\right) = -1 + 2 \frac{j-1}{N-1} \tag{20.9}$$

for $j \neq k$. In addition, for $t \in (t_{k-1}, t_{k+1})$, it holds that $\upsilon(t) \geq f(t)$, with equality only at $t = t_k$.

Let the map Υ now be defined as in Eq. (20.6) and denote the jth component of its image by $\tilde{t}_j$. It then follows that $\tilde{t}_j = t_j = t_j^*$ for $j \neq k$. Moreover, provided that $t_{k-1} - t_k < \delta < 0$, it follows that $t_k < \tilde{t}_k < t_k^*$, and similarly in the case when $0 < \delta < t_{k+1} - t_k$. Repeated application of Υ thus results in convergence of the sequence

$$\left\{ \left(\Upsilon^{(i)} \left(\{t_j\}_{j=1}^N \right) \right)_k \right\}_{i=1}^\infty \tag{20.10}$$

to t_k^*. We conclude that the equidistributed partition is asymptotically stable as a fixed point of Υ under perturbations to individual interior mesh points. ∎

Suppose, more generally, that we are given an interpolation scheme that results in an increasing function υ, for each order-N partition of the interval $[-1,1]$, such that

$$\upsilon\left(t_j\right) = -1 + 2 \frac{g_j - g_1}{g_N - g_1}, \tag{20.11}$$

where

$$g_j = -1 + 2 \frac{f\left(t_j\right) - f\left(t_1\right)}{f\left(t_N\right) - f\left(t_1\right)} + st_j \tag{20.12}$$

for $j = 1,\ldots,N$, and s is some positive scalar. Let the map Υ again be defined by Eq. (20.6). In this case, the unique fixed point of Υ is the partition obtained from the equality

$$g_j = \left(-1 + 2 \frac{j-1}{N-1} \right) (1+s) \tag{20.13}$$

for $j = 1, \ldots, N$. The case considered above, and in Example 20.2, corresponds here to $s = 0$. In contrast, in the limit that $s \to \infty$, the interpolant must satisfy the condition

$$\upsilon\left(t_j\right) = -1 + 2\frac{t_j - t_1}{t_N - t_1} = t_j. \tag{20.14}$$

The unique fixed point of Υ is then the *uniform partition* obtained from

$$t_j = -1 + 2\frac{j-1}{N-1} \tag{20.15}$$

for $j = 1, \ldots, N$. For finite, but nonzero, s, the fixed point partition is a compromise between an equidistributed partition with respect to f and a uniform distribution across the interval $[-1, 1]$.

20.1.2   A moving mesh

Given a function $f : [-1, 1] \times \mathbb{R}^q \to \mathbb{R}$ that is increasing in its first argument, we consider next the zero problem $\Phi_f(u) = 0$, where

$$\Phi_f : u \mapsto x - \begin{pmatrix} f(t_1, p) \\ \vdots \\ f(t_N, p) \end{pmatrix}, u = (x, p) \in \mathbb{R}^N \times \mathbb{R}^q, \tag{20.16}$$

for some order-N partition

$$-1 = t_1 \le \cdots \le t_N = 1. \tag{20.17}$$

Since the $N \times (N + q)$ Jacobian matrix

$$\partial_u \Phi_f(u) = \begin{pmatrix} I_N & \begin{pmatrix} -\partial_p f(t_1, p) \\ \vdots \\ -\partial_p f(t_N, p) \end{pmatrix} \end{pmatrix} \tag{20.18}$$

has full rank everywhere, it follows that every point on the q-dimensional solution manifold is regular. In particular, for a given point

$$u^* = \left(f\left(t_1, p^*\right), \ldots, f\left(t_N, p^*\right), p^*\right) \tag{20.19}$$

on the solution manifold, a basis for the corresponding tangent space is given by the columns of

$$\begin{pmatrix} \partial_p f(t_1, p^*) \\ \vdots \\ \partial_p f(t_N, p^*) \\ I_q \end{pmatrix}. \tag{20.20}$$

Now let υ denote an increasing interpolant on $[-1, 1]$ such that

$$\upsilon\left(t_j\right) = -1 + 2\frac{g_j - g_1}{g_N - g_1}, \tag{20.21}$$

where

$$g_j = -1 + 2\frac{f(t_j, p^*) - f(t_1, p^*)}{f(t_N, p^*) - f(t_1, p^*)} + st_j \tag{20.22}$$

for $j = 1, \ldots, N$, and s is some positive scalar. Given some integer M, we may consider the updated partition

$$-1 = \tilde{t}_1 \leq \cdots \leq \tilde{t}_M = 1, \tag{20.23}$$

where

$$\tilde{t}_j = v^{-1}\left(-1 + 2\frac{j-1}{M-1}\right) \tag{20.24}$$

for $j = 1, \ldots, M$ (cf. Eq. (20.6), where $M = N$).

A modified zero problem is then obtained by substituting $\tilde{t}_j$ for t_j in Eq. (20.16). We find a corresponding initial solution guess by

1. constructing an interpolant χ on $[-1, 1]$ such that

$$\chi(t_j) = f(t_j, p^*) \tag{20.25}$$

   for $j = 1, \ldots, N$ and

2. evaluating χ at each of the new mesh points $\tilde{t}_j$, $j = 1, \ldots, M$.

Similarly, let $d\chi$ denote a vector-valued interpolant on $[-1, 1]$ such that

$$d\chi(t_j) = \partial_p f(t_j, p^*) \tag{20.26}$$

for $j = 1, \ldots, N$. We then obtain an approximate basis for the new tangent space from the columns of

$$\begin{pmatrix} d\chi(\tilde{t}_1) \\ \vdots \\ d\chi(\tilde{t}_M) \\ I_q \end{pmatrix}. \tag{20.27}$$

In both cases, we use information available in an existing chart on the original solution manifold, without the need for further evaluation of the function f (which may, in practice, be computationally expensive).

Example 20.3 From Eq. (20.20), it follows that each vector $v \in \mathbb{R}^q$ corresponds to a vector in the tangent space $\mathcal{T}_{u^*}$ given by

$$\begin{pmatrix} \partial_p f(t_1, p^*) \cdot v \\ \vdots \\ \partial_p f(t_N, p^*) \cdot v \\ v \end{pmatrix}. \tag{20.28}$$

In this case, the corresponding approximate tangent vector

$$\begin{pmatrix} d\chi(\tilde{t}_1) \cdot v \\ \vdots \\ d\chi(\tilde{t}_M) \cdot v \\ v \end{pmatrix} \tag{20.29}$$

to the solution manifold of the modified zero problem may be obtained by first constructing
a scalar-valued function that interpolates the values

$$\partial_p f\left(t_j, p^*\right) \cdot v \tag{20.30}$$

at $t = t_j$ for $j = 1, \ldots, N$ and then evaluating this function at $\tilde{t}_j$ for $j = 1, \ldots, M$. ∎

20.1.3 Problem reconstruction

We seek to apply the procedure outlined above within the context of a continuation problem encoded in a COCO-compatible continuation problem structure. To this end, we restrict
attention to the function f given in Eq. (20.3) and the encoding of the corresponding zero
function Φ_f, shown below.

```
function [data y] = tanh_F(prob, data, u)

x = u(data.x_idx);
p = u(data.p_idx);

y = x-tanh(p*data.t)/tanh(p);

end
```

We first explore the properties of the solution manifold in the absence of adaptive mesh refinements and rely, instead, on an order-7 partition of the interval $[-1, 1]$, which is equidistributed with respect to the identity. The following sequence of commands initializes the
function data structure and initial solution guess, accordingly.

```
>> N          = 7;
>> p0         = 1;
>> data       = struct();
>> data.x_idx = 1:N;
>> data.p_idx = N+1;
>> data.t     = linspace(-1, 1, N)';
>> u0         = [tanh(p0*data.t)/tanh(p0); p0];
```

Example 20.4 In the extract below, we rely on the default 1-dimensional atlas algorithm
to generate a cover of a portion of the corresponding 1-dimensional solution manifold.

```
>> prob = coco_add_func(coco_prob(), 'tanh', @tanh_F, data, ...
     'zero', 'u0', u0);
>> prob = coco_add_pars(prob, 'pars', N+1, 'p');
>> prob = coco_add_event(prob, 'UZ', 'p', [1 4 7 10]);
>> prob = coco_set(prob, 'cont', 'h0', 1, 'hmax', 1);
>> coco(prob, 'run1', [], 1, 'p', [1 11]);
```

STEP	DAMPING		NORMS			COMPUTATION TIMES		
IT SIT	GAMMA	||d||	||f||	||U||		F(x)	DF(x)	SOLVE
0			0.00e+00	2.35e+00		0.0	0.0	0.0
1 1	1.00e+00	0.00e+00	0.00e+00	2.35e+00		0.0	0.0	0.0

STEP	TIME	||U||	LABEL	TYPE	p
0	00:00:00	2.3511e+00	1	EP	1.0000e+00
1	00:00:00	2.3511e+00	2	UZ	1.0000e+00

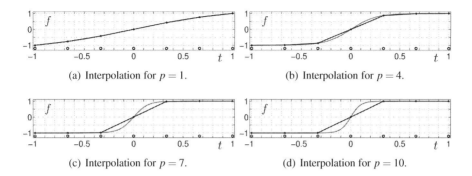

(a) Interpolation for $p = 1$. (b) Interpolation for $p = 4$.

(c) Interpolation for $p = 7$. (d) Interpolation for $p = 10$.

Figure 20.2. *Interpolation of the increasing test function* $f(t, p) = \tanh(pt)/$ $\tanh(p)$ *in Example* 20.4 *on a uniform mesh with seven mesh points for increasing values of* p, *as described in Sect.* 20.1.3. *As seen in the sequence* (a)–(d), *a steep front develops in the center of the graph. This is clearly poorly resolved with a uniform mesh. The circles at the bottom indicate the mesh in* t. *The result of using a mesh that is equidistributed with respect to* f *is shown in Fig.* 20.3.

```
 9   00:00:00   6.1221e+00   3  UZ   4.0000e+00
10   00:00:00   6.7745e+00   4        4.4875e+00
18   00:00:01   1.0194e+01   5  UZ   7.0000e+00
20   00:00:01   1.1604e+01   6        8.0216e+00
26   00:00:01   1.4352e+01   7  UZ   1.0000e+01
29   00:00:01   1.5748e+01   8  EP   1.1000e+01
```

As seen in Fig. 20.2, interpolation along the fixed partition gives only a very crude approximation to the graph of the function f. ∎

In order to accommodate adaptive mesh refinements as well as to generalize the implementation to other choices for the function f, we append several additional fields to the content of the function data structure, as shown below.

```
>> data.N    = N;
>> data.pdim = 1;
>> data.xtr  = zeros(N+data.pdim,1);
>> data.xtr([1 N:N+data.pdim]) = [1 N:N+data.pdim];
>> data.th   = linspace(-1, 1, N)';
>> data.s    = .3;
>> data.HINC = 2;
>> data.HDEC = 0;
>> data = coco_func_data(data);
```

In particular, as we recognize that the discretization order may change, we introduce the N field to hold the current value of N. We imagine making adaptive changes to N when the width of the largest mesh interval exceeds the value stored in the HINC field or falls below the value stored in the HDEC field. We further store an integer array in the xtr field of the function data structure, with nonzero elements referencing those components of the current vector of continuation variables, whose values are invariant under adaptive changes to the partition. We use the th and s fields of the function data structure in the definition of the map Υ. Finally, we convert the function data structure to a coco_func_data class instance in order to support modifications to its content outside of the zero function.

We encode support for adaptive mesh refinements in the `prob` continuation problem structure by including the `'remesh'` optional flag followed by a suitable function handle in the call to `coco_add_func`, as shown below.

```
>> data.t    = [-1; 1/p0*atanh((-1+2/(N-1)*(1:N-2)')*tanh(p0)); 1];
>> u0        = [tanh(p0*data.t)/tanh(p0); p0];
>> prob = coco_add_func(coco_prob(), 'tanh', @tanh_F, data, 'zero', ...
     'u0', u0, 'remesh', @remesh);
```

The initial partition is here obtained from Eq. (20.4). A COCO-compatible encoding of the `remesh` function is shown below.

```
function [prob stat xtr] = remesh(prob, data, chart, ub, Vb)

f  = ub(data.x_idx);
df = Vb(data.x_idx,:);

g  = 2*(f-f(1))/(f(end)-f(1))-1+data.s*data.t;
u  = 2*(g-g(1))/(g(end)-g(1))-1;
t0 = interp1(u, data.t, data.th);
t0([1 end]) = [-1 1];
ua = [interp1(data.t, f,  t0); ub(data.p_idx)];
Va = [interp1(data.t, df, t0); Vb(data.p_idx,:)];

xtr = data.xtr;
N   = numel(t0);
if numel(data.t)~=numel(t0)
  data.N     = N;
  data.x_idx = 1:N;
  data.p_idx = N+data.pdim;
  xtr(end-data.pdim:end) = N:N+data.pdim;
  data.xtr   = zeros(N+data.pdim,1);
  data.xtr([1 N:N+data.pdim]) = [1 N:N+data.pdim];
end
data.t = t0;
prob   = coco_change_func(prob, data, 'u0', ua, 'vecs', Va);

H  = max(abs(diff(data.t)));
N2 = N;
if H>data.HINC
  N2 = min(100, ceil(N*min((H/data.HINC), 1.1)));
elseif H<data.HDEC
  N2 = max(10, ceil(N*max((H/data.HDEC), 0.75)));
end
if N~=N2
  data.th = linspace(-1, 1, N2)';
  stat = 'repeat';
else
  stat = 'success';
end

end
```

Here, the `prob` input and output arguments again refer to the continuation problem structure. The second input argument, `data`, contains a copy of the function data structure of the `tanh_F` zero function. Since we assigned this to a `coco_func_data` class instance, changes to this variable within the body of the `remesh` function persist after execution of this function and are visible to the `tanh_F` zero function.

The `stat` output argument must be assigned one of the strings `'success'`, `'fail'`, or `'repeat'`. In the first case, execution of the COCO finite-state machine continues from the location where a call to `remesh` was first invoked. In the second case, the algorithm terminates with an error. Finally, in the case that the `stat` output argument equals `'repeat'`, the `remesh` function is invoked repeatedly until `stat` equals either `'success'` or `'fail'`, or until a maximal number of calls is reached. In the implementation of the `remesh` function, each change to the discretization order is associated with a modification to the content of `data.th` and a repeated call to `remesh`.

We consider next the third output argument, `xtr`. This must contain an integer array that is equal in length to the cardinality of the index set $\mathbb{K}$, associated with the `tanh_F` zero function, prior to the adaptive change to the mesh. Specifically, nonzero entries provide a translation table between integer indices of the vector $u_\mathbb{K}$ of continuation variables prior to remeshing, and the corresponding elements of the vector of continuation variables after such an adaptive change to the mesh, provided that the corresponding entries of u are invariant to remeshing. In the implementation of the `remesh` function above, we first copy the content of `data.xtr` into the `xtr` variable. In the sequence of assignments

```
N    = numel(t0);
if numel(data.t)~=numel(t0)
  data.N      = N;
  data.x_idx = 1:N;
  data.p_idx = N+data.pdim;
  xtr(end-data.pdim:end) = N:N+data.pdim;
  data.xtr    = zeros(N+data.pdim,1);
  data.xtr([1 N:N+data.pdim]) = [1 N:N+data.pdim];
end
```

we modify the last two elements of `xtr` (since, in this case, `data.pdim` equals 1) to account for a possible change to the discretization order N. We use the identical syntax as during initialization on the command line to update the content of `data.N`, `data.x_idx`, `data.p_idx`, and `data.xtr`, in light of the change in cardinality of $\mathbb{K}$.

20.1.4   An adaptive atlas algorithm

We explore the remainder of the `remesh` function, and the input arguments `chart`, `ub`, and `vb`, after a brief discussion of the modifications to an `AtlasBase` subclass and its class methods that are required in order to support adaptive changes to the problem discretization. To this end, consider a 1-dimensional atlas algorithm obtained by appending handling of computational domain boundaries (as in Sect. 14.1) and event handling (as in Sect. 16.1) to the basic algorithm described in Sect. 12.2. We proceed by introducing two additional settings in the `get_settings` class method, as shown below.

```
function cont = get_settings(cont)

defaults.h      = 0.1;
defaults.PtMX   = 50;
defaults.theta  = 0.5;
defaults.hmax   = 0.1;
defaults.hmin   = 0.01;
defaults.hfinc  = 1.1;
defaults.hfred  = 0.5;
defaults.almax  = 10;
defaults.NAdapt = 0;
```

```
defaults.RMMX   = 10;
cont            = coco_merge(defaults, cont);
cont.almax      = cont.almax*pi/180;

end
```

Here, a nonzero integer for the NAdapt field corresponds to the number of successfully executed continuation steps between each adaptive change to the mesh. The integer assigned to the RMMX field, in turn, equals the maximal number of times that 'repeat' may be assigned to the stat output argument from the remesh function before termination.

The modified encoding of the predict class method, shown below, relies on a call to the coco_remesh utility in order to invoke the remesh function described above.

```
function [prob atlas cseg correct] = predict(atlas, prob, cseg)

chart  = atlas.base_chart;
nad    = atlas.cont.NAdapt;
RMMX   = atlas.cont.RMMX;
if nad>0 && mod(chart.pt,nad)==0
  x0                = chart.x;
  V0                = [chart.t chart.TS];
  [prob chart x0 V0] = coco_remesh(prob, chart, x0, V0, RMMX);
  nv                = repmat(sqrt(sum(V0.^2,1)), [size(V0,1) 1]);
  chart.x           = x0;
  chart.t           = V0(:,1)./nv(:,1);
  chart.TS          = V0(:,2:end)./nv(:,2:end);
end
prcond = struct('x', chart.x, 'TS', chart.TS, ...
                's', chart.s, 'h', chart.R);
th     = atlas.cont.theta;
if th>=0.5 && th<=1
  xp              = chart.x+(th*chart.R)*(chart.TS*chart.s);
  [prob cseg]     = CurveSegment.create(prob, chart, prcond, xp);
  [prob ch2]      = cseg.update_TS(prob, cseg.curr_chart);
  h               = chart.R*chart.TS'*ch2.TS;
  xp              = chart.x+h*(ch2.TS*chart.s);
  prcond          = struct('x', chart.x, 'TS', ch2.TS, ...
                           's', chart.s, 'h', h);
else
  xp              = chart.x+chart.R*(chart.TS*chart.s);
end
[prob cseg]     = CurveSegment.create(prob, chart, prcond, xp);
[prob chart]    = cseg.update_t(prob, cseg.ptlist{1});
[prob chart]    = cseg.update_p(prob, chart);
cseg.ptlist{1}  = chart;
correct         = true;

end
```

The input arguments ub and Vb to the remesh function contain the components of the x0 and V0 input arguments to coco_remesh that pertain to the domain of the tanh_F zero function. The assignments

```
f  = ub(data.x_idx);
df = Vb(data.x_idx,:);

g  = 2*(f-f(1))/(f(end)-f(1))-1+data.s*data.t;
u  = 2*(g-g(1))/(g(end)-g(1))-1;
t0 = interp1(u, data.t, data.th);
```

```
t0([1 end]) = [-1 1];
ua = [interp1(data.t,  f,   t0); ub(data.p_idx)];
Va = [interp1(data.t, df, t0); Vb(data.p_idx,:)];
```

in `remesh` now extract the components of the vector of continuation variables $u = (x, p)$ and the associated array of tangent vectors corresponding to x, and store an updated initial solution guess and array of approximate tangent vectors in the `ua` and `Va` variables. The call

```
prob     = coco_change_func(prob, data, 'u0', ua, 'vecs', Va);
```

makes the necessary modifications to the continuation problem structure. Provided that the `stat` output argument is assigned the value `'success'`, the updated initial solution guess for the entire vector of continuation variables and the associated array of tangent vectors are then returned in the `x0` and `V0` output arguments of the `coco_remesh` function in the `predict` class method.

Finally, we note the need to declare a `'remesh'` signal in the `create` static construction method to the `atlas_1d_min` subclass, as shown below.

```
function [prob cont atlas] = create(prob, cont, dim)

atlas = atlas_1d_min(prob, cont, dim);
prob  = CurveSegment.add_prcond(prob, dim);
prob  = coco_add_signal(prob, 'remesh', 'atlas_1d_min');

end
```

Example 20.5 The extract below now shows the successful application of adaptive changes to the problem mesh during continuation.

```
>> prob = coco_add_pars(prob, 'pars', N+1, 'p');
>> prob = coco_add_event(prob, 'UZ', 'p', [1 4 7 10]);
>> prob = coco_set(prob, 'cont', 'atlas', @atlas_1d_min.create);
>> prob = coco_set(prob, 'cont', 'NAdapt', 1, 'h0', 1, 'hmax', 1);
>> coco(prob, 'run2', [], 1, 'p', [1 11]);
```

	STEP	DAMPING		NORMS			COMPUTATION TIMES		
IT	SIT	GAMMA	||d||	||f||	||U||	F(x)	DF(x)	SOLVE	
0				0.00e+00	2.26e+00	0.0	0.0	0.0	
1	1	1.00e+00	0.00e+00	0.00e+00	2.26e+00	0.0	0.0	0.0	

STEP	TIME	||U||	LABEL	TYPE	p
0	00:00:00	2.2608e+00	1	EP	1.0000e+00
1	00:00:00	2.2840e+00	2	UZ	1.0000e+00
10	00:00:01	3.4946e+00	3		2.0886e+00
18	00:00:02	5.9746e+00	4	UZ	4.0000e+00
20	00:00:02	7.2779e+00	5		4.9575e+00
24	00:00:03	1.0093e+01	6	UZ	7.0000e+00
29	00:00:03	1.4276e+01	7	UZ	1.0000e+01
30	00:00:04	1.5680e+01	8	EP	1.1000e+01

The graphs in Fig. 20.3 show significant improvements in the ability of an interpolant through the solution points to approximate the function f. By assigning 0.3 and 0.2 to the HINC and HDEC fields of the `data` structure, we obtain the results of continuation using an equidistributed moving mesh with a changing number of mesh points. Here, starting with an initial solution guess with $N = 7$, the solution at $p = 1$ is obtained with a mesh of order

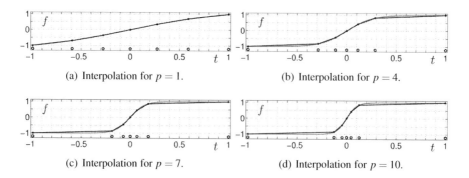

(a) Interpolation for $p = 1$.   (b) Interpolation for $p = 4$.

(c) Interpolation for $p = 7$.   (d) Interpolation for $p = 10$.

Figure 20.3. *Interpolation of the increasing test function* $f(t, p) = \tanh(pt)/$ $\tanh(p)$ *using an equidistributed moving mesh with seven mesh points for increasing values of* p, *as described in Sect. 20.1.3. The circles at the bottom indicate the mesh in* t. *The quality of interpolation of the steep front improves compared with the graphs shown in Fig. 20.2. The result of using an equidistributed moving mesh with variable number of mesh points is shown in Fig. 20.4.*

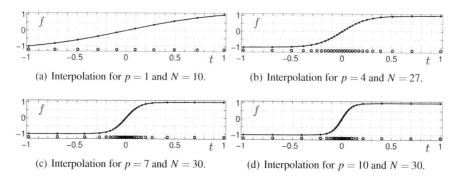

(a) Interpolation for $p = 1$ and $N = 10$.   (b) Interpolation for $p = 4$ and $N = 27$.

(c) Interpolation for $p = 7$ and $N = 30$.   (d) Interpolation for $p = 10$ and $N = 30$.

Figure 20.4. *Interpolation of the increasing test function* $f(t, p) = \tanh(pt)/$ $\tanh(p)$ *using an equidistributed moving mesh with varying number of mesh points for increasing values of* p, *as described in Sect. 20.1.3. The circles at the bottom indicate the mesh in* t. *The overall quality of interpolation improves compared with Fig. 20.3, albeit at the expense of adding mesh points.*

$N = 10$, whereas that obtained at $p = 10$ corresponds to a mesh of order $N = 30$. The corresponding graphs are shown in Fig. 20.4. ∎

20.2   Boundary-value problems

We return to the `'coll'` toolbox from Chap. 18, albeit without the inclusion of the discretization parameters κ among the continuation variables. Instead, we propose a moving-mesh algorithm that updates κ and the discretization order N, following the remeshing paradigm introduced in the previous section.

Accordingly, we leave the encodings of the `coll_isol2seg` and `coll_sol2seg` generalized constructors unchanged. In a similar way, no changes are required to the `coll_arg_check` argument error checking routine or the `coll_read_solution` toolbox extractor. Finally, since the assignments encoded in the `coll_interval` initialization routine are independent of N and κ, this function is identical to the encoding shown in Sect. 18.3.3. The following minimal modifications to the `coll_init_data` and `coll_init_sol` initialization routines accommodate the elimination of κ and λ from the vector of continuation variables.

```
function data = coll_init_data(data, t0, x0, p0)

NCOL = data.coll.NCOL;
dim  = size(x0,2);
data.int = coll_interval(NCOL, dim);

NTST = data.coll.NTST;
pdim = numel(p0);
data.maps = coll_maps(data.int, NTST, pdim);

t  = linspace(0, NTST, numel(t0));
tt = interp1(t, t0, 0:NTST, 'linear');
tt = tt*(NTST/tt(end));
data.mesh = coll_mesh(data.int, data.maps, tt);

end


function sol = coll_init_sol(data, t0, x0, p0)

t0 = t0(:);
T0 = t0(end)-t0(1);
if abs(T0)>eps
  t0 = (t0-t0(1))/T0;
  x0 = interp1(t0, x0, data.mesh.tbp)';
else
  x0 = repmat(x0(1,:), size(data.mesh.tbp))';
end

sol.u = [x0(:); T0; p0];

end
```

In the encodings of the `coll_F` and `coll_DFDU` functions, shown below, we unwind the inclusion of the conservation and mesh conditions in the extended continuation problem, but retain the differentiation of the content of the toolbox data structure in terms of the `int`, `maps`, and `mesh` fields, described in Sect. 18.3.3.

```
function [data y] = coll_F(prob, data, u)

maps = data.maps;
mesh = data.mesh;

x = u(maps.xbp_idx);
T = u(maps.T_idx);
p = u(maps.p_idx);

xx = reshape(maps.W*x, maps.x_shp);
pp = repmat(p, maps.p_rep);
```

```
ode = mesh.fka.*data.fhan(xx, pp);
ode = (0.5*T/maps.NTST)*ode(:)-maps.Wp*x;
cnt = maps.Q*x;

y = [ode; cnt];

end


function [data J] = coll_DFDU(prob, data, u)

maps = data.maps;
mesh = data.mesh;

x = u(maps.xbp_idx);
T = u(maps.T_idx);
p = u(maps.p_idx);

xx = reshape(maps.W*x, maps.x_shp);
pp = repmat(p, maps.p_rep);

if isempty(data.dfdxhan)
  dxode = coco_ezDFDX('f(x,p)v', data.fhan, xx, pp);
else
  dxode = data.dfdxhan(xx, pp);
end
dxode = mesh.dxka.*dxode;
dxode = sparse(maps.dxrows, maps.dxcols, dxode(:));
dxode = (0.5*T/maps.NTST)*dxode*maps.W-maps.Wp;

dTode = mesh.fka.*data.fhan(xx, pp);
dTode = (0.5/maps.NTST)*dTode(:);

if isempty(data.dfdphan)
  dpode = coco_ezDFDP('f(x,p)v', data.fhan, xx, pp);
else
  dpode = data.dfdphan(xx, pp);
end
dpode = mesh.dpka.*dpode;
dpode = sparse(maps.dprows, maps.dpcols, dpode(:));
dpode = (0.5*T/maps.NTST)*dpode;

J = [dxode dTode dpode; maps.Q sparse(maps.Qnum,1+maps.pdim)];

end
```

As before, the content of data.int remains unchanged during continuation, whereas that of data.mesh reflects changes to the discretization during continuation. The distinction between the two implementations of the 'coll' toolbox considered in Sects. 20.2.1 and 20.2.2 below is then found in the absence of changes to the content of data.maps in the first version and the possibility of such changes in the second version. Here, the mesh.fka, mesh.dxka, and mesh.dpka fields are assumed to contain the arrays κ_f, $\kappa_{\partial_x f}$, and $\kappa_{\partial_p f}$, respectively, introduced in Sect. 18.2.

Now suppose that we are given a 1-dimensional array κ of length N, corresponding to the partition

$$0 = t_1 \leq \cdots \leq t_j = \frac{T}{N} \sum_{i=1}^{j-1} \kappa_i \leq \cdots \leq t_{N+1} = T. \qquad (20.31)$$

Let F denote a given increasing 1-dimensional array in $\mathbb{R}$ of length $N + 1$ such that $F_1 = 0$. Finally, given an integer M, denote by ι a monotonically increasing interpolant such that

$$\iota\left(F_j\right) = \frac{M}{T} t_j, \, j = 1, \ldots, N+1, \tag{20.32}$$

from which it follows that $\iota\left(F_{N+1}\right) = M$. The partition

$$0 = \hat{t}_1 \leq \cdots \leq \hat{t}_j = \frac{T}{M} \iota\left(F_{N+1} \frac{j-1}{M}\right) \leq \cdots \leq \hat{t}_{M+1} = T \tag{20.33}$$

is said to *equipartition* the value F_{N+1}. Following Eq. (18.65), the corresponding sequence $\hat{\kappa}_j$ is then given by

$$\hat{\kappa}_j := \iota\left(F_{N+1} \frac{j}{M}\right) - \iota\left(F_{N+1} \frac{j-1}{M}\right) \tag{20.34}$$

for $j = 1, \ldots, M$.

Example 20.6 Consider the monotonically increasing function

$$F : t \mapsto \int_0^t dF(s) \, ds \tag{20.35}$$

for some positive function dF and let $F_j = F\left(t_j\right)$ for $j = 1, \ldots, N + 1$. Denote by F^{-1} the monotonically increasing inverse of F such that

$$\int_0^{F^{-1}(\varphi)} dF(s) \, ds = \varphi. \tag{20.36}$$

Now let

$$\iota = \frac{M}{T} F^{-1} \tag{20.37}$$

such that

$$\iota\left(F_j\right) = \frac{M}{T} t_j. \tag{20.38}$$

It follows that

$$\hat{t}_j = F^{-1}\left(F_{N+1} \frac{j-1}{M}\right) \tag{20.39}$$

or, equivalently,

$$\int_0^{\hat{t}_j} dF(s) \, ds = F_{N+1} \frac{j-1}{M}. \tag{20.40}$$

The partition

$$0 = \hat{t}_1 \leq \cdots \leq \hat{t}_{M+1} = T \tag{20.41}$$

thus equipartitions the integral

$$F(T) = \int_0^T dF(s) \, ds. \tag{20.42}$$

Different choices for the function dF clearly lead to different moving-mesh algorithms. ∎

The change in discretization order $N \mapsto M$ and the transformation $\kappa \mapsto \hat{\kappa}$ associated with the array F correspond to a reparameterization of the extended continuation problem, through a change in the number of constituent elements (equations and unknowns) as well as in their nature. We consider the case when $M = N$ in the next section and return to the fully adaptive algorithm in Sect. 20.2.2.

20.2.1   Fixed discretization order

As in Sect. 18.3.3, the encoding of the `coll_maps` function, shown below, initializes all properties of the problem formulation that depend on the discretization order N but not on the discretization parameters κ.

```
function maps = coll_maps(int, NTST, pdim)

NCOL = int.NCOL;
dim  = int.dim;

maps.NTST = NTST;
maps.pdim = pdim;

bpnum  = NCOL+1;
bpdim  = dim*(NCOL+1);
xbpnum = (NCOL+1)*NTST;
xbpdim = dim*(NCOL+1)*NTST;
cndim  = dim*NCOL;
xcnnum = NCOL*NTST;
xcndim = dim*NCOL*NTST;
cntnum = NTST-1;
cntdim = dim*(NTST-1);

maps.xbp_idx = (1:xbpdim)';
maps.T_idx   = xbpdim+1;
maps.p_idx   = xbpdim+1+(1:pdim)';
maps.Tp_idx  = [maps.T_idx; maps.p_idx];
maps.tbp_idx = setdiff(1:xbpnum, 1+bpnum*(1:cntnum))';
maps.x_shp   = [dim xcnnum];
maps.xbp_shp = [dim xbpnum];
maps.p_rep   = [1 xcnnum];

maps.xtr     = [maps.xbp_idx; maps.Tp_idx];
maps.xtr(dim+1:end-dim-pdim-1) = 0;

rows          = reshape(1:xcndim, [cndim NTST]);
rows          = repmat(rows, [bpdim 1]);
cols          = repmat(1:xbpdim, [cndim 1]);
W             = repmat(int.W, [1 NTST]);
Wp            = repmat(int.Wp, [1 NTST]);
maps.W        = sparse(rows, cols, W);
maps.Wp       = sparse(rows, cols, Wp);

temp          = reshape(1:xbpdim, [bpdim NTST]);
Qrows         = [1:cntdim 1:cntdim];
Qcols         = [temp(1:dim, 2:end) temp(cndim+1:end, 1:end-1)];
Qvals         = [ones(cntdim,1) -ones(cntdim,1)];
maps.Q        = sparse(Qrows, Qcols, Qvals, cntdim, xbpdim);
maps.Qnum     = cntdim;

maps.dxrows   = repmat(reshape(1:xcndim, [dim xcnnum]), [dim 1]);
maps.dxcols   = repmat(1:xcndim, [dim 1]);
```

```
maps.dprows   = repmat(reshape(1:xcndim, [dim xcnnum]), [pdim 1]);
maps.dpcols   = repmat(1:pdim, [dim xcnnum]);

maps.x0_idx   = (1:dim)';
maps.x1_idx   = xbpdim-dim+(1:dim)';

rows          = reshape(1:dim*NTST, [dim NTST]);
rows          = repmat(rows, [bpdim 1]);
cols          = repmat(1:xbpdim, [dim 1]);
Wm            = repmat(int.Wm, [1 NTST]);
maps.Wm       = sparse(rows, cols, Wm);
x             = linspace(int.tm(1), int.tm(2), 51);
y             = arrayfun(@(x) prod(x-int.tm), x);
maps.wn       = max(abs(y));

end
```

We note here the inclusion of the `maps.xtr` field, at first assigned the relative integer indices of the elements of v_{bp}, T, and p, in that order, in the vector of continuation variables. As in Sect. 20.1.3, nonzero entries of this field provide a translation between integer indices of mesh-invariant elements of the vector of continuation variables prior to remeshing and the corresponding elements after an adaptive change to the mesh. Here, we assume a remeshing strategy that holds the boundary values v_1 and $v_{N(m+1)}$, the interval length T, and the problem parameters p fixed. Accordingly, we assign 0 to all elements of the `maps.xtr` array corresponding to interior mesh points of the collocation discretization.

The `coll_mesh` function, shown below, now initializes all remaining properties of the problem formulation that depend on both the discretization order N and on the discretization parameters κ.

```
function mesh = coll_mesh(int, maps, tmi)

mesh.tmi = tmi;

dim  = int.dim;
NCOL = int.NCOL;
pdim = maps.pdim;
NTST = maps.NTST;

bpnum  = NCOL+1;
cndim  = dim*NCOL;
xcnnum = NCOL*NTST;
xcndim = dim*NCOL*NTST;

ka        = diff(tmi);
fka       = kron(ka', ones(dim*NCOL,1));
dxka      = kron(ka', ones(dim*cndim,1));
dpka      = kron(ka', ones(pdim*cndim,1));
mesh.ka   = ka;
mesh.fka  = reshape(fka, [dim xcnnum]);
mesh.dxka = reshape(dxka, [dim dim xcnnum]);
mesh.dpka = reshape(dpka, [dim pdim xcnnum]);

wts       = repmat(int.wt, [dim NTST]);
kas       = kron(ka,ones(dim,NCOL));
mesh.wts1 = wts(1,:);
mesh.kas1 = kas(1,:);
mesh.wts2 = spdiags(wts(:), 0, xcndim, xcndim);
mesh.kas2 = spdiags(kas(:), 0, xcndim, xcndim);
```

```
t  = repmat(tmi(1:end-1)/NTST, [bpnum 1]);
tt = repmat((0.5/NTST)*(int.tm+1), [1 NTST]);
tt = t+repmat(ka, [bpnum 1]).*tt;
mesh.tbp = tt(:)/tt(end);

end
```

We turn next to the introduction of support for adaptive changes to the discretization parameters κ. We note first the a priori initialization of κ in the `coll_mesh` function, using the identical construction as in Sect. 18.3.3. In particular, the `tmi` input argument here contains the sequence

$$\left\{\frac{N}{T}\hat{t}(j)\right\}_{j=1}^{N+1}, \tag{20.43}$$

obtained in terms of the piecewise-linear interpolant defined by Eq. (18.68).

We rely on the a posteriori construction in Example 20.6 for adaptive changes to κ. Specifically, we associate the 1-dimensional array F with the integral of an *error density* such that F provides an estimate of the cumulative discretization error along the solution approximant. The `coll_remesh` function, shown below, implements such a candidate remeshing strategy.

```
function [prob stat xtr] = coll_remesh(prob, data, chart, ub, Vb)

int  = data.int;
maps = data.maps;
mesh = data.mesh;
coll = data.coll;

xtr  = maps.xtr;

u = ub(maps.xbp_idx);
V = Vb(maps.xbp_idx,:);

cp   = reshape(maps.Wm*u, [int.dim maps.NTST]);
cp   = sqrt(sum(cp.^2,1));
cp   = nthroot(cp, int.NCOL);
cpmn = nthroot(0.125*coll.TOL/maps.wn, int.NCOL);

ka = mesh.ka;
s  = data.coll.SAD;
F  = [0 cumsum(((1-s)*cpmn*ka + s*cp, 2)];
t  = [0 cumsum(ka, 2)];
th = linspace(0, F(end), maps.NTST+1);
tt = interp1(F, t, th, 'cubic');

mesh2 = coll_mesh(int, maps, tt);

tbp = mesh.tbp(maps.tbp_idx);
xbp = reshape(u, maps.xbp_shp);
xbp = xbp(:, maps.tbp_idx);
x0  = interp1(tbp', xbp', mesh2.tbp, 'cubic')';
x1  = x0(:);

V1 = zeros(size(V));
for i=1:size(V,2)
  vbp = reshape(V(:,i), maps.xbp_shp);
  vbp = vbp(:, maps.tbp_idx);
  v0  = interp1(tbp', vbp', mesh2.tbp, 'cubic')';
```

```
  V1(:,i) = v0(:);
end

data.mesh = mesh2;

ua = [x1; ub(maps.Tp_idx)];
Va = [V1; Vb(maps.Tp_idx,:)];

prob = coco_change_func(prob, data, 'x0', ua, 'vecs', Va);

stat = 'success';

end
```

Here, we use the mth root of the norm $\|c_m\|$ of the vector of coefficients in front of the σ^m term in each interpolating polynomial as a piecewise-constant value of the function dF on each mesh interval. The assignments

```
s  = data.coll.SAD;
F  = [0 cumsum((1-s)*cpmn*ka + s*cp, 2)];
```

again provide a compromise between equidistributing the integral $F(T)$ for s equal to 1, on the one hand, and a uniform mesh for s equal to 0, on the other hand, parameterized by the 'SAD' setting of the modified 'coll' toolbox. The remaining portions of the encoding mirror the implementation of the remesh function in Sect. 20.1.3. As shown below, the modifications to the coll_get_settings function are immediate.

```
function data = coll_get_settings(prob, tbid, data)

defaults.NTST = 10;
defaults.NCOL = 4;
defaults.SAD  = 0.95;
if ~isfield(data, 'coll')
  data.coll = [];
end
data.coll = coco_merge(defaults, coco_merge(data.coll, ...
  coco_get(prob, tbid)));
if ~coco_exist('TOL', 'class_prop', prob, tbid, '-no-inherit-all')
  data.coll.TOL = coco_get(prob, 'corr', 'TOL')^(2/3);
end
NTST = data.coll.NTST;
assert(numel(NTST)==1 && isnumeric(NTST) && mod(NTST,1)==0, ...
  '%s: input for option ''NTST'' is not an integer', tbid);
NCOL = data.coll.NCOL;
assert(numel(NCOL)==1 && isnumeric(NCOL) && mod(NCOL,1)==0, ...
  '%s: input for option ''NCOL'' is not an integer', tbid);
assert(data.coll.SAD<=1 && data.coll.SAD>=0, ...
  '%s: input for option ''SAD'' not in [0,1]', tbid);

end
```

Here, the default value of 0.95 assigned to the 'SAD' setting gives significant weight to an error-equistributing mesh, while allowing for a uniform mesh in the case of essentially zero estimated errors across parts of the approximant.

From the implementation of the modified predict class method, it follows that all monitor functions are evaluated subsequent to any adaptive change of the mesh. To avoid replacing the error estimate encoded in the coll_err function subsequent to such an adaptive change, we store this estimate as chart data and assign a value to the corresponding element only when none exists, as shown in the following encoding.

```
function [data chart y] = coll_err(prob, data, chart, u)

cdata = coco_get_chart_data(chart, data.tbid);
if isfield(cdata, 'err')
  y = cdata.err;
else
  int  = data.int;
  maps = data.maps;

  cp = reshape(maps.Wm*u, [int.dim maps.NTST]);
  y  = maps.wn*max(sqrt(sum(cp.^2,1)));
  y  = [y; y/data.coll.TOL];
  cdata.err = y;
  chart = coco_set_chart_data(chart, data.tbid, cdata);
end

end
```

The modified encoding of the `coll_construct_seg` constructor ensures that the content of this element of chart data is reset when generating a partial copy of a base chart in the `CurveSegment.create` constructor.

```
function prob = coll_construct_seg(prob, tbid, data, sol)

data.tbid = tbid;
data = coco_func_data(data);
prob = coco_add_func(prob, tbid, @coll_F, @coll_DFDU, data, 'zero', ...
  'u0', sol.u, 'remesh', @coll_remesh);
uidx = coco_get_func_data(prob, tbid, 'uidx');
if ~isempty(data.pnames)
  fid  = coco_get_id(tbid, 'pars');
  prob = coco_add_pars(prob, fid, uidx(data.maps.p_idx), data.pnames);
end
prob = coco_add_slot(prob, tbid, @coco_save_data, data, 'save_full');
efid = coco_get_id(tbid, {'err' 'err_TF'});
prob = coco_add_chart_data(prob, tbid, struct(), struct());
prob = coco_add_func(prob, efid{1}, @coll_err, data, ...
  'regular', efid, 'uidx', uidx(data.maps.xbp_idx), ...
  'remesh', @coll_err_remesh, 'passChart');
prob = coco_add_event(prob, 'MXCL', 'MX', efid{2}, '>', 1);

end
```

We note, finally, the required reference to the `coll_err_remesh` function, shown below, in the second call to the `coco_add_func` constructor.

```
function [prob stat xtr] = coll_err_remesh(prob, data, chart, ub, Vb)

maps = data.maps;

xtr  = [];
uidx = coco_get_func_data(prob, data.tbid, 'uidx');
prob = coco_change_func(prob, data, 'uidx', uidx(maps.xbp_idx));
stat = 'success';

end
```

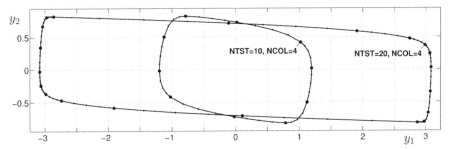

(a) Selected approximate periodic orbits obtained during continuation. The solution with 20 mesh intervals is obtained at $\varepsilon = 20$. The fat dots mark the end points of mesh intervals.

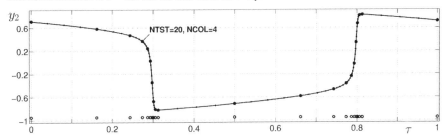

(b) Time profile of y_2 for $\varepsilon = 20$. The circles at the bottom mark the end points of the mesh intervals.

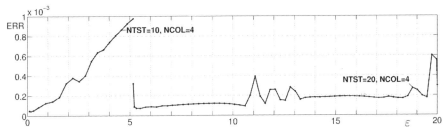

(c) The estimated error for different meshes plotted against variations in the parameter ε. The run with the 'NTST' setting equal to 10 eventually exceeds the desired tolerance of 10^{-3} and terminates.

Figure 20.5. *Continuation of periodic orbits of the dynamical system given by the vector field in Eq. (18.5) using an equidistributed moving mesh without adaptive changes to the discretization order, as described in Sect. 20.2.1. The algorithm allows one to compute the solution family with the same desired tolerance as in Fig. 18.3 with 1/9 of the number of mesh points. Panel (a) shows sample orbits obtained during continuation, and panels (b) and (c) illustrate the effect of adaptation. A refined strategy with varying discretization order is illustrated in Fig. 20.6.*

This ensures that the $\mathbb{K}$ index array for the `coll_err` function is reinitialized following the remeshing action encoded in `coll_remesh`.

Fig. 20.5 shows the result of applying the adaptive 'coll' toolbox with fixed discretization order to the continuation of periodic orbits of the dynamical system given by the vector field in Eq. (18.5). Here, the desired tolerance bounds are accommodated with

1/9 of the number of mesh points used by the brute-force mesh adaptation strategy in Sect. 18.1.2, as shown in Fig. 18.3.

20.2.2 Varying discretization order

We end our development of an adaptive moving-mesh implementation of the 'coll' toolbox by enabling support for adaptive changes to the discretization order N, in addition to the discretization parameters κ. The modified encoding of coll_remesh, shown below, implements such changes in response to variations in the discretization error estimate given in Eq. (18.10).

```
function [prob stat xtr] = coll_remesh(prob, data, chart, ub, Vb)

int  = data.int;
maps = data.maps;
mesh = data.mesh;
coll = data.coll;

xtr  = maps.xtr;

u = ub(maps.xbp_idx);
V = Vb(maps.xbp_idx,:);

cp   = reshape(maps.Wm*u, [int.dim maps.NTST]);
cp   = sqrt(sum(cp.^2,1));
cp   = nthroot(cp, int.NCOL);
cpmn = nthroot(0.5*(coll.TOLINC+coll.TOLDEC)/maps.wn, int.NCOL);

ka = mesh.ka;
s  = data.coll.SAD;
F  = [0 cumsum((1-s)*cpmn*ka + s*cp, 2)];
t  = [0 cumsum(ka, 2)];

NTSTi = min(ceil(maps.NTST*1.1+1),   coll.NTSTMX);
NTSTd = max(ceil(maps.NTST/1.025-1), coll.NTSTMN);
cdata = coco_get_chart_data(chart, data.tbid);
err   = cdata.err(1);
maps2 = maps;
if err>coll.TOLINC && maps.NTST~=NTSTi
  data.coll.NTST = NTSTi;
  maps2 = coll_maps(int, NTSTi, maps.pdim);
elseif err<coll.TOLDEC && maps.NTST~=NTSTd
  data.coll.NTST = NTSTd;
  maps2 = coll_maps(int, NTSTd, maps.pdim);
end
xtr(maps.xtrend) = maps2.xtr(maps2.xtrend);
th = linspace(0, F(end), maps2.NTST+1);
tt = interp1(F, t*(maps2.NTST/t(end)), th, 'cubic');

mesh2 = coll_mesh(int, maps2, tt);

tbp = mesh.tbp(maps.tbp_idx);
xbp = reshape(u, maps.xbp_shp);
xbp = xbp(:, maps.tbp_idx);
x0  = interp1(tbp', xbp', mesh2.tbp, 'cubic')';
u1  = x0(:);

V1 = zeros(numel(mesh2.tbp)*int.dim, size(V,2));
```

```
for i=1:size(V,2)
  vbp = reshape(V(:,i), maps.xbp_shp);
  vbp = vbp(:, maps.tbp_idx);
  v0  = interp1(tbp', vbp', mesh2.tbp, 'cubic')';
  V1(:,i) = v0(:);
end

data.maps = maps2;
data.mesh = mesh2;

ua = [u1; ub(maps.Tp_idx)];
Va = [V1; Vb(maps.Tp_idx,:)];

prob = coco_change_func(prob, data, 'u0', ua, 'vecs', Va);

stat = 'success';

end
```

The encoding makes use of the inclusion in `coll_maps` of the following assignment.

```
maps.xtrend  = maps.xtr(end-dim-pdim:end);
```

In particular, the assignment

```
xtr(maps.xtrend) = maps2.xtr(maps2.xtrend);
```

in `coll_remesh` identifies the translation of the integer indices associated with the mesh-invariant elements of the vector of continuation variables. As shown in the following encoding of the `coll_get_settings` function, default values for the upper and lower bounds of the adaptation window, as well as upper and lower bounds for the number of mesh intervals, are stored in the `'TOLINC'`, `'TOLDEC'`, `'NTSTMN'`, and `'NTSTMX'` settings of the `'coll'` toolbox.

```
function data = coll_get_settings(prob, tbid, data)

defaults.NTST = 10;
defaults.NCOL = 4;
defaults.SAD  = 0.95;
if ~isfield(data, 'coll')
  data.coll = [];
end
data.coll = coco_merge(defaults, coco_merge(data.coll, ...
  coco_get(prob, tbid)));
if ~coco_exist('TOL', 'class_prop', prob, tbid, '-no-inherit-all')
  data.coll.TOL = coco_get(prob, 'corr', 'TOL')^(2/3);
end
NTST = data.coll.NTST;
assert(numel(NTST)==1 && isnumeric(NTST) && mod(NTST,1)==0, ...
  '%s: input for option ''NTST'' is not an integer', tbid);
NCOL = data.coll.NCOL;
assert(numel(NCOL)==1 && isnumeric(NCOL) && mod(NCOL,1)==0, ...
  '%s: input for option ''NCOL'' is not an integer', tbid);

defaults.TOLINC = data.coll.TOL/5;
defaults.TOLDEC = data.coll.TOL/20;
defaults.NTSTMN = min(  5, data.coll.NTST);
defaults.NTSTMX = max(100, data.coll.NTST);
data.coll = coco_merge(defaults, data.coll);

end
```

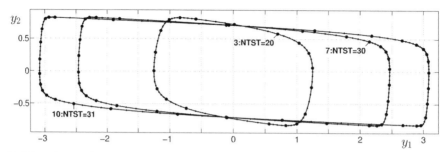

(a) Selected approximate periodic orbits obtained during continuation for increasing ε. The number
NTST of mesh intervals is indicated with the label. The fat dots mark the end points of mesh intervals.

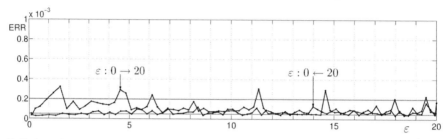

(b) Estimated error during continuation for increasing and decreasing ε. The tolerance is 10^{-3}, and
the adaptation window is indicated by two horizontal lines.

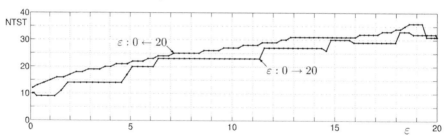

(c) Number of mesh intervals during continuation for increasing and decreasing ε.

Figure 20.6. *Continuation of periodic orbits of the dynamical system given by the
vector field in Eq. (18.5) using an equidistributed moving mesh with adaptive changes to
the discretization order, as described in Sect. 20.2.2. Here, a single continuation run suffices to compute the entire solution family. Due to the conservative definition of the adaptation window, the algorithm uses about 50% more mesh points than the analysis shown
in Fig. 20.5. Panel* (a) *shows sample orbits obtained during continuation, and panels* (b)
and (c) *illustrate the effect of adaptation.*

Fig. 20.6 shows the result of applying the adaptive `'coll'` toolbox with varying discretization order to the continuation of periodic orbits of the dynamical system given by the vector
field in Eq. (18.5). The analysis relies on 50% more mesh points than in the case of a fixed
discretization order shown in Fig. 20.5.

20.3 Numerical comparisons

In this section, we explore the adaptive continuation algorithms described in this and the previous two chapters in the context of two continuation problems that necessitate the greater flexibility in discretization afforded by the adaptive paradigm.

20.3.1 A homoclinic orbit

Recall, from the examples in Sect. 8.2.2, the continuation-based approach to computing an approximation of a homoclinic orbit found at the terminal point of a branch of periodic orbits in the dynamical system given by the vector field in Eq. (8.24) and repeated here for convenience:

$$f(y, p) = \begin{pmatrix} p_1 y_1 + y_2 + p_2 y_1^2 \\ -y_1 + p_1 y_2 + y_2 y_3 \\ (p_1^2 - 1) y_2 - y_1 - y_3 + y_1^2 \end{pmatrix}. \tag{20.44}$$

We consider, in succession, the application to this task of the nonadaptive `coll` toolbox from Chap. 18 and each of the moving-mesh algorithms developed in this chapter.

In Sect. 8.2.2, the construction illustrated in Figs. 8.4–8.5 was performed using the `coll` toolbox with a default discretization order of $N = 10$ and $m = 4$. If, instead, we use the collocation method with error estimation from Sect. 18.1.2 with the default tolerance of 10^{-4}, we find it possible to accommodate the desired error tolerance throughout the continuation run starting at the Hopf bifurcation point with $N = 50$ mesh intervals. The

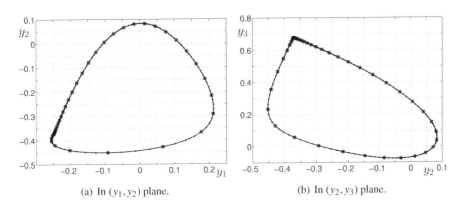

(a) In (y_1, y_2) plane.   (b) In (y_2, y_3) plane.

Figure 20.7. *State-space representations of near-homoclinic approximate periodic orbits of the dynamical system given by the vector field in Eq. (20.44), obtained using continuation with the nonadaptive `coll` toolbox from Sect. 18.1.2 with `NTST` equal to 50 and `TOL` equal to 10^{-4}. Panels (a) and (b) show the terminal low-period solution (gray circles) with 50 mesh intervals found using continuation with a 1-dimensional atlas algorithm starting at a Hopf bifurcation point, as well as the corrected high-period solution (black curve and end points of mesh intervals denoted by $\times$'s) found using continuation with a 0-dimensional atlas algorithm applied to a reconstructed periodic orbit with 1,250 mesh intervals. The estimated error is 4.2532×10^{-5} for the low-period orbit and 4.2531×10^{-5} for the high-period orbit. These results are compared in Figs. 20.9 and 20.11 with computations using moving-mesh adaptation.*

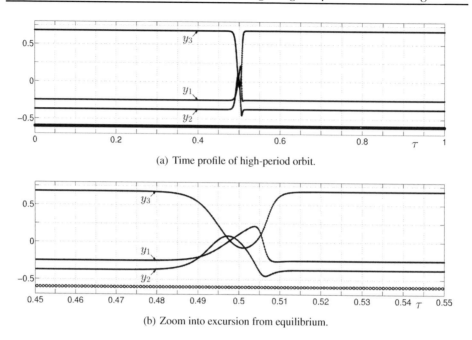

(a) Time profile of high-period orbit.

(b) Zoom into excursion from equilibrium.

Figure 20.8. *Time profiles of the high-period orbit constructed in Fig. 20.7 on a uniform mesh shown over the full period* (a) *and in a zoom into the excursion from the equilibrium* (b). *The distribution of mesh points is indicated with circles at the bottom. In panel* (a), *the mesh is so dense that it is not possible to see individual mesh intervals.*

terminal periodic orbit is represented by the gray circles in Fig. 20.7 corresponding to end points of individual mesh intervals.

Following the methodology of Example 8.4, we proceed to construct a periodic initial solution guess with both the period T and the mesh size N multiplied by a factor of 25. The high-period periodic orbit obtained by applying a 0-dimensional atlas algorithm to correct this initial guess is represented by the black curve in Fig. 20.7. We observe again that the low-period terminal orbit obtained after the initial continuation run and the corrected high-period orbit are virtually identical in state space, including the distribution of mesh points along the excursion from the equilibrium point. The time profile of the high-period orbit and the corresponding uniform mesh are illustrated in Fig. 20.8.

We next repeat the construction illustrated in Fig. 20.7 using the adaptive collocation method with a moving mesh of fixed discretization order from Sect. 20.2.1, with the default tolerance of 10^{-4}. In this case, 20 mesh intervals suffice in order to accommodate the desired error tolerance throughout the continuation run that starts at the Hopf bifurcation point. We proceed to construct a periodic orbit with the period T multiplied by a factor of 5,000, while increasing the 'NTST' toolbox setting from 20 to only 120. As before, the additional mesh segments are inserted at the passage closest to the equilibrium. Due to the nonuniformity of this construction, we cannot expect that the new time mesh will result in an initial solution guess with a small discretization error. Instead, we apply a 0-dimensional atlas algorithm that uses the remeshing map in between every successful application of the nonlinear corrector, for a total of 100 times before returning a final solution point.

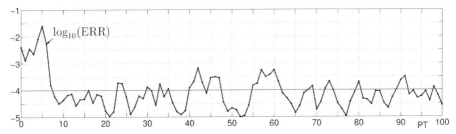

(a) The logarithm of the estimated discretization error during 100 iterations of the remesh-correct cycle of a 0-dimensional atlas algorithm. A mesh with the desired tolerance of 10^{-4} (gray horizontal line) is arrived at within 10 iterations.

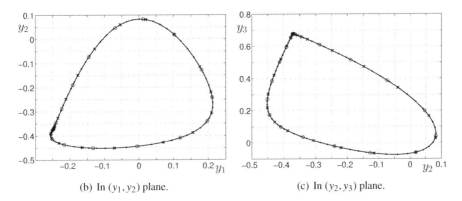

(b) In (y_1, y_2) plane.   (c) In (y_2, y_3) plane.

Figure 20.9. *Mesh adaptation and an approximate near-homoclinic periodic orbit of the dynamical system given by the vector field in Eq. (20.44), obtained using the collocation method with a moving mesh of fixed order from Sect. 20.2.1 with* 'TOL' *equal to* 10^{-4}. *Panel (a) shows variations in the discretization error during 100 iterations of a 0-dimensional remesh-correct cycle applied to the reconstructed initial solution guess obtained by a 5,000-fold increase in the period T and a 6-fold increase in discretization order N. In panels (b) and (c), we compare the corrected high-period orbit (black) with the low-period orbit (gray circles) obtained using continuation from the Hopf bifurcation point. The number of mesh points along the excursion from the equilibrium is nearly identical for both orbits, although the mesh points move somewhat. The estimated error is 2.2619×10^{-5} for the low-period orbit and 4.2620×10^{-5} for the high-period orbit. The adapted mesh and time profile of the high-period orbit are shown in Fig. 20.10. Figs. 20.11–20.13 repeat the analysis with a moving mesh with variable discretization order.*

The dynamics of this process is illustrated in Fig. 20.9(a), where we show the estimated discretization error plotted against the number of iterations of the remesh-correct cycle of the 0-dimensional atlas algorithm. We observe that the algorithm settles onto an acceptable mesh after a short initial phase with large approximation errors. The resultant near-homoclinic orbit, obtained after 10 iterations, is shown in state space in Figs. 20.9(b) and 20.9(c), and as a function of time in Fig. 20.10. In Fig. 20.10(a), we observe the effect of the convex combination between uniform and fully error-adaptive meshes. Along the part with near-constant dynamics, the mesh is nearly equidistributed in time. In contrast,

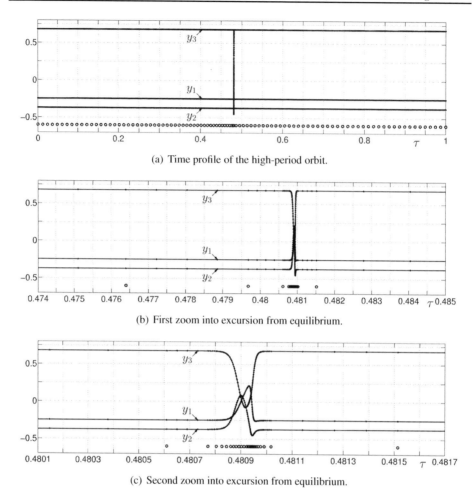

(a) Time profile of the high-period orbit.

(b) First zoom into excursion from equilibrium.

(c) Second zoom into excursion from equilibrium.

Figure 20.10. *Time profiles of the high-period orbit constructed in Fig. 20.9 on a moving mesh shown over the full period* (a) *and in two subsequent zooms into the excursion from the equilibrium* (b) *and* (c). *The distribution of mesh points is indicated with circles at the bottom.*

close to the part that corresponds to the excursion from the equilibrium, the distribution of mesh points is refined due to the localized nature of the discretization error. The ratio between the largest and the smallest mesh interval across the entire orbit is $\max\{\kappa\}/\min\{\kappa\} = 4{,}694$. For comparison, to obtain an orbit of the same period with the same accuracy on a uniform mesh would require $1{,}250 \times 5{,}000/25 = 250{,}000$ mesh intervals instead of the 120 used here.

As a final comparison, we apply the adaptive collocation method with a moving mesh of variable discretization order to the construction of the near-homoclinic high-period periodic orbit of the dynamical system given by the vector field in Eq. (20.44). The graphs of the estimated discretization error and the number of mesh intervals in Figs. 20.11(a) and 20.11(b) show an initial rapid decay in the error and growth in the discretization order,

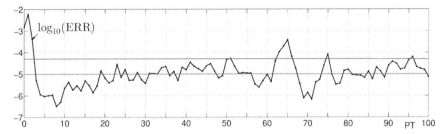

(a) The logarithm of the estimated discretization error during 100 iterations of the remesh-correct cycle of a 0-dimensional atlas algorithm.

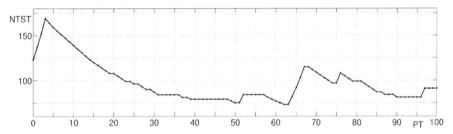

(b) The number of mesh intervals plotted against the number of iterations of the remesh-correct cycle of a 0-dimensional atlas algorithm.

Figure 20.11. *Mesh adaptation for an approximate near-homoclinic periodic orbit of the dynamical system given by the vector field in Eq. (20.44), obtained using the collocation method with a moving mesh of variable discretization order from Sect. 20.2.2 with* 'TOL' *equal to the default tolerance of* 10^{-4}. *Panels* (a) *and* (b) *show variations in the discretization error and discretization order during* 100 *iterations of a 0-dimensional remesh-correct cycle applied to the reconstructed initial solution guess obtained by a 5,000-fold increase in the period T and a 6-fold increase in discretization order N. Again, we observe that the iteration settles onto an acceptable mesh after a somewhat longer transient phase. Although the approximation error drops much faster than in Fig.* 20.9(a), *the dynamics is dominated by the slow reduction of the number of mesh intervals. After* 40 *steps, the solution settles onto a mesh of size N = 79. A comparison of the resulting solutions is shown in Fig.* 20.12. *Fig.* 20.13 *shows the adapted mesh and time profile of the high-period orbit.*

arriving below the desired tolerance of 10^{-4} within four iterations. This is followed by a slow growth of the discretization order as the algorithm reduces adaptively the number of mesh intervals. Throughout this latter phase, the error remains close to the adaptation window represented by the horizontal gray lines. A comparison between the low-period orbit used to construct the initial solution guess for the 0-dimensional atlas algorithm and the corrected high-period orbit, obtained after 40 iterations, is shown in Fig. 20.12. From the time profiles of the high-period orbit in Fig. 20.13, we note the use of a coarser mesh along the part with near-constant dynamics and a doubling of the ratio between the largest and smallest mesh intervals relative to the results obtained with the moving-mesh algorithm with fixed discretization order.

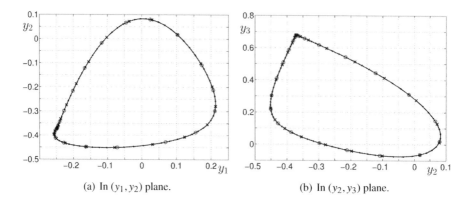

(a) In (y_1, y_2) plane.   (b) In (y_2, y_3) plane.

Figure 20.12. *Comparison of the high-period orbit (black) obtained in Fig. 20.11 with the low-period orbit (gray circles) used for constructing the initial solution guess in two different projections* (a) *and* (b). *Again, the number of mesh points along the excursion from the equilibrium is nearly identical for both orbits, although the mesh points move somewhat. The estimated error is* 1.2506×10^{-5} *for the low-period orbit and* 1.5747×10^{-5} *for the high-period orbit. Fig. 20.13 shows the adapted mesh and time profile of the high-period orbit.*

20.3.2 Canards

As a final example of the use of adaptive discretization, we consider the continuation of the *canard family* of periodic orbits in the *Van der Pol* dynamical system given by the vector field

$$f(y, p) := \left(\begin{array}{c} p_1(p_2 - y_2) \\ y_1 + y_2 - \frac{1}{3}y_2^3 \end{array} \right), \qquad (20.45)$$

where $p = (\varepsilon, a)$. The equilibrium at $y = (-a + a^3/3, a)$ undergoes a Hopf bifurcation at $a = -1$ corresponding to the birth of a family of periodic orbits with angular frequency limiting on $\sqrt{\varepsilon}$ as $a \to -1$. As seen in the bifurcation diagram in Fig. 20.14(a), this family of periodic orbits exhibits a dramatic increase of amplitude over an exceedingly small variation in a, known as a *canard explosion*; orbit 6 in Fig. 20.14(g) is the godfather of the term *canard orbit* as it resembles the head of a duck (*canard* in French).

Fig. 20.15 shows the results of applying the adaptive 'dft' toolbox from Chap. 19 and the nonadaptive 'coll' toolbox from Chap. 18, respectively, to the continuation of the family of canard orbits. In both cases, continuation fails to compute the entire family. The occurrence of oscillations in the solution of the spectral method is a typical artifact of Fourier methods at the limit of convergence. It could be prevented here by using a longer tail for estimating the error, which would trigger the terminal event associated with the estimated error earlier.

In contrast to the adaptive 'dft' toolbox and the nonadaptive 'coll' toolbox, the collocation methods on (co)moving meshes all succeed in computing the complete family with mesh sizes smaller than or equal to the mesh size used in Fig. 20.15. The resulting meshes are illustrated with the canard orbit from Fig. 20.14(g) in Figs. 20.16–20.18. In the

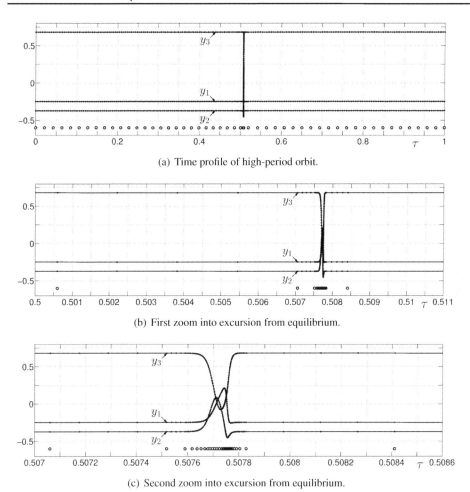

(a) Time profile of high-period orbit.

(b) First zoom into excursion from equilibrium.

(c) Second zoom into excursion from equilibrium.

Figure 20.13. *Time profiles of the high-period orbit constructed in Fig. 20.11 on a moving mesh with variable discretization order shown over the full period* (a) *and in two subsequent zooms into the excursion from the equilibrium* (b) *and* (c). *The distribution of mesh points is indicated with circles at the bottom. A comparison between panel* (a) *and Fig. 20.10(a) indicates that the difference in discretization order of 41 mesh intervals is due mainly to a coarser mesh along the part with near-constant dynamics. The ratio between the largest and smallest mesh intervals,* $\max\{\kappa\}/\min\{\kappa\} = 9245$, *is here almost twice as large as for the moving mesh with fixed discretization order used in Fig. 20.10.*

case of the comoving-mesh algorithm, we replace the assignments

```
data.h = cseg.prcond.h;
```

and

```
data.h = h;
```

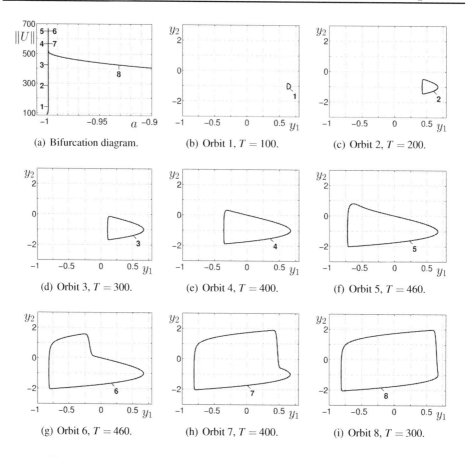

Figure 20.14. *A canard explosion in the Van der Pol dynamical system given by the vector field in Eq. (20.45). Here, a dramatic increase of amplitude (represented by the norm* $\|U\| = \|(u, \mu_{\mathbb{J}})\|$*) along a family of periodic orbits over an exceedingly small variation in a system parameter shows up as a virtually vertical branch in a bifurcation diagram* (a). *Panels* (b) *to* (h) *show members of the canard family. The orbits along the family were selected according to their period, indicated in the caption. We observe evidence for a fold point with respect to period between orbits 5 and 6.*

in `coll_update` and `coll_update_h`, respectively, with the assignments

```
data.h = data.coll.hfac*cseg.prcond.h;
```

and

```
data.h = data.coll.hfac*h;
```

The toolbox setting `'hfac'` here provides an optional scaling of the step size used by the comoving mesh relative to the step size employed by the atlas algorithm. We assign a default value of 1 in the modified `coll_get_settings` function.

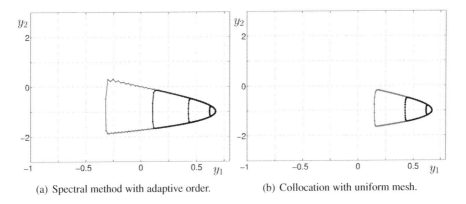

(a) Spectral method with adaptive order.   (b) Collocation with uniform mesh.

Figure 20.15. *Selected approximate periodic orbits obtained during continuation along the canard family of the dynamical system given by the vector field in Eq. (20.45). Both the adaptive spectral method* (a) *from Sect. 19.3 with* 'NMAX' *equal to* 100 *and the nonadaptive collocation method* (b) *from Sect. 18.1.2 with* 'NTST' *and* 'NCOL' *equal to* 100 *and* 4, *respectively, fail early on the canard family, given the desired error tolerance of* 10^{-2}.

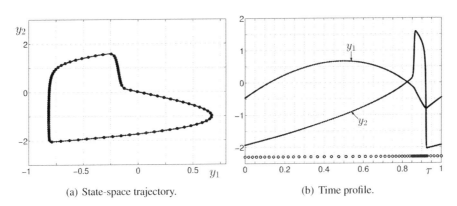

(a) State-space trajectory.   (b) Time profile.

Figure 20.16. *Discretization of the canard orbit in Fig. 20.14(g) obtained using continuation with the comoving-mesh method from Sect. 18.3.1 on a mesh with* 'NTST' *equal to* 100, 'NCOL' *equal to* 4, 'TOL' *equal to* 10^{-2}, *and* 'hfac' *equal to* 5. *The fat dots in* (a) *and the open circles in* (b) *mark the end points of mesh intervals. The estimated discretization error of this solution is* 3.5326×10^{-3}.

A comparison of the condition numbers of each of the three (co)moving mesh strategies during continuation is shown in Fig. 20.19. This graph illustrates the reason why the computation of canard orbits is a difficult problem, not only because of the occurrence of localized errors along an orbit. In particular, for the comoving-mesh method, we observe a window around $PT_{rel} = 0.26$ for which the Jacobian of the restricted continuation problem is singular within the accuracy of double-precision arithmetic. A similar, but much smaller, peak occurs for the moving-mesh method with fixed order. The moving-mesh method with varying order exhibits the best overall performance.

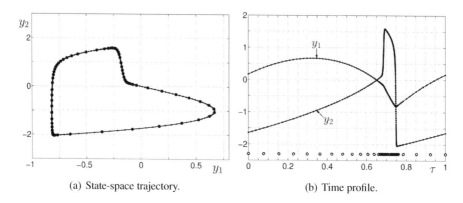

(a) State-space trajectory.   (b) Time profile.

Figure 20.17. *Discretization of the canard orbit in Fig. 20.14(g) obtained using continuation with the moving-mesh method from Sect. 20.2.1 on a mesh with 'NTST' equal to 70, 'NCOL' equal to 4, and 'TOL' equal to* 10^{-4}. *The fat dots in (a) and the open circles in (b) mark the end points of mesh intervals. The estimated error of this solution is* 5.1787×10^{-6}.

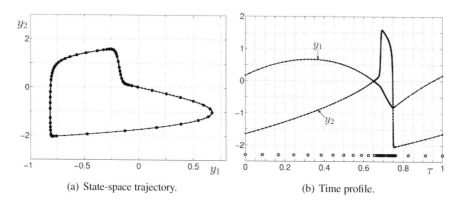

(a) State-space trajectory.   (b) Time profile.

Figure 20.18. *Discretization of the canard orbit in Fig. 20.14(g) obtained using continuation with the moving-mesh method with adaptive discretization order from Sect. 20.2.2 on a mesh with 'NTST' equal to 66, 'NCOL' equal to 4, and 'TOL' equal to* 10^{-4}. *The fat dots in (a) and the open circles in (b) mark the end points of mesh intervals. The estimated error of this solution is* 1.1882×10^{-5}.

20.4   Conclusions

As demonstrated by the numerical results in the previous sections, even relatively simple error estimation and remeshing strategies may successfully navigate highly challenging continuation problems. The approach described in this chapter is thus to be seen only as a first step to more sophisticated implementations that account for problem-specific properties. Even within the family of adaptive discretization strategies developed in this part of the book, there remains great opportunity to explore the sensitivity to various optional

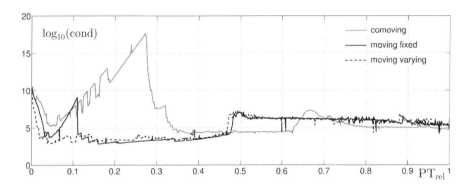

Figure 20.19. *The logarithm of the condition number of the Jacobian of the restricted continuation problem plotted against the point number normalized by the maximum point number during continuation of the canard family of the dynamical system given by the vector field in Eq. (20.45) using the (co)moving-mesh adaptive discretization strategies in Sects. 18.3.1, 20.2.1, and 20.2.2, respectively.*

toolbox settings, including some that have been hard-coded in the versions of the toolboxes shown here.

The power of the task-embedding paradigm formulated in this book is further illustrated by the possibility of segment-specific adaptation strategies for multisegment boundary-value problems. The reader is encouraged to revisit the examples from earlier chapters with the modified adaptive versions of the `'coll'` toolbox. The changes required of the `'bvp'`, `'msbvp'`, `'hspo'`, and `'var_coll'` toolboxes offer further useful practice in the programming paradigm.

Exercises

20.1. Show that the partition

$$-1 = t_1 \leq \cdots \leq t_N = 1$$

is equidistributed with respect to the function

$$t \mapsto \frac{\tanh pt}{\tanh p}$$

provided that

$$t_j = \frac{1}{p} \operatorname{arctanh}\left(\left(-1 + 2\frac{j-1}{N-1}\right)\tanh p\right).$$

20.2. Consider the map Υ in Eq. (20.6) and show that

$$t_j = \left(\Upsilon\left(\{t_k\}_{k=1}^N\right)\right)_j$$

if and only if the partition

$$-1 = t_1 \leq \cdots \leq t_N = 1$$

is equidistributed with respect to the function f, independently of the interpolation scheme.

20.3. Explain why the sequence

$$\left\{ \left(\Upsilon^{(i)} \left(\{t_j\}_{j=1}^{N} \right) \right)_k \right\}_{i=1}^{\infty}$$

in Example 20.2 is expected to converge as $i \to \infty$ to t_k^*.

20.4. Show that the unique fixed point of the map Υ given by Eq. (20.6), where υ is obtained from Eq. (20.11), must satisfy the equality

$$g_j = \left(-1 + 2 \frac{j-1}{N-1} \right) (1+s)$$

for $j = 1, \ldots, N$. Use numerical continuation to explore this family of fixed points under variations in s.

20.5. Does the conclusion regarding the asymptotic stability of the equidistributed partition under iterations of Υ in Example 20.2 generalize to nonconvex functions f? Justify your answer either with an analytical (counter)example or a numerical counterexample.

20.6. For some chosen function f, investigate numerically the properties of any of the maps Υ introduced in Sect. 20.1.1 through the successive iterates $\Upsilon^{(i)}(\{t_j\}_{j=1}^{N})$ for some random partition of the interval $[-1, 1]$ of the form in Eq. (20.1).

20.7. Consider the zero function

$$\Phi_f : u \mapsto x - \begin{pmatrix} f(t_1, p) \\ \vdots \\ f(t_N, p) \end{pmatrix}, u = (x, p) \in \mathbb{R}^N \times \mathbb{R}^q,$$

for some order-N partition

$$-1 = t_1 \leq \cdots \leq t_N = 1.$$

Show that every solution to the corresponding zero problem is regular and that the corresponding tangent space is spanned by the columns of

$$\begin{pmatrix} \partial_p f(t_1, p^*) \\ \vdots \\ \partial_p f(t_N, p^*) \\ I_q \end{pmatrix}.$$

20.8. Explain the construction of an initial solution guess, and a corresponding approximate description for the new tangent space, for the modified zero problem obtained from the updated partition in Sect. 20.1.2. Identify those elements of the `remesh` functions in the remainder of the chapter that implement this construction.

20.9. Verify that the nonzero entries of the `xtr` output argument of the `remesh` function in Sect. 20.1.3 provide a translation between the integer indices, before and after remeshing, of elements of the vector $u_{\mathbb{K}}$ that are invariant under the remeshing operation.

20.10. Provide a complete encoding of a 1-dimensional atlas algorithm obtained by appending handling of computational domain boundaries (as in Sect. 14.1), event handling (as in Sect. 16.1), and support for adaptation (as in Sect. 20.1.4) to the basic algorithm described in Sect. 12.2 and use this to verify the results in Example 20.5.

20.11. In the construction of the initial partition stored in `data.t` and in the `remesh` function in Sect. 20.1.3, we eliminate possible consequences of round-off errors on the values of the end points. Use the atlas algorithm from the previous exercise to provide a numerical comparison between the results in Examples 20.4 and 20.5 and those obtained by removing the explicit assignment of the -1 and 1 to t_1 and t_N.

20.12. Identify those elements of the `coll_remesh` function that implement the reparameterization described in Example 20.6.

20.13. Verify that the nonzero entries of the `xtr` output argument of the `remesh` function in Sect. 20.2.1 provide a translation between the integer indices, before and after remeshing, of elements of the vector $u_\mathbb{K}$ that are invariant under the remeshing operation.

20.14. Reproduce the results in Fig. 20.5 and investigate their sensitivity to the value of the `'SAD'` toolbox setting.

20.15. Comment on the choice of function dF in the implementation of the adaptive `'coll'` toolbox in Sect. 20.2.1 in terms of the mth root of the norm $\|c_m\|$ of the vector of coefficients in front of the σ^m term in each interpolating polynomial. Review the discussion about error equidistribution in [Ascher, U.M., Mattheij, R.M.M., and Russell, R.D., *Numerical Solution of Boundary Value Problems for Ordinary Differential Equations*, SIAM, Philadelphia, 1995] and investigate possible alternative choices.

20.16. Verify that the nonzero entries of the `xtr` output argument of the `remesh` function in Sect. 20.2.2 provide a translation between the integer indices, before and after remeshing, of elements of the vector $u_\mathbb{K}$ that are invariant under the remeshing operation.

20.17. Reproduce the results in Fig. 20.6 and investigate their sensitivity to the values of the `'TOLDEC'` and `'TOLINC'` toolbox settings.

20.18. Follow the methodology of Example 8.4 and reproduce the results of Figs. 20.7 and 20.8. What tolerance is achieved with the discretization order used in Chap. 8?

20.19. What are the benefits of the moving-mesh algorithms to the approximation of the homoclinic orbit of the dynamical system given by the vector field in Eq. (20.44) relative to the uniform mesh algorithm? How large a scaling factor can the latter support during the construction of an initial solution guess to the high-period near-homoclinic periodic orbit? What is the source of any limitations?

20.20. Explore the default 0-dimensional atlas algorithm used in this book. Identify those elements of its encoding that support the remesh-correct cycle relied on in generating the results in Figs. 20.9–20.13. What is the relationship between this implementation and the definition of the Υ map in Sect. 20.1.1?

20.21. Use an appropriately modified implementation of the `var_coll` toolbox to investigate the stability of the canard orbits explored in Sect. 20.3.2.

20.22. Can you continue the entire canard family in Sect. 20.3.2 using the adaptive `dft` toolbox with a larger value of `NMAX` than in Fig. 20.15? How about with the nonadaptive `coll` toolbox with a larger value of `NTST` than in Fig. 20.15?

20.23. The graph of the condition number of the Jacobian of the restricted continuation problem in Fig. 20.19 shows an effectively singular Jacobian for the comoving-mesh method within the available accuracy. What might be the source of this behavior? What are the expected consequences to the numerical predictions? Use comparisons with the moving-mesh algorithms to explore the accuracy of the predicted canard orbit and corresponding parameter values.

20.24. What range of values of the toolbox setting `hfac` in the modified comoving-mesh algorithm support the computation of the entire canard family in Sect. 20.3.2?

20.25. Repeat the analysis of the canard family from Sect. 20.3.2 using the adaptive `coll` toolbox from Exercise 18.22.

Part VI

Epilogue

Chapter 21

Toolbox Projects

It is the authors' experience that a book of this nature, albeit hopefully useful in providing implementable solutions to problems of a common nature, may be particularly valuable inasmuch as it inspires the next generation of applied mathematicians to pursue computational problems of the type considered here. The core functionality of COCO provides a safe context for developing skills of problem formulation and algorithm development. Only time and imagination limit the applicability of this framework to a vast range of continuation tasks available in the existing literature. To stimulate such a pursuit, we leave the reader with a sketch of possible projects for further study, including a list of exercises whose use is encouraged as project assignments, whether for independent study or as part of a formal course.

21.1   Calculus of variations

Consider the problem of finding extremal curves $y = f(t)$ for the integral functional

$$J(f) = \int_0^T L\big(f(t), f'(t), p\big) dt \tag{21.1}$$

that satisfy the boundary conditions $f(0) = 1$ and $f(T) = Y$ for some problem parameters p and Y. This problem clearly includes the catenary functional discussed in Chap. 1. We indicate below a possible discretization scheme corresponding to the third numerical method treated in Sect. 1.3.

Following the methodology of previous chapters, introduce the substitution $t = T\tau$ such that

$$J(f) = T \int_0^1 \tilde{L}\big(\tilde{f}(\tau), \tilde{f}'(\tau), p\big) d\tau, \tag{21.2}$$

where $\tilde{f}(\tau) = f(T\tau)$, $\tilde{f}'(\tau) = Tf'(T\tau)$, and

$$\tilde{L}\big(\tilde{f}, \tilde{f}', p\big) = L\left(\tilde{f}, \frac{\tilde{f}'}{T}, p\right). \tag{21.3}$$

Let $\tau_j = (j-1)/N$ for $j = 1, \ldots, N+1$ such that

$$J(f) = T \sum_{j=1}^{N} \int_{\tau_j}^{\tau_{j+1}} \tilde{L}\left(\tilde{f}(\tau), \tilde{f}'(\tau), p\right) d\tau. \tag{21.4}$$

Then, for $\tau \in [\tau_j, \tau_{j+1}]$, the coordinate transformation

$$\tau = \tau^{(j)}(\sigma) := \tau_j + \frac{1+\sigma}{2}\left(\tau_{j+1} - \tau_j\right) \tag{21.5}$$

implies that

$$J(f) = \frac{T}{2N} \sum_{j=1}^{N} \int_{-1}^{1} \hat{L}^{(j)}\left(\hat{f}^{(j)}(\sigma), \hat{f}^{(j)\prime}(\sigma), p\right) d\sigma, \tag{21.6}$$

where $\hat{f}^{(j)}(\sigma) = \tilde{f}\left(\tau^{(j)}(\sigma)\right)$, $\hat{f}^{(j)\prime}(\sigma) = \frac{1}{2N}\tilde{f}'\left(\tau^{(j)}(\sigma)\right)$, and

$$\hat{L}^{(j)}\left(\hat{f}^{(j)}, \hat{f}^{(j)\prime}, p\right) = \tilde{L}\left(\hat{f}^{(j)}, 2N\hat{f}^{(j)\prime}, p\right) = L\left(\hat{f}^{(j)}, \frac{2N\hat{f}^{(j)\prime}}{T}, p\right). \tag{21.7}$$

Finally, approximate the integral by a weighted sum of the values of the integrand at m suitably chosen collocation nodes on each of the N intervals:

$$J(f) \approx \hat{J} := \frac{T}{2N} \sum_{j=1}^{N} \sum_{k=1}^{m} w_k \hat{L}^{(j)}\left(\hat{f}^{(j)}(\sigma_k), \hat{f}^{(j)\prime}(\sigma_k), p\right). \tag{21.8}$$

Denote by $\hat{L}$ the column matrix whose $m(j-1)+k$th component equals

$$\hat{L}^{(j)}\left(\hat{f}^{(j)}(\sigma_k), \hat{f}^{(j)\prime}(\sigma_k), p\right) \tag{21.9}$$

and consider the diagonal matrix Ω whose $m(j-1)+k$th diagonal entry equals w_k. Then,

$$\hat{J} = \frac{T}{2N} 1_{1,Nm} \cdot \Omega \cdot \hat{L}. \tag{21.10}$$

We now introduce a discretization of the unknown function in terms of a set of base point values υ_{bp} as in Chap. 6 such that $\upsilon_{bp,1}$ and $\upsilon_{bp,N(m+1)}$ may be determined from boundary conditions of the form $f_{bc}\left(\upsilon_{bp,1}, \upsilon_{bp,N(m+1)}\right) = 0$ for some function f_{bc} and such that the vector $\hat{L}$ may be obtained by a vectorized application of $\hat{L}^{(j)}$ to the column vectors $W \cdot \upsilon_{bp}$ and $W' \cdot \upsilon_{bp}$. Given T and using the method of Lagrange multipliers, the zero problem is then given by the derivatives of $\hat{J}$ with respect to the interior base point values, bearing in mind that the continuity conditions $Q \cdot \upsilon_{bp} = 0$ imply that derivatives with respect to coincident points should be replaced by a single term equal to the sum of the two derivatives.

As an example, it is possible to show that

$$\left(\partial_{\upsilon_{bp}} \hat{J}\right)^T = \frac{T}{2N}\left(W^T \cdot \Omega \cdot \frac{\partial \hat{L}}{\partial \hat{f}} + W'^T \cdot \Omega \cdot \frac{\partial \hat{L}}{\partial \hat{f}'}\right), \tag{21.11}$$

where

$$\frac{\partial \hat{L}}{\partial \hat{f}}\left(\hat{f}^{(j)}, \hat{f}^{(j)\prime}\right) = \frac{\partial L}{\partial f}\left(\hat{f}^{(j)}, \frac{2N\hat{f}^{(j)\prime}}{T}\right) \tag{21.12}$$

and

$$\frac{\partial \hat{L}}{\partial \hat{f}'}\left(\hat{f}^{(j)}, \hat{f}^{(j)\prime}\right) = \frac{2N}{T}\frac{\partial L}{\partial f'}\left(\hat{f}^{(j)}, \frac{2N\hat{f}^{(j)\prime}}{T}\right). \tag{21.13}$$

We leave the implementation of a corresponding COCO-compatible toolbox as an exercise.

21.2 Nonlinear boundary conditions

As described in Chap. 10, use of the $Y(0) = I_n$ boundary condition makes the 'var_coll' toolbox susceptible to unbounded growth of the corresponding solution to the variational equation along the solution manifold. To avoid the singularity associated with all such linear conditions, we are led to consider a set of n^2 nonlinear conditions on the entries of $Y(t)$. As an example, consider the complex-valued linear scalar differential equation

$$\frac{dY}{dt} = \lambda Y, \lambda \in \mathbb{C}, \tag{21.14}$$

the nonlinear condition

$$\int_0^1 \left(1 + Y(t)^*\right) \cdot Y(t)dt = 2, \tag{21.15}$$

and the corresponding linearization

$$Z(t) \mapsto \int_0^1 \left(1 + 2Y(t)^*\right) \cdot Z(t)dt. \tag{21.16}$$

Given solutions to Eq. (21.14) of the form

$$Y(t) = e^{\lambda t}c, c \in \mathbb{C}, \tag{21.17}$$

we find functions $Z(t) = e^{\lambda t}$ in the kernel of the corresponding linearization provided that

$$c = -\frac{\left(1 - e^{\bar{\lambda}}\right)\Re[\lambda]}{\left(1 - e^{2\Re[\lambda]}\right)\bar{\lambda}}. \tag{21.18}$$

Substitution into the left-hand side of the nonlinear boundary condition then yields

$$\frac{a(\cos b - \cosh a)}{2\sinh(a)(a^2 + b^2)}, \tag{21.19}$$

where $a = \Re[\lambda]$ and $b = \Im[\lambda]$. As this quantity is everywhere negative, we conclude that the kernel of the linearization is empty for all solutions to the nonlinear conditions.

To obtain an approximate initial solution to the nonlinear condition in Eq. (21.15), we may rely on the following iterative scheme. Let $Y_1(t) = 1$ and solve successively the scaled variational equation

$$\frac{dY_{i+1}}{dt} = \beta_{i+1}\lambda Y_{i+1} \tag{21.20}$$

together with the linear condition

$$\int_0^1 \left(1 + Y_i(t)^*\right) \cdot Y_{i+1}(t)\, dt = 2 \qquad (21.21)$$

for $i \geq 1$ and some finite sequence of values of β, where $\beta_1 = 0$, terminating at 1.

Alternatively, we may apply a nonlinear corrector directly to the original variational equation and the nonlinear condition with some suitable initial solution guess. During parameter continuation, either approach may now be used in order to seek to maintain bounds on the entries of the solution to the variational equation along the solution manifold. We leave the implementation of these ideas in the `'var_coll'` toolbox and the exploration of the numerical properties of each scheme, especially for $n \geq 3$, as a possible research project.

21.3   Connecting orbits

Consider the continuation of connecting orbits between equilibria in smooth dynamical systems. Two examples of this were discussed in Chap. 7. In Sect. 7.3.3 we coupled the `'coll'` and `'alg'` toolboxes in order to allow for the simultaneous continuation of a pair of equilibria, their associated eigenspaces, and the connecting orbit, as would typically be necessary in instances where these were not known a priori.

The two examples considered in Chap. 7 were both planar dynamical systems. In both cases, a connecting orbit was obtained by considering two trajectory segments, glued through appropriate boundary conditions to the appropriate eigenspaces of the two equilibria. The boundary conditions and the system parameters were then varied as needed to ensure that the separation between the intersections of these trajectories with a transversal section was reduced to and held at zero.

Consider, more generally, a case of two equilibria y_* and y_{**} in an n-dimensional state space with stable and unstable manifolds of dimension $\dim \mathcal{W}^u_* = n_*$, $\dim \mathcal{W}^s_* = n - n_*$, $\dim \mathcal{W}^s_{**} = n_{**}$, and $\dim \mathcal{W}^u_{**} = n - n_{**}$. Clearly, in the homoclinic case, $y_* = y_{**}$ implies that $n_* + n_{**} = n$. By the uniqueness of solutions to initial-value problems for autonomous differential equations, it follows that any intersection of $\mathcal{W}^u_*$ with $\mathcal{W}^s_{**}$ must be at least 1-dimensional. Generically, it then holds that the intersection is persistent under parameter variations when $n_* + n_{**} \geq n + 1$ and locally isolated when $n_* + n_{**} = n + 1$. In contrast, if $n_* + n_{**} < n + 1$, as is the case for a homoclinic orbit, a manifold intersection would generically disappear under variations in any single parameter. By releasing $n + 1 - n_* - n_{**}$ additional problem parameters, we may guarantee a persistent connecting orbit in an augmented state space, in which the released parameters are now treated as constant state variables.

Now suppose that Σ is a transversal section at some point on the intersection between $\mathcal{W}^u_*$ and $\mathcal{W}^s_{**}$. As illustrated in Sect. 10.2.2 in the context of a connecting orbit between an equilibrium and a periodic orbit, *Lin's method* implies that unique values for any released parameters may be obtained by imposing a sufficient number of constraints on the separation between the intersections of the manifolds with Σ. In the example in Sect. 7.3.2, we had $n = 2$ and $n_* = n_{**} = 1$ and it became necessary to release one system parameter in order to ensure a persistent heteroclinic orbit. In particular, by letting p_2 vary, it was possible to ensure that the gap $y_2^{(1)}(T_1) - y_2^{(2)}(T_2)$ was reduced to zero. In contrast, in the example in Sect. 7.3.3, we had $n = n_* = 2$ and $n_{**} = 1$, ensuring a persistent heteroclinic orbit without the need to release additional system parameters. We leave the implementation of general toolboxes for connecting orbits between equilibria and for connecting orbits between equilibria and periodic orbits as an exercise.

21.4 Conclusions

Finally, we close with an opening: do not let yourself be held back by the scope of the task, but take one step, however small, at a time. We have tried to encourage this incremental approach throughout this book. Some obvious next steps that constitute active areas of exploration for us are to incorporate adaptation in the expanding-boundary atlas algorithms considered in Part III, to allow for asynchronous collocation schemes in stiff dynamical systems, to explore parallel implementations of the multidimensional atlas algorithms, and to provide support in the COCO core classes for control-based continuation applied to equation-free problems. We hope you see other opportunities and that the steps you have taken to get through this text will be only the beginning of a beautiful journey!

Exercises

21.1. Show that
$$\left(\partial_y \hat{J}\right)^T = \frac{T}{2N}\left(W^T \cdot \Omega \cdot \frac{\partial \hat{L}}{\partial \hat{f}} + W'^T \cdot \Omega \cdot \frac{\partial \hat{L}}{\partial \hat{f}'}\right),$$
where $\hat{J}$ is given by Eq. (21.10).

21.2. Show that
$$\partial_T \hat{J} = \frac{1}{2N} 1_{1,Nm} \cdot \Omega \cdot \hat{L},$$
where $\hat{J}$ is given by Eq. (21.10).

21.3. Derive vectorized expressions for the Jacobians, with respect to the interval length T and the interior base point values, of the zero problem defined by the conditions for a local optimum of $\hat{J}$ in Eq. (21.10) given the continuity conditions $Q \cdot v_{bp} = 0$.

21.4. What is the dimensional deficit of the zero problem formulated in the previous exercise?

21.5. Write a COCO-compatible toolbox for continuing extremal curves $y = f(t)$ for the integral functional given in Eq. (21.1). Demonstrate the application of your toolbox to several classical problems from the calculus of variations, including the catenary problem considered in Chap. 1.

21.6. Verify the conclusions regarding the kernel of the linearization of the nonlinear condition in Eq. (21.15) in the case of an equation of the form in Eq. (21.14). What can you say about the more general case of a matrix equation of the form
$$\frac{dY}{dt} = A \cdot Y$$
for some matrix $A \in \mathbb{R}^{2 \times 2}$? What about for $A \in \mathbb{R}^{3 \times 3}$?

21.7. Write a COCO-compatible implementation of the `'var_coll'` toolbox that relies on an n-dimensional version of the nonlinear condition in Eq. (21.15) in lieu of the $Y(0) = I_n$ boundary condition. Compare this to an implementation that replaces the nonlinear condition with the linear condition

$$\int_0^1 \left(1 + \tilde{Y}(t)^*\right) \cdot Y(t)\,dt = 2I_n$$

for some suitably updated matrix $\tilde{Y}$. Comment on the differences in implementation and possible advantages and disadvantages.

21.8. Write a COCO-compatible toolbox for continuation of periodic solutions of delay differential equations of the form

$$\frac{dy}{dt} = f(y^t),$$

where $y : t \to \mathbb{R}^n$, $y^t \in \mathcal{C}\left([t-\tau,t],\mathbb{R}^n\right)$, and $f : \mathcal{C}\left([t-\tau,t],\mathbb{R}^n\right) \to \mathbb{R}^n$. Here, τ is the *maximal delay* of the system. Refer in your development to [Engelborghs, K., Luzyanna, T., and Roose, D., "Numerical bifurcation analysis of delay differential equations using DDE-BIFTOOL," *ACM Transactions on Mathematical Software*, 28(1), pp. 1–21, 2002] and other publications by these authors.

21.9. Write a COCO-compatible toolbox for continuation of periodic solutions of hybrid delay differential equations of the form discussed in [Barton, D., "Stability calculations for piecewise-smooth delay equations," *International Journal of Bifurcation and Chaos in Applied Sciences and Engineering*, 19(2), pp. 639–650, 2009].

21.10. Write a COCO-compatible toolbox for continuation of solutions to differential-algebraic boundary-value problems of the form discussed in [Ascher, U.M. and Spiteri, R.J., "Collocation software for boundary value differential-algebraic equations," *SIAM Journal on Scientific Computing*, 15(4), pp. 938–952, 1994].

21.11. Use COCO to implement continuation in the context of differential-algebraic boundary value problems arising in optimal control as described in [Fabien, B.C., "Numerical solution of constrained optimal control problems with parameters," *Applied Mathematics and Computation*, 80(1), pp. 43–62, 1996] and demonstrate your implementation using the examples therein.

21.12. Use COCO to implement the methods for continuation of connecting orbits between equilibria described in [Liu, L., Moore, G., and Russell, R.D., "Computation and continuation of homoclinic and heteroclinic orbits with arclength parameterization," *SIAM Journal on Scientific Computing*, 18(1), pp. 69–93, 1997] and demonstrate their use on the examples considered therein.

21.13. Use COCO to implement the method described in [Osinga, H.M. and Moehlis, J., "Continuation-based computation of global isochrons," *SIAM Journal on Applied Dynamical Systems*, 9(4), pp. 1201–1228, 2010] for the computation of isochrones and demonstrate your implementation on the examples considered therein.

21.14. Use COCO to perform continuation of the unstable manifold of the Lorentz attractor through any of the formulations described in [Krauskopf, B. Osinga, H.M., Doedel, E.J., Henderson, M.E., Guckenheimer, J., Vladimirsky, A., Dellnitz, M., and Junge, O., "A survey of methods for computing (un)stable manifolds of vector fields," *International Journal of Bifurcation and Chaos in Applied Sciences and Engineering*, 15(3), pp. 763–791, 2005].

21.15. Use COCO to implement algorithms for the continuation of symmetric periodic orbits described in [Wulff, C. and Schebesch, A., "Numerical continuation of symmetric periodic orbits," *SIAM Journal on Applied Dynamical Systems*, 5(3), pp. 435–475, 2006] and demonstrate your implementation on the examples considered therein, as well as on the continuation problem described in Examples 17.3–17.5.

21.16. Use COCO to implement the comoving-mesh strategy described in [Budd, C.J., Koomullil, G.P., and Stuart, A.M., "On the solution of convection-diffusion boundary value problems using equidistributed grids," *SIAM Journal on Scientific Computing*, 20(2), pp. 591–618, 1998] and reproduce the numerical results presented therein.

21.17. Use COCO to implement the alternative error estimation and remeshing strategies described in [Russell, R.D. and Christiansen, J., "Adaptive mesh selection strategies for solving boundary value problems," *SIAM Journal on Numerical Analysis*, 15(1), pp. 59–80, 1978] and illustrate your implementation on the examples in Part V of this text.

21.18. As an alternative to collocation, use COCO to implement an embedded Runge–Kutta scheme with error control as described in [Dormand, J.R. and Prince, P.J., "A family of embedded Runge-Kutta formulae," *Journal of Computational and Applied Mathematics*, 6(1), pp. 19–26, 1980].

21.19. Use COCO to implement the discretization approach for second-order initial- and boundary-value problems with automatic error estimation described in [Jator, S.N. and Li, J., "An algorithm for second order initial and boundary value problems with an automatic error estimate based on a third derivative method," *Numerical Algorithms*, 59(3), pp. 333–346, 2012] and reproduce the numerical results presented therein.

21.20. Use the idea of an automatic error estimate based on a parallel calculation with half the mesh size from the previous exercise and modify the adaptive `'coll'` toolboxes considered in Chap. 20 accordingly. Repeat the numerical analysis of the homoclinic orbit and the canard family from Sect. 20.3.

Index